Computer-Generated 3D-Visualisations in Archaeology

Between added value and deception

Joyce Wittur

BAR International Series 2463

2013

Published in 2016 by
BAR Publishing, Oxford

BAR International Series 2463

Computer-Generated 3D-Visualisations in Archaeology

ISBN 978 1 4073 1071 8

BAR Publishing is the trading name of British Archaeological Reports (Oxford) Ltd. British Archaeological Reports was first incorporated in 1974 to publish the BAR Series, International and British. In 1992 Hadrian Books Ltd became part of the BAR group. This volume was originally published by Archaeopress in conjunction with British Archaeological Reports (Oxford) Ltd / Hadrian Books Ltd, the Series principal publisher, in 2013. This present volume is published by BAR Publishing, 2016.

Printed in England

BAR
PUBLISHING

BAR titles are available from:

BAR Publishing
122 Banbury Rd, Oxford, OX2 7BP, UK
EMAIL info@barpublishing.com
PHONE +44 (0)1865 310431
FAX +44 (0)1865 316916
www.barpublishing.com

Contents

Please note that the CD referred to throughout the text has now been replaced with a download available at www.barpublishing.com/additional-downloads.html

The password is: BAR246317

List of Figures

List of boxed features

Chapter 1

Preface

The present publication has been submitted as a doctoral dissertation to Heidelberg University.

1.1 Personal motivation

My interest in reconstruction modelling was first raised during my archaeological studies at the University of Southampton (U.K.) where I partook in a master course for 'Archaeological Science: Archaeological Computing', which provided me, among other subjects, with theoretical and practical experience in the field of Computer Aided Design and reconstruction modelling, but also in the use of Geographical Information Systems and a theoretical knowledge of database systems. Critical reflection of these applications was also part of the course. For my master thesis I created a model of the *oppidum* on the Dünsberg (Germany) and used it to assess hypotheses concerning the purpose of the *Strahlenwälle*, a feature which can only be observed at this *oppidum*. After my return to Germany I first started to create 3D-models for the Fraunhofer Institute (IGD) in Darmstadt and later begun working as a tourist guide and educational officer at Lorsch Abbey. At Lorsch a virtual reconstruction of the abbey is displayed in the museum. My experience with the visitors in the museum, my knowledge of the ambiguities inherent to the site and my background in reconstruction modelling led me to questions concerning how best to impart complex archaeological interpretations based on equivocal finds and features to an interested audience, particularly as it was clearly observable that the visitors were interested in the archaeological discussion and were eager to observe, engage and to judge for themselves.

My research on this and related topics culminated in this book which hopefully will provide an impulse for a more judicious use of visualisations and for future research.

1.2 Acknowledgements

Many thanks go to Matthias Untermann, Susanne Krömker and Thomas Meier for their supervision and support for this study, including all the advice, critique and long discussions, which were necessary during its preparation.

I also want to particularly thank Liselotte Saurma, Hans Georg Bock, Michael Winckler and the HGS Math-Comp in general for providing me with a one-year scholarship in the final stages of my work.

Special thanks go to Martin Girschick, Sabine Koch, Tina Jäger and Herbert Pradt for proof-reading, advice and support during all the years, and also to Carole Avison who proof-read the final version of this study.

I want to thank Daniela Scagliarini Corlàita, Antonella Guidazzoli and Antonella Coralini for providing me with literature and answers concerning the Casa del Centenario project; Dirk Callebaut for his valuable information concerning St. Lawrence's church and especially Daniël Pletinckx for showing me Ename, helping me to obtain literature, for letting me use his pictures, for his hospitality and his endless patience regarding my questions. I also want to thank Geert Berings for letting me use his unpublished article, Dirk de Vries for helping me with the definition of the word *'slaper'*, Marie-Claire Van der Donckt for letting me use the pictures from the 'Nieuws uit het Verleden' website and Tina Jäger for helping on my second excursion to Ename. Thanks also go to David Wheatley and Mark Gillings for answering questions and providing additional material on the Negotiating Avebury project and together with Graeme Earl and Joshua Pollard for letting me use the pictures from their publications. I also want to thank Paul Cripps for allowing me to use his unpublished MSc dissertation including the pictures and animations.

Further thanks go to Herbert Pradt for helping with a Latin source translation and Angelika Schineller for translating some of my correspondence with Antonella Guidazzoli into Italian.

I want to thank all of the following persons or institutions for kindly allowing me to use images from their archives or publications: ACM SIGGRAPH, Algemeen Rijksarchief Brussel, Archaeolingua, Bodleian Library Oxford, Dirk Callebaut, Simon R. Davies, Philippe Decaudin, Luc Devliegher, Geschied- en Oudheidkundige Kring Oudenaarde, Mark Gillings, Bruce Gooch, Ulrich Haarlammert, MAS Volkskundemuseum Antwerpen, Michael O'Rourke, Oxford University Press, Pearson Education, Donald H. Sanders, Stefan Schlechtweg, Hermann Schlimme, Thomas Strothotte, Tokiichiro Takahashi, Taylor & Francis, Universität Bonn, Universiteit Gent, Verlag Degener, Verwaltung der Staatlichen Schlösser und Gärten Hessen, Oleg Veryovka, Torre Zuk.

Finally I want to thank the HGS Math-Comp for membership, covering travel expenses to conferences and to Ename and for enabling me to partake in soft-skill courses.

Chapter 2

Introduction

2.1 Study aim

The first computer-generated visualisations in archaeology were created in the early 1980s and the technology has become popular ever since (see Section 4.1). Nevertheless it took almost ten years before the first critical appraisal of the models took place and the discussion concerning applied ethical aspects[1] was initiated (see Sections 4.2 and 5.1) culminating in the allegation that virtual models are potentially deceptive. On the other hand it was recognised that the visualisations could, if prepared appropriately, provide a range of features other media could not, resulting in an added value especially for communication and research applications. The range of applied ethical issues in connection with computer-generated visualisations is wide and encompasses all decision making and modelling stages: from data collection to the final presentation (Chapter 5, see also Chapters 9 and 10).

Despite the critical discussion, most virtual models are created to present complete, preferably photo-realistic visualisations to the public, e.g. on television, in films and computer games and, frequently enough, also in cultural heritage institutions. The impression the observer obtains from these perfect and slick presentations is that no doubt exists about past realities, and indeed that they are knowable. Thereby the fact that research is an ever ongoing process is obscured and the observers are deprived of their right to think for themselves.

These misconceptions could be remedied by indicating the data sources used for the reconstruction, uncertainties in the model and alternative visualisations. Not only lay can audiences profit from these measures, but specialists and future researchers also can benefit, for example, from a well documented project.

[1]For a definition see Chapter 3.

The aim of this book is to point out methods which, if applied, will result in added value to virtual models in archaeology and help avoid deceiving the unsuspecting observer.

This general aim leads to several avenues of enquiry:

- What use have virtual models in archaeology?
- How are they perceived?
- Who is the intended audience?
- Which applied ethical issues exist?
- How can ethical awareness lead to added value?

Bound up with this, is the question of:

- How are these models created?

There is no easy answer to any of these questions and consequently there will be no ubiquitous recipe for building an 'ethically correct' model at the end of this book. The reason for this lies in the fact that every model is built to fit its own application area and will therefore pursue particular aims. Accordingly the methods which lead to a critically informed model will vary.

What this book offers is a collection of possible methodologies for dealing with individual problems concerning visualisation aims. So far, no comprehensive synopsis of different approaches with a critical stance on computer-generated visualisations has ever been attempted.

2.2 Approach taken

This study is primarily concerned with reconstruction models of architecture. Other model types (e.g.

for documentation purposes) are mentioned, but are outside the focus of this work. Architectural reconstructions yield the best basis for the study of critical approaches to visualisations in archaeology because they are generally complex in spatial and temporal concerns (especially when compared to artefacts) and require an ample amount of interpretational input (in contrast to models which record extant remains).

In order to approach the topic of reconstructions with applied ethical objectives, the work takes the form of a comparative study and operates on two levels. Articles from journals, conference proceedings and collections in book form have been studied to provide an overview of the full range of applied ethical issues which have so far been discussed in the literature (Chapter 5). These publications naturally provide no in-depth insight into the projects they describe. In order to achieve a more thorough picture of how virtual modelling projects are approached, three case studies are examined in detail (Chapters 6, 7 and 8).

The projects studied for this book are international, but exhibit a European focus, which is also mirrored by the three case studies. This regional focus is certainly in part due to the available and traceable literature dealing with reconstruction models (see also Chapter 4).

It can be said in general that the publication practice in the field of virtual archaeology has influenced the direction of this study: most reconstruction projects are not accompanied by a publication, while the projects which are presented at conferences or in articles usually fall into the application categories of museum presentation, site or artefact documentation, or research, i.e. application areas which are of interest to an archaeological audience.

Aspects which have so far been neglected in research, e.g. concerning the perception of virtual reconstructions, have been approached by using literature from other disciplines, ranging from psychology to philosophy.

2.2.1 Case studies

The three case studies were selected because of their differences, which will be listed below, but they also have two properties in common: they were all begun at approximately the same time (i.e. around 2001) and they all pay attention to ethical issues. Otherwise an effort was made to find projects which were produced and concerning sites in three different countries: Casa del Centenario in Pompeii (Italy), Ename (Belgium) and Avebury (U.K.). The projects are concerned with architecture from three different periods, i.e. a Roman house, a medieval settlement (with the focus on St. Lawrence's church) and a neolithic monument complex. The projects also had different aims: while the Casa del Centenario was primarily intended as a museum application and as a visualisation tool for the restorers, in the Ename 974 project the reconstructions were to illustrate the work and interpretation in progress for the local population while the church was closed due to excavation and building research. The model of Avebury was designed for research purposes and not intended for public display. According to the various aims, appropriate approaches were chosen.

The chapters concerned with the case studies (Chapters 6, 7 and 8) are all structured in the same way. First, the site is introduced, including its history and the available sources. Depending on the available information, this results in a substantial amount of information. A comprehensive study of the source material has been thought necessary to give a full account of the model basis, such interpretations as were made upon it and, finally, to point out uncertainties inherent to the data. All these points have a significant impact on the resultant model.

Second, the visualisation project is discussed. The aims and ambitions which were pursued by the project members were taken as the starting point because they determine how the model construction was approached. They were also taken as the basis for the later evaluation of the project.[2] Then the approach to model creation is examined, i.e. how are the available materials evaluated, interpreted and used in the project. The question of who was involved and in which capacity is addressed as well. Of special interest also are the modelling process itself and how the model was finally used.

Some of the projects exhibit a very long history, i.e. they have been reviewed or added to, new ideas have been developed by the people involved[3] or a similar endeavour has been taken up by other researchers[4]. These have been included into the case study description.

[2] It seemed inappropriate to evaluate the projects according to principles which were not on the agenda of the model makers. Suggestions concerning additional features, which would have been beneficial to the outcome of the projects are nevertheless made after the evaluation.

[3] For instance documentation or level of confidence issues concerning Ename 974 by Pletinckx (2008) (Chapter 7).

[4] For example, Davies (2009) in case of the Negotiating Avebury project (Chapter 8).

In the end, it was reviewed if the intended aims were indeed reached by the project members. Sometimes additional application areas for the models were discovered in the course of the project. Finally, some supplementary refinements or additions, which would have be beneficial to the project, were suggested by me.

Unfortunately the available information on each of the case studies was very heterogeneous and made adjustments in the chapters' structure necessary. For the same reasons, the weighting of the different chapter parts exhibits considerable differences.

2.2.2 Model creation and preservation

To round off the examination of critical approaches to computer-generated visualisations, a description of the various reconstruction phases is presented in Chapters 9 and 10. The decision to let these two chapters succeed the case studies has been taken because in this way ample reference to the case studies can been made, in order to provide the reader with concrete examples for the different approaches. Chapter 9, Pre-modelling, is concerned with the collection and assessment of the sources and their subsequent interpretation, while Chapter 10, Modelling, describes the decision-making process for the visualisation and its subsequent implementation. The latter chapter includes descriptions of technical approaches while avoiding the use of mathematical formulae, so that non-specialists in the field of computer graphics can follow the explanations. Regard has also been given to indicate what impact the choice of technical approaches will have on the final model and its presentation form.

Lastly, strategies for preserving and maintaining the models and the associated data are introduced. This is a factor which is often disregarded, but archiving and maintaining digital data can cause unanticipated follow-up costs.

2.3 General information

The terms model and reconstruction will frequently be used synonymously; further definitions of words will be given in the text or in Chapter 3. The book does not provide a glossary; instead ample use of hyperlinks was made. To benefit from this feature the PDF version of this document has to be used, which can be found on the accompanying CD-ROM. Links within the document are marked in blue, external links, i.e. to web-pages, in magenta.[5] The print version only provides greyscale images, though the text and the captions refer to the colour images. The latter are accessible via the PDF version on the CD-ROM.*

When page numbers are given in the citations: *f.* means 'the following page', *ff.* 'the following two pages'. Any number of pages beyond the two following pages is given as a range: *x–y*.

*The CD-ROM has now been replaced by a download available from:

http://www.barpublishing.com/additional-downloads.html

The password is: BAR246317

[5]For example, because the reconstruction model of St. Lawrence's church (for Chapter 7) is only available digitally, a link to the model is provided.

Chapter 3

Definitions

The ethical issues discussed in this book fall into the category of *applied ethics*. The Internet Encyclopedia of Philosophy[1] mentions that 'Generally speaking, two features are necessary for an issue to be considered an "applied ethical issue". First, the issue needs to be controversial in the sense that there are significant groups of people both for and against the issue at hand.' The different positions which are presented concerning virtual reconstructions will be described in detail below. 'The second requirement for an issue to be an applied ethical issue is that it must be a distinctly moral issue.' Regarding this point the Internet Encyclopedia of Philosophy provides a list of normative issues in applied ethics,[2] for example: not to deceive others, a person's right to information, or beneficial consequences for society; either of these can certainly be applied to virtual reconstructions too. The Encyclopedia of Philosophy (Borchert, 2006, 239) states: 'In applied ethics, ethical theory is often far less important than moral insight and the defence and development of appropriate guidelines suited to a complex circumstance.' In this regard—with the aid of various published reconstruction projects and the secondary literature—the issues will be discussed and some ways to handle them proposed. The order in which they will be presented follows the sequence of the reconstruction process.

The reconstruction process frequently begins *after* the archaeological interpretation has been completed,[3] so, therefore, the first steps in model creation are often based on archaeological excavation publications. These publications are usually structured in a way which is compliant with an inductive approach, despite this not being the way archaeological research is actually conducted.[4] In this work I will, nevertheless, follow the structure provided by archaeological publications or the sequence of model creation respectively, both in the description of the case studies (Chapters 6, 7 and 8) and in chapters concerned with visualisations in general (Chapters 5, 9 and 10), by providing first the information concerning the data basis, next the model basis itself and then the interpretations which were built upon it, leading up to further considerations concerning model creation.

In my work I will take a 'scientific' concept of reality as a basis, in so far as to accept an external world. This external world is perceived subjectively so that records of the external world (i.e. observations[5]) already constitute interpretations.[6] The data with which the archaeologist has to deal (i.e. the physical evidence from the external world in his or her own subjective perception, as well as records and interpretations by others) is only fragmentary and sometimes conflicting.[7] Therefore under- and overdetermination of theories by the data will occur frequently[8] and make the presentation of alternative reconstructions advisable (Section 5.9.2), together with the indication of (subjective) uncertainties (Section 5.9.1) and the accessibility of the underlying data and/or model basis (Section 5.8.3). In those cases where only one theory has to be selected for visualisation as a reconstruction, subjective decisions have to be made.[9]

---

[1] http://www.iep.utm.edu/ethics/#H3 (Accessed: 24th of January 2012.)

[2] http://www.iep.utm.edu/ethics/#SH3a (Accessed: 24th of January 2012.)

[3] A fact which has been criticised, e.g. by Gillings (2005, 226).

[4] cf. http://www.britarch.ac.uk/publications/puns/punsrep2.html (Accessed: 13[13]th of March 2012) especially sections 2.2 and 2.5.

[5] See boxed feature 4 on page 207.

[6] Various authors cited in this work hold different views on this topic.

[7] See also Vatanen (2004, 5 on CD).

[8] See Section9.2, Shanks and Hodder (1995, 10, 29) and http://plato.stanford.edu/entries/scientific-underdetermination/ (Accessed: 6th of March 2012.).

[9] This raises the question as to which basis the 'inference to the best explanation' is made and whether it can be made at all. For a discussion of this problem see http://plato.stanford.edu/entries/scientific-realism/#SceAboInfBesExp and http://www.iep.utm.edu/explanat/#SH4e (Both accessed: 6th of March 2012.).

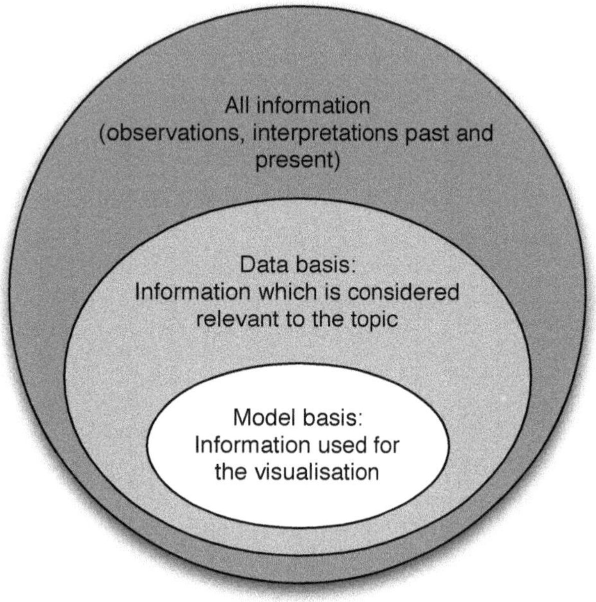

Figure 3.1: The relation of different groups of information: especially the data basis and the model basis. (Drawing by Joyce Wittur)

Interpretations based on the external world (i.e. physical remains, events, experiences, *etc.* including those from past realities) have been considered 'legitimate' data or information for the construction of a virtual model in this work.[10] They constitute the pool from which the *data basis* is drawn (see Fig. 3.1). In Fig. 3.1 the data basis is defined as 'information which is relevant to the topic'[11] whereby the decision as to which data is considered to be relevant to the topic is yet again a subjective choice *and* will be limited to information which is accessible and known to the researcher(s).[12] Lastly, not all of the information which has been reviewed for the interpretation and visualisation process will finally make its way into the model(s).[13] The information which will be used in the construction of the visualisation(s) is the *model basis*.

[10]This does not necessarily reflect the opinion of other authors cited in this book.

[11]This information consists not only of physical, textual, pictorial or numerical matter pertaining to the site or object under research itself but also includes analogies.

[12]cf. Hodder (1999, 43) and http://www.iep.utm.edu/explanat/#SH4e (Accessed: 6th of March 2012.): 'We are unable to compare proposed explanations to others that no one has yet thought of, [...].'.

[13]See Section 9.1 and as an example Section 7.2.7 together with Fig. 7.27.

Chapter 4

Trends in the application of virtual models in archaeology

Almost 30 years have passed since the first computer-generated archaeological 3D-models were created. During this time the areas of application, opinions and expectancies respecting this technology have changed.

In the following an attempt will be made to point out trends in the development. It is not possible to structure the history of research into unequivocally identifiable phases—every attempt must remain subjective and to some extent arbitrary—indeed, movements within the scientific community are rather likely to be found to overlap. Neither is a complete overview of all the virtual models feasible as it would go beyond the scope of this book, especially as many of the archaeological models have not been accompanied by publications,[1] which are a very important source for evaluating the methods of work and the objectives of the projects.

4.1 The beginnings: 1983 – 1993

The first years of using 3D-modelling software for archaeological projects are characterised, as is often the case when a new technology is adopted, by optimism and inventiveness. A comprehensive overview is presented by Reilly (1992, 149–158). His article mirrors the optimistic attitude during these years towards computer-visualisations and stylises their development from mere publicity instruments towards analytical tools (Reilly, 1992, 156). This 'prediction' should be proved untrue in the following years.

The first[2] computer-generated archaeological model, the Roman temple of Sulis Minerva and the adjoining buildings in Bath, was produced in 1983/84 (Woodwark, 1991, 19). There was interest on both participating sides: the computer experts were looking for challenging material to try out new modelling techniques while the archaeologists thought of the advanced presentation methods for a television broadcast. However, there existed no true collaboration, as the archaeologists were merely handing over their research results and were later presented with the finished product, which had been programmed for them (Reilly, 1992, 149).

Due to the public success of the reconstruction, more models were to follow in rapid succession: a reconstruction of the civic baths of Roman Bath (Reilly, 1992, 150 f.), the baths complex of the Roman Fortress at Caerleon (Wales) (Woodwark, 1991, 19) and the Saxon minster at Winchester (created in 1985) (Burridge et al., 1989, 562; Reilly, 1992, 152).[3]

Also, more sophisticated models for presentation purposes existed. For example, a hypermedia system developed for parts of Pompeii showed reconstructed views of the houses, but also gave additional information (text, pictures, *etc.*) on parts of a house (Reilly, 1991, 134 f., Reilly, 1992, 161 f.). The functionality, i.e. airflow, lime and coal transportation of the Langcliffe lime kiln (Wood and Chapman, 1992, 141) was shown in another model designed for the public,

[1] See Earl (2005, 206), who identifies a clear bias concerning publications about models (with a focus on museums and research) in comparison to the total of reconstructions made (including models for television, games and movies).

[2] At least to my knowledge.

[3] Besides the projects described here several other models for presentation purposes were generated during this time and shortly after: the reconstruction of Cluny III (Grellert, 2007, 175), the Stabian baths in Pompeii, S Giovanni in Sardinia, the Athenian Acropolis, the Pyramids at Giza in Egypt (Miller and Richards, 1995, 19), a reconstruction of Paris in 1789 (Collins et al., 1995, 19), the Dresden Frauenkirche (Collins et al., 1995), and many more.

thus providing more insight into the processes on site (Wood and Chapman, 1992, 142).

All these reconstructions were used to present existing data and were not used for research purposes. But such applications existed as well. In the projects concerning Malew18 (Reilly, 1988, 194-216) and Mathrafal (Arnold and Huggett, 1995; Arnold et al., 1989) topographical survey data were used to generate three-dimensional surfaces with or without geophysical prospection results mapped onto them, enabling the identification of features (Reilly, 1988, 199, 206; Arnold et al., 1989, 150) and the acquisition of new insights[4]. For Navan[5] different reconstruction possibilities were tested on the basis of surveys and excavation data (Burridge et al., 1989, 562; Reilly, 1992, 155 f.). Two models were built: one based on the assumption that the post-hole rings found on site once supported a thatched roof and belonged to a building; the other hypothesis being modelled was that the posts once belonged to a wooden henge monument and alignment to celestial bodies was suggested.[6] Another early model (presented at the CAA[7] in 1993) was intended for research, this time concerning the natural lighting conditions in the monastery complex of Rievaulx Abbey. However, the model could not be finished in time for the testing phase (Kemp, 1995, 249, 253).

In St. Veit-Klinglberg the generation of a three-dimensional excavation record from standard 2D documentation was attempted (Reilly and Shennan, 1989, 158), with limited success (Reilly and Shennan, 1989, 160). A slightly different endeavour had been undertaken by the Universities of Southampton and York with the 'Southampton-York Archaeological Simulation System' (SYASS).[8] The system was designed for students without excavation experience who were faced with a virtual site on which they could try out different excavation and survey methods, comparing the results from each (Reilly, 1991, 133).

Two further ways to reconstruct should be mentioned here. First, a stone-by-stone model of Furness Abbey, whereby it is questionable why this approach was chosen as the resultant reconstruction is a schematic one, not based on the photogrammetric measurements which were taken on site (Delooze and Wood, 1991, 141; Wood and Chapman, 1992, 128). The stones were idealised so that they could be easily copied and pasted (Wood and Chapman, 1992, 130). An even more automated approach is pursued by Ozawa (1993) who lets the computer reconstruct Japanese Keyhole tombs from only seven measurements and some additional information which can be entered by archaeologists (Ozawa, 1993, 417). In a second project he intended the reconstruction of Japanese villages, with prefabricated house models which only had to be placed on the footprints unearthed by the excavation (Ozawa, 1993, 420).[9]

Most possible application areas of 3D visualisations in archaeology have their roots in these early attempts. The articles which describe the projects frequently focus on the technical side of the reconstruction process, but tend to leave the reader in the dark concerning the archaeological side. Information regarding, for example, the state of preservation of the sites, the degree of cooperation between the archaeologists and the computer experts, the data basis for the reconstructions and how the archaeologists arrived at the reconstruction from the model basis is often missing. Also the research value of many of the models seems to be overestimated, as frequently the hope is expressed that in the future more useful research applications will be found; but still one of the main uses seems to be to produce 'pretty pictures' for the public. Nevertheless, it was shown that there was a great variety of ways to use visualisation in archaeology in these early years. Individual approaches to the different problems were found and, maybe hindered by the still young technology, were bound to arrive at fewer conclusions than was hoped for.

4.2 Critical voices: 1994 – 2000

The years from 1994 – 2000 can be identified as the core years of critique, although applied ethical aspects have been discussed at earlier and later points as well. Two prominent outliers are certainly the first critique by Reilly (1992)[10] presented on the CAA in

[4]See Reilly (1988, 203, 203 Fig. 103, 204 Fig. 104); Arnold et al. (1989, 152); Reilly (1989, 577) and Reilly (1992, 155).

[5]The descriptions given in the texts best matches Navan Site B (cf. Mallory and Lynn, 2002, 533, 537 f., 539).

[6]The discoveries and interpretations are quite similar to the Sanctuary (Section 8.1.3) in the Negotiating Avebury case study (Chapter 8).

[7]Computer Applications and Quantitative Methods in Archaeology conference.

[8]See also Section 5.5.4.2.

[9]This approach naturally leaves a lot of problems unresolved: while the individuality of the location and the surrounding landscape has been noticed (Ozawa, 1993, 419 f.) the individuality of the buildings or tombs has not. Through such an automated approach, the real questions concerning the sites certainly remain unrecognised.

[10]See also Section 5.2.

1991 and the London Charter[11], which was drawn up in 2006.[12]

In 1994 Miller and Richards (1995) presented, in a seminal contribution at the CAA, a harsh and partly exaggerated critique concerning computer visualisations. Some of their arguments echo Reilly's (1992) views[13], but also move beyond them:

- Miller and Richards (1995, 19 f.) state that computer visualisations could be employed as research tools in archaeology, but would be mainly used to present 'existing data to the public' focussing on Roman and medieval architecture because their principles are well understood.[14]
- They admonish that archaeologists are not in control of the projects (Miller and Richards, 1995, 20), but they also warn the archaeologists against naively creating computer graphics themselves, if they are unaware of the possible pitfalls which may in the worst case culminate in misleading results (Miller and Richards, 1995, 21).
- Another problem relates to the three-dimensional nature of the reconstructions: unlike as in drawings, uncertainties cannot be hidden by using smoke and mist, like Alan Sorrell. Instead all parts of the 3D-model are open to inspection (Miller and Richards, 1995, 20). Accordingly, all parts of the reconstruction appear to be knowable and seem to possess the same 'likeliness'. Museum visitors are being patronised through confronting them with a finished model and little or no additional information on alternatives or uncertainties in the underlying data, consequently prohibiting them from judging for themselves (Miller and Richards, 1995, 20). This could mislead the 'data-naive public' into thinking that the models represent a past reality (Miller and Richards, 1995, 21). To prevent such misconceptions there should be links between the initial data (the data basis) and the reconstruction as well as the possibility of indicating problems with the data and the visual representation (Miller and Richards, 1995, 21).

In short, Miller and Richards (1995) argue that virtual representations should not be reduced to 'pretty pictures' but should rather be used in research; that collaboration between archaeologists and technicians is essential to avoid meaningless or misleading results; and that in order to avoid deception it is necessary to confront the recipients of the models with uncertainties, alternative reconstructions, and the relationship of the data basis with the reconstruction.

In the following years a number of authors[15] had a critical look at the application of virtual models in archaeology. In general they follow Miller and Richards (1995) in their condemnation of misleading or merely spectacular images.[16] Instead they see the application areas for the models to be in communication and research, whereby the interactive and analytical possibilities the models offer for exploration, interpretation and hypothesis testing/assessment are highlighted.[17]

General consensus concerning the necessity both to indicate uncertainties, and to reveal the relationship between the extant remains or other sources and the reconstruction, is observable,[18] while the provision of alternatives is less frequently named.[19] The main goal is to emphasize that the model is not a faithful representation of a past reality, but only one possible interpretation. How this should be achieved is variously discussed. Among the recommendations is to

[11] On the London Charter see also Section 5.7.

[12] http://www.londoncharter.org/introduction.html (Accessed: 1st of December 2010.)

[13] Reilly (1992) addresses issues like: *using virtual models as research tools instead only for presentations to the general public* (Reilly, 1992, 156), whereby research is simply seen as being superior to mere representational purposes. He also recommends the provision of *alternative reconstructions or dynamic models* (see e.g. Roberts and Ryan, 1997) and the indication of a *level of confidence* to show the hypothetical and uncertain character of the reconstruction in order to inform the 'uncritical viewer' that the model does not depict a past reality (Reilly, 1992, 159).

[14] To support this notion Miller and Richards (1995, 19) mention several projects, among them the previously described reconstructions of Bath, Caerleon, Winchester and Furness Abbey, while others, which are more research oriented (e.g. Malew 18, Navan), are left out.

[15] For example, in 1996: Reilly (1996); in 1997: Pringle and Moulding (1997), Roberts and Ryan (1997), Daniels (1997) and Gillings (1999); in 1998: Barceló (2000), Gillings (2000) and Forte (2000); in 2000: Eiteljorg (2000), Bateman (2000), Huggett and Guo-Yuan (2000), Niccolucci (2002a) and Frischer et al. (2002).

[16] This goes actually so far that some researchers feel compelled to add justifications for using photo-realistic output in their projects (see Chalmers et al., 1995, 225)!

[17] See Reilly (1996, 38–40, 42), Pringle and Moulding (1997, 23, 27), Gillings (1999, 247–250, 253), Daniels (1997, Introduction, sections 1.1, 1.2, 1.3, 3.1, 3.2 and 4), Gillings (2000, 59 ff.), Barceló (2000, 9 f., 29), Forte (2000, 247, 249–252) Niccolucci (2002a, 4, 6) and Frischer et al. (2002, 15).

[18] See Reilly (1992, 159), Ryan (1996, 96), Pringle and Moulding (1997, 23), Kanter (2000, 47), Forte (2000, 249), Isenberg (1999), Eiteljorg (2000), Huggett and Guo-Yuan (2000, section 5), Frischer et al. (2002, 10, 13–15) and Niccolucci (2002a, 4).

[19] See Pringle and Moulding (1997, 23), Roberts and Ryan (1997), Eiteljorg (2000, section: Attractive Images or Good Information?), Huggett and Guo-Yuan (2000, section 3.2), Niccolucci (2002a, 4) and Frischer et al. (2002, 13).

raise the visualisation literacy of the target audience (Reilly, 1996, 39 f.) or to include information in text format which makes it clear that the reconstruction depicts just one possible scenario (Ryan, 1996, 105).

Also, the call for more collaboration between archaeologists and technicians is frequently made.[20]

Other criteria, like the possibility to make updates[21] and to incorporate metadata[22] as well as the listing of interpretational steps[23], are seldom mentioned.

Concerning other topics, opinions diverge. Some authors argue that **accuracy** is an important feature (e.g. Frischer et al., 2002, 14), besides the fact that there is some difference concerning what people think the term accuracy pertains to. It can range from the exactness of the measurements[24] (Martens et al., 2000, 206; Addison, 2001, 350) to how well the reconstruction represents the underlying interpretation (Hermon and Niccolucci, 2004, 1) or to the quality of the interpretation (Pletinckx et al., 2001, 198). Accuracy is also often perceived to be bound up with 'authenticity'[25] (Roussou and Drettakis, 2003, 52).[26]

This opens up the questions of what exactly makes a model accurate (Addison, 2001, 350) and whether models can be accurate (Pringle and Moulding, 1997, 22) (or authentic (Gillings, 1999, 252)) at all. And if they can, how to measure or ensure this accuracy? In answer to the latter question Addison (2001, 354) as well as Frischer et al. (2002, 7, 10) suggest peer reviews. But who would be suitable to review reconstructions? And does this not imply that the validity of interpretations which diverge from the reviewer's interpretation are cast into doubt?[27]

The question of whether models have to be accurate to be useful depends on their intended use. If they are conceptualised as surrogates[28] they do have to be (cf. Martins and Bernandes, 2000, 353), but for other applications—for example, to experience a place—this is not necessary (cf. Gillings, 2000, 59).

Equally debated is the notion whether virtual models are **objective** (Barceló, 2000, 28; Earl and Wheatley, 2002, 5).

One of the greatest discussions is centred on the question whether models should look **realistic/naturalistic**[29] or **non-realistic**[30]? Are non-realistic models ugly (cf. Kirchner and Jablonka, 2001, 237)? What does the audience expect ('photo-realism' according to Roussou (2002, 98)) and should these expectations be satisfied while at the same time running the risk of deception?

The best way to treat uncertain parts also remains unresolved. Some authors argue for leaving them out (Niccolucci, 2002a, 4; Kanter, 2000, 48; Frischer et al., 2002, 14). Some plead for completed views, especially for lay audiences (Kanter, 2000, 48), with (Eiteljorg, 2000, section: Attractive Images or Good Information?; Frischer et al., 2002, 14) or without marking the supplemented features.

The expectancies of finding appropriate application areas for virtual models (possibly beyond communicating archaeological information to lay audiences) are optimistic. Application areas are seen in Processual (e.g. Daniels, 1997) and Post-processual (Gillings, 2000) research, in low level theory,[31] but also concerning ritual and human experiences (Chalmers and Stoddart, 1996, 85; Barceló, 2000, 10) which are the particular focus of Post-processual approaches (e.g. Gillings, 2000, 60 f.). The application areas also depend on 'what the models are seen to represent' (see Section 5.3). Hence a surrogate model can be handled and examined instead of the original in order to keep the original safe, or it can be used as a record of an endangered original (Martins and Bernandes, 2000, 353). A simulation model can be employed to run different scenarios (e.g. with artificial life forms) in order to evaluate which one comes

[20]See Reilly (1992, 149), James (1997, 23), Frischer et al. (2002, 13), Forte and Borra (2000, 239), Borra (2000, 260 ff.), Addison (2001) and Scagliarini Corlàita et al. (2003, 237) concerning the Casa del Centenario case study.

[21]Pringle and Moulding (1997), Frischer et al. (2000, 158), McGregor (2000, 7), Messika (2002, 149) and Callebaut (2002, 180).

[22]The calls for incorporating meta data into the models are all rather late: Frischer et al. (2002, 13), D'Andrea and Niccolucci (2001, 215), Ryan (2001, 247) and Niccolucci and Cantone (2003, 58).

[23]See Reilly (1991, 135), Barceló (2000, 28), Vote et al. (cf. 2001, 315), as well as the more recent articles by Vatanen (2003, 70-74), Vatanen (2004).

[24]Including the question what exactly are we measuring (Lock, 2003, 4 f.).

[25]'[A]ccuracy and authenticity [...] are two sides of the same coin. Accuracy pertains to the data and metadata; authenticity to the user's experience of the data and metadata.' (Frischer et al., 2002, 14).

[26]The concept of authenticity is discussed in Section 5.4.

[27]This is basically a theoretical question: is there a general and quantifiable truth (New Archaeology) or is a plurality of opinions allowed or even desired (Post-processualism)?

[28]See Section 5.3.1.

[29]Ozawa (1993, 415), Collins et al. (1995, 21 f.), Chalmers and Stoddart (1996, 86), Burton et al. (1999, section 4), Martins and Bernandes (2000, 352) and Bohuslav et al. (2002, 209, 212).

[30]Daniels (1997, section 3.1), Pringle and Moulding (1997, 22), Eiteljorg (2000, section: Fantasies may seem very real), Gillings (1999, 247, 249 f.), Gillings (2000, 59), Goodrick and Harding (2000, 116), Kanter (2000, 47) and Freudenberg et al. (2001).

[31]Reilly (1992, 170), Kanter (2000, 48) Barceló (2000, 10) and Daniels (1997).

closest to the recorded evidence on site (Forte, 2000, 252 f.).

Unfortunately the participants in the critical discussions and the protagonists in innovative application areas remain a minority, especially as more and more models are created. Frischer et al. (2002, 12) state: 'official archaeology and VR seem to have marched on separate if not divergent paths.' The same problems are mentioned by D'Andrea and Niccolucci (2001, 215 f.) as well as Goodrick and Earl (2004, section 1.1): 'Concepts of authentication and visual critique are increasingly referenced, if not pursued, [also] theoretical questions receive little attention.' The problem continues to exist and not even the publication of the London Charter in 2006 had a significant effect on this development.

4.3 The boom years: 1996 – 2002

The boom years overlap largely with the phase of the first critical appraisal, but, as mentioned above, the critique had little influence on most of the projects.

The copious increase in virtual models can be attributed to a decrease in the price of the necessary hard- and software (Grellert, 2007, 175) and to the fact that the software became more user-friendly. This led to an increase in publications[32] and presentations at conferences[33] concerned with virtual models, whereby the number of unpublished reconstructions for television, cinema, exhibitions, games, research and the internet now exceeds the number of published reconstructions by far (cf. Earl, 2005, 206).

A large number of 3D-visualisations are concerned with documentation, be they of excavation data or of extant remains; the advent of the 3D-laser scanner especially seems to have contributed to this development.[34]

Studying the publications named above (see Footnotes 32 and 33) it can be observed that many of the papers concerning virtual reconstructions address several topics, i.e. technical questions, research and presentation issues, as well as some points of applied ethics,[35] wherefore it is noticeable that technical aspects, research and presentation (usually for museum applications) are represented in almost equal measure throughout the whole of the considered time span.[36] However, the total number of publications for each year varies considerably.[37]

The optimistic appraisal of Earl and Wheatley (2002, 6) concerning the uptake of virtual reality modelling in archaeology cannot be retraced with confidence: Earl and Wheatley (2002, 6) state that the development had moved from adoption and uncritical application, through a phase of examination and critique, and will, finally, reach a state of equilibrium when research has moved beyond applied ethical issues and other problems.

In reality the development seems rather to have bypassed most of the former.

4.4 End of the discussion? Since 2003

It can be observed that since 2003 there has been a decline in publications dealing primarily with virtual reconstructions. Models for documentation purposes are still very popular and there is a clear trend towards documentations in combination with information systems[38] or Geographical Information Systems in combination with virtual models, where the latter play a subordinate role (e.g. Davies, 2009). Many articles also deal with technical issues of information presentation.[39]

There are hardly any articles whose primary focus rests on the critical examination of virtual models any more[40] and their contents do not exhibit the quality of

[32]For instance articles in *Archaeologia e Calcolatori* or *Internet Archaeology*, but also in collections like *Imaging the Past*.

[33]For example, the *Computer Applications and Quantitative Methods in Archaeology* (CAA), *Archäologie und Computer* and the conference on *Virtual Reality, Archaeology, and Cultural Heritage* (VAST).

[34]Due to the focus of this study, i.e. reconstructions which pay heed to applied ethical issues, models which fulfil recording or documentation purposes have been largely excluded from further research.

[35]The article by Gillings (2000) is such an example: applied ethical, theoretical, research and technical issues are well presented in this paper.

[36]This result is based on an admittedly crude sampling process, which certainly introduced some bias into the analysis: an article which describes mainly technical aspects, but mentions some research goals and some (not well-thought-through) applied ethical aspects, has simply been noted as addressing the three aforementioned topics disregarding the quantity and quality of the presented statements.

[37]The sample includes thirty publications for 1998 and only two for the year 1999. Presentations at conferences were attributed to the year the conference was held, which often diverges from the year the conference proceedings were published.

[38]For instance (Bobowski et al., 2008) and (Ferschin et al., 2008).

[39]For example, Fritz et al. (2005) or Luyten et al. (2006).

[40]Some exceptions exist, e.g. Gillings (2005). An interesting point of view is provided by Thomas (2004). Both positions are examined in Section 5.3.

the early critique, neither in their theoretical understanding nor in their argumentative depth.[41] Instead *specific* topics are addressed, for instance, relating to metadata and persistency,[42] how uncertainty can be quantified and made visible,[43] or concerning documentation strategies (e.g. Vatanen (2004) and Pletinckx (2008)). In many articles the applied ethical aspects have only a minor role (e.g. in Viti, 2004) thereby reducing them to mere 'trimmings'. As before, most authors do not consider theoretical or applied ethical perspectives at all.

This book hopefully aids in reopening the discussion.

[41] For example, the texts by Knoll et al. (2004), Monod and Klein (2005) or Tost (2008).

[42] For instance Lancaster (2005) or Christodoulou et al. (2006).

[43] Hermon and Niccolucci (2004), Zuk et al. (2005) and Sifniotis et al. (2006) constitute examples.

Chapter 5

Some ethical problems

Ethical aspects of computer-generated visualisations make the difference between added value and deception. They encompass many stages in the creation of virtual models, starting from the conception of the enterprise and ending with the perception of the final product. This chapter will provide a detailed discussion of the different applied ethical issues, which will be presented following the sequence of the reconstruction process.

It will become clear that the way in which models are seen (as surrogates, simulations, hyperreal or mimetic models, or simple communication devices; see Section 5.3) has an effect on which aims can be achieved by employing them. Additionally the general question of whether the virtual models are authentic or objective has to be addressed (Section 5.4). This also has bearing on their possible application areas.

The modelling applications or aims (Section 5.5), in turn, will have an influence on how the modelling process will be approached (Sections 5.6, 5.7, 5.8 and 5.9) and determine the form of the outcome and therefore, how the models will be perceived (Section 5.10).

A clear concept of what 3D models are and what can be accomplished with them, will help find appropriate areas of application, preferably in the categories of documentation, communication, front-end and analysis (Sections 5.5.4, 5.5.5, 5.5.6 and 5.5.7) as they are the categories with the greatest impact on the dissemination of archaeological research and information.

Close collaboration between the different individuals concerned with the model production will ensure a satisfying end-product (Section 5.6) while documenting all the stages of the modelling process will both guarantee transparency regarding the sources, interpretation stages and implementation, and facilitate changes and updates of the visualisation (Section 5.7). Storage and retrievability of sources used for the visualisation (preferably from within the visualisation) help to make the model comprehensible and assessable for experts and laypersons alike (Section 5.8) and can at best even enable reinterpretations with new conclusions.

The fragmentary record of the past combined with the ambiguities which are inherent to any source lead inevitably to uncertainties and potential alternative interpretations (Section 5.9). These should be made visible in order not to deceive the observer into believing that he or she is faced with irrevocable facts. Instead, it would be desirable to encourage the observer to raise questions and thus to actively engage him- or herself with the visualisation.

The visualisation aim can sometimes dictate the manner of representation. Where this is not the case, the choice of the representation can have a considerable impact on how the visualisation is perceived (Section 5.10), e.g. whether the importance, certainty or other relevant properties of objects can be judged from a purely visual standpoint, without recourse to the documentation or the data basis[1] of the model. Particularly in this respect, a great amount of research is still pending.

However, it should be noted that the added value, which is gained by considering ethical issues like 'honesty in representation' or 'transparency' and taking measures to implement them in the models, comes at the price of an increased workload, whilst also avoiding deception.

Regarding nomenclature, it has to be remarked that the objective of this chapter is to address the various ethical questions related to computer-generated 3D-models in general, albeit most of the quoted texts here are explicitly referring to *virtual reality* (VR)

[1] See Section 9.1.

models[2]. Most of the following text is equally applicable to non-VR models. The term *virtual reality* is in itself highly contested, resulting in a number of other terms used by different authors (e.g. hyper-reality (Gillings, 2002), three-dimensional models (Cripps, 2001), virtual archaeology (Forte, 2000), virtual models (Viti, 2004), computer visualisations (Bateman, 2000), virtual reconstructions (Martens and Peter, 2004), and virtual reproductions (Viti, 2004)) which also capture a multitude of different meanings.[3]

5.1 Awareness of ethical issues

Over the last decades a vast amount of archaeological visualisations have been created, but for only a few of the projects is background information or information about their creation available in published form. Even among these, ethical/theoretical questions are rarely addressed (D'Andrea and Niccolucci, 2001, 213–216). Many of the authors seem to be completely unaware that such issues exist. But for all that, there are a number of texts which do deal with ethical questions. Unfortunately they frequently are not linked to a practical implementation (e.g. Frischer et al., 2002) or if they are, their example has no impact on other visualisation projects. Additionally, little testing of how applications are used and how successfully they perform their tasks has been carried out (Jablonka et al., 2003, 15).

As mentioned in Chapter 4 the first computer-generated architectural models in an archaeological context were created in the early 1980s (Woodwark, 1991, 19). Since then the areas of application and the critique these models elicit have hardly changed. While effort has gone into overcoming technical problems, theoretical issues and questions of collaboration between model creators and their clients (i.e. archaeologists, curators, *etc.*) remain often unaddressed (D'Andrea and Niccolucci, 2001, 215 f.)[4] despite the fact that some of them have been identified quite early (Reilly, 1992, 157 ff., 168 ff.).

5.2 Ethical and theoretical problems

Reilly (1992, 159) is one of the first to mention ethical problems which can be encountered during the creation and consumption of three-dimensional reconstructions:

A *degree of confidence* about reconstructed parts needs to be indicated, for example, by using a colour coding scheme or transparency in order to indicate how sound the data basis is for them.

Indicating the data or model basis for the visualisation. This can be achieved with augmented reality (AR) by overlaying the extant/documented remains (for example, an excavation photograph or a live-stream from a camera) with the reconstruction.[5] Additionally or alternatively it is possible to use hyperlinks to connect the reconstructed parts to their model basis (i.e. analogies, finds, texts, *etc.*)[6] or even to the data basis, which comprises information which has not been included in the reconstruction.

The model is not a representation of a past reality and this has to be clearly demonstrated to a lay audience. It is necessary to point out which data are available and that several reconstruction hypotheses are possible.

In later years Reilly's principles were elaborated by other authors, but the focus remained the same. Some additional topics, which have to be considered, were brought up in the discussions:

Virtual reality models are often under-theorised (Goodrick and Gillings, 2000, 56) because they are divorced from technical and theoretical discussion in their fields (Miller and Richards, 1995, 20). To use virtual reality effectively in archaeology it is first necessary to understand what it is (and what it is not) and starting from there, suitable applications can be developed (Goodrick and Gillings, 2000, 41–48).

[2] Virtual reality applications have been defined as being *immersive, interactive and real-time* (Frischer et al., 2002, 10). The majority of the so called 'virtual reality' models in archaeology do not fulfil these requirements.

[3] In the case studies the terms chosen by the various authors were retained, the terms *model* and *reconstruction* were used as alternatives.

[4] D'Andrea and Niccolucci (2001, 213–216) discuss ethical aspects which should be considered and in this light analyse the presentations given at the Computer Applications and Quantitative Methods in Archaeology conference in 2001 (Burenhult, 2002) particularly in respect of which questions were addressed in them. The result is devastating: only two presentations view virtual reality applications in a critical light, six are concerned with specific topics and therefore do not mention ethical aspects, and eight should be mentioning them, but avoid the topic.

[5] An example of this approach can be found in the Ename 974 project, Section 7.3.2.1.

[6] This approach has been taken in the Casa del Centenario project, see Fig. 6.6.

The intended use of the models should be specified before the project is started (which is unfortunately not always the case) (Gillings, 2000, 59). It would be desirable to employ virtual reality models not only for presentations to a lay audience, but also as explanatory tools with a concrete research goal in mind (Barceló, 2000, 29; Miller and Richards, 1995, 19).

The archaeologists need to have control over the modelling process for the reconstructions to be archaeologically useful (Barceló, 2000, 28 f.).

Quality control of the models is necessary (Miller and Richards, 1995, 20). This can only be achieved by documenting the reconstruction process, e.g. by adding metadata[7] to the reconstructed parts (Ryan, 2001, 247 ff.) or by some other documentation scheme[8]. Reconstructions could then be certified by a committee (Ryan, 2001, 247).

Think more about how people are seeing these models, especially if they are designed for a lay audience in the cultural heritage sector. Are people media-literate enough to question the images presented to them? Are they aware that this is just *one* possible reconstruction and that there are uncertainties (Miller and Richards, 1995, 20 f.)?

One of the most critical stances was taken by Miller and Richards (1995). They point out the danger of deceiving the observer with ill-documented and poorly presented computer models, and go so far as to write in their conclusion 'that computer graphics should carry a health warning.' (Miller and Richards, 1995, 21).

5.3 What are these models?

Following the suggestion by Goodrick and Gillings (2000, 43 ff.), an overview of what kind of relationship the models are thought to have to reality is presented here. This relationship to reality is important, as it determines in which way (if at all) the virtual models can be useful for archaeological applications.

The different views are presented in chronological order, i.e. the order in which they have been applied to virtual models in archaeology, which is by no means identical to the order in which the underlying philosophical concepts have been developed.

5.3.1 Surrogates?

Sometimes virtual models are seen as surrogates for the real sites. They are meant to 'preserve' the monuments they represent as accurately as possible for the use of future generations,[9] in the hope that they can be used for further research when the real monuments have gone. A similar standpoint is taken by some authors, who promote the use of virtual excavation documentations (e.g. Reilly, 1991, 133, 135), which then enable the archaeologist to re-excavate the site and to 'search for evidence which escaped attention during the actual dig' (Reilly, 1991, 135).

This may work in cases in which, by chance, a feature was documented which later proved to be essential for interpretation. But documentation is a matter of choice as well as interpretation. What has not been documented will inadvertently be lost; choices made on site (e.g. where does one feature end and the next begin) may be irreversible—so that in the end a true re-evaluation of the excavation data is impossible.

Surrogates are documentations of existing matter (at the time of documentation), but it is possible that the models are supplemented with additional material to form a 'reconstruction'[10]. These added parts (being interpretations) are not part of the surrogate (which is taken for 'almost' objective) itself.

5.3.2 Simulations?

The models can be seen as simulations (Daniels, 1997, abstract), i.e. simplified versions of the real world, which are based on assumptions. Detail is deliberately removed in order to achieve a simpler version of 'the real world' which helps to explain certain phenomena. This also means that the model has an underlying theory (Daniels, 1997, section 1.1). Simulations can be used for experimentation (e.g.: what

[7] See Section 5.8.2.

[8] A 'philological' approach has been proposed by Frischer et al. (2002, 13 ff.). Other, more elaborate ways of documentation, which also include the underlying archaeological interpretations, were suggested by Pletinckx (2008) and Wittur and Krömker (2009, 88 ff.). See Section 5.7.

[9] Statement by Harrison Eiteljorg II in an interview with Sims (1997, 13).

[10] Visualising an excavated area is also a reconstruction in the sense that the original is no longer there, even though all the data which goes into this kind of reconstruction has previously been recorded in the field. Nevertheless, a reconstruction can also include parts which complement fragmentary extant remains, e.g. reconstructing a building from a ruin.

happens if parameters are changed?) (Barceló, 2000, 9).

Barceló (2000, 9) uses statistical terms to describe what information can be gleaned from the models: the archaeological data, which constitutes the model basis for these simulations, is but a sample. The simulation, i.e. the visualisation of the sample, enables the archaeologist to draw inferences about the population[11]. It can also be used to explain properties of the data set, whereas human action can only be seen in the *effects* it has on the data set (Barceló, 2000, 10); this means, in fact, that in Barceló's opinion, visualisation is only suitable for low level theory. He also demands that the simulation should strive to be close to reality and should not include information which is solely based on interpretation.[12]

5.3.3 Communication devices?

Creating understanding by conveying an interpretation can also be seen as an important function for virtual models, whereby 'visual coding' can facilitate and accelerate comprehension (Hermon, 2008, 41), not only regarding observable features but also concerning abstract concepts (Hermon, 2008, 40).[13]

Pringle and Moulding (1997, 22 f., 31) see the relation of virtual reality to reality as being identical to an archaeological illustration and reality. Neither the illustration nor the model is meant as a surrogate. Their use lies in conveying information and/or an interpretation which is interesting and useful to archaeologists. They need not be 'realistic' for this purpose. In fact Pringle and Moulding (1997, 22) argue that realism can actually obscure more important information. Also the level of detail (i.e. the accuracy of the image) should support the imparting of relevant aspects rather than leading to a clutter of unnecessary features. Instead, the creator of the models has to identify the requirements of the users and to create a model that elucidates the important points. This requires interpretation (and also selection). The representation is already an interpretation and its function is to convey this interpretation successfully to the user (Pringle and Moulding, 1997, 31). Pringle and Moulding (1997, 22) emphasise the necessity first to identify the problem and then to determine whether a virtual reality model is the best solution, and not in the reverse order as is often the case.

The next three definitions refer to postmodern concepts of reality.

5.3.4 Hyperreal?

Gillings (1999, 250) discerns between virtual reality models which represent existing remains of a site and other models, which have hardly any 'tangible referent' (he gives the Peel Gap model[14] as an example) and therefore cannot be tested against reality. The latter reconstruction (or rather most of it) would then be a simulacrum as defined in Section 5.3.5 (cf. Goodrick and Gillings, 2000, 43). Instead of using the word simulacrum, Gillings prefers hyperreality (Gillings, 1999, 251). This term is based on the definition by Baudrillard, who states that hyperreal models are 'generated of a real without origin or reality' (Baudrillard, 1983, 2), and that 'hyperreality is more real than real' (Baudrillard, 1983, 146). Gillings (1999, 250) equates this definition to Noel Gray's second definition of virtual-reality, i.e. of it being *manufactured intensity*.

Gillings (1999, 252) claims that the—so-to-say 'imperfect'—virtual models can be used as 'catalysts for exploration and interpretation', but that the focus will have to shift from attempting to recreate outward appearance towards process. Process and the relationship between people and their world have been identified as constituents of authenticity[15] (Gillings (1999, 252) following Dovey (1985)).

5.3.5 Simulacra?

Thomas (2004, 198–201) views virtual reality models as simulacra, following the definition of Jean Baudrillard, i.e. that they bear no relation to reality. He takes the stance that the virtual reality models are

[11]The population is the complete set of individuals in a special group, for example, *all* the houses in a city, or *all* children between 8 and 10 years in a country. Performing a statistical analysis it is usually impossible to use the whole population, so a sample is picked, of which it is hoped that it represents the population well. How well an archaeological sample fits the population is always difficult to determine because unaccountable factors (e.g. preservation conditions, et al.) have a great influence on the composition of the sample.

[12]'Nothing should be wrong or "imaginary" in a virtual reconstruction, but should follow what we know [...]' (Barceló, 2000, 28 f.).

[13]Hermon (2008, 41) calls these models 'expressive tools'.

[14]Cf. Gillings and Goodrick (1996, section 2.5.1).

[15]At the same time Gillings (1999, 252) states that a virtual representation can never be authentic. See also Section 5.4.

only 'dots and lines'.[16] Thomas (2004, 199) points out that virtual reality models give primacy to the visual sense (which is a modern Western concept) and help to establish a subject-object divide because the observer is looking *at* the model. In addition, the observer is biased by his or her sociocultural background, which is (with a high probability) a modern Western, analytical, objectifying (male) view. These points have already been mentioned by Goodrick and Gillings (2000, 42, 54) and Earl and Wheatley (2002, 7)[17] in relation to the Negotiating Avebury project (Chapter 8). Thomas (2004, 199 f.) points out that some studies[18] attempt to shift the focus towards a subjective view that would see the body as a mediator between the physical world and the mind (i.e. adopting a phenomenological approach similar to Tilley's (Tilley, 1994, 11–14; cf. Brück, 2005, 47 f.)). Notwithstanding, several problems emerge from this approach: again the problem of the subject-object divide ensues and, in order to experience the landscape as did the people in the past, the meanings of the landscape would have to be 'transparent'. But they are not. To understand the landscape in the same way as people did in the past, we would need to know its context, but this is lost to us. Thomas (2004, 200) argues that the difficulties in finding the meanings of a landscape with the aid of virtual reality models are increased by the fact that the models themselves have no context. The conditions of manufacture would also be occluded,rendering them unsuitable for an experiential archaeology (Thomas, 2004, 201). This line of reasoning leads inevitably to the conclusion that with or without virtual reality models nothing can be said about past meanings, because we lack the contextual background to interpret them.

5.3.6 Mimesis?

The process of creating a virtual reality model is far removed from creating a real object, and the relationships of past peoples towards their objects are usually unknown to us. Hence no tangible connection exists between the virtual re-creation and the authentic object or site. But for the virtual models to be useful, they need to relate to reality in some way. Gillings denies that such a connection exists but unconsciously assumes it to be there. Exploring or interpreting entirely fictional settings would hardly constitute archaeology. There would also be no need for the approaches to be problem-driven (cf. Gillings, 1999, 252 f.). It rather appears that he assumes that the virtual models are at least partly simulations, though they are actually more than reduced models of reality: they are not only selective, but also creative (Goodrick and Gillings, 2000, 43).[19] This perspective becomes particularly clear in his latest article on this subject (Gillings, 2005, 234 ff.), in which he views virtual models as mimetic according to Aristotle's definition. This means for Gillings (2005, 235) that they consist of part imitation and part creativity or fiction which elicits feelings from the observer. He believes that virtual reality can evoke emotional responses, particularly by presenting unexpected, challenging and ambiguous content, which it is hoped will provoke a desire to engage oneself with the subject of the virtual reality presentation (Gillings, 2005, 231).

Actually Aristotle's understanding of mimesis as a part of poetry goes further: the fictional element in mimesis is not entirely fictional but stays within the boundaries of what is possible[20] by drawing on universal themes, in order to entertain and to educate the observer.[21]

In the construction of virtual reality models in an archaeological context, recourse to existing information is made. This information can consist of observations and interpretations made on the basis of extant remains, modern or historical texts, images, measurements and analogical material. This data already contains 'creative' elements from the outset, and the creative component is increased due to the way the data is used by different individuals, also which information is selected and how it is combined with other

[16] Thomas (2004, 200): '[...] the representation of the world as a geometrical structure composed of points, lines, surfaces and solids, picked out by pattern of contrast and luminance [...]' Thomas (2004, 201): 'Digital technologies reduce the past to a pattern of pixels [...]'

[17] See also Section 8.2.1.4.

[18] For example, the Negotiating Avebury project, though this is not explicitly mentioned by Thomas.

[19] In this context Goodrick and Gillings (2000, 43) refer to the article by Pringle and Moulding (1997) discussed in Section 5.3.3.

[20] 'all imaginative literature [...] is mimetic, in the sense that it describes not what really happened but the sort of thing that could happen.' '[...] the world portrayed by poetry is not a fictional world, but a recognizable version of our own real one: poetry does not show us what actually happened, but it does show us the kinds of things that could have happened and might happen. [...] the possibilities it describes are real, not fictional, ones.' (Craig, 1998, entry: Mimēsis).

[21] 'As we engage in more advanced forms of mimēsis, imitation gives way to representation and depiction, where we need not be regarded as attempting to copy anyone or anything in any narrow sense of the term. For tragedy does not set out merely to copy what is the case, but rather, as we have seen in Aristotle's differentiation of tragedy from history, to speak of what might be, to engage universal themes in a philosophical manner, and to enlighten an audience by their depiction.' http://plato.stanford.edu/entries/aristotle/ (Accessed: 19th of March 2012.).

data from other sources. In the case of *reconstructions* entirely speculative parts are frequently added to supplement the often fragmentary data basis.
The question concerning which application areas are best suited for computer visualisations *in general* cannot be answered easily. As Gillings (2005, 234 ff.) or Pringle and Moulding (1997, 33) point out, the models should be employed to resolve *specific* problems which arise from the archaeological context and can only be tackled in this distinct way. This approach appears to be the obvious way to employ any tool, but as Gillings (2005, 234 ff.) as well as Pringle and Moulding (1997, 33) articulate correctly, this has seldom been the case with virtual reality applications.

5.4 Authenticity and objectivity

As already mentioned in Section 5.3.4 virtual models can never be authentic, especially if authenticity does not refer to (a visual approximation of) form alone, but also to process, biography and embeddedness (Goodrick and Gillings, 2000, 44 f.). This definition is related to the definition of *aura*; the latter is supposed to be inherent to an original (in contrast to a reproduction) (see Nerdinger, 2010, 478).[22] The term *authentic* was originally used to convey that a written document had been copied faithfully and was authorised for use. This explains the first two definitions of *authenticity* in the Oxford English Dictionary: '1. as being authoritative or duly authorized. 2. as being in accordance with fact, as being true in substance.' (OED, 2010). Authenticity also denotes the original (in contrast to a copy *etc.*), which can be gleaned from the third and fourth definition: '3. as being what it professes in origin or authorship, as being genuine; genuineness. 4. as being real, actual; reality.' (OED, 2010). This definition, in parts contradictory, (i.e. one pertaining to a copy the other to the original) makes the use of this word difficult. But even if the second dictionary definition is used, not much of the substance of a virtual model (material, work processes, use and even form) can be seen as corresponding to the original.

Some authors use the term 'authentic' synonymously to 'lifelike'[23] so that some reconstructions aim to be very lifelike by including people and/or objects (e.g. furnishings), even though the presence of the latter cannot be deduced from the archaeological record (see e.g. Fig. 5.1). These models are often intended for public consumption and it is felt that by showing only what can be archaeologically proven the public might conclude that the past was, for example, devoid of implements made from perishable materials if they are not explicitly shown (Kanter, 2000, 48). The decisions about whether to include or exclude certain objects (plants, furnishings *etc.*) from the model are already a manipulation of the reconstruction and would therefore be contrary to its 'authenticity' (see also Eiteljorg, 2000, section: Towards an Honest and Complex View of the Past). Instead it has to be accepted that a model is a projection of a hypothesis (Barceló, 2001, 242) and that the model is 'wrong' (James, 1997, 25). The degree to which the model is 'wrong' is the degree to which the data do not fit the hypothesis (cf. Hodder and Hutson, 2003, 239), keeping in mind that the data basis is already an interpretation in the first place (see Section 9.1). Nevertheless, computer reconstructions in particular are assumed to impart a sense of authority and objectivity (Eiteljorg, 2000, section: Reconstructions or Fantasies?), which is predicated on their increased level of detail (in comparison to drawings) and by their mode of creation (Eiteljorg, 2000, section: Fantasies Can Seem Very Real)[24] even though the choice of elements which are depicted (in greater detail) is still the subjective choice of the creators (Bateman, 2000, section 4).

5.5 Model applications

From the discussion in Section 5.3 the conclusion becomes obvious that what the proposed models *are* (or are believed to be) is closely dependent on what the models can be used for.

This book is primarily concerned with reconstruction models, but in this chapter a short look at models for documentation purposes will also be presented (see Section 5.5.4).

5.5.1 Modelling aims

Reconstructions differ depending on the attempted goals of their makers. These goals can fall into six

[22] A very short critique concerning the concept of *aura* and *authenticity* can be found in Nerdinger (2010, 478).

[23] This is obviously not the case in Goodrick and Gillings (2000, 44 f.).

[24] Eiteljorg asserts that computers are thought to be machines which produce 'images based on logical process without artistry', although we are at the same time aware that the computer only implements the user's input.

Figure 5.1: Example of a lifelike reconstruction with the depiction of persons and objects (e.g. chair) for which there is no direct archaeological evidence: view of the Vari House (Attica, Greece; *c.* 300 BC) courtyard rendered from the Learning Sites virtual world. (Copyright 2010 Learning Sites, Inc.; image used with permission)

categories: technical feasibility, entertainment, documentation, communication, front-end for another application and analyses. These categories have been sorted in ascending order according to what I subjectively believe is their profitableness for archaeological applications.[25]

The following text provides a short introduction to these six categories and other classification schemes before the former are discussed in detail in Sections 5.5.2, 5.5.3, 5.5.4, 5.5.5, 5.5.6 and 5.5.7.

1. Objects or places, which were reconstructed to show the **technical possibilities**, are nothing more than mere staffage. They could easily be substituted by other challenging and fascinating sceneries, which inspire the audience with a similar feeling of awe.

2. In **entertainment** archaeological sites often serve as a backdrop, they are used to situate[26] the observer, i.e. to capture the imagination, but they are usually not meant to convey any further scientific information about the sites they present.

3. Three-dimensional models of extant remains or excavated features for **documentation** or simulation purposes have been among the popular application areas from the start. Extending the documentation into the third dimension and being able to manipulate the visualisation can help viewers to understand and interpret the site and site formation processes. Efforts to use excavation simulations for university teaching have also been made.

[25]Opinions on this matter may obviously vary.

[26]For the different functions of images see the Boxed Feature 5 on page 212.

4. For **communication** purposes, the creators of the models are intent on telling something about the objects and sites. This communication can work at different levels, though most of the virtual models are used for public display: in museums, on site, on television, in video games, *etc.* Their aim is to communicate information and/or to entertain.[27] The advantages of virtual models lie in the possibilities which reach beyond the ones of traditional reconstruction drawings. The computer generated models offer methods to alter the presentation of an object or monument easily, in order to indicate, for example, uncertainties or the underlying data basis and thus offer additional information to the observer.
5. A three-dimensional model can be used as a **front-end** to a database or a Geographic Information System (GIS)[28]. The entries in the database are then linked to parts of the model and can thus be accessed. This connection can work both ways. It is possible to display selected items (by way of a database query) in the model, or parts of the model can be clicked to display the associated entry in the database. Such software combinations can prove to be valuable research tools.
6. **Analysis** is deemed to be the most important application area for virtual models. It will therefore be elaborated in detail by presenting a number of application areas. Models can, for example, help researchers with structural inquiries, perception analyses, simulations, interpretation of excavation results, *etc.* In these cases the models need to be well planned to fulfil all the necessary functions in an analysis. Note should be taken that not every kind of modelling technique is sufficient for every kind of research question. In some cases photo-realistic models will be necessary, for others reduced models are more appropriate (see Section 5.10). The same accounts for surface and solid models: if cuts through objects are necessary, or building phases of a wall aer to be shown, a solid model is required. For displaying murals, a surface model may suffice.

Reconstructions need not have a single use only; some models will have to meet various purposes, for example, to serve as means for research but also communicate the research results to the public.

[27]This book is primarily concerned with models which have an aim beyond mere entertainment or serving as a backdrop. Therefore such models will not be discussed in great detail.

[28]A GIS has also a database attached, which could be accessed via the model.

It should not go unmentioned that some other classifications have been proposed, for example, by Kanter (2000, 48): research, pedagogy (i.e. teaching students at university) and public consumption (here computer games and the internet are mentioned). Forte (2000, 250) names communication, education, simulation and research. Sanders (2000, 37) mentions the following objectives as frequent areas of applications: background for games, demonstrations of computer modelling prowess, visualisations for research or presentation. As further uses he adds: publication, education (teaching in schools), research (e.g. the possibility of bringing dispersed artefacts from one site back together again), sharing excavation results (full excavation reports could be issued, which would otherwise be too costly to produce in paper format).

5.5.2 Technical feasibility

Computer companies and also other institutions use 3D reconstructions as showcases for demonstrations of what is technically possible.[29] These reconstructions are designed to impress visually and are therefore often used for public display. Unfortunately these models predominantly ignore archaeological questions, because they usually lack the involvement of archaeologists (Ryan, 1996, 107)[30]. This issue will be discussed in detail in Section 5.6.

As feasibility studies provide attractive views of archaeological sites, they are therefore mostly used on television or in museums. But besides their engaging appearance they often have little to offer, as they are not designed to be communication devices.[31]

Additionally, technical problems during their presentation can lead to further confusion: Forte (2000, 247) mentions the model of a Roman Forum presented by Silicon Graphics on the SIGGRAPH '99, where wrong scaling, lack of an appropriate navigation system and insufficient marking of building phases resulted in disorientation and thus in a loss rather than in a gain of information. Frischer et al. (2002, 11 ff.) report similar effects on the visitors of the Pompeii models by the Simlab of Carnegie Mellon University.

[29]For example, the reconstruction of Pompeii at the ETH Zürich: http://www.g-o.de/wissen-aktuell-8094-2008-04-16.html (Accessed: 1st of February 2012.). The primary aim is to develop and test a device which enables people to move through a virtual world. That the virtual world is, in fact, a reconstruction of Pompeii, is secondary.

[30]See also the example in Frischer et al. (2002, 12).

[31]See also the arguments by Pringle and Moulding (1997, 22) and Section 5.3.3.

Using archaeological models as mere presentations of the latest computer technologies is certainly in part maintained by the European Union's funding policy, which includes computer-generated modelling within Information Technology and degrades the cultural heritage sector to a provider of content (Frischer et al., 2002, 11 ff.; Monod and Klein, 2005, section 1). This tendency is mirrored in the conference contributions in which ethical questions hardly ever find consideration (D'Andrea and Niccolucci, 2001, 216).[32] There are several opinions as to why computer scientists have an interest in modelling archaeological sites and objects and where exactly the benefits lie:

Miller and Richards (1995, 20) find that the involvement, i.e. investment, of large companies in these projects makes a very clever publicity stunt through attracting public interest in a 'politically safe' area, where third party funding is always welcome.

Reilly (1992, 149) on the other hand sees benefits for both sides in other areas: the archaeologists have a chance to get illustrations of monuments, while, during the modelling process, the computer scientists get a chance to overcome technical problems. These included, at the time of Reilly's writing, the quest for photo-realism for which archaeology seemed to be a good application (Reilly, 1992, 157)[33]. However, the purpose for which the archaeologists need the resulting illustrations of monuments appears to be second rate.

In this light, the appeal of Ryan (1996, 107) to museum staff and archaeologists seems appropriate in asking them to 'strive to retain a clear understanding of the purpose of their models', which lies in communication instead of showing 'the latest computer graphics effects.'

This should not lead one to the belief that there exist only examples of projects where collaboration between the archaeologists and computer scientists did not work out, but those that did work out are unlikely to be found under the heading of technical feasibility.

5.5.3 Entertainment

As already mentioned archaeological reconstructions act as backdrops for films and games. Often these reconstructions do not possess a great scientific quality (but usually a very good visual one). Martinez (2001, 10) thinks that in films or games more correct archaeological details should be presented and that in this way a wider audience could be reached (Martinez, 2001, 15). It would certainly be beneficial to see more scientifically accurate representations of the past in films and games, but it is doubtful whether the audience would expect to be presented with probable versions of the past. Films and games do reach a wider and in many cases a different audience than, for example, museums, but even a film which is not really based on fact can still spark the interest of people in finding out more about the historical setting.

In the section 'The Double-Edged Sword' Eiteljorg (2000) calls attention to another problem concerning films:

> My point here, of course, is that fiction can be as easily conveyed as fact in a good movie. Indeed, most movies are entirely fiction, and good fiction presents no problem. It is a problem, however, when fact and fiction are confused, because an effective movie will overwhelm viewers with the image and sound of apparent reality, leaving the viewer with a permanent and strong memory that cannot easily be dislodged.

By using this opening example on films Eiteljorg wants to demonstrate that graphical representations can be deceptive for the viewer. Fact and fiction are hard to tell apart if they are not indicated or the observer has a sound knowledge of the topic. This insight can be transferred to the realm of reconstructions in general, as they consist of both fact (primary data) and fiction (interpretations, scientific analyses, educated guesses)[34]. The resulting problems will be mentioned in Section 5.9.

5.5.4 Models of excavations and extant remains for documentation, simulation and research

A detailed discussion dealing with 3D-models of extant remains, preserved artefacts or excavation trenches for documentation purposes does not fall within the scope of the book. A short mention has to be made, because these models can be part of the model basis for more elaborate visualisations (see

[32] This matches my experience when looking for relevant articles for this book.

[33] i.e. the Archaeologists were acting as guinea pigs for technical applications. See also Daniels (1997, sections 3.1 and 4).

[34] Concerning *primary data* and *interpretation* see Boxed Feature 4 on page 207.

Section 9.1.2.1). It shall suffice to mention at this point that these records are neither objective nor should be treated as surrogates.

5.5.4.1 Excavation reconstruction

There have been a number of attempts to generate 3D models of excavated features. One of the first was made of a pit in Bavaria[35] with the aid of software (FRODO) originally designed for modelling protein crystallography. Besides modelling the layers, the program was able to display find positions and was meant to help the archaeologists in interpreting the stratigraphic sequence and the use of the pit which was recut several times (Ottaway et al., 1986). The aim of this, and many other reconstructions of archaeological features, is to be used as a cognitive tool (Forte, 2000, 250):

- to analyse the stratigraphy in 3D (Reilly, 1989, 571) in order to provide new views on features and artefacts (Viti, 2004, 527) by being able to see the excavated layers from all sides (Reilly and Shennan, 1989, 158).
- to explore the data to find patterning (Reilly and Shennan, 1989, 157), e.g. by comparing artefact densities to test hypotheses and to develop new ones (Lukesh, 1996, 246 f.), by visualising the distribution of artefacts in 3D space (Avern, 2002, 158), or by identifying specific layers[36] (Reilly, 1989, 570 f.; Reilly and Shennan, 1989, 159).
- to understand formation processes (Reilly and Shennan, 1989, 157, 162)[37], or to reconstruct post depositional processes[38] (Viti, 2004, 527).

Different paths were taken to obtain the data for the models. Some are based on traditional recording methods, i.e. plans, section drawings and photographs,[39] especially if they were dealing with old excavation data which had not been specifically obtained for the purpose.[40] In other instances 3D measurements were taken in the field (Vote et al., 2002, 84), e.g. with a total station (Cosmas et al., 2001, 297, 300)[41].

The models which were created also used very different modelling techniques: solid modelling (Reilly and Shennan, 1989, 158 f.), wireframe (Cosmas et al., 2001, 303), or voxels (Lieberwirth, 2008a). The models were usually connected to a database[42] and some of them are GIS-based[43]. The approach by Lieberwirth (2008a, 2) claims to be a GIS, because it uses a GIS-software (GRASS) to create the model, but it cannot be so, because no statistical analyses of the model are possible (Lieberwirth, 2008a, 7).

For the modelling of two grave examples (Avern, 2002; Viti, 2004) another approach was taken. The position of the finds taking precedence over the stratigraphy. In both cases the finds (bones, stones, artefacts) were modelled in 3D and placed in the appropriate position inside the grave; no database was necessary.[44] Another difference between these two reconstructions and the ones mentioned above are the more-or-less clearly defined research goals (especially in the case of the chariots (see Footnote 40), even though the problem could not be resolved in this way (Avern, 2002, 158 f.)). This is in contrast to the general aims of 'looking for patterns' or 'analysing how the strata relate to each other' prevalent in the other examples. It has also been suggested that the 3D records were suited for letting 'future archaeologists revisit the site in virtual reality in order to make their own interpretation of the finds.' (Cosmas et al., 2001, 297).

[35]The name of the site is never mentioned in the article!

[36]In the case of the Potterne midden, layers containing datable material (to build up a chronological sequence) or in which the bone fragmentation is very high, were singled out (Reilly, 1989, 570 f.). In St. Veit-Klinglberg the presence or absence of copper objects and bronze age pottery was depicted (Reilly and Shennan, 1989, 159).

[37]Reilly and Shennan (1989, 157) warn that such an approach can have drawbacks, because the human brain has the tendency to identify patterns where there are none.

[38]A grave was modelled in which a person had been buried and his/her body were covered with stones. After the decomposition of the corpse the stones splintered the bones (Viti, 2004, 527).

[39]For instance Reilly and Shennan (1989, 158 f.) and Viti (2004).

[40]For example, Lieberwirth (2008a, 2) describes the reconstruction of stratigraphy from plans and sections made in Akroterion in Kythera (published in 1972) (Lieberwirth, 2008b, 85, 89). In the case of two Iron Age chariot burials (Avern, 2002, 158) the excavator, P. Bonenfant, hoped that his records would in the future help to create a 3D model of the chariots which would specifically lead to insights about the chariot's superstructure. Therefore he recorded in 1967 and 1969 all objects with 3D coordinates.

[41]Cosmas et al. (2001, 300 f.) also envisaged that they would in future be able to calculate geometry from videos taken on site, a process they call 'shape from video'.

[42]For example, Reilly and Shennan (1989, 159); Lukesh (1996, 249); Cosmas et al. (2001, 298); Vote et al. (2002, 84).

[43]For instance Bobowski et al. (2008, 2 f.) and Vote et al. (2002, 85). These two applications had problems with the representation of their data. The strata could not take complex forms which hampered usability (Bobowski et al., 2008, 3; Vote et al., 2002, 85).

[44]In the previously mentioned examples single finds were usually displayed as abstract shapes, information relating to their shape could be retrieved from the databases.

Besides a number of technical problems[45] the display mode itself also posed difficulties. Both Reilly and Shennan (1989, 158) as well as Vote et al. (2002, 84) encountered the problem that an opaque set of layers which is intended to represent the strata (for example, when the colour of the strata indicates the amount of bulk finds present in it) blocks much of the view and makes analysing the stratigraphy impossible, especially if one is dealing with a complex site. Vote et al. (2002, 84) reverted to showing the layers as transparent units but arrived at technical problems when the bulk finds had to be added to this (Vote et al., 2002, 85). Vote et al. (2002, 84) also noted that displaying the stratigraphy in a realistic manner is not the best way for data analysis.

Considering this, Bobowski et al. (2008) present an interesting approach by trying to display not only the stratigraphy[46] but rather the Harris Matrix in 3D, so that its related features can be visualised in 3D space. For this the Harris Matrix is displayed in the shape of cylinders which are positioned in the middle of their stratigraphic context. The cylinders' height does not correspond with the height of their stratigraphic contexts; instead they show the relative sequence of strata. Relations between units are indicated by lines (Bobowski et al., 2008, 4, 4 fig. 2).

Sometimes it is disregarded that the data displayed in excavation reconstructions is a selection and that even a great amount of measurements taken in the field will not generate a surrogate site, which can be re-excavated by computing new section drawings.[47] Even though Forte (2008) meant the following as a critique of 3D stratigraphies generated from conventional recording methods instead of 3D measured features in the field, it is also applicable for those: 'In the best of circumstances, 3D is used to represent data collected by hand and in two dimensions and/or integrating spatial information that is either not correct or simply invented.'[48] (Forte, 2008, 23).

5.5.4.2 Excavation simulation

In the late 1980s and the early 1990s 3D models of (fictitious) excavations were also used for simulation and training purposes: Reilly (1989, 570 f.) suggests using an elaborately recorded Saxon pit from Hamwic (Southampton) for training and evaluation studies. He argues that the model could be used to be 're-excavated', this time not as a whole, but only in parts. By using different excavation or sampling strategies and comparing their outcome with the results from the complete excavation, it is hoped to find out, e.g. if there are cheaper methods, which might provide (almost) the same results.[49] Reilly (1989, 571) proposes that these methodologies could be used in the training of archaeology students as well, by letting them try out various approaches and comparing these also in connection with budget, personnel and time restraints. None of these trials would lead to the destruction of real archaeological records.

This was also the intended application field for the Southampton-York Archaeological Simulation System (SYASS) and its follow up SyGraf: teaching archaeology students the decision-making process in planning and carrying out archaeological excavations (Molyneaux, 1992, 312; O'Flaherty, 1988, 491, 495 ff.; Reilly, 1991, 133). The program is 'plan-based', i.e. two-dimensional,[50] and is supposed to help the students to identify how decisions affect excavation strategies and interpretation (Molyneaux, 1992, 313).[51] The decisions are: where to excavate and to what depth, which excavation tools are used and dependent on this, in what detail the finds are sampled and the features are recorded. The program then searches the database and retrieves the 'excavated' contexts and the finds and calculates the budget (O'Flaherty, 1988, 495 f.). It is certainly beyond

[45]Problems occurred specifically with older models, which could not be interactively viewed, or with systems based on GIS-software which have trouble displaying the complex shapes of the stratigraphic layers.

[46]This was without much success, but it also was not the primary aim of the project.

[47]For example, Lieberwirth (2008a, 2) tried to merge the section drawings from different trenches on the same site into a stratigraphy of the whole excavated and unexcavated area. She found that this was not possible because stratigraphic layers were intersecting each other (Lieberwirth, 2008a, 4). Consequently she reverted to using only one section as an example and the method presented a coherent result (Lieberwirth, 2008a, 4). Instead of questioning the methodology used to obtain the result (which would have been advisable seeing that the first attempt of merging failed, indicating that the interpolation does not produce reliable output - not to speak of 'an objective point of view [...] provide[ing] the user with further options for spatial analysis.' (Lieberwirth, 2008a, 1)), she blames the old section drawings for the failure (Lieberwirth, 2008a, 8). Stratigraphic layers are not always straight and may have unexpected features, which can hardly be interpolated if they have not been observed in the field.

[48]With 'invented' he means interpolated data. But as long as we do not start to measure every grain of sand, the data between two points of measurement has to be interpolated.

[49]The question remains whether the results gained from these comparisons are applicable to other sites.

[50]The excavation results are presented to the user as a series of plans and a number of retrieved finds (Molyneaux, 1992, 316).

[51]Whereby it is not meant as a tool to evaluate how good the decision making processes of the individual students are, but it should rather stimulate discussions about the site and experimentation with different approaches (Molyneaux, 1992, 318 f.).

dispute that this is quite removed from an actual excavation situation. In order to emphasize the usefulness of the approach Molyneaux (1992, 314 f.) argues that the simulation is capable of providing the same perceptual experiences as the real world, because[52] a stimulus to the eye will always provide the same experience, whether real or artificial.

Regarding the abstraction of the displayed finds and features, I would strongly disagree with Molyneaux's notion, because of the important fact that only visual stimuli are sent out, added to the inherent simplification in respect of reality which is inborn to any simulation.

Reilly (1992, 163) also remarks critically on the project: 'SYASS is not simulating an archaeological site, it is simulating what is known in Britain as Level 2 or 3 information.'[53] The point he is intending to make is that mere two-dimensional representations (equivalent to drawings made in the field) are not the same as a three-dimensional stratigraphy with embedded finds. Despite the arguments by Molyneaux (1992, 320) a site is not perceived as a sequence of plans (as which it is stored and displayed in SyGraf) during excavation, but in 3D. A three-dimensional model may depict the stratigraphy more realistically, because not only plans, but also profiles can be generated. Additionally, the decision when to draw a/the next plan is subjective and the drawing itself a time-consuming task which also has consequences on budgeting, *etc.* This does not mean that simulations for teaching decision-making processes—so that students can learn to judge costs, necessary materials and efforts—are not useful, but that in order to let students make these decisions and find out the consequences, a 3D representation would be more appropriate (cf. Reilly, 1992, 163).

5.5.4.3 Re-assembling sites

Some archaeological sites have been dismantled in the past and their parts distributed all over the world. 3D-modelling of the still-existing parts and virtually re-assembling them is often the only way to show them in context. Sanders (2000, 39 f.) mentions the relief decoration from the Northwest Palace of Ashurnasir-pal II in Nimrud, as an example which has been treated in this way.[54]

5.5.5 Communication

Communication is a wide field and a prevalent application area for virtual models. It incorporates, among others, museum displays, models on the internet and on CD-ROM, which aim at informing their audience about various topics.

One could argue that some parts, which fall here under the heading of communication, could as easily be termed pedagogics or teaching, however, the common goal in all these application areas is to impart an understanding of the underlying concepts to the observers.

Communication can take place at museums, schools and universities and at home (via the internet or a CD-ROM) and will address different audiences ranging from laypersons to professionals. Its form will also vary in informational content and on the insight gained during production (cf. Boxed Feature 1 on page 27).

Frequently, computer-generated models are used in the same way reconstruction drawings would be (Goodrick and Gillings, 2000, 42), with the sole difference that films can be generated from them. This underrates the potential of these models, especially because most of these reconstructions are also completely uncritical regarding their underlying archaeological and interpretational basis (Barceló, 2000, 28). They are mere end-products presenting archaeological 'results' to the general public without having produced new knowledge during their construction (Miller and Richards, 1995, 19 f.).

Instead, as communication devices, they should fulfil the objectives mentioned by Pringle and Moulding (1997, 22 f., 31), regardless of whether they are meant to address archaeologists or a lay audience. That is, they should identify the needs of the users, judge whether a virtual reality model is the best way to address those needs, and then create a representation which conveys the information appropriately.

[52] Here he refers to J. J. Gibson's *The Senses Considered as Perceptual Systems* (1966), p. 228.

[53] In a footnote Reilly (1992, 170) explains that: level 1 is the physical data, level 2 the site archive, level 3 the archival interpretation, level 4 the synthesis.

[54] This is one of the rare examples where such a reconstruction actually took place, though this application field is often mentioned in the literature. One of the problems certainly lies in the difficulties involved in gaining permission to record the archaeological remains and to travel to all the collections to which they have been distributed.

Excursus 1: Automatic reconstructions

Automated reconstructions are specific examples of 'reconstructions' which are meant for public display and do not have any impact on archaeological research (and upon which archaeological research undoubtedly also has little impact).
Some examples are mentioned by Barceló (2000, 20)[a]. For instance, Ozawa (1993) used a computer to calculate the shape of Japanese Keyhole tombs based on seven measurements.[b] In a second project villages of the Yayoi period were automatically reconstructed by placing pre-prepared house models on the post-holes identified by the archaeologists. Both approaches negate the monuments' individuality and aim at the quick and easy generation of reconstructions.
Related to the topic of automated reconstructions is the demand for a library of architectural elements (e.g. in Reilly (1992, 157) and D'Andrea and Niccolucci (2001, 216)) with the aim of reusing the elements in other reconstructions.
In another example the north portico of Arles' forum is compared to an automatically created counterpart based on *De architectura* by Vitruvius (Florenzano et al., 1999). The column shafts were found to be inconsistent with Vitruvius' rules (as had been already assumed by the archaeologists before the model was built). Consequently, Florenzano et al. (1999) surmised that column shafts may have been reused from another building. It is interesting to note that from the outset the authors of the article were aware that the probability of receiving disparaging results between the actual monument and the 'norm', as established by Vitruvius, was very high. This knowledge begs the question why this painstakingly complicated methodology was adopted in the first place.
That the trend for automated reconstructions, especially in creating city models, remains unbroken is demonstrated by Flack et al. (2001), Arnold (2002) and Dikaiakou et al. (2003). Often these models are made without a concrete goal in mind, only with the vague idea that they could be presented to the general public or be used as backdrops.

[a]Some of the references in his article are faulty, e.g. Roberts and Ryan (1997) and Ryan (1996) did not write about automatic generation of temples but about displaying alternatives utilizing the example of a Roman theatre.
[b]A paper similar in content was presented at the VAST 2006 conference: see Ozawa (2006).

Museum displays and similar applications often fall short of these expectations. The reason for the shortcoming is because they are only presenting stunning and often photo-realistic images to the audience. But communication must go beyond this. Indeed, departing from a 'realistic' representation in order to clarify a point is often advisable (Pringle and Moulding, 1997, 22; Weidenmann, 2002a, 88 f.). Besides the subject of the illustration, 'control codes'[55] can be included which help the observer to process the contents of the representation: i.e. to guide the observer's attention and to highlight or enlarge important parts of the subject (Weidenmann, 2002a, 89).

Virtual models in the cultural heritage sector can accomplish even more, for instance, by offering alternatives, showing levels of confidence or the underlying data and model basis.[56] By providing this supplementary content, the models offer great advantages in comparison to conventional illustrations where such effects are difficult or time-consuming to achieve (cf. Mainka-Mehling, 2008, 245 f.). Additionally, reconstructions which honour these principles can offer the public the opportunity to think and judge for themselves (Eiteljorg, 2000).

Also, the way these reconstructions are presented can have an effect, whether they are static images, a film or an interactive experience.

Fly-throughs (e.g. Terras, 1999, section 4) do provide an overview of a site but are unquestionably not the way a monument was experienced in the past (or is in the present (Krasniewicz, 2000, 164)). Still, elevated views on reconstructed sites are very popular and can help observers to understand the layout of a site. The down-to-earth variant, walk-throughs[57], provide a more realistic perspective of the site (in relation to the viewpoint of the observer) and should therefore be preferred for exploration.[58]

[55]The German technical term is 'Steuerungscode' (see Weidenmann, 2002a, 88 f.).

[56]A good example for the depiction of the model basis is described in Chapter 6.

[57]For example, Kadobayashi et al. (1999, 2000) provide an interactive example of a village, which can be explored by intuitive gestures while standing in front of a huge screen.

[58]A more theoretical note on why plan-views should be avoided

The interactive experience appears to be the best solution but it has been reported that people feel lost when exploring a (large) model on their own (Frischer et al., 2002, 11; Jablonka et al., 2003, 16 f.) and they may miss the 'important parts'. Some type of guidance is advisable, for example, by providing hot spots as in the reconstruction of Troy (Jablonka et al., 2003, 17). An interactive approach also gives the user the possibility of accessing additional information or features incorporated into the model (e.g. as in the application described in Chapter 6).

Displaying the reconstruction as a conventional film provides guidance but leaves the user with no choice about what is displayed. Terras (1999, section 11) mentions that the users of 'The Virtual Tomb of Sen-nedjem' model preferred the fly-through to the VRML[59] model which could be experienced interactively. Similar observations have been made by other researchers, leading to the conclusion that the presentation method which is most popular with the visitors is a linear exhibition guide, because it is seen as non-threatening (Lock, 2003, 226).[60] In an educational context, however, it has been demonstrated that film sequences can lead to an information overload[61] (Weidenmann, 2002a, 95) whereas providing supplementary content, which could be accessed interactively by the learners, helped in avoiding this overload (Weidenmann, 2002a, 96). Audio input in combination with a picture or object is also a successful combination in learning situations (Weidenmann, 2002a, 94). The time users take to look at the image/object is extended and their attention can be guided, but the matching and synchronisation of both the visual and the auditory contents is necessary. Otherwise the result would be an information overload (Weidenmann, 2002b, 53 f.).

Audio and video is also very suitable to catch the attention of people and to provide introductory, superficial information (Lock, 2003, 223). Lock (2003, 226) suggests using audio-visual displays to attract people into engaging with an information system and to provide in-depth information in an interactive, non-linear way. The latter is better suited for experienced and motivated learners (Haack, 2002, 134) for the simple reason that it takes more time and effort to access the information (Lock, 2003, 226). Interactivity also increases the learning efficiency (Weidenmann, 2002b, 57). Nevertheless learners usually prefer structured learning approaches (Haack, 2002, 134).[62]

A compromise between passive and interactive presentations would be to provide short film sequences which offer a choice of consecutive film sequences that lead along different routes (Lock, 2003, 153) or deal with different aspects of the monument (similar to TimeScope 3 in Section 7.3.5).

Different user groups (e.g. school children, adult laypeople or archaeologists) have different questions regarding the site or objects they encounter. Consequently, an information system's usefulness depends on whether it can be adapted to suit the target audience or even provide user-specific content.[63] Different needs will have to be addressed: students may need motivation; the information presented should be relevant to their interests, appropriate for their age and in accordance with their curriculum. Laypersons have diverse interests, needs and no common prior knowledge of the subject, although it can be expected that, for example, museum visitors are already motivated to learn. Scholars, in contrast, are likely to have far-reaching questions, and they may want to know in depth about details and why the makers of the model chose a certain way to reconstruct. This does not rule out the fact that the same reconstruction can be used for all three audiences. Nevertheless, changes in handling the information provided and the model displayed, for example, may be necessary to offer the observers an adequate level of the detail needed to answer the different questions the user groups will have concerning the reconstruction.

in favour of a more phenomenological approach, which includes exploring the site by walking through it, is discussed in Chapter 8.

[59]VRML stands for Virtual Reality Modeling Language. For a specification see http://www.web3d.org/x3d/specifications/vrml/ (Accessed: 1st of February 2012.).

[60]A survey conducted by Baumeier (2008, 5) had slightly different results. Ninety-three senior citizens were interviewed on which kind of presentations they would like in a museum. The preference was films with real scenes (42%) over 3D virtual presentations (34%). Although 80% of the people wanted to navigate actively, using predefined paths (56%) instead of completely passive presentations (5%). Also, completely free navigation was rather unpopular (16%). Pre-rendered paths provide guidance and a more 'linear approach' which, in turn, could be seen in keeping with the findings reported in Lock (2003, 226). It is important to note that the survey by Baumeier (2008, 5) asks questions about the *intent* of users, whereas the study referred to in Lock (2003, 226) looks at people actively *using* the different systems.

[61]Weidenmann (2002a, 95) suggests several remedies: slowing down the animation, 'advanced organising' i.e. preparing the learners for what they will see, repetitions, stopping the animation at crucial points and structuring the animation by inserting explanatory text and/or using audio comments.

[62]This has also been mentioned by Tergan (2002, 109), who ascribes the problem to the additional cognitive effort (besides the learning effort) which is necessary to find the relevant information and to structure it appropriately (Tergan, 2002, 108 f.).

[63]This had been planned for the project described in Chapter 6.

Excursus 2: Media, encoding and modality

By *media* the kind of object which presents the information is meant: e.g. a book, a television set, a paper print of a photograph or a computer. *Encoding* relates to the way the information is represented: text, pictures or numbers. The third category, *modality*, describes the way in which we perceive the information, e.g. by hearing or seeing (Weidenmann, 2002b, 45 ff.).

Several ways to present interest- or audience-specific content have already been explored: for example, Tobler et al. (2006, 322) enable the user to choose from different categories of content (history, music, literature, *etc.*). Luyten et al. (2006, 223 f.) provide different interfaces for adults and children.[64] Users can also state their interests in advance or during the visit. Additionally, user behaviour is taken into account (long stays in a certain place, requests for more information on a specific topic, *etc.*) to build up a user profile. Kadobayashi et al. (2000, 135, 142) provide two very different user interfaces for their application: VisTA is supposed to be used by archaeologists for research and hypothesis testing (for example, by querying the underlying database and visualising hypothetical scenarios), whereas museum visitors can use the gesture-controlled VisTA-walk to navigate through the reconstructed village and can access some background information about objects by pointing at them.

More important than the presentation medium is the way in which the presentation is incorporated into the learning context (Lock, 2003, 226; Weidenmann, 2002b, 59 ff.). In addition computer literacy has been observed to play a role in whether the information is accessed—80% of the people who use interactive media in a museum context have had previous experience with computers (Lock, 2003, 226).

5.5.6 Front-end

Pringle and Moulding (1997, 28) already suggest that three-dimensional reconstructions could be used as a front-end or as an appendix to other applications, that is, multicodal[65] content (in the form of a database (DB)), GIS and the Internet. These views have been variously echoed in the literature[66], though actual applications are rare.[67]

The Casa del Centenario project described in Chapter 6 has some additional multicodal content which can be accessed via the virtual model.

A combination of GIS and virtual reality (though without explicit query facilities) is used in the Negotiating Avebury project (Chapter 8).

The added value of combining different applications lies in the simplified access to additional data (in the case of the database systems) or in the possibility of analysing spatial data in relation to the virtual model (in the case of the GIS). Both paths can provide useful insights into the structure and the data of the site and therefore facilitate research.

At the moment a number of problems are still hampering the creation of GIS-VR-DB combinations: so far no real 3D GIS system exists, which allows queries in three-dimensional space. A possible solution could be provided by working with voxels. Nigro et al. (2002) describes an excavation model created with voxels which has been connected to a GIS coupled with a database. Unfortunately this system is unable to deal with complex three-dimensional queries (Nigro et al., 2002, 118).

An example of a combination of a multicodal database with a GIS is presented by Fronza et al. (2003). It is possible to query the GIS for similar structures[68] from within the multicodal database (Fronza et al., 2003, 151). From within the GIS it is possible to access the content of the multicodal database (Fronza et al., 2003, 150). No three-dimensional content has been included.

[64] cf. Schedl (2003, 4 f.).

[65] The expression frequently used in this context is 'multimedia', which is incorrect. See Boxed Feature 2 on page 29.

[66] A combination of GIS and virtual reality has, for example, been proposed for the Troy-Project (Kirchner and Jablonka, 2001, 237). Combinations of virtual models and database systems were, for instance, suggested by Earl and Wheatley (2002, 6) for the Negotiating Avebury project, Forte et al. (2003, 54) for the Aksum project, and in a completely general manner by Sanders (1999, section 5.1).

[67] Meyer et al. (2006, 134) present an example where 3D-models can be used as user-interface to access the documentation data for Vianden castle, Luxembourg. The granularity of the model and the linked data can neither be judged from this nor from a related article (Meyer et al., 2007).

[68] Further search parameters can also be entered.

5.5.7 Analysis

Several authors have demanded that virtual models should be used for research (e.g. Earl and Wheatley, 2002, 8; Miller and Richards, 1995, 19), but the number of research projects which yielded (published) results is small in comparison to the number of models used for communication. Besides these few examples of published research results, there are also a couple of papers in which the authors state a vague intent of using their models for some sort of (usually undefined) research, or in which possible application areas are pointed out—mainly without a concrete example of how this could be done. They are therefore adding only little to the corpus of analysable projects.

Some of the application areas are listed below, whereof the classification provided here being necessarily a subjective one; indeed, the boundaries between several of the subdivisions are fluid. As there is no reason to believe that specific research areas are more important than others and different approaches have been attempted at various points in time,[69] the research topics have been treated like 'nominal data' and are thus presented in a more or less arbitrary order.

Applications lie either within the realm of low level theory (Reilly, 1992, 170) (e.g. structural analysis), some are meant as interpretational aids, by visualising what can otherwise not be seen, others can be used as simulations (e.g. for astronomical phenomena). The models are, due to their nature, best suited for visual analysis. Modelling of sound has been attempted (see Pope and Chalmers, 2000), but has found no widespread use. Some models have multiple application areas, e.g. the Santa Maria Maggiore project (Frischer et al., 2000, 158), where the questions range from the interrelationship between the basilica and the surrounding buildings, lighting conditions, furnishings and interior décor, comparing several reconstruction hypotheses, *etc.*

There are also critical voices: Krasniewicz (2000, 164) thinks that reconstructions are unsuitable for research: their focus lies on the reconstructed buildings, while they are at the same time devoid of debris and artefacts, which are the main archaeological sources (i.e. contexts and artefacts). She heralds instead the use of QTVR[70] and other virtual reality applications which recreate the context of the archaeological excavation (cf. Section 5.5.4). The critique that the main focus of reconstructions often lies on the buildings and less on other artefacts and features is certainly true. It could be added that most reconstructions do not include human beings and animals. The question is where the main focus of the model and the research lies. If the focus of attention lies on people then they should be included in the reconstruction (Hodgson, 2001, 10). But one should be aware that their depiction can raise a number of problems concerning gender roles, activities, costumes *etc.*, which have been discussed in connection with life scenes (e.g. images or dioramas) in museums or books (cf. Jud and Kaenel, 2002; Mainka-Mehling, 2008; Moser, 1998). They will also invariably become the centre of the observer's attention (Hodgson, 2001, 11), though are often the most uncertain factor in the reconstruction and may detract attention away from other important features (e.g. architecture). In case animals should be included, special attention relating to their behaviour and—for domestic animals—the appearance of early breeds and seasonal changes has to be given (Hodgson, 2001, 11).

5.5.7.1 Structure

Several authors suggested that solid modelling could help to investigate questions belonging to the realm of low-level theory (Reilly, 1992, 170), for example, concerning the *physical properties* of extant remains or reconstructed buildings and artefacts. Daniels (1997, section 3.2) recommends the use of solid modelling (see Chapter 10) because it is possible to calculate from these models 'volume, mass (using finite element analysis), centre of gravity, moments of inertia, radii of gyration and stresses.' To know volume, mass and stresses is certainly worth while,[71] but it is questionable whether the remaining values have any application in archaeological research.[72] Knowledge about volume and mass can lead to questions about labour cost and construction methods (Daniels, 1997, section 3.2). Additionally, structural features which are usually invisible can be included into the models, e.g. the way in which two walls join (Daniels, 1997, section 3.2).

A stone-by-stone reconstruction is described by Wood and Chapman (1992) and Delooze and Wood (1991).[73] It is a reconstruction of Furness Abbey

[69] Besides the fact that this collection is only a sample and does not aspire to completeness.

[70] QuickTime Virtual Reality, i.e. bubble worlds created from panoramic photographs.

[71] For instance to find out how much raw material was needed, or whether an object could be fragile in certain places and may need support (Daniels, 1997, section 3.2).

[72] See also Gillings (2000, 59).

[73] To my knowledge Furness Abbey is the only stone-by-stone reconstruction attempt, probably due to the enormous amount of

church of which only ruins remain (Delooze and Wood, 1991, 141; Wood and Chapman, 1992, 126, 127 fig. 11.1). Two-dimensional reconstruction drawings were used to aid the modelling process. As most of the church has disappeared, the reconstructed parts can obviously not show the stones in their former size and position in the masonry. The modelling process for these parts is described in Delooze and Wood (1991, 143) and Wood and Chapman (1992, 130). It consists mainly of copying or mirroring already created parts to accelerate the modelling process. Neither of the texts elaborates whether the extant parts of the church are modelled according to the existing remains or whether they are also produced schematically. Little benefit is to be expected from this painstaking methodology in a purely schematic approach. Showing construction phases (as had been planned (Delooze and Wood, 1991, 144 f.)) would only make sense for a model of standing remains where these phases can be discerned, and even then, it might be unnecessary to model each and every stone. Blocks of masonry, corresponding to the individual phases, would suffice—they could, for example, help in matching (possible) building phases in adjacent rooms or buildings. If more detailed information about the masonry needs to be stored, e.g. tool marks, material and damage, a stone-by-stone model may become necessary, as in the example of the St. Peter's Bridge in Prague (Pavelka et al., 2006, 164).[74]

Solid modelling allows for the simulation of structural stresses (Delooze and Wood, 1991, 144), to test the structural feasibility of reconstructions (Hermon, 2008, 39) and to simulate distortions (Reilly, 1989, 578) (for example, of damaged metal artefacts).

Despite the possibility of doing these kinds of analyses, they seem to have hardly ever been performed, even if questions concerning the statics have been raised as in the example of the in the Santa Maria Maggiore (Rome) project (Frischer et al., 2000, 159): the form of the apse has been reconstructed on the basis of liturgical sources, but after viewing the reconstruction, the experts were in doubt whether it would have been statically feasible (Frischer et al., 2000, 159 ff.). There is no indication that the question was resolved.[75]

work and computational power which has to go into such an endeavour and the little added value it provides.

[74]The bridge was modelled as a wireframe, and not as a solid model.

[75]There was no follow-up to this article and also a search of the World Wide Web brought no further results.

That statical analysis can create interesting results is demonstrated by Levy and Dawson (2008), who researched the stability of Thule whalebone houses when tested with dead loads (like skin, sod and snow) and life loads (like strong winds). The analyses had to be based on assumptions because the statical properties of whale bones are largely unexplored, but it became evident that the houses could withstand a great amount of dead loads though could be prone to collapse in very strong winds. On the other hand it was observed that some bones (crania and maxilla) used for the houses were insufficiently strong on their own and had to be supported by additional bones (mandibles, which are the standard building material), which therefore made the crania and maxilla actually superfluous. Levy and Dawson (2008, 137 f.) explained this redundancy by suggesting that the weaker bones had symbolic character, especially so because they have been inserted in the entrance areas to the houses.

5.5.7.2 Surface models combined with additional data

One of the very early 3D modelling examples already makes use of this technique: a mediaeval castle in Mathrafal, Wales was surveyed in 1988 (Arnold et al., 1989, 149). The topographic, magnetometer and resistivity surveys were all executed using the same grid. Then the topographical data was used to create a wireframe surface (with exaggerated height values, to emphasize the irregularities in the terrain) and the model was artificially lighted (Arnold et al., 1989, 150). Among other visual analyses of the three survey data sets, the resistivity and magnetometer survey data was mapped onto the terrain model and thus allowed the interpretation of features, for example, the continuation of a path was partly visible in the terrain and partly in the resistivity survey. Other data was only visible in one of the surveys (Arnold et al., 1989, 152; Reilly, 1989, 577).

Barceló (2001, 232) suggests combining three-dimensional terrain models with data concerning the soil type or hydrography (this would be useful for landscape research), and using the same technique on the artefact level, for example, projecting use-wear patterns upon an artefact. Sometimes even legacy data can generate new results (Forte, 2008, 28): an aerial photograph of Aksum (Ethiopia) taken in the 1960s was mapped on a terrain model. The scene was artificially lighted and a low altitude flight was simulated. Several so far unknown man-made sites could be discovered through this exercise. Forte (2008, 28 f.)

attributes this success to the combination of glancing light, the dynamic viewing and the interactivity which the model provided. This made it possible to view the landscape from different angles and to discover the correlations between the terrain data and the photograph.

5.5.7.3 Explanation or comprehension of data

Usually archaeological data is only represented in 2D plans and elevations, adding the third dimension helps to conceptualise it, and understand it both better and quicker (Daniels, 1997, section 1.2). Daniels (1997, section 1.3) attributes this to the way our brain works. This is supported by research in how graphical representations can help in problem solving.

Diagrams can be used as external memories or as memory aids (Scaife and Rogers, 1996, 193 ff.). They can help in searching, recognising and inference-making because information that is included in the diagram may not to be kept in mind ('computational offloading'). This makes the problem easier to solve.[76] In order to receive the best results the representations should be designed so that the attention is directed to the key components.[77] The best cognitive results are achieved when the diagrams are prepared by the person who wants to solve the problem (Scaife and Rogers, 1996, 207 f.). Scaife and Rogers (1996, 210) point out that animations and virtual reality have so far not exhibited the same learning benefits as static representations. They believe that this could be due to the time constraints that animations and virtual reality place on the viewer.

This argument is certainly valid for the animations, but virtual reality environments should not place time constraints on the user. Maybe a different explanation might be that the amount of unweighted information contained in the virtual reality is too much to be processed easily so that no 'computational offloading' can be achieved.

[76]Hermon (2008, 37) has suggested the use of VR and 3D modelling to express alphanumerical data visually in order to express thoughts and ideas. In another article Hermon and Fabian (2002, 103) emphasise the cognitive, theoretical side of archaeological models as information visualisations: Hermon and Fabian (2002, 104) follow the Intelligence-Design-Choice model for decision making by Newell and Simon (1972) and remark on the problem that usually not all of the possible alternatives are explored, it can be observed that 'standard' solutions to problems are preferred (Hermon and Fabian, 2002, 106). The same sadly applies to the visualisation, where normally only one of the possible alternatives is presented (see Section 5.9.2).

[77]Note the similarities to the article by Pringle and Moulding (1997).

The combination of data from different disciplines into one model can be helpful (Hermon, 2008, 38).[78] Hermon (2008, 38) cites an example where geomorphological studies, palaeoclimate and palaeobotanic studies, as well as palynological analyses were combined with ethnographical data to reconstruct a landscape and to estimate the changes which occurred there over time. In conjunction with a simulation, the amount of arable land, labour used for gaining new agricultural areas and the possible population size could all be estimated.

3D-models of architectural remains can also help to show how buildings relate to each other and where gaps in hypotheses can be found. Abernathy and Johanson (2005, 224) describe the Roman Forum project in which the buildings of the forum were modelled from the available data (usually each building was recorded according to an independent coordinate system). The buildings were then combined into a single model which allowed one to see the buildings in their context with neighbouring buildings and to check whether the coordinates given in the old excavation documentations could be trusted. Problems with the height of buildings and the adjacent streets presented themselves, temporal changes aggravated the problems, and dependencies between buildings, streets, *etc.* became apparent. It stands to reason that without the three-dimensional modelling many of these problems would not have been identified.

Similarly, the virtual model of Salagassos (Turkey) provides indications that the town planners of the city made use of the specifics of the existing terrain to construct visually attractive townscapes (Martens et al., 2000, 209). For example, monumental buildings were constructed on several terraces in order to be widely visible and to attract the attention of citizens and visitors alike (Martens et al., 2000, 208 fig. 5, 209).

5.5.7.4 Change

The already-mentioned Roman Forum project (Abernathy and Johanson, 2005, 224) deals in some ways with change: destruction and rebuilding of houses leads to new situations and to new contexts for neighbouring houses and places. Kadobayashi et al. (1999) introduce with the VisTA system a tool which allows researchers to study possible building sequences in a Japanese village. Based on the excavation plan the locations of the wooden houses and outbuildings

[78]See also the examples by Forte (2008, 28 f.) and Arnold et al. (1989, 152).

have been stored in a database; the archaeologists can now create and test hypotheses of temporal change within the village (Kadobayashi et al., 1999, sections 1 and 2.1).

Another kind of change is investigated by Sanders (2007, 565). A shipwreck near Kyrenia, Cyprus was modelled according to the archaeological findings. By virtually reassembling the wreck the researchers hope to find out how and where the cargo was stored in the vessel, for example, by simulating the wreckage and investigating where the cargo from specific points in the ship comes to rest on the bottom of the sea.

5.5.7.5 Astronomy

It has been suggested for some monuments that they could have been used to observe celestial events. While observations concerning the movement of the sun can still be made in the field (if the condition of the site allows it) virtual models provide the unique opportunity to test hypotheses concerning the night sky. They can also be used to study inaccessible or poorly preserved sites.

Stonehenge does not count among the truly poorly preserved sites[79], but it nevertheless features in a reconstruction which is meant to investigate the play of light and shadow on the sarsen trilithons at the time of the summer solstice (Pásztor et al., 2000, 111). From the observations made using their model, they assume that the top of the largest trilithon may have reflected light back into the stone circle, while the rest of the space inside the sarsen circle would still be left in darkness (Pásztor et al., 2000, 112).

The time of the summer solstice is also the focus of the studies relating to the Island of the Sun (Abernathy and Johanson, 2005, 225), Lake Titicaca, Bolivia. Chroniclers have recorded that people came to specific locations on the island to see the sun set. The viewing locations stand in connection with 'solstice markers' through which the sun seems to set. The research endeavour was to identify so far unknown possible viewing locations (Abernathy and Johanson, 2005, 225). For this, a model of the landscape and its architectural features was built and viewsheds were generated, from which possible new viewing locations could be calculated, to be later explored in the field (Abernathy and Johanson, 2005, 226).

It has been suggested that features in the Thornborough monument complex (Great Britain), could be aligned towards stellar constellations[80] (Goodrick and Harding, 2000, 116). Instead of looking at the three aligned henges, a double pit alignment[81] was chosen, which lies close to the southernmost henge. The pits had once held posts which were included in the model. Finally, a representation of the neolithic night sky was draped over the model (Goodrick and Harding, 2000, 117). The whole endeavour is rather disputable from the outset. The night sky changes in the course of one night, also in the course of the year and from year to year (Goodrick and Harding, 2000, 118). A very exact date for the construction of the monument would be necessary to detect any alignments. Additionally the monument would only be aligned to the specific event (rising or setting of a constellations or star) for a short time. No specific date could be assigned to the double pit alignment under examination (Goodrick and Harding, 2000, 118), so that, according to the hypothesis that Orion's Belt may have influenced the layout and the ritual practices in the monument complex[82], possible dates were investigated for alignments with this star constellation. Two possible dates (3000 BC and 2200 BC) for different parts of the pit alignment were identified, which may in one case coincide with the date of erection for this particular monument part (Goodrick and Harding, 2000, 118). This study leaves a lot open to question, not least, whether it is at all sensible to look for stellar alignments, due to the rapidly changing night skies, even if reasonably good dating evidence is available.

5.5.7.6 Lighting conditions

Studies of lighting conditions are part of several projects,[83] but usually only play a minor role regarding the research goals.

In the research concerning Rievaulx Abbey, lighting conditions had a central part. It has been hypothesized that the wish for more sunlight may have led to the church and the chapter house being orientated to the south instead of to the east (Kemp, 1995, 249). That light may have had an influence on the choice of orientation was suggested by the unusually large

[79] The state of 'accessibility' may be open to debate.

[80] Alignments with landmarks, solar and lunar events had to be dismissed (Goodrick and Harding, 2000, 116).

[81] http://thornborough.ncl.ac.uk/double_pit_alignmentC_mainplan.htm (Accessed: 1st of February 2012.)

[82] See http://thornborough.ncl.ac.uk/vrml.htm (Accessed: 1st of February 2012.). The three henges mirror the position of the stars in Orion's Belt.

[83] For example, the Santa Maria Maggiore project (Frischer et al., 2000, 158, 161) or the model of the Northwest Palace, Nimrud (Iraq) (Sanders, 2007, 569).

windows in the chapter house and also by the fact that to the east of the abbey grounds lies higher terrain which would have blocked some of the sunlight (Kemp, 1995, 249, 251 fig. 37.4). Due to time constraints the research project was not finished at the time when the article was published and further testing remained to be done (Kemp, 1995, 250).

More results could be presented by Peterson et al. (1995)[84] who were researching activity areas in a pithouse in Keatley Creek site, British Columbia, Canada. Light could enter the building only through a central smoke hole (Peterson et al., 1995, section: Light Availability). The illuminated area was then matched with the identified activity areas inside the dwelling. Heavily retouched scrapers were predominant in areas which were lit by the midday sun—it was assumed that producing them required good lighting conditions (Peterson et al., 1995, section: Results).

5.5.7.7 Ritual

Chalmers et al. (1995) and Chalmers and Stoddart (1996) propose a photo-realistic reconstruction of the Ġgantija temple on Malta to evaluate ritual. Lighting conditions (with artificial light sources) and sight lines of participants (Chalmers and Stoddart, 1996, 87), the possible positions both of people actively involved in the rituals and of observers, (Chalmers et al., 1995, 226) as well as the influence of smoke, dust or fog on the scenario, are the main points of interest in this research example (Chalmers et al., 1995, 225). Photo-realism in combination with interactivity is deemed essential to test various hypotheses about how the ritual space could have been used (Chalmers et al., 1995, 226). Besides visual 'realism', the sound impressions gained from the simulation should also be realistic (Pope and Chalmers, 2000). The researchers argue that sound provides additional information about the space in which the user is acting (Pope and Chalmers, 2000, 105), e.g. size or position in the room, which is especially helpful if the virtual world can only be experienced on a computer screen instead of being immersive. To this end Pope and Chalmers (2000, 107) propose a voxel-based acoustic model for the site. Unfortunately no research results are presented for this specific example, the articles remaining statements of intent only.

[84] http://www.saa.org/Portals/0/SAA/publications/SAAbulletin/13-4/SAA18.html (Accessed: 1st of February 2012.)

In the Santa Maria Maggiore project the relation between liturgy and the structure of the apse (especially the question whether there was really a space behind the apse as suggested by literary sources) was one of the many questions which were to be researched with the help of the virtual model (Frischer et al., 2000, 156, 158).

5.5.7.8 Hypothesis testing

This term is often associated with Processual Archaeology, but may also have its place in Post-Processual research. Chadwick (2004, 21-22) suggested the 'assessment' of hypotheses instead of their 'verification', i.e. to give them a value of their 'likeliness', e.g. expressed as a percentage. This would still leave the possibility of different hypotheses to be valid, but with a differing degree of probability. There are various examples where research has aimed at hypothesis testing or evaluation.

Hypothesis testing was at the core of the above mentioned model of the Ġgantija temple (Chalmers and Stoddart, 1996; Chalmers et al., 1995).

In the Santa Maria Maggiore project the evolvement of the building over time was modelled, as well as several conflicting interpretations in order to evaluate them, for example, different entrance situations to the nave or the disputed existence of an atrium (Frischer et al., 2000, 158, 160).

Testing hypotheses of temporal change in a Japanese village was the focus of the VisTA project by Kadobayashi et al. (1999).

Earl (1999, 260) built different models of the earthwork at Danebury on the basis of different interpretations of the excavation data. These were tested and the model was modified where necessary (Earl, 1999, 262).[85]

5.5.7.9 Visibility and/or phenomenology

Visibility studies[86] are often coupled with phenomenological approaches[87], as in the case of the Negotiating Avebury project (see Chapter 8) or in the

[85] This is similar in approach to the Negotiating Avebury project, see Chapter 8.

[86] They often incorporate GIS for viewshed analyses.

[87] Monod and Klein (2005, section 3.3) plead for the concepts of 're-enactment, embodiment, hermeneutics and phenomenology' to be part of the research agenda for cultural heritage projects in order to make past events experienceable. It is questionable whether their concepts can in any way be realised. Re-enactment to under-

case of the Peel Gap study by Gillings and Goodrick (1996, section 2.5.1). The latter is concerned with a Roman tower on the line of Hadrian's wall, whose place seems not to conform with the usual assumptions made concerning optimal defensive positioning. The tower is located near a pass instead of on higher ground and is therefore an exception in the positioning of the Roman watchtowers. A viewshed (Gillings and Goodrick, 1996, fig. 14) and an animation created from a 3D-model (Gillings and Goodrick, 1996, fig. 15) show that the view along Hadrian's wall is heavily restricted, but that on both sides of the wall the pass is well visible. To give the observers of the model more degrees of freedom the model was exported as VRML (Gillings and Goodrick, 1996, fig. 16) which could then be altered, provided that the user had enough knowledge on how to change the VRML code! In that case, for example, the height of the wall or the tower could be modified and its effect observed in the VRML-model.

For Salagassos, the question was whether the visibility of certain locations had an influence on town planning and on the specific way the buildings were constructed (Martens et al., 2000, 208). The hypothesis was that in a large town visual recognition of public buildings could be essential for the orientation of visitors and citizens alike. It was questioned whether during the town planning stages special care had been taken to make use of the terrain and of specific architectural features in ortder to ensure this visibility (Martens et al., 2000, 209). Indications supporting this hypothesis could be found with the help of a 3D model (Martens et al., 2000, 208 fig. 4 and 5, 209, 210 fig. 6).

5.5.7.10 Human activity, site function

Reconstructions are often free of any trace of human life.[88] This has been criticised by some authors (e.g. Krasniewicz, 2000, 164), because human usage is what gives the places meaning.

Crowd simulation can help to assess the effectiveness of building designs, for instance how well a military camp can be defended, or other tasks can be performed within the structure (re the example of Avdat in Israel see Hermon (2008, 38)).

Similarly a small amphitheatre at Petra (Sanders, 2007, 571) was reconstructed and 'populated' in order to see how entering and exiting the theatre could have taken place and how many people would fit into it. 'Autonomous individuals' were programmed and some of their behaviour was based on information from those ancient texts which are concerned with theatre scenes. The simulations resulted in the insight that 200 people could easily be fitted into the theatre, that it takes them 7–8 minutes to enter the building and take their seats, and 5 minutes to exit. Besides that, the bottlenecks caused by the theatre's entry and exit design-system could be identified.

Simulation[89], in combination with 3D models could provide a very interesting outlook for hypothesis assessment. Murphy (2007, 18) describes a project where agents were programmed to cooperate and compete along a water canal system modelled on actual findings from Arizona. The point was not to recreate the past, but to see how a social system can work, how stable it is against outside influences, and how much central control is necessary to build and maintain the canal system. Not only the social components were of interest, however, but also environmental factors, which might sometimes challenge the system (long periods of drought, choice of cultivated plants, *etc.*). The simulation allows 'what if' questions to be asked to test the hypothetical scenarios. This could be a very useful concept for testing hypotheses about the use of archaeological monuments as well, for instance, in evaluating different possible ritual practices (for example, in Santa Maria Maggiore or Ġgantija).

Bonde et al. (2009) try to reconstruct monasticism. For this they studied the architectural layout and in particular the abbey customary of Saint-Jean-des-Vignes (Soissons, France). The virtual models play a subordinate role in the project. Frequently only the abbey plans are used, to show which daily routines are prescribed in the customary or to which areas brothers, lay-brothers and visitors to the abbey had access (Bonde et al., 2009, 369-374).

A completely different problem is posed by 'simulating' individuals: i.e. by populating reconstructions with computer controlled agents (Anderson, 2008). The application here lies less in research than in communication, which poses additional problems. There

stand the motives of past agents would necessitate a way to escape our modern views (which is not possible). Neither does virtual reality fulfil the requirement for true bodily experience (as yet), making phenomenology impossible. For a more detailed discussion of phenomenological concepts see also Section 8.2.1.4.

88 Including humans, animals, plants and other objects can increase the degree of uncertainty in the representation and is therefore often avoided. See Section 5.4.

89 Murphy (2007, 20 f.) provides in his article a very clear outline of what can and cannot be achieved with simulations and where their strengths and weaknesses lie.

are historical records of how certain people acted in specific situations, but they will often not suffice to construct a general code of behaviour for any time and any group of individuals. Therefore it is problematic to place virtual agents in a game-like scenario to be encountered by avatars of, for example, museum visitors.[90]

5.5.8 Conclusion: Implications of model applications

Models have to suit their purpose and their intended user group (Niccolucci, 2002a, 4).

With reference to the intended use of the model, the model type (surface or solid model[91]) best suited to the purpose has to be chosen. If displaying murals is the main purpose then a surface model may suffice, but if structural analyses have to be made a solid model is necessary. In some instances also an idealised (Lucet, 1997, see for example) or reduced model (Gillings and Goodrick, 1996, 2.5.1) is suitable for the intended purpose, while in other circumstances a great amount of detail appears to be necessary for the intended analysis (Chalmers and Stoddart, 1996; Chalmers et al., 1995).

The choice of application area has influence also on the expectations of the model users. It would, for example, be desirable if a model for entertainment purposes were designed with great attention to archaeological and historical detail, but on the other hand it is doubtful whether the audience would expect so much diligence going into such a model, if it were not explicitly stated. The opposite is certainly true for educational software, models presented in cultural heritage sites (Terras, 1999, section 2.7) and especially models built for research.[92]

5.6 Problems of collaboration

The question of who is working on the models, in what capacity and during which stages, has a considerable influence on the outcome of any visualisation project.

The archaeologists involved in the production of the earliest models (Bath, Caerleon, Winchester) often provided only the necessary data for the reconstruction, usually plans and sections, which depicted the already reconstructed object on paper. Otherwise they had no further part in the modelling process (Reilly, 1992, 149, 157)[93],[94] leaving the model makers (often persons without training in archaeology) to interpret the material. By providing finished interpretations in the form of sections and plans, i.e. reconstructions in their own right, the archaeologists colluded with this process so that the resulting model could only be an end-product—a 'pretty picture'. That this lack of collaboration between the two parties, namely cultural heritage professionals and technicians, has so far not been overcome can be gleaned from comments of professionals in the museum sector[95].

Miller and Richards (1995, p. 20) criticised the above mentioned practice. They see the archaeologists divorced from their data because the control over it was taken out of their hands by the technicians[96]. The results are 'expensive gadgets', which 'exploit' cultural heritage (Frischer et al., 2002, p. 13), so that visitors of heritage sites are provided with more data, but not with better content.

But the problems do not only affect the visitors to museum exhibitions, they can have an effect on archaeological research issues, too. Models constructed in this manner are only representations of what is already known (Miller and Richards, 1995, 19), and restrict the use of the model for explorative purposes. This is the reason why some authors have recommended that the archaeologists should build the models themselves (Gillings, 2000, 61). Nevertheless his method may also have its drawbacks because archaeologists often may not have the necessary know-how to represent the information in a readily understandable way, rendering the resultant models—in the worst case—meaningless or even misleading (cf.

[90]On the other hand, if sufficient information was available on a specific individual from the past, this approach could be seen as a possible chance for creating a hermeneutic circle, helping the visitor to better understand the past (cf. Monod and Klein, 2005, section 3.3).

[91]See Chapter 10.

[92]Therefore I would argue against Terras (1999, section 2.7), it is not the fact that the reconstruction was created with the help of a computer which creates the sense of authority, but the issuing or exhibiting institution.

[93]This statement stands in contrast to the 'collaborative affairs' and 'symbiotic relationship' between archaeologists and computer scientists which Reilly (1992) claims to exist on page 149.

[94]See also Barceló (2000, 28 f.).

[95]Personal comments by Hermann Schefers (concerning permanent exhibition at Lorsch Abbey Museum (2001)) and Verena Türck (concerning Staufer exhibition in the Reiss-Engelhorn Museum 2010).

[96]It has to be remarked that before the advent of computer modelling the draughtspersons, who were drawing the reconstructions, were also standing between the archaeologists and their data.

Miller and Richards, 1995, 21). The only way to overcome both sets of problems is a close collaboration between archaeologists and technicians (see Miller and Richards, 1995, 20 f.). In a collaboration it should be the archaeologists, who drive the project instead of the technicians (Addison, 2001, 346) because in the end the goal is to resolve an archaeological question (cf. Stork, 2009, 3).[97] In the case of models for museum exhibitions it may also be beneficial to include a person proficient in didactics on the team.

There are exceptions to the above mentioned practice of strictly dividing the area of work between archaeologists (collecting and interpreting) and technicians (modelling). Among them are the three case studies presented in this book[98], and also the 'Troy VR' project (Jablonka et al., 2003, 16), in which the archaeologists took the role of the modellers and the technicians were optimising the results.

In cases where collaboration did take place, it was observed that communication between the two groups could prove difficult, due to differences in vocabulary and thought patterns (Brandl et al., 2002, 25), problems which have to be overcome first. It was also noticed that the model itself became a valuable communication device between the different researchers (see Section 6.4.1).

5.7 Documentation

In order to create added value for any model it is necessary to document the decisions which have been taken before and during its creation, i.e. the pre-modelling and modelling stages. Pre-modelling (see Chapter 9) starts with data collection (and inadvertently selection) in the field or archives (see also Boxed Feature 4 on page 207) and ends just before the decisions for the visualisation are taken (Fig. 5.4) (Wittur and Krömker, 2009, 89, 90 fig. 7). During the modelling process (see Chapter 10) additional choices have to be made with regard to which methods should be used for modelling and how the result is to be displayed (see also Fig. 5.4).

A thoroughly documented reconstruction process is very helpful if additions or changes to the model have to be made in the future (Pletinckx, 2008, 5). It also enables validation of VR models (Vatanen, 2003, 70). Additionally, the documentation can be used to record intermediate results (together with the various iterations through which the model has passed during its creation (Frischer et al., 2000, 158))[99] or possible alternative interpretations. Before and during the modelling process, documentation can help to structure thoughts and modelling approaches (Beacham et al., 2009, section 4). An effective documentation should make note at every point of which decisions were taken and on what basis. This way changes can be implemented more easily because dependencies become obvious and decisions are traceable, no matter whether a source has to be re-evaluated, inferences become unlikely, new information is available or the visualisation needs to be revised for transferral into a new storage format. Information concerning intellectual property rights and privileged information can be stored with the sources or the inferences they affect. Likewise, metadata can be connected to the items (e.g. sources or model elements) they pertain to (Wittur and Krömker, 2009, 89).

It is necessary to decide on structuring principles before starting with the reconstruction (Martens and Peter, 2004, 2 on CD). There are different ideas on how the the documentation data should be structured. While Jablonka (2004, 4 on CD) proposes an object-oriented approach, Vatanen (2004, 4 on CD) argues that a structure which follows along the lines of argumentation (rather than being content driven) would help in the retrieval of relevant information. Besides documenting the data and model basis as well as the different levels of interpretation, it is also useful to document the materials, textures, *etc.* used during the modelling process (Martens and Peter, 2004, 4 on CD).

Suggestions on how documentation should be carried out have been proposed by AHDS (2002, sections 5.2 and 5.4) in *Guides to Good Practice: Creating and Using Virtual Reality* and the *London Charter* (Beacham et al., 2009, sections 3 and 4) which both aim at giving guidelines for virtual reality modelling.

The sections in *Guides to Good Practice: Creating and Using Virtual Reality* (AHDS, 2002, sections 5.2 and 5.4) concerning documentation primarily see the need to provide metadata[100] with the virtual reality application: title, creator, subject, ...source, ...images/textures, ...interpretative objects, *etc.* This does not constitute a full documentation of the modelling project, but more a technical description of the end-product. It is more suitable for finding, retrieving and using a virtual reality

[97] Addison (2001, 346) argues that 'technology should be in the service of society and culture'.

[98] See Chapters 6, 7 and 8.

[99] This has been the case for St. Lawrence's church in Ename, Chapter 7.

[100] See Section 5.8.2.

model from the internet or a database, and also to give the user some indication on how 'trustworthy' the reconstruction might be (on the basis of the institution in charge of the model). But it is not suitable as a guide to understanding the model and the underlying interpretation, nor for verification or altering the model.

More detailed information on what documentation should include, and for what end, is given by the *London Charter*, though this document leaves completely open how the process of documenting should be done. For instance, the section concerning the *Research Sources* (Beacham et al., 2009, section 3)[101] recommends that the relevant sources which were used, considered or have influenced the project should be identified. Source evaluation, including the identification of possible factors biassing their contents, should be executed 'with reference to current understandings and best practice within communities of practice' and documented also.
This description leaves open what constitutes a 'relevant' research source—which may be a very subjective choice. In some cases it would also be useful to keep track of the seemingly 'irrelevant' sources, because they may become relevant in the changing light of new research. There is also no explanation concerning the way the source evaluation should proceed or how exactly the documentation should be carried out.

The same is true for the section entitled *Documentation*[102] (Beacham et al., 2009, section 4). The list of which parts of the pre-modelling and modelling process should be documented is very comprehensive:

1. What does the model represent? Where is the border between fact and fiction?
2. Documentation of all the research sources used.
3. Documentation of 'paradata', i.e. all the interpretational and creative steps which have been taken in the composition of the model. They should show how the sources are linked to the model and what dependencies exist.
4. Documentation of methods, including why a virtual representation was chosen and which methods were used in the visualisation process. (The authors are requested to write an easily understandable document.)
5. Dependencies between elements (i.e. parts of the model, hypotheses, *etc.*) should be made explicit.
6. The documentation should be disseminated including documents in encodings[103] (images, text, *etc.*) best suited for the purpose.

As this short summary shows, the guidelines given here are open to be implemented in various ways. This is certainly intended by the authors of the London Charter because different projects may find different ways leading to a suitable documentation.

Nevertheless, there are more concrete suggestions on how to approach documentation issues. An early example is presented by Frischer et al. (2002, 13 f.) who propose a philological approach. This has been followed in the 'Progetto *Insula* del Centenario (IX, 8) and is therefore explained in detail in Chapter 6 (Boxed Feature 3 on page 61). The philologic approach pertains mainly to data and metadata in relation to the virtual model (Frischer et al., 2002, 14); inferences are dealt with, but only marginally.

Vatanen (2004, 4 on CD) criticises that such a content-driven approach[104] would not be efficient. Instead he suggests an argumentation-driven approach[105], which would be better suited for information retrieval. Vatanen (2003, 71) argues rightly that metadata, especially in linear text format,[106] is on its own not enough for the purpose of model validation. He suggests the use of a metadata network: a system of 'descriptor nodes'. These descriptors should be linked to the model, but also linked to each other.

> [. . .] apart from describing only one entity in the virtual model the descriptor nodes should also contain information on the probability of information, its probable, definite and possible relation to other descriptors and entities, characteristics, and information on how to deconstruct the current document structure and rearrange it according to a new set of premises [. . .]. (Vatanen, 2003, 73)

In a later article this idea is elaborated, integrating the need to allow for different subjective views in the in-

[101] http://www.londoncharter.org/principles/research-sources.html

[102] http://www.londoncharter.org/principles/documentation.html

[103] See Boxed Feature 2 on page 29.

[104] Parts of the model are annotated with metadata and other information pertaining to them.

[105] The lines of reasoning are the relevant parts of this scheme.

[106] Encodings (e.g. images, text) as well as format (linear/nonliner) should be appropriate to the application area and to the sensory impressions they are best suited to make (Vatanen, 2003, 71).

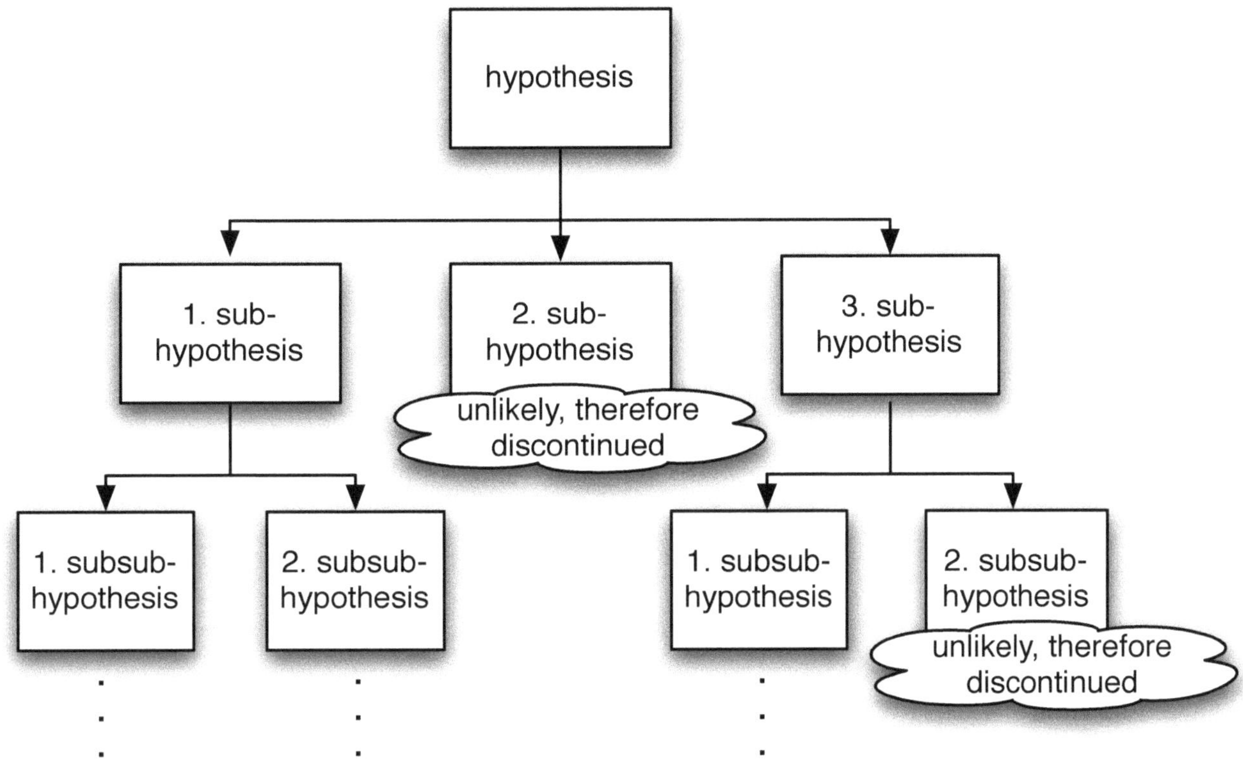

Figure 5.2: Hypothesis tree as described by Pletinckx (2008, 13 f.). In the end only the most likely hypothesis 'branch' will be modelled. (Drawing by Joyce Wittur)

terpretation and representation (Vatanen, 2004, 5 on CD).

> This presents two important demands on the practical realisation of the infrastructure: (1) that the practical infrastructures consist of co-existing, interrelating, controversial, conflicting and incomplete structures that still complete each other. (2) Infrastructures have to be able to be evaluated, not by absolute values, but on relative, explicitly political [read: subjective] scale that relates itself to relative yet established categories. (Vatanen, 2004, 5 on CD)

The approach recommended by Vatanen (2003, 2004) contains some very useful ideas. For one, it is insufficient to provide only information about the model basis regarding visualised parts (which in fact does nothing much to *explain* the reconstruction) or metadata if it is structured in a hierarchical way: i.e. each element in the reconstruction is linked to its own set of metadata. Information which can be extracted from the sources is often interconnected, pertaining to several parts in the model; likewise interpretations take dependencies between different parts of the model[107] into consideration. These should be expressed in the documentation, and, as a consequence, the documentation has to be non-linear to allow for a network structure.

Different researchers will come to different (subjective) conclusions if confronted with the same material, through subjective evaluation of sources and differences in making inferences. Therefore it is reasonable to provide a means of evaluating 'infrastructures' on a relative scale[108] to find out how 'probable' an interpretation/model seems to be.

Two problems stand out: first, how to integrate information on how to deconstruct and rearrange the document (see Vatanen, 2003, 73); second, how to make a coherent whole including controversial and conflicting information (see Vatanen, 2004, 5 on CD)?

To tackle the first problem, all the *probable* (in contrast to all the *possible*—i.e. everything can be linked with everything) links between the pieces of information and all possible rearrangements of the data would

[107]Usually in the form of 'if ... then ...' relationships.

[108]Cf. Chadwick (2004, 22) and his concept of hypothesis assessment.

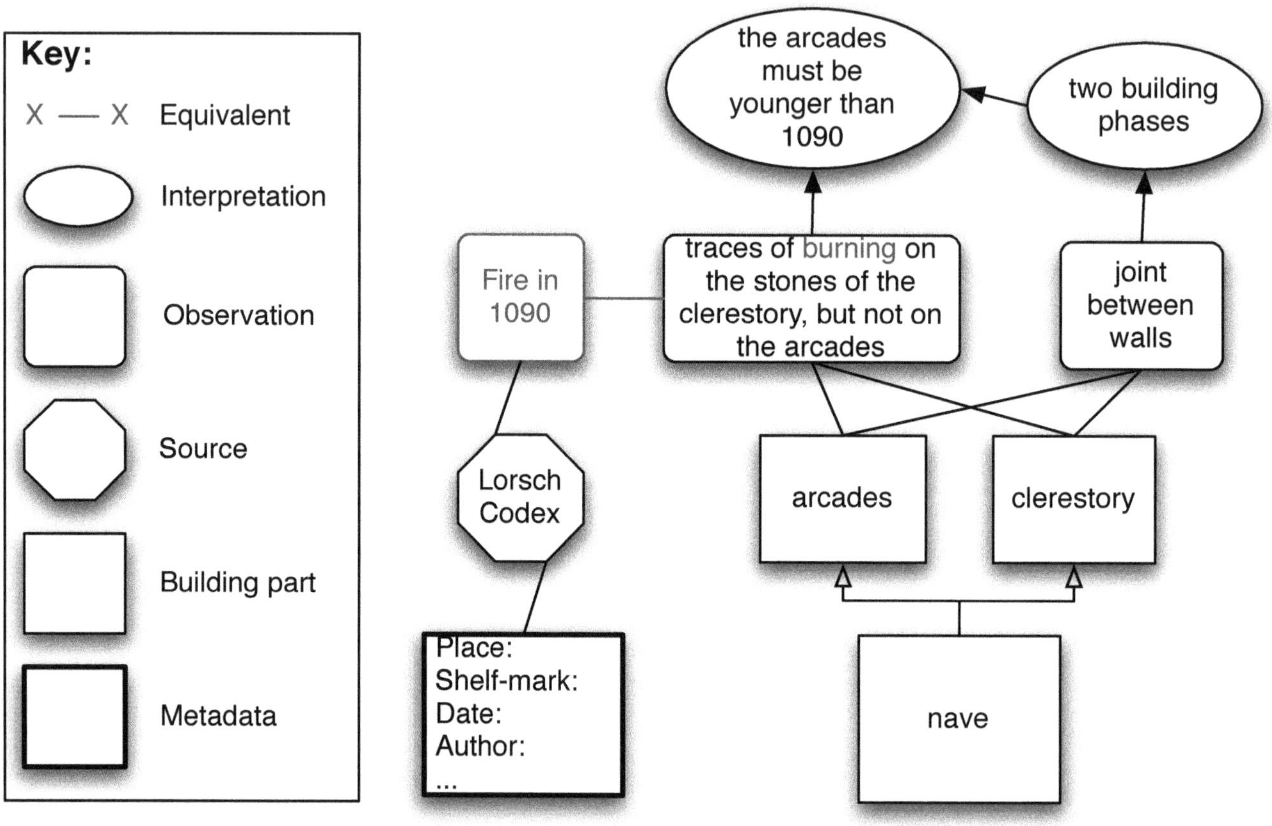

Figure 5.3: Example of an Argumentation Network, NB the reading direction is from the bottom to the top. One nave wall consists of the arcades and the clerestory. It can be observed that a joint exists between these two wall parts which leads to the conclusion that they belong to two different building phases. It can also be observed that there are traces of burning on the stones of the clerestory, but not on the arcades. For 1090, a fire in the church is reported in the Lorsch Codex. This fire is taken to be equivalent with the burn marks on the stones of the clerestory. This, and the conclusion that there are two building phases, leads to the further conclusion that the arcades must be younger than 1090. (Shortened argument taken from Platz (2005, 2007b).) (Drawing by Joyce Wittur)

have to be known beforehand—a rather unlikely situation. In any case it is necessary to have an understanding of the whole available information and argumentation (and the information dependencies in particular) to see how deconstruction and rearrangement would effect the final outcome. Dependency indicators can help in the task of understanding the information structure, but cannot provide ready-made solutions to the problem for the researcher.

The second problem can be tackled more easily, by simply not making a coherent whole of all the available information. It is important to collect all the information on the subject matter, and if a piece of it has, with good reason, been deliberately excluded from or cannot be fitted into the interpretation, stating the fact and the reasons for it will suffice to establish a link.

Documentation strategies applied to real archaeological data are presented by Pletinckx (2008, esp. 26–33) and Wittur and Krömker (2009).

The approach described by Pletinckx (2008) is discussed in detail in Section 7.3.6 and presents a top-down (or 'content-driven') approach. Starting from general hypotheses about the whole building (cf. Fig. 5.2[109]) these will be broken down into sub-hypotheses about individual building parts and evaluated. If a (sub-)hypothesis seems to be unlikely further evaluation will be discontinued. Only the most likely hypothesis will be modelled, the modelled parts (each corresponding to a sub-/subsub-/...-hypothesis) will be linked to the corresponding documentation part. Despite starting with the arguments—formulated as hypotheses—this is a

[109]In Fig. 5.2 only one 'hypothesis' has been illustrated as an example; alternative hypotheses will be treated the same way.

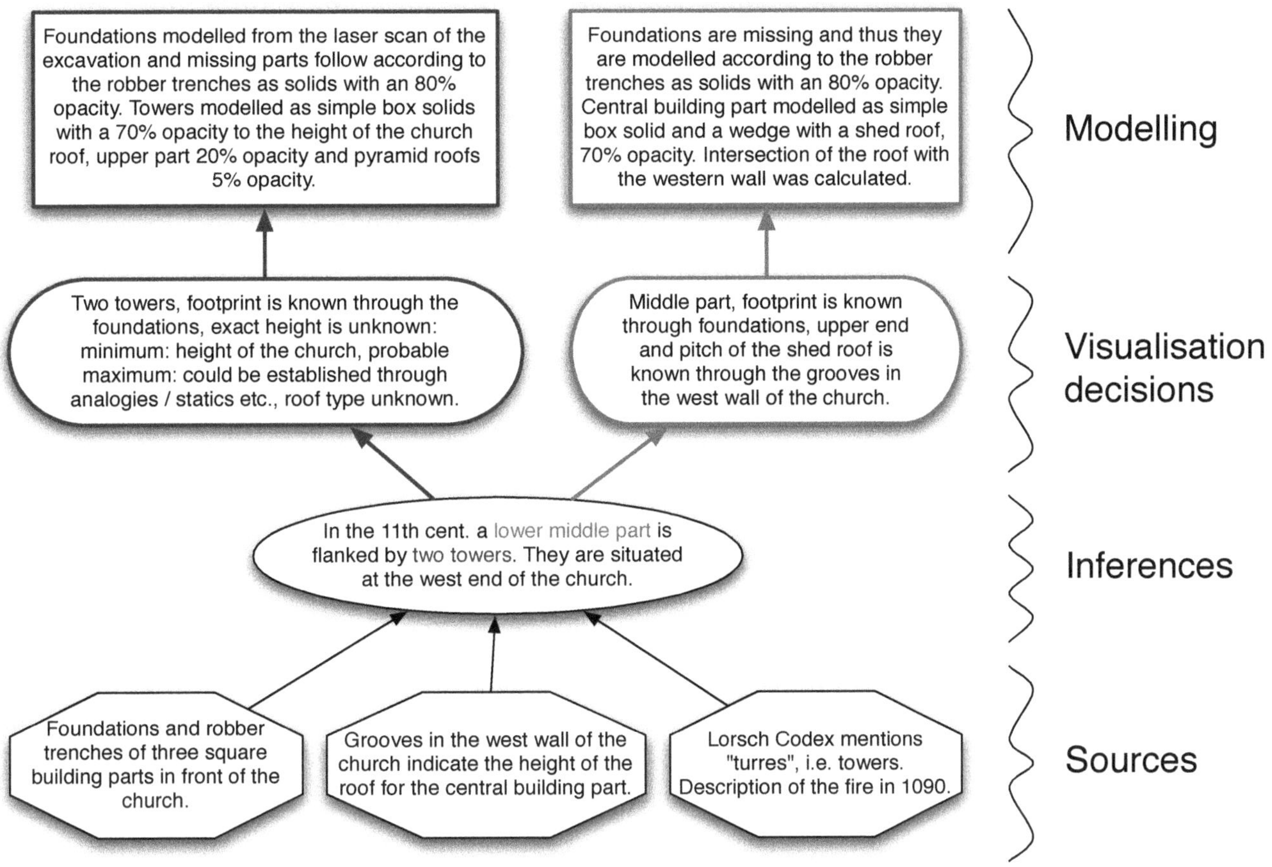

Figure 5.4: A simplified example of how Argumentation Networks can be used to document the sources, inferences, visualisation decisions and the modelled parts in a reconstruction (Wittur and Krömker, 2009, 90 fig. 7). (Drawing by Joyce Wittur)

content-driven approach because the hypotheses are closely tied up with the resultant model, and it is top-down because it starts with the whole structure (building) and is then broken into smaller parts.

I have proposed a bottom-up (or 'argumentation-driven') approach (Wittur and Krömker, 2009, 88-90) which can be used to document already existing interpretations as well as new ones.[110] A bottom-up approach was chosen, because this follows the way in which archaeological reports are usually structured: beginning with the excavation results hypotheses will be introduced. The reports are written in a linear form, but the arguments used often go back to earlier observations or draw on other source material (historical texts, images, *etc.*) so that the resultant diagram forms an irregular network.

Several stages starting from the primary data to the final reconstruction have been identified. As mentioned in Boxed Feature 4 on page 207 the primary data (e.g. the excavated remains or the historical event) have to undergo a stage of interpretation, here called *observation*, to be defined and described. These observations (e.g. a feature recorded in an excavation report or a historical text) constitute the *data basis* and can be annotated with metadata. On the data basis a number of further inferences are based (these can also lead to decisions to exclude some pieces of information from the model basis). Further, the inferences build upon each other and in some cases it becomes necessary to draw on additional sources (e.g. observations so far not considered or texts by other researchers) which are thus added to the data and/or model basis. An example is shown in Fig. 5.3. The argumentation part depicted in Fig. 5.3 is in respect

[110]The initial aim had been to retrace interpretations made by Adamy (1891) and Behn (1934a) which resulted in reconstruction drawings of Lorsch Abbey church by the respective authors. The written publications were analysed according to which sources were used, which inferences were made and which conclusions drawn in order to understand the arguments behind the reconstructions. I treated two texts by Platz (2005, 2007b) in the same way though they stood not in connection with any reconstruction. The results of these analyses were then transcribed into diagram form (Argumentation Networks), constituting a proof of concept.

of its interpretation stages equivalent to the two bottommost stages in Fig. 5.4 labelled *sources* and *inferences*.

While the part labelled *inferences* in Fig. 5.4 gives an impression of what the archaeologist assumes to have been there, the description in the diagram is too imprecise to be rendered into a visualisation. An elaboration which meets the visualisation requirements takes place in the next step: *visualisation decisions*. Yet another step is necessary during the actual *modelling* process: decisions about how to display uncertainty, which rendering methods to use, *etc.*

This approach enables the user to document all the sources used, all the inferences which were drawn from them and also how the results were incorporated into the model. It would be possible to attach a degree of confidence to the sources, which would then have an impact on the interpretations based on them. The resulting degree of confidence for the interpretations should not be static—it is quite possible that, based on two sources possessing a high degree of confidence, an interpretation is made in which the author of the interpretation (or another researcher) places no great confidence at all. Therefore it is necessary that subjective degrees of confidence can be placed alongside the calculated ones[111]. This way interpretations can also be evaluated by other researchers holding different opinions.

5.8 Data basis

Any model must cope with data[112] and it is important to be aware that all kinds of documents—be they excavation records, find descriptions (Bateman, 2000, section 1; Vote et al., 2002, 83) or mediaeval text sources—are interpretations. They are (unintentionally) affected by cultural and personal experiences of their creators (Bateman, 2000, section 2) and therefore only as 'objective' as their publisher (Hermon and Fabian, 2002, 106). This is especially true in the case of reconstructions because they are interpretations (Zhukovsky, 2001, 297) on the basis of interpretations.

[111]Which are subjective in their own right.

[112]The matter of the data basis is elaborated in Section 9.1. The term is defined in Chapter 3.

5.8.1 Data

Many models created for communication (esp. for museum exhibits) are not designed for subsequent changes. Once created they are not updated, but remain in their original state.[113] If a model is to be used for a longer period of time (which should include updates of the contents (Messika, 2002, 149)) or for research purposes, it is important to consider what data format should best be used, especially if different reconstructions are to be combined or complemented (Frischer et al., 2002, 15).

Also, transferring models from one computer system or display device to another often meets with great obstacles. In order to resolve these problems, standard systems need to be established (Goodrick and Earl, 2004, section 3.1). Regarding the ephemerality of the software used to create the models and the often proprietary data formats (which should best be avoided) it produces, it is of paramount importance to pay attention to means of data preservation (AHDS, 2002, section 6.2)[114] as well (see Section 10.5).

5.8.2 Metadata

Metadata are often described as 'data about data'.[115] The use of metadata is essential for finding and retrieving objects or pieces of information from the model, the documentation or an affiliated database (AHDS, 2002, section 7.2.1)[116]. They are therefore used in cataloguing and archiving as well as in content and technical descriptions (AHDS, 2002, section 7.2.1). The usage of metadata should best follow an established standard, for example, Dublin Core (DCMI, 2010)[117]. The latter is well suited for use with the data basis of virtual models, but does not go far enough to be considered a sufficient documentation tool for the model itself (cf. Section 5.7). Additional, important information pertaining to the virtual model concerns its underlying structure, its content as well as the sources and methods employed in its construction (Ryan, 2001, 249).

[113]This is usually due to funding policies. Money is made available for a specific purpose, e.g. creating a temporary or permanent exhibition including a virtual model, but no funds are allocated for its future maintenance.

[114]http://guides.archaeologydataservice.ac.uk/g2gp/Vr_6-2.

[115]A useful example of metadata are the contents of a library catalogue: the datum would be the book, while the metadata would comprise the information concerning shelf mark, author, title and year, but also condition of the book, price, *etc.*

[116]http://guides.archaeologydataservice.ac.uk/g2gp/Vr_7-2

[117]http://dublincore.org/ (Accessed: 1st of February 2012.)

Most of the modelling software and 3D file formats do not support metadata, which means that if metadata are used they have to be stored in external documents and then linked to the respective model parts (Ryan, 2001, 255). One of the rare exceptions is X3D (Extensible 3D)[118]. Trials using this format for archaeological purposes have already been made (Niccolucci and Cantone, 2003).[119] X3D allows metadata to be associated with parts of the model or the whole X3D world,[120] but allows only for metadata in the following formats: strings, single-precision and double-precision floating point values, and integers,[121] which can be roughly 'translated' as alphabetic characters and numbers. Metadata are in general represented in text format disregarding the item they describe.

Virtual models have also been compared with text[122] and some of them (e.g. VRML and X3D) can be output as text. However, the data basis and the interpretations behind them are non-linear and would therefore require another approach, which among other things allows for browsing and starting/ending at any point (Vatanen, 2003, 71 f.). Vatanen (2003, 71) also argues, with reason, that metadata in text format only is not enough because words cannot express everything—in particular when transferring information from one encoding to another (e.g. image to text) some informational content will be lost. Additionally, the brain processes different sensory input in different ways. Depending on the code and mode of the information, it can arrive at deviating conclusions (see also Barceló, 2001, 221). However, different metadata encodings can lead to query problems. It is already difficult to find specific content in 3D models and their associated metadata (Vatanen, 2003, 72) even though searching for a particular term in a textual database is comparatively simple[123]; but to search for particular shapes in images or even films is a much more complicated task.[124] The search mechanisms required for the task are not available yet (Vatanen, 2003, 72). Some ideas on how to approach these problems (albeit with another focus[125]) have been proposed by Ragia and Leopold (2003). In the meantime mere textual metadata will have to be used.

5.8.3 Accessing the data basis

Connecting the data basis with the model is an essential part of reconstruction assessment especially so that a level of confidence (see Section 5.9.1) based on the reliability of the sources may be obtained (Hermon and Nikodem, 2008, 1). Several authors[126] have suggested that the best solution would be to link the data basis, or at least the model basis, (e.g. in the form of database entries) directly to the individual objects in the model, so that it is possible to use the model not only as a visualisation of the interpretation, but also as a front-end for queries of the data underlying it. Nevertheless, it is wrong to assume that the 'ruins [. . .] will speak for themselves' (Lock, 2003, 155). Additional information is therefore necessary.[127]

Some critical voices have remarked that it would be impossible to include all the available sources in a model because of problems resulting from the sheer amount of data (Roussou, 2002, 98)[128] combined with the necessity to establish relations between archaeological finds and features, taking their spatial location into consideration, *etc.* for which conventional databases are not well equipped (Vote et al., 2001, 313 f.). Yet, this kind of information is essential for creating, evaluating and communicating interpretations (see also Section 5.7).

So far no such complete and structured documentation approach has been fully implemented. Some information systems for communication in the field of didactics have been established (for example, see Chapter 6 or Bonfigli and Guidazzoli (2000)), or

[118] X3D is the XML (Extensible Markup Language)-based successor of VRML. See also http://www.web3d.org/about/overview/ (Accessed: 1st of February 2012.).

[119] They were met with some technological problems, because the format had not been sufficiently developed for the purpose by then (Niccolucci and Cantone, 2003, 59).

[120] http://www.web3d.org/x3d/specifications/ISO-IEC-19775-1.2-X3D-AbstractSpecification/Part01/components/core.html#Metadata sections 7.2.4.3 and 7.2.4.4. (Accessed: 14th of March 2011.)

[121] http://www.web3d.org/x3d/specifications/ISO-IEC-19775-1.2-X3D-AbstractSpecification/Part01/components/core.html#Metadata section 7.2.4.2. (Accessed: 14th of March 2011.)

[122] See Frischer et al. (2002, 13 f.) and their philological approach to documentation, Section 5.7 and Boxed Feature 3 on page 61.

[123] Even though synonyms, different languages, spelling, *etc.* can create obstacles (cf. Ragia and Leopold, 2003, 103).

[124] There are research efforts to overcome these problems, see for example, http://hci.iwr.uni-heidelberg.de/COMPVIS/research/se/ (Accessed: 1st of February 2012.).

[125] The intention is to match place descriptions from ancient texts with (modern) geographic places. Starting from the identification of relevant text passages and the identification of possible sites with the aim of matching them (Ragia and Leopold, 2003).

[126] Scagliarini Corlàita et al. (2003, 252), see also Section 6.2.1.1, Forte et al. (2003, 54) and Ryan (2001, 246).

[127] See section 3.2 and 3.3 of the London Charter (Beacham et al., 2009) as well as Vatanen (2003, 73) and Pletinckx (2008, 7–13).

[128] Selection of the data included in the model is certainly necessary. This can be achieved if a clear research question is kept in mind so that superfluous data will be excluded.

databases containing sources linked to the models, but without any additional structured interpretation (e.g. Cosmas et al., 2001; Jablonka et al., 2003).

5.9 Interpretation of the data basis

After the data basis has been evaluated further interpretation stages are necessary to come to conclusions about the structure and workings of the reconstructed object.

To arrive at a complete or even partial picture it is necessary in order to depart from the observations of the remains and to advance into more complex interpretations. The further the interpretations advance the more uncertainties enter the model. The question is how visually to separate uncertain or reconstructed parts from finds which are still *in situ* or otherwise elicit a higher degree of confidence, and how a degree of confidence in general can be expressed (Section 5.9.1).

Similarly, different researchers[129] will come to diverging interpretations of the primary data and thence arrive at different observations, interpretations and finally alternative reconstructions. How shall such alternatives be treated? Is visualising only one interpretation sufficient or should alternatives be shown, considering the cost factor which would be involved (Section 5.9.2)?

5.9.1 Displaying uncertainties and reconstructed parts

In the restoration of cultural heritage, whether it is a building or an artefact in a museum, it is common to mark additions, for example, by using a different colour or other materials[130]. It is therefore surprising that the call for similar procedures in virtual reconstructions (see for example, Barceló (2001, 240) or Ryan (2001, 246)) is rarely heeded.[131]

In order to mark reconstructed or uncertain parts in a virtual reconstruction, different methods have been employed:[132]

- using transparency to mark uncertain parts (Viti, 2004, 7 on CD fig. 7; Schedl, 2003, 6; Goodrick and Gillings, 2000, 54 fig. 6, see also Section 8.2.2.1)
- using augmented reality (AR) to differentiate between the still existing parts and the virtual reconstruction (Vlahakis et al., 2001, 131, 136 fig. 6; Pletinckx et al., 2001, 199 fig. 4, see also Section 7.3.2.1)
- using greyscale for uncertain parts, while certain parts are displayed in colour (Scagliarini Corlàita et al., 2003, 246 f. see also Section 6.2.1.1) or using colour-codes (Reilly, 1992, 159; Eiteljorg, 2000, fig. 8 a, b)
- using non-photo-realistic rendering (NPR) (Freudenberg et al., 2001, 97; Strothotte et al., 1999, 36, Lehmann and Haarlammert, 2010, 174) or wireframes (Hermon and Nikodem, 2008, 4, 5 fig. 3) to mark uncertainty.

Employing transparency, greyscale or NPR also helps to remove the impression of 'realism' from the reconstruction, showing that it is only a hypothesis.[133]

Blurring can also be a solution to indicate uncertainty. The blurred textures used in the Santa Maria Maggiore project performed the double purpose of indicating that murals covered the walls, but at the same time prevented the identification of any particular motif (Frischer et al., 2000, 160).

Standardising the representation method for displaying uncertainty or reconstructed parts has been suggested (Vote et al., 2002, 83) especially in regard to representing different *levels of confidence* which should be distinguishable to the observer.

Establishing a level of confidence will always be based on a subjective assessment (Hermon and Niccolucci, 2004, section 1). Therefore the degree assigned will belong to the ordinal measurement

[129] Even the same researcher can come to multiple conclusions.

[130] For instance the 'pink cement approach' in Britain (Cornforth et al., 1992, 224).

[131] Even more surprising is the fact that life scenes in images seem to lag even further behind in this respect (cf. Mainka-Mehling (2008, 246) who proposes an approach for archaeological images following the principle of anatomy books in which transparent sheets can be overlaid to show different aspects or parts of the human body).

[132] Several more methods are mentioned by Kensek (2007) and Zuk et al. (2005, 103 table 1), but not all of them are applicable to virtual models (for example, tactile cues (Kensek, 2007, section 4)) or have been utilised in the archaeological field outside of feasibility studies (e.g. uncertain objects move (Zuk et al., 2005, 104 fig. 6)).

[133] The augmented reality approach may still make the model appear rather realistic to an observer.

scale[134]. The degree can either be assigned to the individual parts of the reconstruction or can be based on the evaluation of the data basis (Solmi and Vecchietti, 2002, 201; Coralini, 2007, 25) and interpretations which underlie the reconstructed part (cf. Pletinckx (2008, 5) and Reilly (1992, 159), see also Section 5.7). Both outcomes will be subjective, though the latter approach will perhaps result in a more structured process of evaluation.

Finding a sufficiently suitable scale for measuring a degree of confidence has been the problem in the Roman Forum Project. In the end only one building, the Curia Julia, was displayed with three levels of confidence (Abernathy and Johanson, 2005, 222 f.). The endeavour was discontinued for the rest of the Roman Forum.[135] For the Curia Julia a graded display has been attempted by using colour for parts still *in situ*, greyscale for data which was later lost, and transparency for areas where nothing remains. These different grades were based on the available source material.[136]

As can be seen from the above and the case studies in Chapters 6 and 7 three levels of confidence have been preferred in these contexts.[137] Other authors have suggested the use of finer gradients, especially if a degree of confidence is to be calculated. Both fuzzy logic (Vatanen, 2003, 72; Hermon and Niccolucci, 2004; Hermon and Nikodem, 2008, 2 ff.) and probability theory (Sifniotis et al., 2006) have been recommended for this use. The mathematical calculations will inevitably result in fine grades of confidence. These are easy enough to display[138] but small differences are not easily discernible[139] to the human eye (see also Eiteljorg, 2000, section: Attractive Images or Good Information?). In order to overcome this problem, thresholds can be employed to reduce the number of styles displayed (e.g. complete or wireframe) (Hermon and Nikodem, 2008, 4).

In some cases the spatial extent of a feature is unknown either because it is only mentioned in written sources to which no archaeological features correspond (any more) or because the size of the feature cannot (yet) be established by means of archaeological methods or surveys. Jablonka (2004, 6, 7 fig. 4) presents a way of displaying this data in a GIS: by giving a maximum and minimum extension of the feature and indicating the site type (settlement, cemetery, *etc.*) and period to which it belongs, by a colour- and shape-coded symbol at its centre. The same approach could be used in a virtual reconstruction by marking the respective areas in the 3D-scene.

5.9.1.1 Time

Visualising temporal change or temporal uncertainty has rarely been tackled. It poses slightly different problems from displaying spatial uncertainty, but can be of essential help in the interpretation and reconstruction process, e.g. in 'validating' the reconstruction of a particular time period, by comparing it to models of previous and following periods (Pletinckx, 2008, 13).

Some applications incorporate a time slider or some other device which lets the user choose a particular time period. By using it, an automatic query in a database is launched which leads to objects from the requested period being retrieved and displayed (see e.g. Bonfigli and Guidazzoli (2000, 143), Cosmas et al. (2001, 297) or Meyer et al. (2006, 133)).

However, for most archaeological finds no absolute creation and destruction date[141] exists, so most model parts can only be placed in a relative chronology (cf. Holst, 2001, 159; Zuk et al., 2005, 99 f.).

A method for recording building sequences[142] was presented by Holst (2001), which allows for making note of all the relative chronological relations between structures. From these records it is possible to develop (usually several) potential building sequences (Holst, 2001, 160 f.). The outcome of these records would be an input ideally suited for the visualisation system envisaged by Zuk et al. (2005). Their approach uses timelines of objects (fig. 5.5, left), which can also include some uncertain values (fig. 5.5, right). In this way it can be established and displayed whether the artefact in question existed,

[134] The values will be in an ordered scale: ranging e.g. from 'very high confidence' to 'very low confidence'.

[135] Instead, it was preferred to include links to the source material and to offer alternative reconstructions.

[136] Similar to the *Insula* del Centenario project, Section 6.2.1.1.

[137] The number three seems to lend itself readily for categorisation into 'sure/*in situ*', 'medium' and 'hypothetical/speculation'.

[138] The modelling software allows for any value between 0-100% opacity.

[139] They may not be discernible at all, or, even with a key at hand, may result in reading mistakes due to optical illusions: see example in Ware (2004, 75 and fig. 3.8).

[140] Even more detailed information about the *degree* of certainty/uncertainty could be included if the timelines took the shape of violin plots.

[141] Whereof the destruction date may lie in the future (Zuk et al., 2005, 100).

[142] The method is aimed at recording the overlapping footprints of iron age and mediaeval wooden buildings and suggesting possible construction and destruction sequences for a settlement (Holst, 2001, 159).

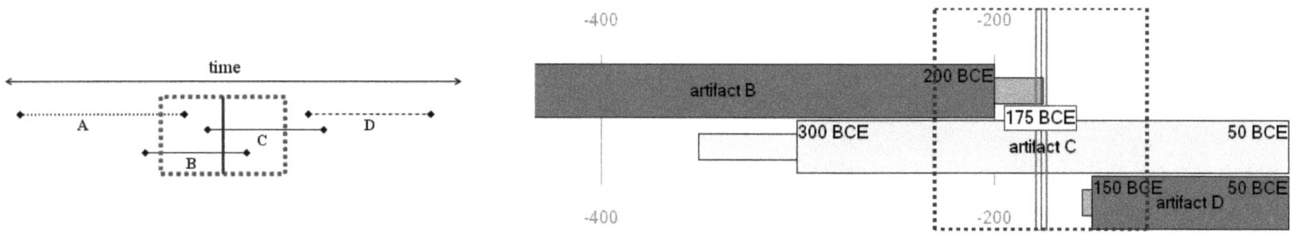

Figure 5.5: Timelines and time windows. Left: four timelines for four objects (A–D) whose existence overlapped in parts. The software can display objects which existed at a certain point in time (B and C, crossed by the red line) or if there is uncertainty about the exact point in time, a 'time window' may be viewed which also includes previous and later points in time (yellow box, which also includes A). In the visualisation of the latter case B and C would be displayed as 'certain' while A would be displayed as 'uncertain'. Right: the schema in the left image is elaborated to show that uncertainty about the object's timeline can also be included. The timelines of artefacts B, C and D have lighter and thinner extensions attached to them (reminiscent of box-and-whiskers plots)[140] which indicate that it is uncertain whether the artefacts were in existence during these times. If these extensions should fall in the chosen time frame, the artefacts would also have to be marked as uncertain in the visualisation. (Zuk et al., 2005, 100 fig. 1, 102 fig. 3).

may have existed or did not exist at a certain point in time (i.e. whether the artefact timeline coincides with the red date line in figs. 5.5) (Zuk et al., 2005, 100). Additionally it is possible to look left and right of the date line (i.e. to create a time window) and to include—with uncertainty—artefacts whose timeline reaches into the window, but not up to the date line (Zuk et al., 2005, 100). The two uncertainty factors (inherent to the timelines and the time window respectively) can be regarded separately or combined (Zuk et al., 2005, 100). The resultant visualisation is connected to a time slider so that the user can explore the changes over time interactively (Zuk et al., 2005, 101).

5.9.2 Alternative reconstructions

There are few authors, who argue for the use of alternative reconstructions (e.g. Ryan (2001, 246); Huggett and Guo-Yuan (2000, section: The Lack of Alternatives) or Eiteljorg (2000, section: Towards an honest and complex view of the past)) and even less who present (e.g. Viti (2004, 5 on CD fig. 5, 6 on CD fig. 6) or Niccolucci and Cantone (2003, 61 fig. 1)) or plan to construct them (Jobst, 2004, 6 on CD).[143]

In some cases the creators of the models seem unaware of the problem, have the impression that they have to pick only one hypothesis for visualisation (Hermon and Fabian, 2002, 104; Pletinckx, 2008, 19 f.) or are afraid to overtax their audience by presenting several choices (cf. Eiteljorg, 2000, section: Towards an honest and complex view of the past). But virtual model representations should not be too undemanding either. Lay audiences should be informed about what is archaeologically known and what is not, and that different scholarly opinions exist and why (cf. Eiteljorg, 2000, section: Towards an honest and complex view of the past). A survey by Baumeier (2008, 5) might even show that laypeople *want* to be informed about different hypotheses instead of being presented with only one solution.

Alternative reconstructions can actually be used for different purposes. Abernathy and Johanson (2005, 222) kept older versions and alternative models in order to show the stages and outcomes of different argumentation processes. Alternatives can also be used in an endeavour to express uncertainty (James, 1997, 31 f., 30 fig. 2.3, 31 fig. 2.4, 32 fig. 2.5): through the preparation of three reconstruction drawings of a house—two of them presenting the extremes of the possible spectrum (with the 'truth' lying somewhere in between)—it was intended to encourage discussion among other researchers.[144] In the Negotiating Avebury project (see Chapter 8) the visualisation of different hypotheses (between which the user can toggle) is used to evaluate them (Earl and Wheatley, 2002, 11). In the proposal for the Lorsch Abbey Information System, previous reconstructions, intended to be linked to their model basis[145] and the underlying interpretations, are meant to help create a 'history of research' for the site (Wittur and Krömker, 2009, 87).[146] A more playful and creative approach

143This may be due to time and budget constraints.

144To the regret of James (1997, 32) no discussion concerning the feasibility of the two extremes followed the publication.

145Where identifiable rejected information should be recorded as well.

146See also http://www.iwr.uni-heidelberg.de/groups/ngg/Lorsch/aims.php?L=E (Accessed: 1st of February 2012.).

in the form of an interactive model of the Roman theatre in Canterbury is presented by Roberts and Ryan (1997, fig. 8). Within the borders of what is known of Roman theatre architecture (cf. Ryan, 1996, 106) the user is allowed to change the number of seats, their height and their depth as well as the height of the building. It is also possible to add a walkway between the seats. Roberts and Ryan (1997) tested three further possibilities in order to publish alternative reconstructions/interpretations with varying degrees of interactivity. With the degree of interactivity rises the level of knowledge the user needs to have about the subject matter if he or she wants to manipulate the model within reason (cf. Roberts and Ryan, 1997, section 2.4.3), while in the low interactivity example the authors see the risk of biasing the user's viewpoint (Roberts and Ryan, 1997, section 2.1.3).

In any case, if alternative reconstructions are shown, it is important to point out their differences in order to guide the observer (cf. Weidenmann, 2002a, 90) and to enhance his or her understanding by explaining the underlying interpretations (cf. Roberts and Ryan, 1997, section 2.1).

5.10 Perception

In relation to computer generated models, very little attention has been given to aspects of perception and cognitive theory. However, the fear of misleading the observer with technically perfect images is great (see for example, Cornforth et al. (1992, 224); Miller and Richards (1995, 20) and Eiteljorg (2000, section: The Double-Edged Sword)). These fears are usually underpinned with personal experiences of the authors, for instance Eiteljorg (2000, section: The Double-Edged Sword) uses his experience in relation to documentary-style motion pictures to proclaim that; if fact and fiction are mingled and the images are very vivid, the fiction may in the end be comprehended as fact.

The realistic pictures and the adoption of highly developed technical equipment for their production are identified as the main problems: they are thought to provide an impression of objectivity and authority. Laypeople especially are likely to be unaware that the models do not depict a past reality (Eiteljorg, 2000, section: Reconstructions or Fantasies?).

5.10.1 Level of detail

In cases where there is insufficient material to support a reconstruction, it should not be attempted. The result could be misleading and dishonest (James, 1997, 26).[147] If there is a sufficient basis for a reconstruction the level of detail and the manner of representation should be chosen according to the intended audience (Mainka-Mehling, 2008, 245 f.), but also according to the intended impact (James, 1997, 26). It can be preferable to present a schematic drawing instead of a complex scene (James, 1997, 26) because the important parts are pointed out (Weidenmann, 2002a, 85 f.). Besides, the more detail is included in a scene the more convincing it becomes to the observer, even though the opposite is actually the case (James, 1997, 26): the complexity of the scene is, according to James (1997, 25), inversely proportional to its accuracy.

Hodgson (2001, 5), whose main concern lies with reconstruction drawings, recommends not cluttering the picture with unnecessary detail and instead focussing on one main and some subsidiary aspects or functions of the site to be shown. The selection should bear reference to the context in which the reconstruction will be presented, e.g. in an exhibition. In case many factors have to be represented, it is advisable rather to consider a series of themes instead of just one picture (Hodgson, 2001, 5). In contrast to an image, a virtual model often has no 'main viewpoint' because a 3D model can (theoretically) be viewed from all sides. Exceptions are presented when, e.g. only still images of the model are created, if an animation only shows a certain perspective, or if an augmented reality application constrains the viewer to a single viewing location on site (see e.g. TimeFrame™/TimeScope 1 in Section 7.3.2.1). Therefore the focus of the visualisation has to be indicated by other means, for example, by enhancing important parts and excluding or visually reducing any superfluous information. On the other hand, presenting different themes can be achieved within one and the same display system by hiding or displaying certain parts of the reconstruction (cf. the different representations of the northern part of the West Kennett Avenue in Section 8.2.2.2).

[147]Other means of indicating that something has been there can be used, for example, marking an area on the ground (see 5.9.1).

5.10.2 Line drawings

James (1997, 39) and Eiteljorg (2000, section: Fantasies Can Seem Very Real) agree that coloured images or computer models have more impact on the observer than (coloured) line drawings and that they will eventually be more convincing and stay longer in the observer's mind (Eiteljorg, 2000, section: Fantasies Can Seem Very Real).

Strothotte et al. (1999, 36) attribute this effect to the impression that a photo-realistic image aspires to appear as if it had been taken with a camera (a snapshot of reality), while a line drawing has no such aspirations.

It could be shown that realistic images are not well suited for illustration purposes (Weidenmann, 2002a, 89 f., 90 fig. 6.4) and that hand-drawn[148] line drawings in particular stimulate discussion. They can also guide the observer's attention and depict the right amount of relevant detail (Strothotte et al., 1999, 36 f.). These images can be of high aesthetic quality, but poor images also can have their use. They are often considered untrustworthy and can therefore be used to mark interpretations as potentially dubious (James, 1997, 26 f.).

5.10.3 Photo-realism or non-photorealistic rendering

In the end, the rendering process determines how a virtual model will look, while the level of detail has been determined during model creation. Photo-realism is employed to make a reconstruction seem 'real' (whereby Gillings (2005, 228) argues that **photo**realism is not real).[149] Whether realism is a desirable attribute for a virtual model is highly debated. On the one hand Roussou (2002, 98) claims that museum visitors expect photo-realistic displays[150] and Chalmers and Stoddart (1996, 86) use photo-realism for research on rituals. On the other hand Ryan (1996, 103) and Hodgson (2001, 17) argue that a non-photorealistic presentation would enhance the model character of a reconstruction.[151]

It is certainly true that for researching ritual, especially in connection with the effects of smoke, dust and fog (Chalmers et al., 1995, 225) on visual perception (Chalmers and Stoddart, 1996, 86), a photorealistic approach has to be taken. There are also other research-related topics which do benefit from a photo-realistic approach, e.g. the production of cave art or the colouring of mediaeval pottery under artificial lighting circumstances (Goodrick and Earl, 2004, section 2). Forte (2008, 30) also claims that photorealism acts as a 'perceptual-evocative catalyst', but not because it is 'real' but rather because our brain can process it without our prior conscious attention (Ware, 2004, 149)[152], which in turn helps us to learn, visualise and interpret (Forte, 2000, 254).[153] Vatanen (2003, 70) follows Forte's argument, stating that a detailed and attractive model would aid the comprehension of how objects were seen in the past. To show uncertainty, it would be necessary to attach metadata to the model—not to reduce its visual realism. If viewers know that what they are seeing is not 'the past' (and it can be argued that both the archaeologists and the general public have become critical enough to scrutinise computer images (Goodrick and Earl, 2004, section 2)) then it follows that good models do not have to be ugly (Niccolucci and Cantone, 2003, 57).

The question of aesthetics will be brought up again below, but it can be observed that there is a tendency to associate photorealistic rendering with 'pretty pictures' in both the positive and the negative sense. The above-cited authors have no qualms concerning the possible deception of the observer (cf. Miller and Richards, 1995, 20) when using photo-realistic images. However, images are more easily remembered than text (written or spoken) (Weidenmann, 2002b, 49 ff.) or metadata (for which the user has to actively enquire). Embedding the uncertainty into the image seems to be an effective way to ensure that it will be later remembered.[154] It can be argued that most of the authors above are concerned with research applications for experts, who are familiar with the underlying sources and interpretations. In this context photo-realistic representations are a useful tool.

Roussou and Drettakis (2003, 58 f.), on the other hand, recommend (after studying the assets and

[148]This probably includes non-photo-realistic rendering techniques which imitate hand-drawings.

[149]Rendering methods which achieve more or less photo-realistic output are explained in Section 10.3.1.2.

[150]Jablonka et al. (2003, 15) argue along similar lines.

[151]A selection of non-photo-realistic rendering methods is provided in Section 10.3.1.3.

[152]Forte (2008, 30) refers to Gibson and his concept of 'affordance', which I take to be very similar to what Ware (2004, 149–154) calls *preattentive processing*.

[153]The question is whether a simplified approach would not be more suitable than a complex, 'real' scenario.

[154]As already remarked in Section 5.5.5: if image and text harmonise and are presented at the same time, the effect on learning the information is greatest (Weidenmann, 2002b, 52). In this context this would mean that both image and text would need to depict/mention uncertainty if this is the relevant information.

drawbacks of photo-realistic and non-photorealistic rendering) the use of photo-realistic models for the general public—to give them a 'window into the past', while non-photo-realistic rendering could 'suffice' for some research applications, or be preferable because it imitates 'traditional methods of depiction'.[155] I would argue rather for the reverse in relation to the general public and follow Goodrick and Gillings (2000, 46) in their pragmatic approach of choosing between realistic or non-realistic representations depending on the research topic.

Authors arguing against the unrestrained use of photo-realism point out that, despite often being seen as the 'default option', it is an unnecessary feature for many applications (Daniels, 1997, section 4; Kanter, 2000, 48). It also impedes the indication of uncertainty in the image (Traser, 2007, 589) and the separation of the 'real from the restored' (Eiteljorg, 2000, section: Attractive Images or Good Information?). Eiteljorg (2000, section: Attractive Images or Good Information?) advises that if 'compelling' (i.e. photo-realistic) images are to be used it is paramount *first* to confront the viewer with the extant ruins before showing the reconstruction, in order to moderate the persuasiveness of the reconstruction.[156]

The fundamental aim of a reconstruction should not lie in presenting compelling images, but in the visualisation of information and ideas (Barceló, 2000, 29; James, 1997, 23) in this respect aesthetics are secondary, but they should not be disregarded (James, 1997, 23). That realism has indeed nothing to do with aesthetics[157] is easily proved by regarding the reconstructions by Kieven et al. (2000), Lehmann and Haarlammert (2010), Traser (2007) (see Fig. 5.6) and others.

The three examples presented in Fig. 5.6 manage to convey a sense of 'unreality'. In the case of Kieven et al. (2000) the aim was to construct a building which had never existed. The geometry was kept simple and was textured with the line drawings taken from the original architectural plans (Kieven et al., 2000, 6). The model was created so that it would later run on the Unreal engine[158], which allows the observer to move through the model. The model has been used by the researchers to explore the architecture (Kieven et al., 2000, 6–9) and to compare the views it offers to stage-set design in theatres or to wooden models by the same architect.

Lehmann and Haarlammert (2010) as well as Traser (2007) employ a mixture of photo-realistic rendering for extant parts and drawing styles for the reconstructed parts in their models, which set apart the two levels of confidence well. Lehmann and Haarlammert (2010, 174 f.) go further in indicating which reconstructed parts can be deduced from the remains *in situ* (for example, the patterning of the apse floor) and also, where possible, indicate colour trends for these parts in muted colours. Further codes employed include using red colour for structures reconstructed on textual sources, and grey employed to indicate later changes to/reconstructions of the physical building. For the user to fully understand the intentions of the model makers and the intended effect of the visualisation, it is necessary to explain which methods were used; simply giving a key is insufficient for this purpose (Weidenmann, 2002a, 90).

5.10.4 Cognitive theory

So far no research has gone into assessing what visualisation and presentation styles are best suited for conveying information with reconstructions. Even concerning more simplistic diagram styles there is no established best practice[159] because many factors play a role in how effective a representation is (Spence, 2001, 69). Research into which effect different encodings and their combinations have on learners has been undertaken, but the results are not readily applicable to the questions at hand: for example, 'what must be done to aid understanding without overwhelming the senses'? (Eiteljorg, 2000, section: Attractive Images or Good Information?). By trying to display spatial and temporal uncertainty (maybe even including several levels of it) too many different cues might be present in the model for anyone to distinguish them. Some display methods may prove entirely unsuitable: for example, indicating find densities by using colour coded strata (Vote et al., 2002, 84). In Vote's example the problems were mainly due to occlusion, but it is generally inadvisable to use

[155] Reading between the lines you could get the impression that Roussou and Drettakis (2003, 57) think that archaeologists live in the past (i.e. pre-photographic days) and are quite unwilling to adopt 'new' kinds of pictorial representation—much different from the general public.

[156] For the related question concerned with how much has to be reconstructed for different audiences (i.e. do laypeople or students really require 'complete' reconstructions (Kanter, 2000, 48; cf. Pletinckx, 2008, 17)?) see Section 5.4.

[157] This equation has been set up by Vatanen (2003, 70).

[158] http://www.unrealengine.com/ (Accessed: 1st of February 2012.)

[159] And, indeed, there might be none, see Schnotz (2002, 81).

Figure 5.6: Examples of non-photo-realistic rendering. Top left: Kieven et al. (2000, 21 fig. 18) visualised a never realised project by architect Filippo Juvarra and used the hand-drawings as textures for the model (http://wissensgeschichte.biblhertz.it/3d-bridge-html/lectures.html Accessed: 21st of November 2011). (Copyright: Elisabeth Kieven, Hermann Schlimme, Günter Eger, Bibliotheca Hertziana - Max-Planck-Institut für Kunstgeschichte, Rom). Top right: Traser (2007, 588 fig. 24) integrates extant remains in a hidden wireframe reconstruction of Pilis Abbey fountain. Bottom: Lehmann and Haarlammert (2010, 174 f., 172 fig. 6) depict *in situ* remains at Chimitile/Nola in a photo-realistic manner (floor and lower wall in the apse) while reconstructed parts are shown in line-drawing style with further information visually encoded into them (for example, the parts shown in red are reconstructed after a textual source). (Copyright: maßwerke GbR).

colour to indicate quantitative differences (Schnotz, 2002, 76), in addition to the difficulties in identifying a specific hue (even if a key is present) (Ware, 2004, 75 and fig. 3.8). Ware (2004, 205) shows that using a semi-transparency (in his example in a GIS) can also cause problems, because viewers cannot decide on which layer the feature lies; several semi-transparent layers can also lead to occlusion.[160] This is just to mention some of the problems involved.

So far, virtual models almost exclusively address vision, while the other senses are neglected (Earl and Wheatley, 2002, 5). Especially in research which focuses on embodiment (see e.g. Chapter 8) it would be most important to address the other senses too, for they are equally involved in the perception of our surroundings (Knoll et al., 2004, 6 on CD). There is already some research in acoustics, e.g. Pope and Chalmers (2000) (see Section 5.5.7.7).

5.11 Conclusion

This chapter has shown that by paying attention to applied ethical issues a visualisation can become more than a 'pretty picture' (which in the worst case deceives the observer) and indeed can generate added value in a multitude of forms for its creators and its users. It has also become clear that there are many ways in which such a visualisation can be approached, which form it can take and for which uses it can be employed. Nevertheless there is still a great amount of research to be done concerning how best to utilise the available methods in creating and displaying the models to the best advantage for all involved parties.

[160] Instead, the use of differently pattered regions is recommended (Ware, 2004, 205).

Chapter 6

Progetto *Insula* del Centenario (IX, 8)

The first case study is concerned with the reconstruction of a part of Casa del Centenario in Pompeii. The reconstruction work started while the field work (in this case documentation of the remains, surveys and small excavations) were still in progress. The excavation of the *insula* to which the Casa del Centenario belongs had already taken place in 1879–1880 (see Section 6.1).

As depicted in Figure 6.1 the available data about the site was the basis for the reconstruction. Obviously, as in all other cases, the creators of the model had to find a way to deal with missing and insufficient information. The aim was for the resulting reconstruction to be used for communication and other predetermined purposes (see Section 6.2.1), but it yielded some unexpected applications too (see Section 6.4.1). New findings, i.e. changes in the data basis, led to an updated reconstruction, but neither did the reconstruction trigger new research aims nor did the use of the model lead to changes in the reconstruction or to new research efforts—unlike the 'Excavation Model' and the 'Research Model' (Chapters 7 and 8).

This chapter starts with an introduction to the site, and is then concerned with the reconstruction project itself. The reconstruction is part of two projects with different goals: first, 'Pompei – *Insula* del Centenario (IX, 8)' which aims at documentation, restoration and the education of students through hands-on experience (Scagliarini Corlàita et al., 2003, 237; Scagliarini, 2006, 9). Second, the MUSE project, in which it was planned to develop a system for site management and site exploration leading up to the Whyre® system[1] (see Section 6.2). This explains why the goals range so widely (Section 6.2.1). Besides these primary aims, prerequisites concerned with applied ethical issues were set by the members of the projects (Section 6.2.1.1); in this context an article by Frischer et al. (2002), which recommends a 'philologic' approach, greatly influenced the work. The chapter proceeds by looking at the methods employed in the making of the reconstruction, from the point of data gathering to the modelling work, and the uses of the model (Section 6.3). The final part is concerned with the question of whether the aims set out by the teams have been reached (Section 6.4).

[1]Whyre is a contraction of 'why' and 'ware' (Cinotti et al., 2007, 41), because the system shall answer the questions of the user.

6.1 *Insula* del Centenario

Pompeii lies in the region of Campania in Italy. The city was buried during the eruption of Mount Vesuvius in AD 79. 1800 years later the excavation of the *insula* began and it was named 'del Centenario' to mark the anniversary of the eruption (Scagliarini Corlàita et al., 2003, 237). The excavation began on the 16th of May 1879 in the south-west of the *insula* and proceeded to the north[2], then the eastern side of the *insula* was excavated from north to south. Over a hundred workers were involved in the task.

Of special interest to the excavators were the frescos, which were described in great detail (Capoferro Cencetti et al., 2001, 3), while only selected finds were preserved: especially, bronze and glass objects survive in the Museo Archeologico di Napoli today (Scagliarini, 2006, 15). In September 1880 the excavation ended, though in the south about 20 per cent of the *insula* remains unexcavated.

The *insula* lies in *regio* IX, bordering to the north on Via di Nola, one of the *decumani*, i.e. main east-west roads, of the town (Fig. 6.2). The excavated area

[2]For convenience' sake Via di Nola is taken to be lying in the north and Vicola di Centenario as lying to the west, though in truth the *insula* does not really lie on a north-south axis (see Fig. 6.2).

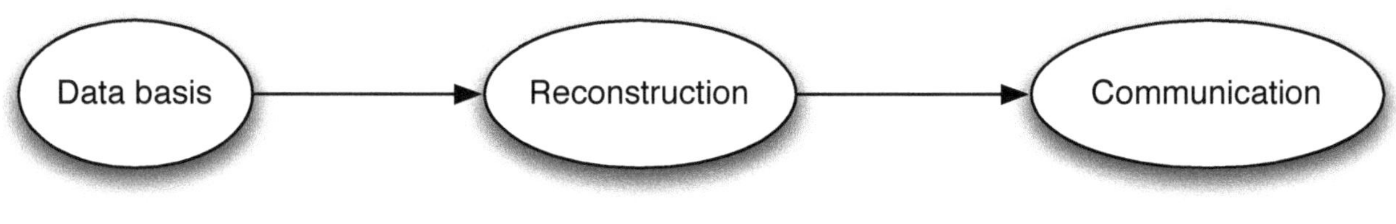

Figure 6.1: Museum Model as demonstrated in the Casa del Centenario Project.

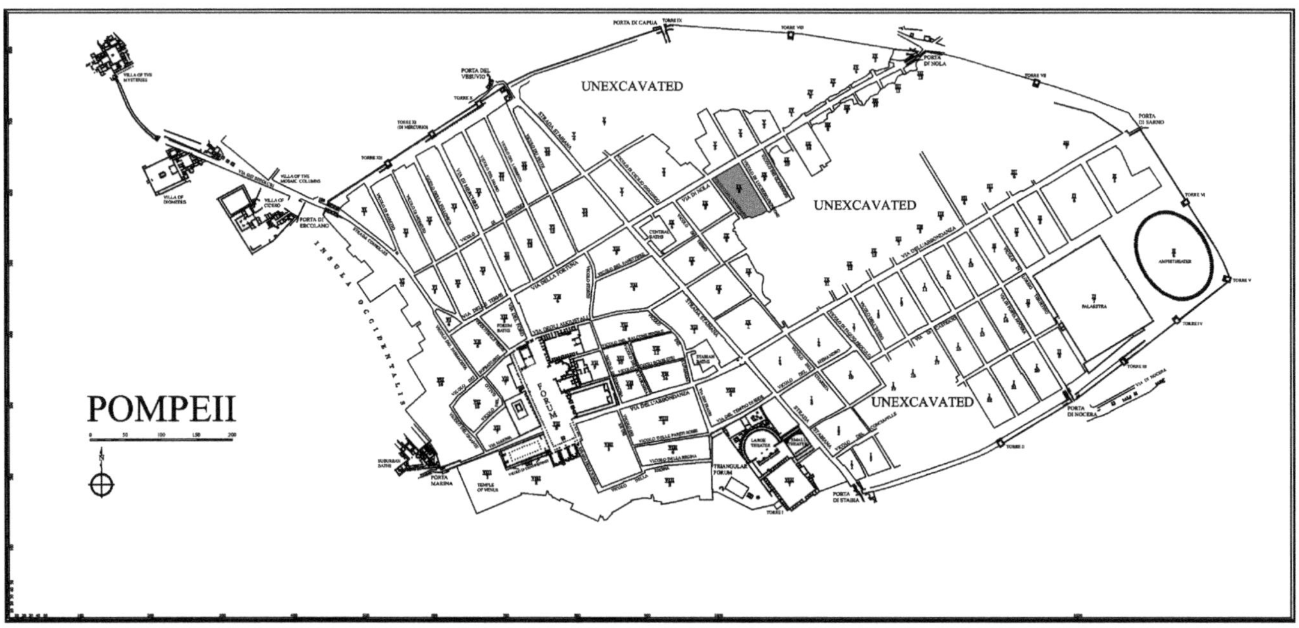

Figure 6.2: Pompeii (after Dobbins and Foss, 2007, CD-ROM), *Insula* del Centenario is marked in red.

(Fig. 6.3) is comprised of three houses, four shops (entrances 1, 4, 5 and 7, all on Via di Nola) and a 'philosophy school' (entrance 2). The latter was identified as such due to a painted frieze on the eastern wall depicting philosophers (Scagliarini, 2006, 18)[3]. The biggest house (to which the shops also belonged), Casa del Centenario, has three entrances: two on Via di Nola (3 and 6) and one on Vicola di Centenario (a). The two smaller houses, *hospitium Hygini Firmi*, so called because of a black inscription discovered during excavation on the north wall[4] of the vestibule, and the house with a peristyle, both have their entrances on Vicola del Centenario (b and c respectively)[5].

Casa del Centenario had at the time of its destruction two atria and a peristyle; its size is the result of merging several houses into one. A preliminary report on the possible events is given by Santoro et al. (2005, 249–251):

Some parts of the building could belong to the first quarter of the third century BC; these parts are situated around the main atrium (2), in the service area in the south-east and the border between it and the *hospitium* (Santoro et al., 2005, 250). This is also possibly the time when the two atrium houses in the north of the *insula* (atria 2 and 16) were built, with their fruit and vegetable gardens lying to the south. The wall between the gardens can be dated to the first half of the second century BC, and the atrium house to the east may have had a portico to the south[6], indicat-

[3]A description of the 12 figures can be found in Mau (1882b, 31), who also mentions landscape paintings on the remaining walls. I am not convinced that depictions of philosophers must necessarily signify that the rooms were used as a philosophy school.

[4]According to Mau (1882b, 116) it was found on the door-jamb.

[5]The house with the peristyle is not fully excavated and may have more than one entrance.

[6]Photographs in Capoferro Cencetti et al. (2001, 4 fig. 5, 13 figs. 3 and 4) show that some of the columns in the north-east of the peristyle possess channelled shafts, which were later covered with plaster, while the columns in the north-west, west and south are made of plastered masonry. These three channelled columns could belong to the first portico and were later incorporated into the peristyle (Santoro et al., 2005, 250).

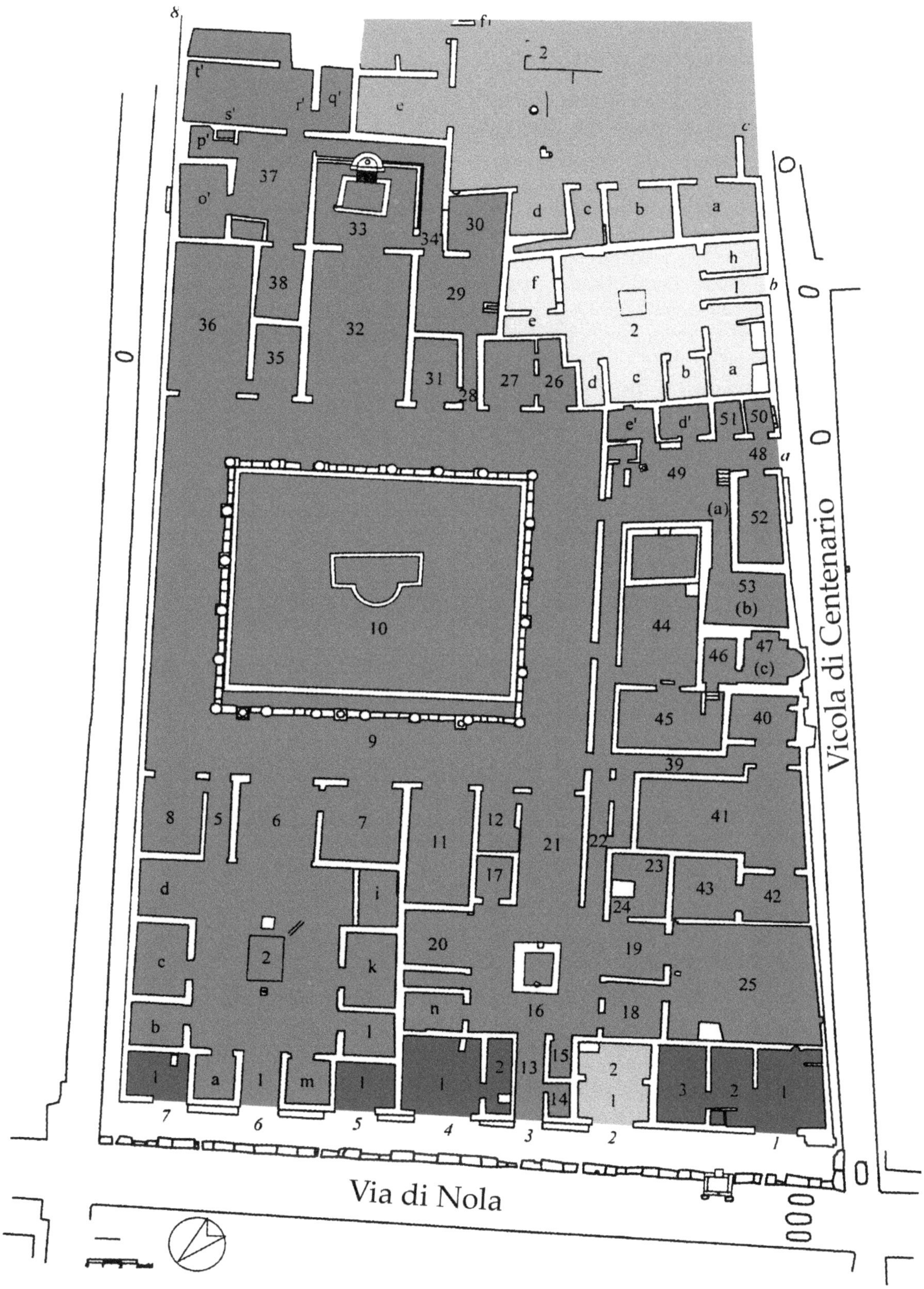

Figure 6.3: *Insula* del Centenario (after Custodi et al., 2006, 191 fig. 3 and Capoferro Cencetti et al., 2001, 3 fig. 2). Blue: Casa del Centenario, yellow: Hospitium Hygini Firmi, green: house with a peristyle, red: shops, and orange: philosophy school. (Copyright: 2000–2002, Progetto 'Pompei – Insula del Centenario IX 8', Università di Bologna, Dipartimento di Archeologia e ARCES; CINECA, VisIT Lab; Soprintendenza archeologica di Pompei).

ing that the fruit and vegetable garden has become a 'pleasure garden'. A round basin in *opus incertum* was added at some point (Santoro et al., 2005, 250).

Differences in floor levels are taken by Santoro et al. (2005, 250) to indicate different properties, and indeed that adjusting the levels shows that the properties were joined[7]. The floors of the rooms south of the garden (32) behind the eastern atrium house (2) were raised in late Augustean times to the same height as the garden itself, being followed by a raised floor level of room 33 in the first half of the first century AD. Several other minor changes concerning the *fullonica* in the north-west of the *insula* (entrance 1) and the *fullonica* in the west of the *insula* (entrance a), which becomes a service area for the atrium house (16), occur in late Augustean times (Santoro et al., 2005, 250).

Finally, after the earthquake in AD 62 the two atrium houses are merged into one property: Casa del Centenario. The dividing wall in the garden is demolished resulting in the building of a new peristyle (9, 10), which is now the centre of the property, the round basin in its middle is substituted by a rectangular basin with an apse. This basin and the *thermae* in the west are connected to the water supply. By closing the communicating doors between some of the shops to the main house they are rendered independent (Santoro et al., 2005, 250).

In its last phase Casa del Centenario is composed of several areas. Entrance 6 leads into the main atrium (2), the adjoining *cubiculi* and *ala* i were fitted with storage shelves for the new owner, destroying wall paintings (room b, *ala* i) in the process (Capoferro Cencetti et al., 2001, 14), while the atrium walls themselves retained the murals depicting scenes from the theatre more suited to the reception room it once was (Scagliarini, 2006, 18).

Entrance 3 leads into atrium 16, the secondary atrium. Already Mau sees it as minor to atrium 2, because it is smaller, less regular and its decorations are inferior, in realisation and materials, in comparison to atrium 2. Nevertheless the adjoining rooms of atrium 16 are partly redecorated in the fourth style[8] (rooms n, 17, 18) while on the other hand the rooms leading off the main atrium (the exception is room b with a decoration in the fourth style) possessed simply a white intonaco (Mau, 1881, 122–128). This actually indicates that the rooms surrounding atrium 16 are much more representative than the storage rooms around atrium 2; it hints also that atrium 2 was only a room for passing through, while the rooms around atrium 16 may have been used in the usual way by the occupants of the house.

The focus of the house was the peristyle (9, 10) with the rooms surrounding it (Santoro et al., 2005, 250 f.). Two *oeci*, open rooms with fine murals on a black (8) and white (7) ground respectively, open southward to the peristyle. At the centre of its southern side is a *triclinium*[9] (32) (the spacing in the columns of the peristyle is altered to fit the opening of this room) and at its far end another open space: a *viridarium* (33) with a *nymphaeum*[10] (Capoferro Cencetti et al., 2001, 7). From the *viridarium* a passage (*cryptoporticus* 34) leads to a service area.

Another service area is situated in the west next to entrance a. Special interest was aroused by the *lararium* in atrium 49 which is visible from the entrance. The mural, showing Bacchus (uniquely dressed in grapes), Mount Vesuvius and a serpent, is today in the Museo Archeologico di Napoli; it was removed at a very early stage and is therefore very well preserved.[11] The other wall is decorated with two lares (i.e. gods of the land) (Capoferro Cencetti et al., 2001, 14).

Further to the north are the *thermae* with an *apodyterium* (45), i.e. a room to undress, a *tepidarium* (46) and a *caldarium* (47), the warm and hot baths, and a *frigidarium* (44), the cold bath (Mau, 1881, 229). Private *thermae* exist only in very few (luxury) houses, and even before the Casa del Centenario was excavated in the 19th century the *thermae* had been plundered (Capoferro Cencetti et al., 2001, 7). Robber trenches through the roof and wall gave access to the rooms, which were then stripped of their marble revetments. Otherwise the surviving vaulted roofs provided shelter for the murals inside the *tepidarium* and the *caldarium*, which has preserved their vivid colours until now (Capoferro Cencetti et al., 2001, 7).

[7] While bringing adjacent rooms onto the same floor level can be taken as a sign that they belonged to the same estate, different floor heights need not indicate different properties.

[8] The four decoration styles were first described by Mau (1882a) and are still valid today.

[9] A triclinium is a room used for meals. It should be kept in mind that rooms in a Roman house may be used for different purposes and that the labels attached to them, like *cubiculum*, *oecus*, *exedra* or *triclinium* (due to proportions, size of the openings, whether they are reached via an atrium or a peristyle and so on) may have no connection with their real usage. Obviously other rooms are clearly recognisable through their installations (*thermae*, kitchen, latrines, *etc.*).

[10] A garden with a fountain.

[11] This painting in connection with the shelves in the main atrium lead to the conclusion that the last owner of the *domus* must have been a rich wine grower and trader (Capoferro Cencetti et al., 2001, 14), which I find not entirely convincing.

Adjoining these rooms in the north is a 'private apartment' (Capoferro Cencetti et al., 2001, 3) with a *triclinium* (41) and three *cubiculi*[12] (40, 42, 43). One of the rooms (43) has erotic scenes painted on the walls; they may have been the reason why some of the rooms of this private apartment were covered with roofs after the excavation, and they were certainly the reason for installing a door. It is both a sign of prudery (only a few people were allowed to enter the room for years) and a prevention from theft (small erotic scenes especially are often removed from the walls) (Capoferro Cencetti et al., 2001, 8, 16). Room 41 presents another speciality: here some of the older central pictures were taken from the walls and substituted by newer pictures with mythological scenes (Scagliarini, 2006, 19).

Scagliarini (2006, 19) tries to give a picture of the last owner of Casa del Centenario: he may have been a man called Aulus Rusticus Verus (re election inscriptions on the facade of the building) who ran for the highest office of the city twice as well as promoting other persons for political offices. He may have come from a family of freed slaves, arriving at Pompeii some decades earlier. The family had obviously had some economic success—apparent in the size of the house—which brought political influence with it, though the decoration and architecture show, according to Scagliarini (2006, 19), a certain lack of taste.

The state of preservation is very heterogeneous, but mostly rather poor (Capoferro Cencetti et al., 2001, 14, 16). After the excavation in 1879–80 the building remains were left exposed to the elements, only some of the rooms being covered with roofs (b, 7, 8, 9 in the west, 12, 33, 34, 36, 41, 42, 43). Elsewhere individual pictures were provided with little roofs, unfortunately to no avail and at the cost of destroying the plaster in the place they were attached. Moisture rising from the ground has also affected the paintings (Capoferro Cencetti et al., 2001, 16).

Only three paintings were completely removed from the site: the afore-mentioned picture of Mount Vesuvius and Bacchus from the *lararium*, and a picture each from *ala* d and *ala* 20 (Capoferro Cencetti et al., 2001, 14). The remaining pictures are washed out, sometimes so much that nothing remains of them. Other, better preserved, images have been stolen by thieves removing the picture with the plaster from the wall. In both cases only the early documentation of the site can help to reconstruct the original decoration (Capoferro Cencetti et al., 2001, 22).

[12]A cubiculum is a bedroom, but see footnote 9.

The mosaics are also endangered. They have warped after the eruption in AD 79 especially where cavities (cisterns, hypocausts or channels) lie underneath. Additionally moisture has lifted some areas so that the tesserae have fallen off. Attempts to repair this immediate damage have been made in the past (Solmi and Vecchietti, 2002, 198, 202).

Another problem was the effect of tourists who could visit the house until 30 years ago (Coralini, 2007, 19), nowadays the house is closed to the public, though one of the aims of the project is to reopen it.

6.2 The reconstruction project

The reconstruction of a part of Casa del Centenario (Pompeii, *Insula* IX, 8) is the result of a joint project with a more technically oriented side, the MUSE Project, and a preservation side, the project 'Pompei – *Insula* del Centenario (IX, 8)' (Scagliarini et al., 2001, 83). MUSE[13] aims at creating a system which can be used to enhance the experience of cultural heritage, while 'Pompei – *Insula* del Centenario (IX, 8)'[14] aims at documenting, studying and revaluing the *insula* by working out a restoration plan.[15]

6.2.1 Aims and ambitions

As mentioned above, several projects are involved in the reconstruction of a part of the *Insula* del Centenario, resulting in manifold goals.

The reconstruction model itself should accomplish several things:

[13]Boconsult IdS, University of Bologna, CINECA and Sinet, funded by the Ministerio dell' Università e della Ricerca Scientifica e Tecnologica (Scagliarini et al., 2001, 83).

[14]A cooperation between the Dipartimento di Archeologia dell'Università di Bologna and the Soprintendenza Archeologica di Pompei (Capoferro Cencetti et al., 2001, 5).

[15]In the project 'Pompei – *Insula* del Centenario (IX, 8)' several other institutions and researchers are involved. Capoferro Cencetti et al. (2001, 5) name: Dipartimento di Ingegneria delle Strutture, dei Trasporti, delle Acque, del Rilevamento del Territorio (DISTART); Consorzio Interuniversitario per il Calcolo Automatico dell'Italia Nord Orientale (CINECA); Centre d'étude des peintures murales romaines (CEPMR), and researcher from: Research Centre of Applied Mathematics (C. I. R. A. M.), Università di Bologna; Facoltà di Architettura, Universtà di Ferrara; Dipartimento di Chimica, Università di Modena e Reggio Emilia; Dipartimento di Biochimica 'G. Moruzzi', Università di Bologna; Dipartimento di Scienze della Terra, Università di Parma; Dipartimento di Chimica, Università di Parma; Dipartimento di Scienze della Terra, Università di Roma 'La Sapienza'; Istituto per le Tecnologie Applicate ai Beni Culturali (ITABC), Consiglio Nazionale delle Ricerche (CNR).

- Supporting the conservators in their work. They have the opportunity to try out several approaches and judge their effects as a whole. During the phase of data gathering, on-site adjustments can be made to the model in order to include the new results. This is in line with one of the aims of 'Pompei, *Insula* del Centenario (IX, 8)', namely, to compile guidelines for the restoration of the *insula* (Scagliarini Corlàita et al., 2003, 244 f.).
- Helping the archaeologists to verify their interpretations. As gaps and defects become obvious, different reconstruction alternatives can be tested (Scagliarini et al., 2001, 83).
- Giving the general public the chance to see (light, shade, colours) and understand the site, i.e. the results of the archaeological work. Several ways are proposed for this: the model can be viewed in a Virtual Theatre, television recordings can be made with actors 'in' the model using the Virtual Set, on site the visitor can explore the ruins with a hand-held device, and also large screens can show the model within a part of the site (Scagliarini Corlàita et al., 2003, 244, 247 f.). To enhance the understanding of the site the models are interactive; the users can navigate them freely and get additional information regarding the location.
- Didactics appear to be considered by some of the authors as being different from 'touristic' enterprises; in particular, the Virtual Set is often mentioned in this context (Scagliarini Corlàita et al., 2003, 251; Guidazzoli, 2007, 86). Maybe the focus is now seen to be shifted away from experiencing the site to being confronted with educational matter concerning the location in a non-interactive way.

The project is also concerned with applied ethical issues. One of the main foci lies on the interactivity of the model: giving the user the possibility to move freely and to explore the reconstruction. On the other hand the model is meant to be 'transparent', giving the user the opportunity to validate the reconstruction as well as to be made aware of uncertainties (Coralini, 2007, 25 f.).[16]

The reconstruction was not meant to encompass the whole *insula*, but only part of it. It was considered that the main axis of Casa del Centenario would be best suited for this purpose (Scagliarini Corlàita et al., 2003, 240): the blue area in Fig. 6.4 shows the rooms which were modelled; they lie in the line-of-sight when entering the house and would have been open and visible to visitors received by the owner of the *domus*: the *fauces* (entrance, 1), the main atrium (2) with the *alae* (d, i), through the *tablinum* (an open room lying opposite the *fauces* of the atrium, 6) into the peristyle (9, 10), from there the two *oeci* (7, 8) flanking the *tablinum* can be approached; on the other end of the peristyle lies a *triclinium* (32) and a *viridarium* (33) with *cryptoportius* (34).

6.2.1.1 Tackling the applied ethical issues

In order to make the reconstruction verifiable three different models were created (Fig. 6.5). One shows the ruins as they are today (this model is not necessary when the system is used on site, as the extant ruins are visible). The two other models show the Casa del Centenario at around AD 79[17] but only in two aspects (Coralini, 2007, 19, 18 fig. 2): the 'didactic' reconstruction shows the site with all the reconstructed parts in colour to give a good impression of what the site may have looked like: decorative program and architecture. The 'philological' reconstruction (see Boxed Feature 3 on page 61) shows only the more probable reconstructed parts in colour, while parts, for which a certain amount of proof does not exist, are shown in greyscale (Capoferro Cencetti et al., 2001, 28).

The reconstructed elements are classified according to the reliability of their documentation basis (Coralini, 2007, 17). The resulting three levels are visible either in the model which shows the Casa del Centenario today (level 1) or the 'philological' reconstruction (level 2 is displayed in colour and level 3 is displayed in greyscale) (see Fig. 6.5 left and centre).

The three levels are distinguished according to the following criteria (Scagliarini Corlàita et al., 2003, 246 f.):

Level 1 Remains which are still *in situ* today. (See Fig. 6.6, top left and top right.)

[16] Concerning these applied ethical issues the authors refer to Reilly (1991) (Scagliarini Corlàita et al., 2003, 249), Reilly and Rahtz (1992) and Kanter (2000) (Coralini, 2007, 19).

[17] As Scagliarini Corlàita et al. (2003, 240 f.) mention it is assumed that the state of the Casa del Centenario in AD 79 is greatly mirrored in the state of the house excavated in 1879, so that the reconstruction is to a considerable part a reconstruction of the condition uncovered in 1879; to this model the lost structures (roofs, *etc.*) are added.

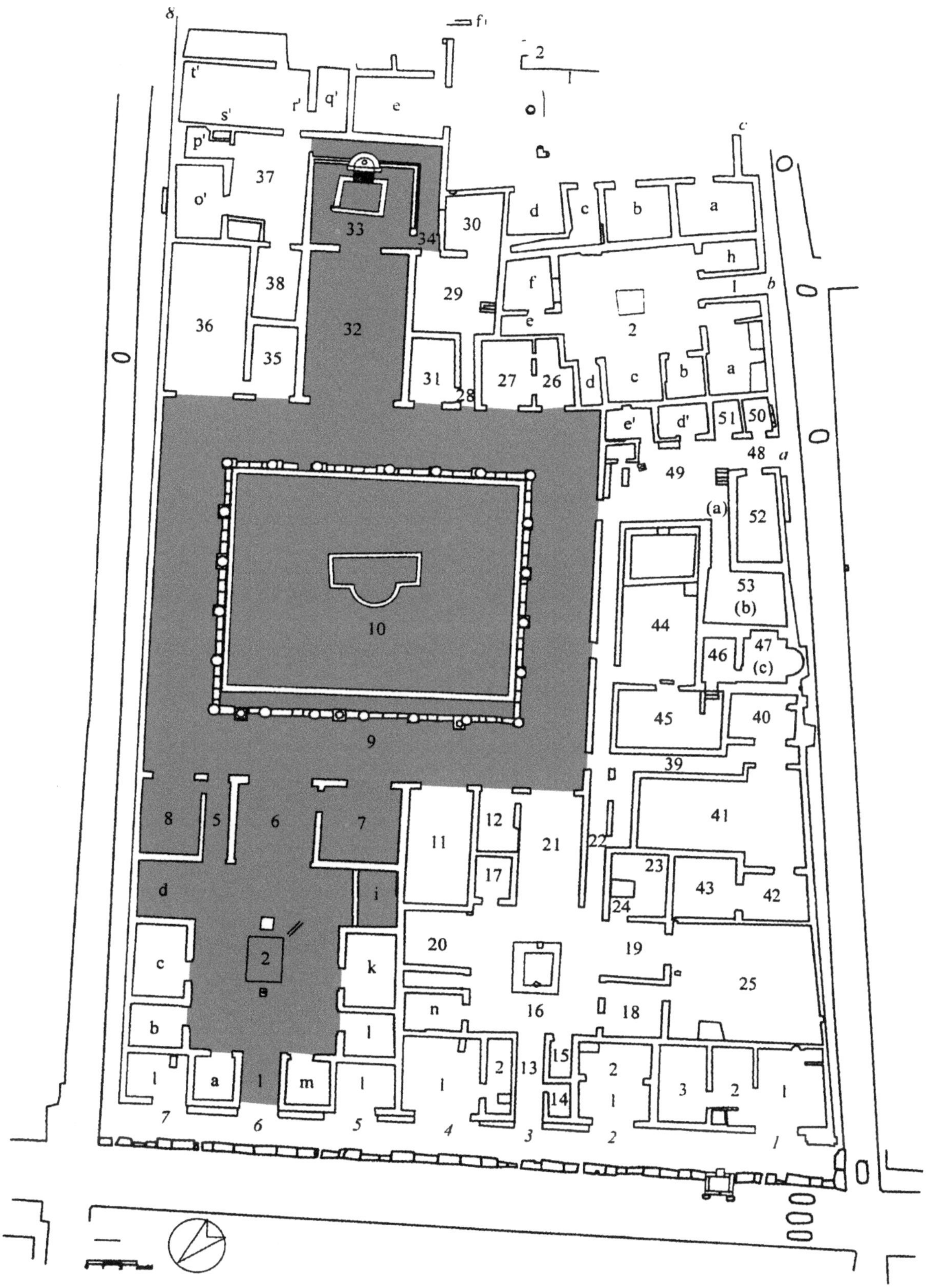

Figure 6.4: Casa del Centenario (after Custodi et al., 2006, 191 fig. 3 and Coralini, 2007, 17 fig. 1). Blue: main axis of Casa der Centenario, denoting the rooms which were reconstructed for the project. (2000–2002, Progetto 'Pompei – Insula del Centenario IX 8', Università di Bologna, Dipartimento di Archeologia e ARCES; CINECA, VisIT Lab; Soprintendenza archeologica di Pompei).

Figure 6.5: Main atrium in three VR models. From left to right: model of the modern ruins, 'philological' model which shows uncertainties, complete model for giving the best impression of what the room may have looked like (Coralini, 2007, 18 fig. 2). (Copyright: 2000–2002, Progetto 'Pompei – Insula del Centenario IX 8', Università di Bologna, Dipartimento di Archeologia e ARCES; CINECA, VisIT Lab; Soprintendenza archeologica di Pompei).

Level 2 Additions to the model which are highly probable because:

- the original elements are preserved elsewhere, e.g. in museums (detached pictures, *etc.*).
- the elements are repeated or mirrored and can be deduced from remains which are still *in situ*.
- the elements are visible on photographs which were taken while the *domus* was in a better state of preservation. (See Fig. 6.6, bottom left.)
- the elements are shown on drawings, watercolours, etchings, prints, *etc.* which were made while the *domus* was in a better state of preservation. Beforehand, possible mistakes, misunderstandings, unintentional additions or omissions have to be ruled out through comparison (if several images exist) or evaluation of the antiquarian and iconographic plausibility.[18] (See Fig. 6.6, centre left.)
- the elements are the results of physico-chemical or optical analyses (Raman-Analysis, diffraction, spectroscopy, photographs taken in glancing light, *etc.*). (See Fig. 6.6, bottom right.)
- they are structural or statical elements, which can be deduced from constructions still *in situ*.

Level 3 The elements (doors, windows, railings, architectonic and decorative characteristics as well as—with great caution—furniture) are known from other contemporary buildings of the Vesuvius region, either from structure, decoration or furnishings.

The reconstructions can be further validated by looking at the sources and the reconstruction hypotheses (Scagliarini et al., 2001, 83). Additionally the three models present a diachronic element to the reconstruction, as the changes through time (Casa del Centenario as it may have looked in AD 79 and how we can see it now)—in this case the degradation—are made apparent (Coralini, 2007, 17).

The validation is made possible by viewing the sources in relation to the reconstruction (Scagliarini et al., 2001, 83 f.). One of the sources is the site in its present state; this is made available either as a model of the site, which can be watched side by side with the reconstruction (Fig. 6.5), or by viewing the reconstruction on site, having the extant ruins before ones eyes. The user is made aware of what is still there and which parts have been added in the reconstruction. Further information is made available through interactivity (Scagliarini et al., 2001, 84). The user has the opportunity to select reconstructed parts and receive background information about them. This is illustrated in Fig. 6.6 where the interplay between modern documentation and archival material, dating from the 19th and early 20th century, lead up to the reconstruction (Scagliarini Corlàita et al., 2003, 242 f.; Coralini, 2007, 19) (see also Section 6.3.1). However not all of the available sources are shown but only a selection and only those which agree with the reconstruction (Scagliarini Corlàita et al., 2003, 243), i.e. the model basis.

Though the aim is 'to show the reconstructive method and verify the entire reconstructive process, from the original data to the final product'

[18] Presumably by trying to find out who is a reliable draughtsperson and seeing whether the element depicted fits into the all over decorative scheme.

Excursus 3: Philologic approach

The idea of philologic reconstructions was first expressed by Frischer et al. (2002, 13–15) remarking on a text by Forte (2000), who demanded 'transparent' models in respect of the initial data, and the use of multiple alternative reconstructions instead of just the one which is supposed to be the 'most probable'. Frischer et al. propose to tackle the problem like a philologist editing a text. Concerning the methodology, they refer to a text by P. Maas (1958)[a] about textual criticism. Maas's steps towards editing the text are as follows: 1.) Find out what possibly has been transmitted of the original text (*recensio*). 2.) Establish which of the transmitted pieces may provide the original text (*examinatio*). 3.) If the pieces do not transmit the original text then attempts must be made to reconstruct the original by presumption (*divinatio*) or at least by excluding the corruption.
Additionally the preface to the edition should describe all the sources, show the relationships between them (in a stemma[b]), describe the quality of the original, and remark on spelling and dialect.
In the text itself added <>, deleted [ ] and corrupted † passages are marked with the respective symbols while underneath the text passage an *apparatus criticus* shows departures from the original text, rejected variants of words or passages, sub-variants, identical readings of otherwise varying texts and doubts re correctness.
Frischer et al. state that this methodology cannot be emulated in exactly the same way for 3D reconstructions but may be useful as an analogy. Some experimentation to this end was done by the University of California, Los Angeles (UCLA)[c].
Uncertainties (additions, lost data, *etc.*) can be flagged in the model (as they are in the texts). Previous reconstructions by other scholars can be seen as 'alternative readings', i.e. they may be displayed as alternative models. This can make the model more credible in the eyes of the observer—in comparison to models where uncertainties are not pointed out—but first a common visual language has to be developed.
In the Casa del Centenario project, and beyond it in the Italian archaeological computing literature, the word 'philologic' became a hypernym, and is used for the same concepts as the term 'applied ethical' in this dissertation.

[a]See Frischer et al. (2002, footnote 43) for the reasons why this text was chosen.
[b]This is not dissimilar to a genealogical tree, describing which text was copied from which other text.
[c]http://www.cvrlab.org/ (Accessed: 5th of January 2006.), for example, Frischer and Favro (2005).

and 'to reveal to the public the reconstruction process' (Scagliarini et al., 2001, 84) the authors themselves already have some doubts whether this can be achieved with the reduced sources displayed to the user. On the other hand they surmise that no lay person will ask to be confronted with a clutter of information more relevant to specialists (Scagliarini Corlàita et al., 2003, 243).

Regarding the latter problem, i.e. confronting the users of the models with unwanted information—or (expressed in a positive way) giving the different kinds of users the information they want and need—reference to several levels of interest is planned according to whether a specialist or a lay person is using the system (Coralini, 2007, 26). Thanks to the interactive approach the user is also free to choose from the pieces of information offered by the system.

The authors (Solmi and Vecchietti, 2002, 201) are aware that virtual reality itself cannot act as a surrogate because it cannot reproduce the original in sufficient detail so that its history can be read from it. Rather they see the potential of virtual reality in enhancing the scientific and educational potential. Paying attention to the latter is an especially efficient way to prolong the duration of a visit to the site (Guidazzoli, 2007, 78).

6.3 Arriving at the model

As stated by Capoferro Cencetti et al. (2001, 28) the value of the model is dependent on the quality of the archaeological data on which it is grounded—a good result would be a model which is scientifically certified.

To arrive at such a model an interdisciplinary group of 28 researchers[19] from the fields of archaeology, archaeobotany, pigment analysis, plaster and mortar analysis, microgravity survey, georadar survey, spectrophotometry, computer graphics and virtual

[19]The researchers are mentioned in Scagliarini Corlàita et al. (2003, 270 f.) and Scagliarini (2006, 10) with only a few variations.

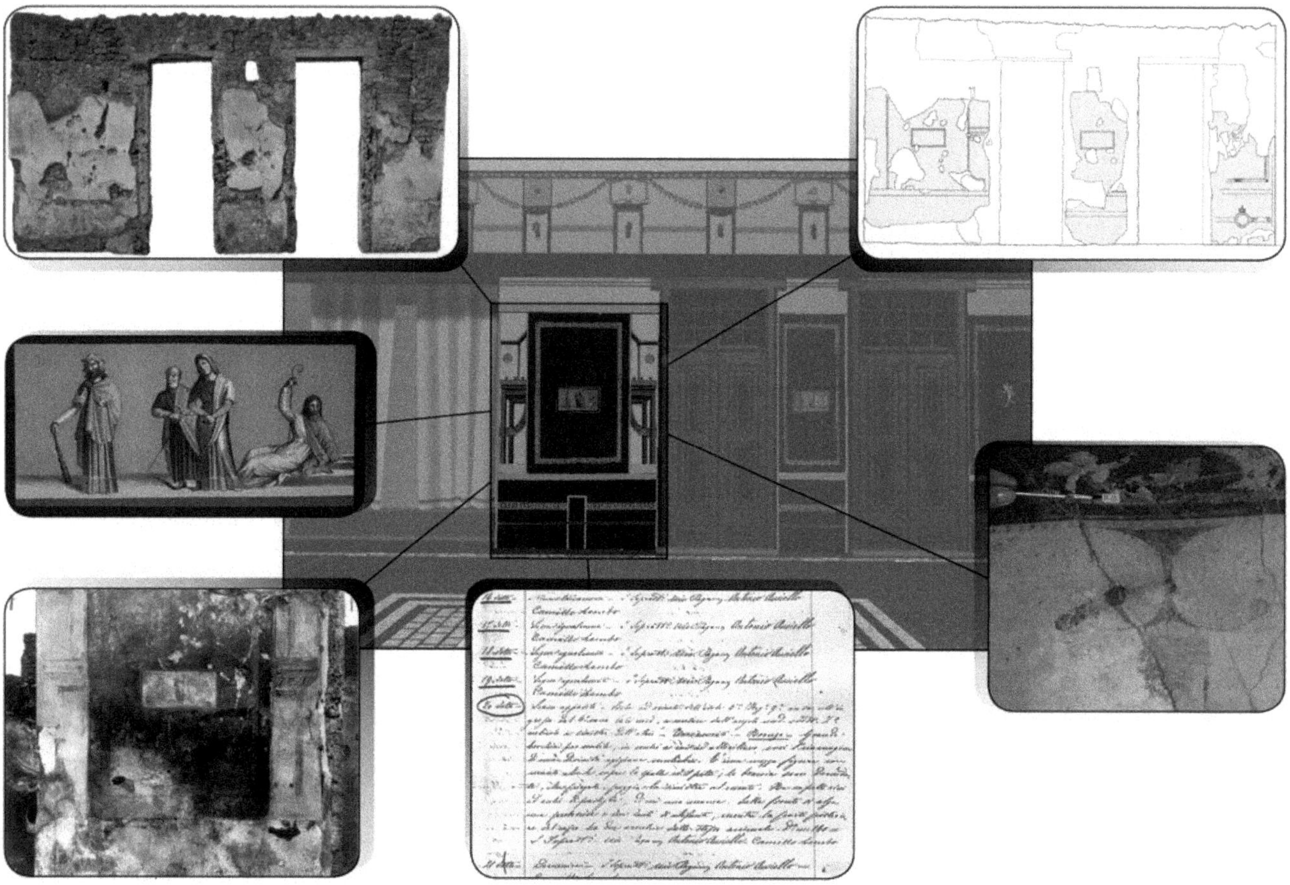

Figure 6.6: Here the sources which were used for the reconstruction are shown. Starting from the top left: photogrammetric survey, plan 1:20 (hand drawn on site), chemical analysis of the pigments, literary sources from the time of the excavation (*Giornale dei Soprastanti*), archival photographic documentation and archival graphic documentation (watercolour by G. Discanno) (Coralini, 2007, 19 fig.3, 20, 22). (Copyright: 2000–2002, Progetto 'Pompei – Insula del Centenario IX 8', Università di Bologna, Dipartimento di Archeologia e ARCES; CINECA, VisIT Lab; Soprintendenza archeologica di Pompei).

archaeology, interactive multimedia systems, photogrammetry and survey, structural analysis, mosaic analysis and archival research, together with other specialists and students, all in all around 120 people, were involved in the project (Scagliarini, 2006, 9 f.). The project was also used to train students who were involved in all the phases necessary for model creation, starting from the field work to collecting and selecting data, compiling a data base, modelling, up to texturing and writing didactic texts (Scagliarini Corlàita et al., 2003, 245). For a successful project the communication between the different groups of professionals was essential (Guidazzoli, 2007, 85) (see also Section 6.4.1).

6.3.1 The documentation

Before the model could be created it was necessary to collect all the available data, starting from the documentation of the excavation in 1879 to modern surveys on site. This collection helped to determine for each part of the model whether it is still *in situ*, has disappeared but is documented, is preserved in another place or can be reconstructed with the help of archaeological science (e.g. colours) (Scagliarini, 2006, 13). Unfortunately this was not an easy task, as the early documentation is rather scarce and 120 years of restorations and additions to the original structures on site have complicated the picture (Scagliarini, 2006, 12).

Textual evidence comes in the form of the unedited Giornale dei Soprastanti, the field diary, where the progress of the excavation is described, as well as

Figure 6.7: Elevation and elevation with thematic overlay (Santoro, 2006, 32 fig. 4). (Copyright: 2000–2002, Progetto 'Pompei – Insula del Centenario IX 8', Università di Bologna, Dipartimento di Archeologia e ARCES; CINECA, VisIT Lab; Soprintendenza archeologica di Pompei).

from the early publications (e.g. Notizie degli Scavi di Antichità, 1879, 1880, 1881; Memorie dell'Accademia dei Lincei, 1879, 1880, 1881; Bullettino dell'Instituto di Corrispondenza Archeologica di Roma, 1881, 1882). Besides this, the inventory, Registri Inventariali, of the Museo Archeologico Nazionale di Napoli gives information about some of the finds from Casa del Centenario (Scagliarini Corlàita et al., 2003, 238, 241; Coralini, 2007, 20; Scagliarini, 2006, 12).

The graphic documentation created during or shortly after the excavation by archaeologists, artists and architects is neither abundant nor complete, but rather a loose collection of images in different media[20] of varying quality and by diverse artists (Capoferro Cencetti et al., 2001, 8, 10; Coralini, 2007, 20). Still, it is essential for research, because many of the depicted paintings, mosaics, *etc.* are lost or in a severe state of degradation (Capoferro Cencetti et al., 2001, 7). This is also the reason why the original settings of some of the images in the paintings are hard to find, because the originals have been washed out completely (Capoferro Cencetti et al., 2001, 8; Coralini, 2007, 20).

To identify mistakes in the paintings and drawings they were compared with other images; the best sources are contact drawings made on tracing paper. They also show fissures in the plaster which have helped in some cases to locate the site of an image even though nowadays no trace of it remains (Capoferro Cencetti et al., 2001, 8).

In the course of the project the images were catalogued. The researchers tried to mark their exact place in the building (and decorative complex) and to add supplementary information, e.g. the artist. The images were also compared to others with the same subject matter in an effort to validate their contents and also to evaluate the work of different draughtspersons, i.e. whether their representations are more or less likely to reproduce the originals exactly (Capoferro Cencetti et al., 2001, 8, 10).

Even though at the time of the excavation of the *insula* del Centenario photographic documentation became a popular medium, there are few photographic records. These photographs show the fully excavated rooms rather than the work in progress and focus on the well-preserved areas: the main atrium (2), the white and black *oeci* (7, 8), the peristyle (9, 10) and the nymphaeum (33) (Capoferro Cencetti et al., 2001, 12). Later photographs often document restoration efforts and are therefore concerned with murals or floors (Coralini, 2007, 20). The only complete photographic documentation was made in 1977[21].

A selection of finds discovered during the excavation as well as three murals (see Section 6.1) from the *insula* del Centenario were transferred to the Museo Archeologico Nazionale di Napoli (Capoferro Cencetti et al., 2001, 14). Other murals were stolen; some of them can still be identified using the early documentation, for example, the picture of two pygmies harvesting grapes, which was taken from the white oecus (7) (Capoferro Cencetti et al., 2001, 16).

The 'Pompei – Insula del Centenario (IX, 8)' project included the documentation and survey of the remains still *in situ* (over 20,000 m^2) (Capoferro Cencetti et al., 2001, 20). Besides being relevant for the reconstruction, the results are necessary to establish the construction and decoration history of the *insula*. The documentation is comprised of several stages. The

[20] Among the media are watercolours, tempera, pen and pencil drawings. Nowadays the images are dispersed and can be found in Pompeii, Naples, Rome and Paris (Scagliarini, 2006, 12).

[21] Istituto Centrale per il Catalogo e la Documentazione (ICCD).

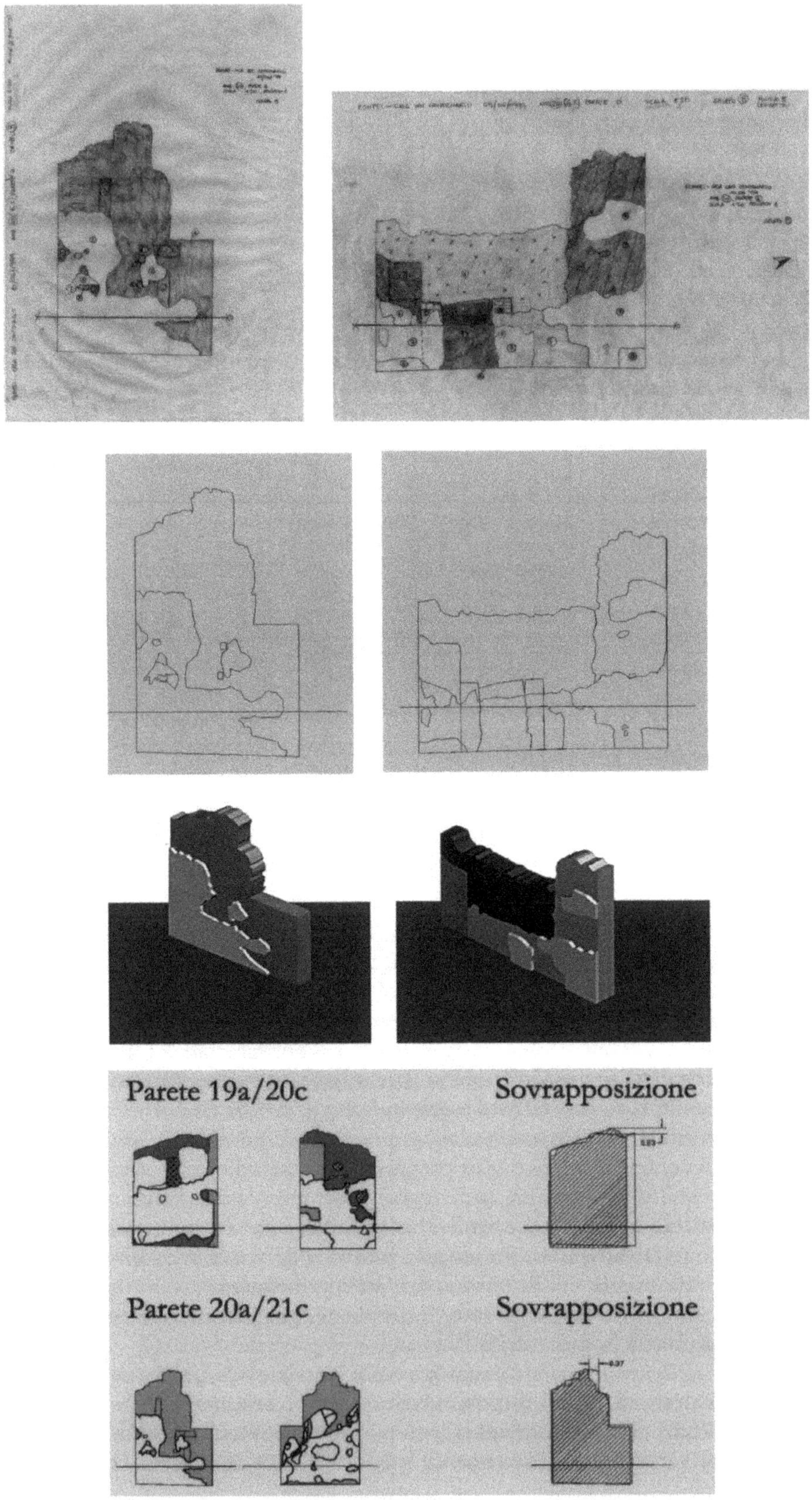

Figure 6.8: From the elevation to the model. Top: original drawings of elevations; upper centre: vector drawing of the elevations; lower centre: 3D models of the walls; bottom: evaluation of the accuracy of the drawings, drawings of both sides of a wall are compared (Santoro, 2007, 142 fig. 2–5). (Copyright: 2000–2002, Progetto 'Pompei – Insula del Centenario IX 8', Università di Bologna, Dipartimento di Archeologia e ARCES; CINECA, VisIT Lab; Soprintendenza archeologica di Pompei).

insula and the surrounding streets are surveyed using a total station and the data is transferred into a CAD[22] environment (Coralini, 2007, 22). The graphical documentation is drawn by hand on site. Details are documented in scale 1:1 using transparent films which are placed directly on the original. Other parts are recorded in scale 1:20 on paper. The images are later scanned and digitised as vector graphics (Scagliarini Corlàita et al., 2003, 241). The photographic documentation (analogue and digital), including a rule and a colour chart (Scagliarini Corlàita et al., 2003, 242), also records all the surfaces of the *insula*. Additionally a photogrammetric survey was undertaken using a camera capable of making orthophotos[23] (Zanfini and Vecchietti, 2004, 850; Santoro, 2006, 27). The various documentation methods and formats are digitally combined to form thematic maps, photomontages, elevations comprised of orthophotos (with or without a thematic overlay, Fig. 6.7 and Fig. 6.8) and 3D models, complemented with analytic sheets which depict the walls as a stratigraphic sequence (Santoro, 2006, 27 ff.).

Having the survey and documentary data in digital format also makes it possible to sort it according to research questions, for example, the sequence of building, renovation and restoration events (Capoferro Cencetti et al., 2001, 20, 22; Scagliarini Corlàita et al., 2003, 241 f.; Zanfini and Vecchietti, 2004, 850; Coralini, 2007, 20–22).

In order to study the remains of the murals several non-destructive multispectral and colorimetric methods were used (Capoferro Cencetti et al., 2001, 24): photography in the ultra-violet reflective and the visible fluorescence band, reflectography, spectrophotometry and telephotometry as well as laser scans. The results from the multispectral and colorimetric research helped to monitor the state of preservation and the usefulness of the conservation work, to identify materials, to differentiate between original and restored surfaces and the physico-chemical properties of materials. Two laser scanners were used to record the whole *insula* (Custodi and Sciortino, 2009, 626 fig 4). Besides that, some analyses, which required the extraction of sample matter, were used: the Raman spectroscopy and the analysis of pigments and binders[24] (Scagliarini Corlàita et al., 2003, 242). These analyses gave information about the colours used by the artists and their original hues (unaltered by the heat of the eruption and the later degradation through the elements). Other methods used on material from the *insula* del Centenario include X-ray scattering techniques, optical microscopy, infra-red spectroscopy, gas chromatography and mass spectrometry. Also a colorimetric examination of the glass tesserae (Scagliarini, 2006, 14 f.) as well as plaster and mortar analyses were started (see Bonazzi, 2006).

Further research was undertaken after the original reconstruction model of the Casa del Centenario was created (e.g. a georadar survey (Piro, 2007) and a microgravimetric survey (Di Filippo et al., 2007)), though some of this research has made use of the data acquired for it (e.g. the structural analysis (Custodi et al., 2006)).

Thus it is possible to draw on a number of sources for most parts of the reconstruction. Vecchietti (Solmi and Vecchietti, 2002, 201 f.) proposes that there is a hierarchy to the sources which corresponds to their 'reliability'. In descending order, the types of sources are: the modern documentation, old photographs, watercolours and written sources. She draws on the example of the main atrium floor which consists of multiple elements, for each of which a number of sources exists. For every element of the floor (e.g. black strips on the outer edge of the floor) she recorded which sources refer to it and ended up with a table (Table 6.1). The amount and the reliability of the sources then provide a 'degree of authenticity'[25].

The proposed methodology is a useful way to establish a level of confidence for parts of the reconstruction. First it is necessary to establish a degree of confidence for the sources. In the example the sources are simply graded by kind, but it will often be more useful to distinguish further between them. For example, Capoferro Cencetti et al. (2001, 10) analysed the drawings which documented the site shortly after its excavation to this end (see Section 6.3.1). Such a table would make it easy to understand why a reconstructed part is seen to be probable or improbable—and even to what degree, but the difficulty would be shifted from the evaluation of the reconstruction towards the evaluation of the sources.

The combined modern documentation yields a basis for the reconstruction onto which the documentation from the archives is added to fill in the missing parts of the decorated surfaces. If no sufficient data from the site exist, analogies are employed to fill gaps in the records. These additions are graphically marked to show that they are 'uncertain' (Capoferro Cencetti et al., 2001, 22).

[22]Computer aided design.

[23]Photographs without lens distortion.

[24]In the analyses no organic binders could be detected.

[25]The term 'grado di autenticità' is used by Solmi and Vecchietti (2002, 202) for each element. A discussion concerning the use of the word 'authenticity' can be found in Section 5.4.

	Modern documentation	Period photographs	Water-colours	Written sources
Border of black tesserae	×	×	×	×
Frame consisting of two white strips	×	×	×	×
Black field regularly interspersed with big white tesserae	×	×	×	×
Frame consisting of two black strips	×	×	×	×
Frame of the *impluvium* (white rectangles on black ground)	×	×	×	×
Revetment of the *impluvium* (Lunesian marble)				×

Table 6.1: The decorative elements of the main atrium floor and the sources recording them. After Solmi and Vecchietti (2002, 202, table 1).

6.3.2 Modelling

Coralini (2007, 22–24) sees the modelling process in two separate stages: the actual modelling (i.e. creating the geometry) and the rendering (i.e. dressing the geometry in textures) (Fig. 6.9).[26]

For the first stage, the data acquired with the total station on site and the vectorised scale drawings of all the necessary walls are used (Coralini, 2007, 22). From the profiles the relative thickness of the walls is deduced and the walls are then extruded; finally the resulting polygons are placed in their exact position on the floor plan. The result is a wireframe model (Solmi and Vecchietti, 2002, 199). For the reconstruction model, which is meant to show one hypothesis for the *insula* in AD 79, further points have to be considered, for example, the walls have not retained their original heights after the eruption of Mount Vesuvius so that they have to be estimated now. Also the roof structure of the atrium is unknown, so the atrium of Casa di Octavius Quartio was used as an analogy, because of its similarity in size and typology (Coralini, 2007, 23).

The fact that the walls are extruded means that the original irregularities of the surfaces and in the thickness of the walls are not actually portrayed in the model, as an extrusion always creates objects with an even thickness.[27] Using measurements made with the total station instead could have resulted in uneven walls in the model representing the irregularities encountered on site more accurately.

The wireframe model is then covered with textures (Coralini, 2007, 23). These need different preparation phases. For the model of the modern site the rectified photographs taken for the documentation can be used as textures. Due to varying weather and lighting conditions when these photographs were taken, or because objects may have obstructed the view, the photographs might have to be adjusted or retouched (Scagliarini et al., 2001, 85). The texturing of the hypothetical model of the *insula* at around AD 79 is more difficult. Thanks to the good preservation conditions of the site at the time of its recovery, the early documentation is of great importance, especially regarding the poor condition the site is in today (Coralini, 2007, 23).

The modern documentation is once again still the starting point for the reconstruction. Areas which have been washed out or are otherwise lost, have to be filled with the information gained from old photographs, drawings and sketches (Solmi and Vecchietti, 2002, 201). Further aid comes from the physico-chemical analyses, especially where the original colours have to be estimated. Where this information is lacking, analogies have to be used (Coralini, 2007, 25).

The interpretations behind this reconstruction process are described by Coralini (2007, 26–31); she presents several examples of various objects in the Casa del Centenario. The simplest example is the floor of the *fauces* (1). It is well preserved and the sources in the archive do not indicate any conservative measures so that, simply, the orthophoto from the modern documentation of the site (Fig. 6.10.b) was mapped as texture onto the polygon marking the floor (Coralini, 2007, 27).

The second example is the mosaic floor of the main atrium (2) which is in a worse condition (Fig. 6.10.c) and therefore needs a different treatment.

[26] Capoferro Cencetti et al. (2001, 30) mentions that, among other software, Maya by Alias Wavefront was used for modelling. No other information on the software involved is provided.

[27] See Section 10.2.1.3.

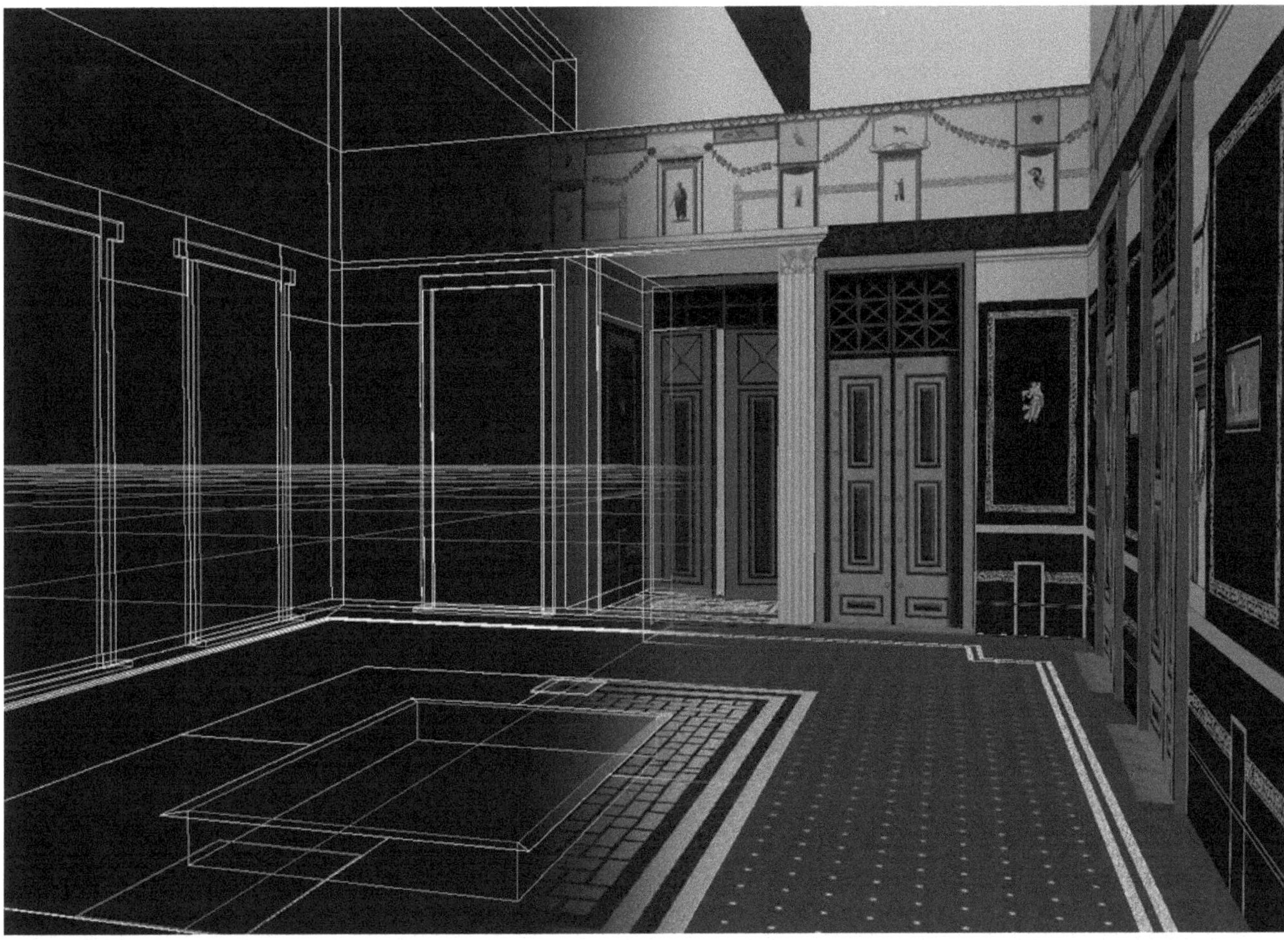

Figure 6.9: Wireframe and textured model. Left: wireframe model of the main atrium; right: the 'philologically' textured model (Zanfini and Vecchietti, 2004, 855 fig. 2). (Copyright: 2000–2002, Progetto 'Pompei – Insula del Centenario IX 8', Università di Bologna, Dipartimento di Archeologia e ARCES; CINECA, VisIT Lab; Soprintendenza archeologica di Pompei).

During the ongoing documentation of Casa del Centenario it was necessary to prevent further damage to the already disintegrating mosaic. It was decided to cover the floor with a protective layer of sand and cloth, which in turn prevented the researchers from documenting the floor as a whole. Only small parts of it could be photographed at a time (Solmi and Vecchietti, 2002, 198). Another characteristic distinguishes this floor from the *fauces*: it is composed of several repetitive patterns (the *fauces* shows a figured scene) which make the reconstruction of the mosaic easier (Fig. 6.10.a). The simple patterns also made it possible to make use of tiling[28]. As the mosaic is comprised of several patterns, each pattern is modelled as a separate plane in the wireframe model (Coralini, 2007, 27). These planes are then filled with the respective tiled patterns (Coralini, 2007, 28). A black and white border (Fig. 6.10.e, .f) frames the floor, which, in turn, is made up of large white tesserae set at intervals into a black background (Fig. 6.10.d). Closer to the *impluvium* there are further borders followed by a frame of rectangular white shapes separated by black 'joints' which enclose an *impluvium* revetted with Lunesian marble (Fig. 6.10.g). Of the latter nothing remains; the marble revetments are only described in the written excavation records (see also Table 6.1) (Coralini, 2007, 28). The only part depicted in greyscale is the opening to a cistern; what once covered this hole in the ground is unknown. The authors claim that the tiled textures worked so well that the number of tesserae in the model matches the number of tesserae in the physical mosaic (Coralini, 2007, 28).

[28]Similar to tiles on a floor a small part of the pattern is repeated over and over to make up the whole of the pattern. This exercise helps to reduce the size of the model.

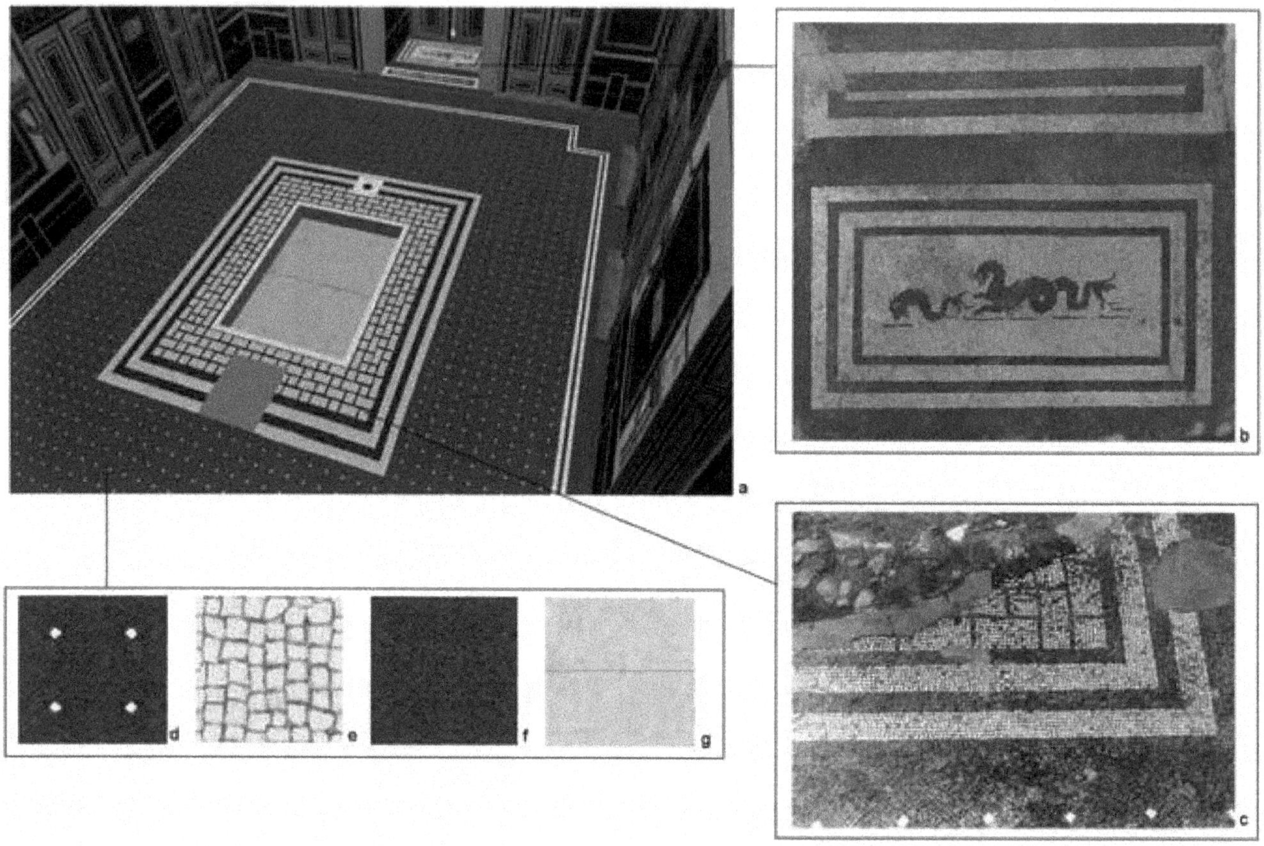

Figure 6.10: Reconstruction of the main atrium and *fauces* floors: a: finished reconstruction; b: orthophoto of the *fauces* floor; c: photograph of the border around the impluvium; d–g: textures used for tiling (Coralini, 2007, 25 fig. 8). (Copyright: 2000–2002, Progetto 'Pompei – Insula del Centenario IX 8', Università di Bologna, Dipartimento di Archeologia e ARCES; CINECA, VisIT Lab; Soprintendenza archeologica di Pompei).

Using tiling for the texturing of the floor elements, appears to have been beneficial for the file size of the model, because only small textures have to be stored. Additionally there was almost no loss of information, if the floor patterns are really as uniform as the authors claim and if the count of the tesserae is identical to the original. Indeed, the question would be, what consequences would inaccuracy in those details have had regarding the use of the model—probably none.

Less coherent seems to be the fact that despite the scarcity of sources the revetments of the *impluvium* are not shown in greyscale. Mau (1881, 123) writes that the impluvium must have been revetted with marble, but that it either remained in construction or was robbed of the revetments in antiquity. So other sources must describe the Lunesian marble. Written sources do not appear in the catalogue of criteria for level 2, i.e. coloured elements in the 'philological' reconstruction—or indeed for any other level (see Section 6.2.1.1), rendering their assigned status as witnesses unclear. Thus it remains questionable whether the revetments should really be displayed in colour.

The third example is the western wall of the main atrium and was chosen because it was well documented during and shortly after the excavation (Coralini, 2007, 28). The wall is divided into three parts by the doors to the adjacent rooms. It is decorated in the fourth style with a socle in black and a middle zone in red; the upper part of the wall is destroyed. In the centre of the middle zone are small pictures showing theatre scenes. The main layout of the design scheme is still visible and was recorded in 1:20 and 1:1 drawings (Coralini, 2007, 30). These drawings were scanned and then vectorised; images from drawings and photographs from the old documentation were integrated where the originals were lost. Finally the vector images were reverted to raster images.[29] The colours were reconstructed accord-

[29]It seems a bit tedious to first scan the drawings (raster image)

ing to the results of the pigment analyses. Where this was not possible—because the pigments have been washed out completely—the colours were taken from period drawings or refashioned in keeping with the descriptions written by the excavators (Coralini, 2007, 30). Ornamental borders, which could not be reconstructed otherwise, were shaped according to borders found elsewhere in the Casa del Centenario. The topmost zone of the room posed the greatest problems (Coralini, 2007, 30). Only one photograph from the beginning of the 19th century shows a little part of it and it seems as if there might have been a white stucco border. Therefore, it was necessary to draw on an analogy, whereby the Pompeian house Casa degli Amanti seemed to be the best choice, but further analogies were used for the decoration of the upper zone[30]. These parts are shown in greyscale in the reconstruction. Additionally the researchers made sure that the added elements in the upper zone were kept as neutral as possible: to indicate that there was decoration in this part of the room, but not to draw unnecessary attention to it, in order to avoid misleading the observer (Coralini, 2007, 30). Finally the doors to the adjoining rooms were fashioned after 'false', i.e. painted, doors found in several Pompeian houses (Coralini, 2007, 30 f.). Just as for the floor of the fauces, tiling of the textures was not possible. Instead huge raster images were used, so that even from close up the walls looked acceptable. This obviously had its effect on the fluidity of the navigation inside the model (see below) (Coralini, 2007, 31).

The last example (Coralini, 2007, 31) deals with the railings in the peristyle (Fig. 6.11). That there had been wooden railings could be proven through fittings in the columns and traces in the plaster covering them, therefore—it is argued—they are displayed in colour. The wood from which the railing was constructed did not survive the eruption and therefore other sources had to be found. Railings fortunately have an almost standardised form which can also be found in paintings in the fourth style (Coralini, 2007, 31). A bigger problem was to find out which kind of wood was probably used in their making. A study of the wooden remains in Herculaneum by Mols (1999, 2002) acted as the starting point for the research. Mols took Roman written sources as well as archaeological finds, especially from Herculaneum but also a few finds from Pompeii, into consideration. Pliny the Elder names the Silver Fir as one of the most commonly used kinds of wood for furniture and fixtures. This is confirmed by the wood analyses from Herculaneum, where most of the wooden furniture was made from Silver Fir. Finally pollen analyses from Pompeii and other Vesuvian sites show that in this region Silver Firs grew in large numbers and were therefore easily obtainable during the period in question (Coralini, 2007, 31). Therefore the probability is quite high that the wood used for the railings was Silver Fir and the texture was chosen accordingly. For this purpose a Silver Fir dowel was photographed, the resulting image turned out to be suitable for tiling and the process was used for texturing the railings. To give the railing a weathered look, the wood texture was darkened a little and a bump map[31] employed to simulate granularity (Coralini, 2007, 31).

Even though there is proof that the railings existed, their form is actually unknown. It is only an assumption that they followed the 'standardised' form known from antique paintings and other sources—they are by definition only analogies, because no wooden remains or other clues were found *in situ*. Therefore the railings should have been marked as level 3 in the 'philologic' model (see Section 6.2.1.1). Indeed their case is equivalent to that of the doors leading to the rooms adjoining the atrium. These also must have existed, but they were displayed in greyscale, because their design was based on analogies.

Besides wanting to present a 'transparent' model, the reconstruction also aimed at being interactive (Scagliarini, 2006, 15; Coralini, 2007, 17). The model was constructed around 2000–2002 and at that time processing power could still be a great limitation. Too many polygons, huge textures or the calculation of the illumination could slow down the navigation inside the model considerably (Coralini, 2007, 24).

Therefore the textures were not only used to give the objects their original colour, but also to convey a sense of light and shadow (Coralini, 2007, 24, see also Scagliarini Corlàita et al., 2003, 224). For the reconstructed area of the Casa del Centenario an ideal lighting was calculated and from this the areas of light and shadow were noted. The shade was painted on the object texture while the shadow the objects cast on other surfaces was modelled as a polygon to which a transparent grey tone was assigned. Through this, the model seemed more realistic, but the size of the model did not increase much, still

then redraw them as vector images and then convert them into raster images again.

[30]Coralini (2007, 30) names rooms from Casa di Pinarius Cerialis, Casa delle Danzatrici, Casa di Nettuno and Casa degli Amorini Dorati.

[31]For the definition of a bump map see Section 10.3.4.4.

Figure 6.11: Two views of the 'philological' reconstruction of the peristyle with the railings in place (Coralini, 2007, 29 fig. 11). (Copyright: 2000–2002, Progetto 'Pompei – Insula del Centenario IX 8', Università di Bologna, Dipartimento di Archeologia e ARCES; CINECA, VisIT Lab; Soprintendenza archeologica di Pompei).

allowing for a smooth navigation inside the model (Coralini, 2007, 24).

Photo-realistic representations of the objects in the model were meant to symbolise the idea that the probability for particular reconstructed parts is high (Coralini, 2007, 24)[32]. As a great number of polygons and very detailed textures require a lot of processing power, objects were produced in varying levels of detail (closer objects are displayed with a greater level of detail than the ones further away), to keep the size of the model small (Guidazzoli, 2007, 84) and navigation fluid.

6.3.3 Using the reconstruction model

The model was specifically designed to be interactive, allowing the user to navigate freely, to interrogate the objects in the model concerning their sources and thus to make the reconstruction transparent. For the reconstruction to be used in this way it has to run on a sufficiently fast computer and needs a user interface (Scagliarini Corlàita et al., 2003, 247, 249 ff., 253).

As mentioned in Section 6.2.1 the reconstruction is meant to be used on several device types: a 'Virtual Set' where it can be used as a backdrop for TV productions[33], a 'Virtual Theatre' with a large semicircular screen giving the user the impression of being immersed in the scene, large monitors on site for a similar experience and a portable device, 'Whyre®', to be used on site offering information during the visit (Scagliarini Corlàita et al., 2003, 255; Coralini, 2007, 26). The different media all have specific characteristics which have to be considered during the planning stages, because ideally there will be one central system delivering the relevant and adequate contents to the different user platforms. Scagliarini Corlàita et al. (2003, 254) call this a multichannel system; it holds a common information basis, but the contents are specifically conditioned for each 'channel'. The channels correspond to the different devices on which the model can be used (Whyre®, Virtual Set, *etc.*) and the contents as well as the user interface have to be adjusted according to their characteristics (size of the display, data held locally or transmitted via WLAN) (Scagliarini Corlàita et al., 2003, 254). Also, the way the information is conveyed may differ accordingly: from the immersion into the model, through still pictures to animations, many different ways of presentation are possible. Scagliarini Corlàita et al. (2003, 257) think that communication specialists should decide which kind of interaction and navigation is suitable, while web designers should look after the graphic aspects of the interface.

The information held in the system should have different depths to it, so that specialists, students or lay persons get the right amount of information (neither too little nor too much) and can choose which pieces of information they are interested in (Coralini, 2007, 26; Scagliarini Corlàita et al., 2003, 256).

[32]To avoid confusion, it has to be added that there are no non-photo-realistic elements in the model, except the parts displayed in greyscale.

[33]Such a TV production was created in collaboration with RAI, the program can be viewed at http://streaming.cineca.it//raieducational/ (Accessed: 12th of January 2012.) (Scagliarini Corlàita et al., 2003, 251).

Scagliarini Corlàita et al. (2003, 251–268) discuss some of the necessary steps towards a reasonable application of the models for educational and communication use. The goal is to enhance the understanding of the model or of the remains on site by giving access to two-dimensional data. In this way the model becomes an interface to the underlying databases (e.g. GIS or multimedia databases)[34] (Scagliarini Corlàita et al., 2003, 252).

As a specific example, the use of the model on site is considered. Two partly contradictory aims are pursued. On the one hand the devices used on site should not distract the visitor from the original remains, on the other hand the natural curiosity about the history and the background of the site should be satisfied. The authors propose that a small device carried by the visitor during the stay on site would fulfil these prerequisites best, reacting to queries from the visitor or signalling that interesting information may be accessed (Scagliarini Corlàita et al., 2003, 253).

To this end, in the context of the MUSE project[35], a new portable device called Whyre® was created. It is based on a prototype by Intel and was designed according to the most important specifications made by the team of researchers (Scagliarini Corlàita et al., 2003, 256 f.).[36] Additionally a stationary installation on site is proposed, where large monitors show the reconstruction which can be navigated and manipulated with the hand-held device Whyre®. And finally the visitors will also be able to take their experiences of Pompeii home by recording information and making pictures on site using Whyre®. The data collected thus will then be transferred to CD-ROM to be taken home as a souvenir (Scagliarini Corlàita et al., 2003, 255).[37]

The mobile system Whyre® allows only for a little interaction, as the attention of the visitors lies in the surroundings (Scagliarini Corlàita et al., 2003, 258). A number of sensors in the hand-held device help to localise the visitors and find out in which direction they are moving and looking, i.e. establishing the focus of attention. By this means a choice of information only relevant to a particular position will be displayed (Cinotti et al., 2004, section: MUSE Main Features) via a small enough number of options for easy selection (Fig. 6.12). Thus the system is enhancing the total experience of the visible remains (Scagliarini Corlàita et al., 2003, 258; Coralini, 2007, 26). Consideration was given to the modality[38] of the information best suited for on-site usage. As the eyes of the visitors should be focused on the original remains, audible information seemed to be the best choice, complemented with maps, animations, films and 3D models (Scagliarini Corlàita et al., 2003, 258).

Scagliarini Corlàita et al. (2003, 259) suggest that pieces of information could be assigned a general 'degree of interest' according to which (and according to the place where the visitor stands) a hierarchy of the contents could be built. In practice this means that when a visitor arrives at an area a short introduction is presented to him or her followed by a choice of further information which is only presented on demand.

Creating the information contents has several stages. It begins with the data collection. The sources found during research have to be digitised and are then included in what the authors call a 'digital habitat' (Scagliarini Corlàita et al., 2003, 260 f.)—by this they mean an information basis which is already so arranged that it is impossible to get lost in it or to use the data in an incorrect way[39]. The 'digital habitat' is comprised of a centralised database storing a catalogue and the layout of the contents, a Geographical Information System (GIS), a multimedia database, and a distribution system which sends the contents on demand via radio waves to the hand-held device (Scagliarini Corlàita et al., 2003, 260). It also encompasses the media content (texts, images, animations, *etc.*). These contents should not only be prepared by the researchers in the relevant fields but also by communication specialists (Scagliarini Corlàita et al., 2003, 251), and people acquainted with theatrical productions are needed as well to present the information in an engaging way (Scagliarini Corlàita et al., 2003, 259).

Another decision which has to be taken is whether to present the contents in a non-interactive (film, slide show, animation, *etc.*) or a slightly interactive way (films where decisions have to be taken, or additional information can be accessed); the latter one poses the problem that the narrative will be interrupted and extra measures have to be taken to remedy this (Scagliarini Corlàita et al., 2003, 262 f.).

The idea of including communication specialists, web designers and performing arts specialists (the

[34] See also Santoro (2006, 28) for an application within research.

[35] See Cinotti et al. (2007) regarding the project.

[36] The prototype still had some drawbacks, for example, concerning the weight of the device (Scagliarini Corlàita et al., 2003, 256).

[37] See also Cinotti et al. (2004, section: MUSE Main Features).

[38] The modality refers to the part of the sensory system addressed: hearing, sight, *etc.* See Boxed Feature 2 on page 29.

[39] Valacchi, 2001, http://www.dssg.unifi.it/_storinforma/Ws/archivi2/Fonti%20archivistiche,%20risorse%20digitali%20e%20digitalizzazione.ppt (Accessed: 28th of April 2011.).

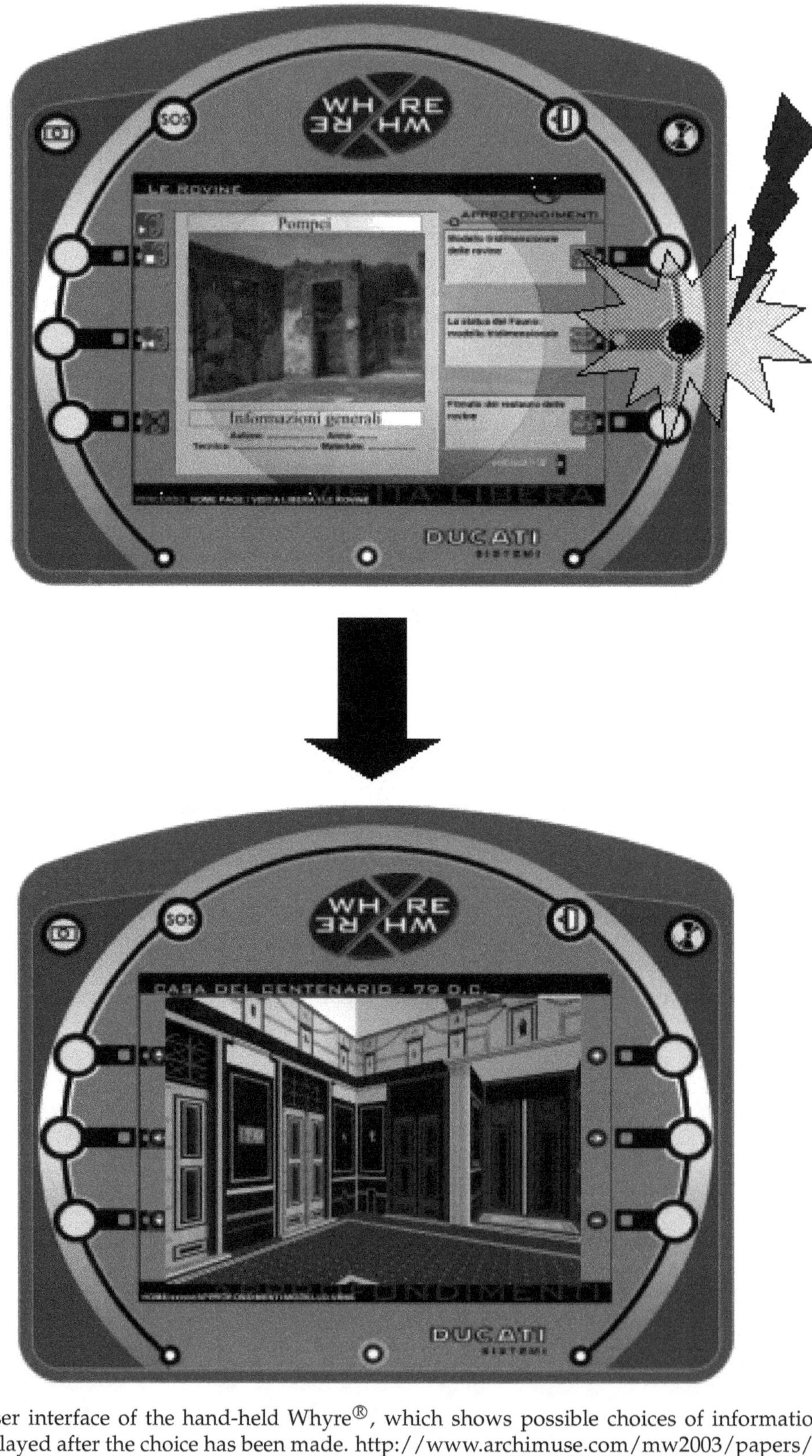

Figure 6.12: User interface of the hand-held Whyre®, which shows possible choices of information for the user and the contents displayed after the choice has been made. http://www.archimuse.com/mw2003/papers/garzotto/garzotto_Garzotto.fig8.JPG (Accessed: 12th of January 2012.). (Copyright: 2000–2002, Progetto 'Pompei – Insula del Centenario IX 8', Università di Bologna, Dipartimento di Archeologia e ARCES; CINECA, VisIT Lab; Soprintendenza archeologica di Pompei).

latter two professions fall into the category of communication design, the first may also comprise communication psychology) is highly recommended. It will help to ensure user-friendliness regarding the man–machine communication, i.e. an easy-to-use user interface that can be handled without long introductions. Additionally, the contents delivered by the system should be designed to fulfil the user's needs: they should be comprehensible, interesting, as ell as answering the most important questions the user might have (and should not only tell what the curator wants to express); ideally it would provide different contents according to the user profile (specialists, students, interested lay persons, children and so on). Terefore, it is not enough simply to include these specialists in the design phase of the system, it is also necessary to test whether the application really is appropriate for its use. All these points have been considered in this project.

In the example of the use of Whyre® presented in Scagliarini Corlàita et al. (2003, 263–268) it becomes clear that this way of using the reconstruction and the information basis at least has some limitations regarding the interactivity and the choice of supporting data which can be accessed. The path which the visitor has to follow from the Porta Marina through the Forum to Casa del Fauno and finally to Casa del Centenario is predetermined, likewise the accessible features, so that it is, for example, not possible to view the sources of all the reconstructed parts but only some selected ones.

Scagliarini Corlàita et al. (2003, 264) argue that the long way from the Porta Marina to Casa del Centenario will enable the users to familiarise themselves with Whyre® and its user interface.

The contents presented by the system, from which the user can choose, appear to be well thought-out, namely a mixture of reconstructions, which give a better understanding of what the site may have looked like, even with temporal developments (for the forum) (Scagliarini Corlàita et al., 2003, 264 f.), a reunion of the decorative parts of the Casa del Fauno, whose mosaics reside today in the Museo Archeologico Nazionale di Napoli, and some detailed information about the different stations of the visit (Scagliarini Corlàita et al., 2003, 266). For Casa del Centenario as the focus of research, apparently even more information is accessible (Scagliarini Corlàita et al., 2003, 266 ff.). Having the whole documentation and all the preliminary results available—and not just the reconstruction as an end result—is not only useful during the progress of the project, but has the potential to show the visitors how archaeological work and reconstruction processes are put into practice. The visitor is made aware of the many tedious analytical processes necessary during archaeological work and the degree of arbitrariness which is present in the added-on parts of the reconstruction. However this information appears to be limited to the main atrium, probably because a good documentation basis exists for it (see Section 6.3.2).

Further features deal with the production stages of a 3D-reconstruction, where all the phases of the virtual reconstruction process are shown and explained[40], and likewise the archaeological profession itself[41] (Scagliarini Corlàita et al., 2003, 267 f.).

6.3.3.1 An evaluation of Whyre®

Cinotti et al. (2004)[42] published an evaluation of the Whyre® device. The test was not set-up in Pompeii but in the Museum and Charterhouse of San Martino[43] in Naples, another location which can be visited with the Whyre® system.

Despite the difference in site and contents, some of the results are independent of the location and can help to shed light on the possible use of the device in the Casa del Centenario. The proposed channels of use are similar to the ones planned in Pompeii: 'an on-site mobile context-aware channel, an on-site stationary channel, a memories channel' (Cinotti et al., 2004, section: MUSE Main Features) (see Sections 6.2.1 and 6.3.3). The evaluation deals with the on-site mobile context-aware channel.

This channel has three modes of use, of which two are interactive. The first gives access to web pages with multimedia content independent of the user's location. The second is context aware and offers information according to the standpoint and line-of-sight of the visitor. A third mode is a non-interactive guided tour (Cinotti et al., 2004, section: MUSE Main Features).

The evaluation is comprised of two parts, a user evaluation and an expert evaluation (Cinotti et al., 2004, section: Our Approach to Evaluation). The user evaluation is more important in this context because the

[40]This information feature is linked in Whyre® to the peristyle.
[41]This information is available in the *viridarium*.
[42]http://www.archimuse.com/mw2004/papers/salmon/salmon.html (Accessed: 28th of April 2011.).
[43]Il Museo e Certosa di San Martino.

expert evaluation has its focus primarily on the testing of the technical aspects of the system (Cinotti et al., 2004, section: Usability Inspection, plus the following subsections):

'Multimodality[44] effectiveness': Can the user find out in which mode the system is operating, what the restrictions of this mode are and whether is it easy to swap between modes?

'Synchronicity': Does the system play the relevant information according to the position and line-of-sight of the user, and what happens if the user changes position or the line-of-sight during play back?

'Fault evidence': Does the tracking of the user position and the line-of-sight work flawlessly? If errors occur, is the user informed about them?

'Recovery/Task Resumability': If a system failure occurs can the user reset the system?

The user evaluation, on the other hand, is concerned with the 'user satisfaction', the 'multimedia content design' and the ergonomics of Whyre® (Cinotti et al., 2004, section: User Testing Sessions). 32 visitors to the museum (mostly Italians from the region around Naples—due to the winter season) were asked to fill out questionnaires after their visit to the museum. Additionally 17 persons (mainly students from cultural heritage related fields of study) were invited to test the system and to fill out the same questionnaires. In the statistical evaluation phase these groups were treated separately, though Cinotti et al. (2004, section: User Testing Sessions) state that the results were very similar.

The questionnaire contained 28 questions targeting seven topics:

The first aim was to identify the user profile and with that the museum audience (Cinotti et al., 2004, section: Set 1 Responses: Museum Visitors). The results are certainly not representative, because the sampling period was short and took place at a time of low visitor numbers (November 2003–January 2004, i.e. in winter); tourists especially were missing from the evaluation, though only Italian speaking tourists could have contributed, because the questionnaires were in Italian[45]. The number of persons approached is also small, but at least men and women are almost equally represented. The visitors predominantly belong to the middle class and have a higher education, their age ranges between 21–65 years, and for most of them it is not their first visit to the Museum and Charterhouse of San Martino. Also, a large number states that they are interested in technical devices—this can be relevant when looking at the user performance with the hand-held Whyre®.

To evaluate 'the novelty of the proposed approach' visitors were asked whether they have seen or used a similar device before. None of them had (Cinotti et al., 2004, section: Set 1 Responses: Museum Visitors).

The third and the seventh aim, i.e. to evaluate 'the capability of the approach to modify visitors' expectations, and from there, its potential to create new demand and a new market' and to understand 'how much a visitor could afford to pay for Whyre® services' Cinotti et al. (2004, section: User Testing Sessions), are obviously marketing related. Nevertheless it is interesting to know that the visitors stated that they would like to use similar devices on other sites.[46]

The fourth topic was concerned with whether 'the proposed development approach fits the specific museum nature' (Cinotti et al., 2004, section: Set 1 Responses: Museum Visitors).

Here, for the first time, the article offers some insight into the nature of the questions on the questionnaires and the predetermined, possible answers (five grades ranging from 'excellent', 'good', 'do not care/no answer' and 'disappointing' to 'very disappointing'). The questions

> Question 21: 'Did you find the visit to the "Immagini e Memorie" section more appealing with WHYRE?'

and

> Question 22: 'Should WHYRE be available in other museum sections, would you expect a better visit quality?'

are unfortunately leading in nature.[47] A neutral question would be: 'How did you find the visit to the

[44]Cinotti et al. (2004) do not use 'multimodality' in relation to the senses, but rather imply that it means 'many kinds' of use. The three modes are described above.

[45]Cinotti et al. (2004) do not mention, whether Whyre® offers other languages than Italian.

[46]The answer to the question about how much people would be prepared to spend on such a service, is not disclosed in the article.

[47]The article (Cinotti et al., 2004) is written in English and it is at least *possible* that the questions were not leading in the original Italian questionnaires. The latter are not published.

"Immagini e Memorie" section with Whyre®?' More suitable answer categories would be: 'very appealing', 'appealing', 'neutral', 'disappointing', 'very disappointing', 'I did not visit the section/I did not use Whyre® there', no answer. This answering scheme would allow for distinguishing between visitors who are unable to answer the question (i.e. the question is inapplicable), because they did not visit the section or did not use Whyre® in that section; visitors who see no advantages or disadvantages; and visitors who simply chose not to answer the question.

The visitors' reactions to Whyre® were mostly positive (whether this is due to the leading questions remains unclear), though there were also some negative answers, e.g. 'the device is uncomfortable' or 'it took too long to understand how it works' (Cinotti et al., 2004, section: Set 1 Responses: Museum Visitors).

The visitors were also asked which exhibit they enjoyed most, but they were not asked whether this was due to/influenced by the background information provided by Whyre®, in which case this would have been an essential question. Notwithstanding, Cinotti et al. (2004, section: Set 1 Responses: Museum Visitors) speculate, whether the visitors' choice could have been influenced by the information provided by Whyre®.

The fifth topic is concerned with 'the Whyre® device ergonomy (audio quality, screen visibility, shape and buttons design, weight)' (Cinotti et al., 2004, section: User Testing Sessions). The visitors thought the video and audio output to be good, but the device was often judged to be quite heavy.

The sixth aim was to evaluate 'the effectiveness of the multimedia contents'. Intermixed with this question seems the desire to know how easy it is to use the device (i.e. with regard to the user interface). All of the users answered that the contents were clear to them (Cinotti et al., 2004, section: Set 1 Responses: Museum Visitors), unfortunately the phrasing of the question was leading again. Concerning the information format (coding) visitors preferred a wider range (audio, video and written text) to audio-visual contents only. Half of the users said they would prefer written instructions on the use of the device, but 90% of the users answered 'very easy' and 'easy' when asked 'Is WHYRE easy to use?'. Is this openly contradictory result due to the influence of the questions?

Despite the fact that the results of this evaluation should be used with some caution, it seems to show that hand-held devices for use in museums and on site do appeal to the visitors and that the additional information is well received by the users.

Further research into how information is received and influences visitor behaviour is necessary. A representative sample of visitors (including tourists and first-time visitors) and also a control group would be necessary, e.g. to judge whether the additional information results in a prolonged stay and/or results in higher visitor satisfaction. A refinement and context-broadening of the questions goes without saying.

6.4 Have the aims been reached?

The first aim, supporting the conservators in their work, appears to have been reached; to what extent exactly cannot be established through the available literature. Only the following example is cited by Scagliarini et al. (2001, 84, 88 fig. 5): the red colour of the central fields in the main atrium (2) had been very bright in the first reconstruction attempts. Later the Raman analysis brought new results on the red pigments used in the murals and the red colour was proved to be much more attenuated than previously thought. The colouring could easily be changed to the new shade and with that the appearance of the room altered (Fig. 6.13 left).

The second aim, helping the archaeologists to verify their interpretations, is even less traceable. Though the authors think that the model will help them in this context, none of the texts makes reference to an actual occurrence.

The modelling of a curtain separating *ala* i and the main atrium (2) from each other can be judged as an attempt to model a reconstruction hypothesis (Fig. 6.13 right). Scagliarini et al. (2001, 89) see it as as an attempt to show everyday life in the domus. There is no proof that the curtain in Fig. 6.13 (right) has been there; the hypothesis is solely based on the assumption that the owner of the house may not have wanted visitors to see the shelving in this area.

Other representations of furnishings are sparse, due to a lack of sufficient sources which could tell something about them. Rust stains marking the place where a chest once stood (Scagliarini et al., 2001, 87) are, for example, not enough to say anything about the latter's appearance.

[48]There might have been a mistake in Scagliarini et al. (2001, 88): the colour on the right side of the image appears rather harsh instead of attenuated.

Figure 6.13: Integrated analyses and hypotheses. Left: reconstruction of the main atrium (2) with partly integrated results of the Raman analysis (attenuated shade of red on the right[48]). Right: reconstruction hypothesis with a curtain to close off *ala* i (in the last phase of the *domus* used as a storage room) (Scagliarini et al., 2001, 88 fig. 5, 89 fig. 8). (Copyright: 2000–2002, Progetto 'Pompei – Insula del Centenario IX 8', Università di Bologna, Dipartimento di Archeologia e ARCES; CINECA, VisIT Lab; Soprintendenza archeologica di Pompei).

The third aim, giving the public the opportunity to see and understand the site, is well thought-out. Great care has been taken to follow recommendations from authors concerned with a critical examination of virtual archaeology applications (e.g. Frischer et al., 2002) and many practical solutions have been found. Among these the indication of uncertainties in the model through the use of greyscale is certainly the most laudable (see Scagliarini Corlàita et al., 2003, 246). The solutions found for estimating the probability for certain objects or decorations are commendable (see Solmi and Vecchietti, 2002, 201 f., 202 table 1):

The three levels of probability (which are mirrored in the display of the models: now and AD 79 in colour or greyscale) are clearly defined with the necessary characteristics and kinds of sources needed to support them (see Section 6.2.1.1). The only problem is the missing definition as to which level belong objects which are only mentioned in written sources (as in the example of the marble revetments in atrium 2). There also seem to be some inconsistencies as to whether an object should be represented in colour or greyscale when it is based on analogies and various clues in the remaining ruins (for example, the doors in the atrium and the railings in the peristyle). Although, following the classification of the levels in Section 6.2.1.1 to the letter, such objects should be depicted in greyscale.

The method for estimating the probability on a finer scale by using a table to draw up how many and what kind of sources support an object or decoration has proved useful (see Solmi and Vecchietti (2002, 201 f.) and Table 6.1).

Besides the visual cues in the model the researchers intended to make the reconstruction verifiable by displaying the sources and hypotheses underlying its component parts, thus enabling the visitors of the site to use the data provided with the model to verify the reconstruction themselves and to arrive at their own hypotheses (Coralini, 2007, 26).

Unfortunately exactly this is prevented by the fact that the sources displayed with the reconstruction are only the ones supporting it. Without alternative or contradictory material the user will certainly not arrive at alternative conclusions. Additionally only the *sources* are displayed, the difficult reasoning behind the reconstruction stays concealed. A good demonstration that such reasoning is essential is presented by the explanations of Coralini (2007) (see Section 6.3.2) which disclose much more than the mere sources could ever convey, but still do not give a full report concerning all the interpretations which are behind the reconstruction. For example, it is never really explained why certain houses were chosen as analogies for parts of the Casa del Centenario.

Despite 'promoting new hypotheses' being an aim of the reconstruction only one of them ('the most likely') is displayed; there are no alternative models.

It can be argued that the situation in Pompeii is interpretationally simpler than on many other sites and it is therefore easier to arrive at a very likely reconstruction: the remains are in a way well preserved (unlike buildings of which nothing remains, except perhaps the foundations); the attempt is to reconstruct the last phase of the *domus* (the situation is much more difficult if later alterations have almost completely obscured the original layout of the building). Changes in layout of the buildings in the *insula* and repairs in antique (and modern) times have been noted, but were not built as separate models, which, on basis of the evidence, would have been a much more difficult task than the reconstruction attempted here. Nevertheless, it was considered by Scagliarini Corlàita et al. (2003, 245) that the reconstruction of previous phases would be a beneficial addition to the model, and also that the further the model reaches back in time the more difficult it would be to fill the gaps.

Concerning the different devices or 'channels', on which the model can be used, it is not always quite clear what really has been implemented. The TV-program in collaboration with RAI educational shows that the model was used with great effect on the Virtual Set. The implementation on the Virtual Theatre is also working and can be booked on demand[49]. How far the Whyre® technology has advanced is not so clear. The 'guided tour' proposed in Scagliarini Corlàita et al. (2003) gives the impression of being planned but not implemented yet, though the implementation clearly may have taken place in the years following the publication of the article.[50] What can be learned from the named text, is that the information on the relevant sources for the reconstruction is not available for all the sites encountered. This clearly comes from the wish not to overwhelm the visitors with information that is maybe not relevant to them, but this also takes away their chance of judging for themselves.

Another anticipated channel are the monitors installed on site which are meant to give better and more detailed access to the model. This and the opportunity to burn a CD-ROM (Scagliarini Corlàita et al., 2003, 255) with the impressions of the day, to take home as a souvenir, are probably not implemented (or if so: are not open to regular use) as the Via di Nola and with it Casa del Centenario are still closed to the public.

6.4.1 Unexpected uses for the model

Besides fulfilling the aims which were set out at the beginning of the project, some uses came as a bonus to the researchers.

For example, the model was used by the stress analysts for surveying the walls and for the analysis of the masonry in order to draw a map re the criticality of the structure. Besides that, proposed roofings were inserted into the model to be tested for statics and aesthetics[51] (Scagliarini Corlàita et al., 2003, 244 f.).

It was very important to the project that the model acted as a lingua franca for communication between the different members of the multidisciplinary team—even more so than for communicating with the visitors (Scagliarini Corlàita et al., 2003, 240).

Finally, having the whole documentation and all the preliminary results available and not just an end result was very useful during the progress of the project (Scagliarini Corlàita et al., 2003, 266).

6.5 Appraisal

Besides minor drawbacks, this is an excellent project demonstrating that applied ethical issues can be addressed and that levels of confidence as well as supporting information (e.g. sources) can be incorporated into the model. Guidelines have been drawn up for a uniform handling and assessment of the sources. A great effort has been made to include specialists from various fields to provide a sound foundation for the reconstruction and its presentation. Even though this leads to a number of different aims for the model, the result is a coherent end-product. In particular the achievement of presenting an interested audience with engaging and helpful information on the Casa del Centenario, thus enhancing their experience of the site, is very well thought-out.

[49] Antonella Guidazzoli, personal comment.

[50] Cinotti et al. (2004) announced that 'soon it [Whyre®] will be demonstrated in Pompeii.'

[51] Presumably this had no been anticipated from the beginning.

Chapter 7

Ename 974

The second case study is concerned with the Ename 974 project, which encompasses several reconstructions in Ename: St. Saviour's church, which later becomes an abbey, the castle, the village and St. Lawrence's church, which is the focus of this chapter.

Several information systems for the different sites were realised, to give the local population—but also other interested individuals—the ability to understand the remains and the history of the sites as well as the research undertaken there. In at least one case the model of St. Lawrence's church did influence the direction of research (Fig. 7.1, see also Section 7.3.3.3). Data concerning the site was collected and interpreted, before models of the different phases were constructed. In the case of St. Lawrence's church the model was built while the excavation and building research was still ongoing. New finds, features and other sources discovered during research led to frequent updates of the models. The models were used for communication purposes in the museum, kiosks and the internet. Older versions of the models are still available and together with their updates give unique insights into the interpretation and revision process during the project.

This chapter starts with the historical background of Ename and the evolution of the site. Then the archaeological background is examined and finally the multiple parts of the Ename 974 project are introduced. The chapter ends with a discussion about whether the aims of the project have been reached and also an appraisal of the work.

7.1 Historical background of the region

The focus of this section lies on the reconstruction of the St. Lawrence's church in Ename. To understand the building and its background it is necessary to look at the development of the castle, the settlement and the abbey.

Ename is situated east of the River Scheldt in Belgium, around 22 km south of Ghent and 50 km west of Brussels. In the Ottonian period the River Scheldt marked the border between the Holy Roman Empire and the Kingdom of France and was secured by three margraviates (Ename, Valenciennes and Antwerp; see Fig. 7.2). Besides this it was also the scene for various, more local conflicts (Franz, 1940). Franz (1940) provides useful information, which helps in understanding the sequence of events leading up to a shift in the domains involved (see Fig. 7.3). Milis and Callebaut (1990) give the best description of the historical sources concerning Ename in particular.

In 958 Reginar III of Mons was banished from his lands by Brun of Cologne, his sons Reginar IV and Lambert I returned from exile and killed Werner and Reinard of Hainaut, upon which Otto II banished them again and gave Hainaut, in 974, to Godfrey of Verdun and Count Arnulf to govern. Arnulf's prior concern was with Valenciennes (he is referred to as *comes Valentianensis* in two deeds (Nonn, 1983, 130)), while Godfrey established himself in Ename (Bethmann (Ed.), 1844, 1846).

A late 9th century Ghent source mentions a man named 'Radin' and his brother, who lived in *'villa Ehinham'* (Gysseling and Koch, 1950, 138). This is the first known reference to Ename though Milis and Callebaut (1990) believe that it does not refer to modern Ename, but rather to the rural settlement of Nederename because another Ghent source dating from 1064 names St. Vedastus' church in Nederename as the mother church of the Ename churches[1]. Considering that St. Vedastus' church belonged to a rural

[1] *'Mater quoque ecclesia, id est beati Vedasti de Inferiori Eiham'* (Gysseling and Koch, 1950, 270).

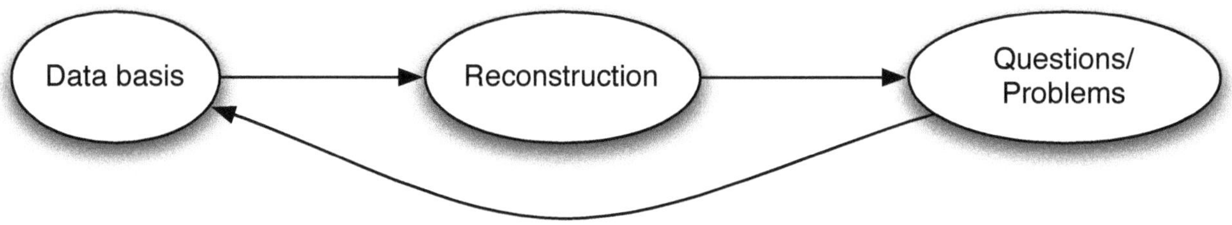

Figure 7.1: Excavation Model as demonstrated in the Ename 974 Project.

settlement, it has rather large dimensions (see Section 7.2.2).

The year when the margraviate of Ename was established is unknown[2]. It is assumed that it was created by joining two counties (Biest and Chièvres), which constituted the parts into which the margraviate of Ename was split up 1033 (Franz, 1940, 241).

Two text sources tell, in retrospective, of the castle, the pre-urban settlement and the owners of Ename (see Fig. 7.4). The first is the *Gesta Episcoporum Cameracensium*:

> ***De villa Iham.*** *Est etiam locus super Scaldum fluvium, quem dicunt Iham, ubi modernis temporibus honorabilis vir comes Godefridus et uxor sua Mathildis, matrona videlicet memorabilis—erat enim suum predium, suis usibus oportunum—castro quidem munito, navigium, mercatum, teloneum, ceteraque negotia statuerunt; infra castrum vero monasterium in honore sanctae Mariae deputatis canonicis, fundaverunt. Extra autem Herimannus filius duo monasteria struxit, unum Sancto Laurentio, alterum vero sancto Salvatori. Nunc igitur locus, utpote noviter instructus, ex omnia sufficientia floret, et tamen esset uberior, nisi crebro hostili incursione quateretur, quod maxime ab inimicis Dei patitur pro stabilitate regni et fidelitate imperatoris.* (Bethmann (Ed.), 1846)[3]

The second text is the *Auctarium Affligimense*:

> ***1005.*** *Iste dux Godefridus dictus est Eihamensis. Erat autem Eiham oppidum et castrum munitissimum, et sedes principalis ducatus regni Lotharici. Qui dux Godefridus duxit Mathildem, filiam Herimanni ducis Saxonum, viduam relictam Baldewini comitis, qui erat filius Arnulfi magni marchionis. De qua idem Godefridus genuit tres filios, Godefridum, Gozcelonem, Hezcelonem. Hezcelo comes, post mortem ducis, castrum Eiham cum provincia Brabantensi suscepit et diu tenuit. Hic enim genuit filium nomine Herimannum, et filiam nomine Berthildem; qui dum adhunc iuvenes essent, defuncti sunt, et in ecclesia apud Felseka sepulti. Qui postea multis miraculorum signis claruerunt, sed a Verdunensibus monachis furtive ablati sunt. Alteram quoque filiam tradidit nuptui Reginero, Montensi comiti, simul cum tota provincia Brabantensi. Deinde cum omnia sua ad votum ordinasset, relicto in manus Regineri castro et comitatu, apud Verdunum effectus est monachus.* (Bethmann (Ed.), 1844)[4]

[2]Nonn (1983, 221) believes it to be established already during the reign of Otto I. Franz (1940, 251 ff.) thinks that the margraviates may have existed since 965, she also discusses several opinions of other authors in her text.

[3]'There is a place over the River Scheldt, which is called Ename, where in recent times that honourable man Count Godfrey and his wife Mathilde, certainly a memorable woman—because it was her own personal property, being suitable for the use—have installed a fortified castle, a harbour, a market, a toll, and other businesses; indeed, inside the castle, they have founded a monasterium in honour of St. Mary, given to canons. Outside, moreover, their son Herman has built two monasteria, one for St. Lawrence the other, indeed, for St. Saviour. Now then this place, being newly set up, has prospered from all possessions, yet it would have been richer, if it had not been hit by frequent hostile attacks, for it suffers the most from God's enemies for the steadfastness of the reign and faithfulness of the emperor.' A *monasterium* is a church with several clerics.

[4]'This Duke Godfrey was called "from Ename". Ename was, however, a town and a strongly fortified castle and principal seat of the duchy of Lorraine. This Duke Godfrey married Mathilde, daughter of Hermann, Duke of Saxony, widow of Count Baldwin, who was a son of margrave Arnulf the Great. With her this Godfrey had three sons, Godfrey, Gozcelo and Hezcelo. Count Hezcelo, after the death of the duke, took over the castle Ename with the province of Brabant and held them for a long time. He fathered a son named Hermann and a daughter named Berthilde; they died when they were young and were buried in the church at Velzeke. They later became famous for many miracles, but were furtively removed by monks from Verdun. Another daughter he gave into marriage with Reginar, Count of Mons, together with the whole province of Brabant. Then, after he had ordered everything according to a vow, leaving castle and county in the hands of Reginar, he became a monk in Verdun.' Hezcelo is another name for Herman of Ename.

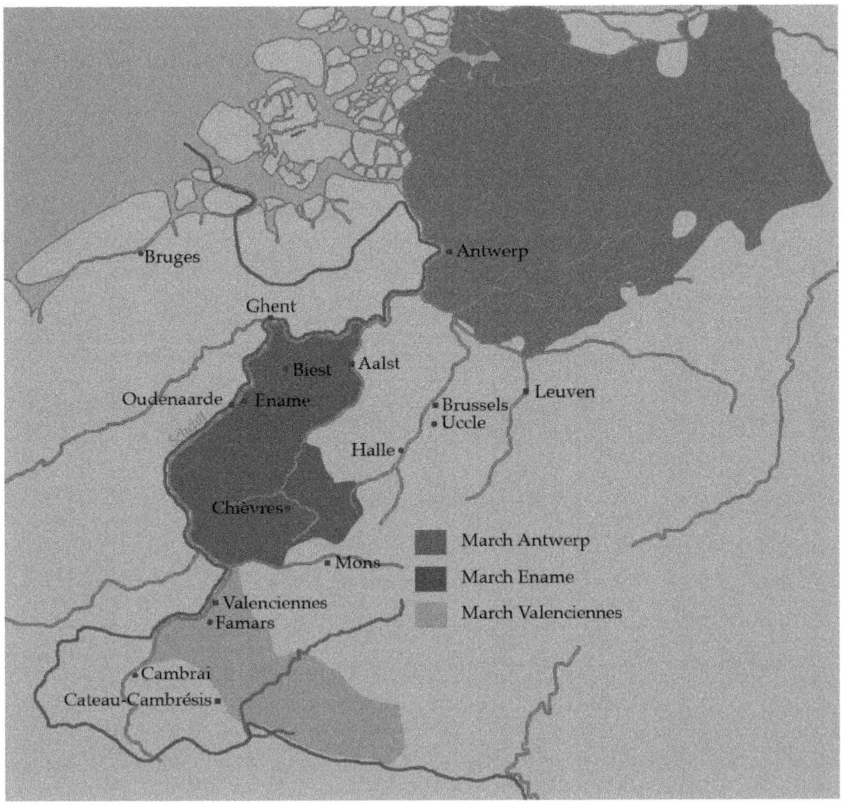

Figure 7.2: The march of Ename and the contemporary marches of Antwerp and Valenciennes (redrawn after Franz (1940, 233)).

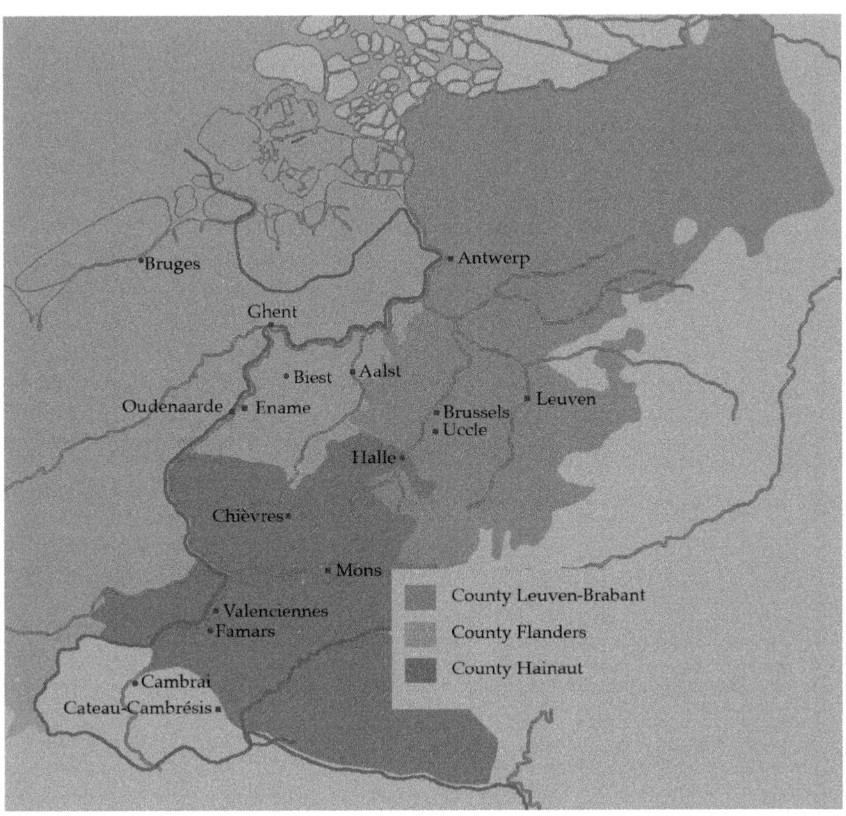

Figure 7.3: The resulting counties after the marches had broken down (redrawn after Franz (1940, 233)).

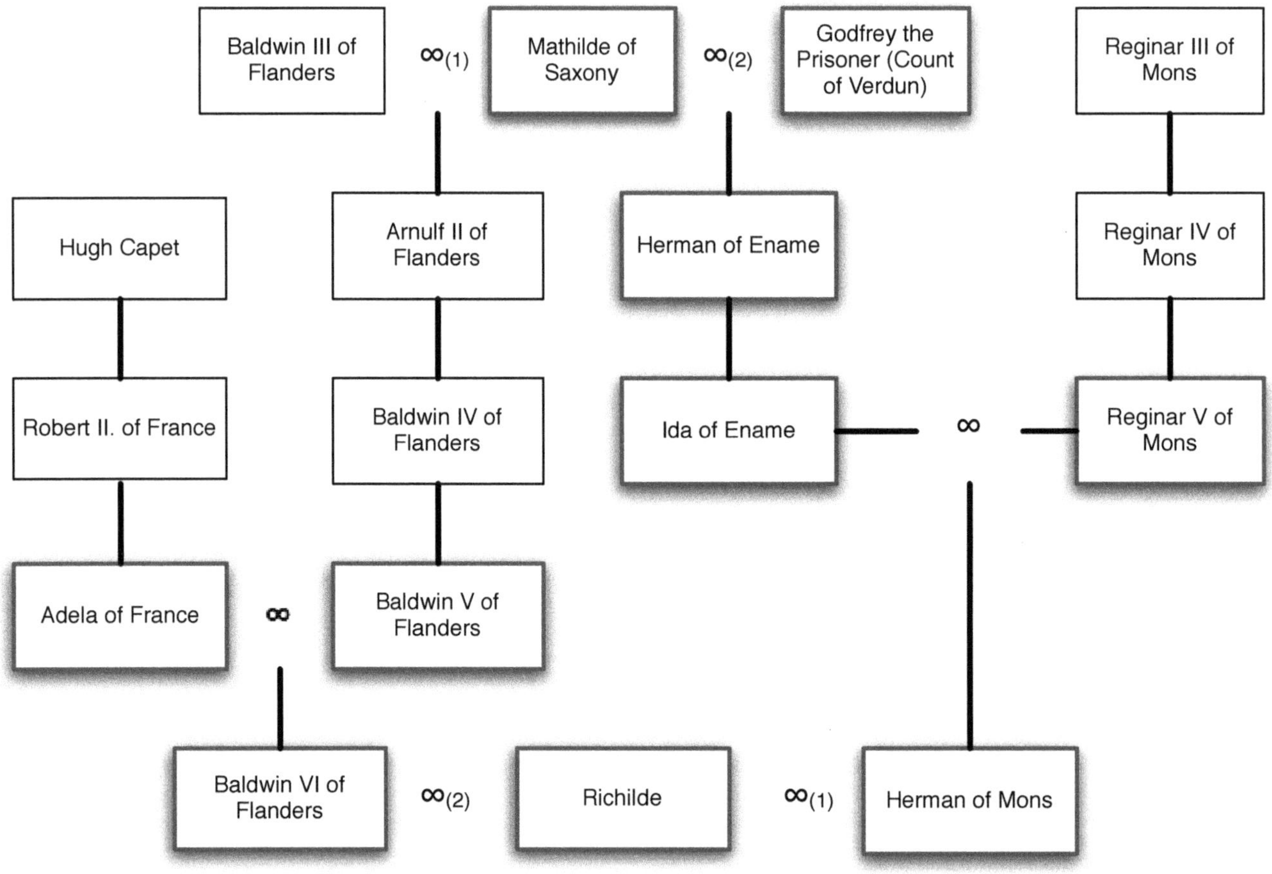

Figure 7.4: Relationships between the houses of Verdun, Flanders and Mons. The persons who were in possession of Ename are framed in red and shaded.

The *Gesta Episcoporum Cameracensium* was written in 1024/25 and mentions that on grounds (*praedium*) possessed by Mathilde of Saxony a *castrum* was built[5] and that she and her husband Godfrey the Prisoner organised navigation, trade, tolls and other affairs. These events can have only taken place after 962 and before 1005/09. The first date marks the death of Baldwin III of Flanders, who was Mathilde of Saxony's first husband. After his decease she married Godfrey I of Verdun, who is mentioned by the *Gesta Episcoporum Cameracensium*. The last time Godfrey appears in written sources is in 998 (Milis and Callebaut, 1990, 463 ff.), but this does not necessarily mean that he died that year or soon after. The latest possible date for establishing the settlement could be set either to Mathilde's year of death (1009) or the year 1005, for which the second source, the *Auctarium Affligimense* (written before 1164), tells of an *oppidum et castrum munitissimum, et sedes principalis ducatus regni Lotharici* at Ename. The question is how reliable is the date given by the *Auctarium Affligimense*; especially considering that the text was compiled with the help of other texts (Gorissen, 1952, 52 ff.). The *Gesta Episcoporum Cameracensium* mentions a castle built under Godfrey and Mathilde including a church of St. Mary, which was used by canons. The *Auctarium Affligimense* adds that it was strongly fortified and also the

[5]The fact that the castle, which is the administrative centre of the march of Ename and a stronghold defending the border, i.e. an 'imperial castle' is built on Mathilde's *allodium*, is taken by Milis and Callebaut (1990, 469) as a sign for 'strong identification between state and family interests'. But it could also mean the opposite. Originally Velzeke, a former Roman *vici* (Ameels, 2003), was one of the administrative centres for the region. It also has a 10th century church which is comparable in size to that of the first design for St. Lawrence in Ename. That Velzeke held some importance during the Ottonian Period is evidenced by the fact that Herman of Ename buried his two children there. The choice of building a new castle, administrative centre and the foundation of a harbour settlement with a toll and a market on private land and for private profit, hints at a diversion of power to a personal domain, away from the Emperor's reach. The undertaking was successful. Similar events have taken place, for example, in Freiburg, Germany. (Matthias Untermann, personal comment.) The allodium on the other hand, must have come from the property of the Counts of Flanders. Mathilde of Saxony would have no possessions in this region, but could have gained them by marriage to Baldwin III.

principal seat of the duchy of Lorraine[6] and, therefore, an administrative site (which may indicate a second reason for the canons to be at the castle: to see to administrative tasks).

Another source mentions a *gynaeceum*:

> ***1012, Oct. 1.*** *Ex ginentio comitis Heinrici[7], de portu scilicet Einhamma, erat puella nomine Oda caeca per annorum tempora. [...]* (Holder-Egger (Ed.), 1888, 616)[8]

A *gynaeceum* is usually a work-house for women's work (weaving) and part of a socage. Therefore, it is probably not located inside the castle.

The *Gesta Episcoporum Cameracensium* also refers to trade-related issues (a harbour, a market and a toll booth), which were established during the time of Godfrey and Mathilde. They apparently belonged to a trading-station (*portus*), which appears to have been quite successful, as confirmed by the *Gesta Episcoporum Cameracensium* because Godfrey's and Mathilde's son Herman built two churches big enough for the number of inhabitants and certainly for prestige[9], namely St. Saviour's church and St. Lawrence's church. When exactly the churches were built is once again unclear. The relevant time span is between the date Herman of Ename takes over from his father Godfrey (so we have to assume a date after 998) and before he himself retires to the monastery in Verdun, probably in 1025 (Milis and Callebaut, 1990, 469). The assumption because of their dedications that work on St. Lawrence's and St. Saviour's church had started between 999 and 1002 (Milis and Callebaut, 1990, 484), seems rather daring. The authors are certainly right to point out the connections between the dedications of the Ename churches and those of famous imperial and papal buildings (Aachen Cathedral, Lateran). These were certainly intended, as was also the connection to Otto I's victory on the Lechfeld.

The *hostili incursione* mentioned by the *Gesta Episcoporum Cameracensium* can be imagined when looking at the further events in the region (cf. Franz (1940)):

[6]This title was inherited by Godfrey of Verdun's son Godfrey.

[7]For the difference in name see (Holder-Egger (Ed.), 1888, 616, footnote 3).

[8]'From the gynaeceum of Count Herman [of Ename], from the harbour namely Ename, was a girl named Oda, blind for years.'

[9]This seems a likelier explanation, of why two churches were built at (nearly) the same time, in contrast to the hypothesis by Callebaut (1992, 461), who assumes that the churches may have been built to accommodate a *Festkrönung* of Henry II. Having two churches may have facilitated such an event, but it is unlikely that they were built in anticipation of it in the first place.

Together the two margraves, Godfrey and Arnulf, were able to hold Hainaut against another attack from the house of Reginar in 976. However, in 977 Otto II gave the house of Reginar their allodia back. The county including Mons remained under Godfrey at first until Reginar IV was able to conquer it in 998 (Franz, 1940).

In 978 Lothair of France started an assault on Aachen, which was answered by an attack on Paris by Otto II in 979. In 985 Lothair of France conquered Verdun and took Godfrey of Verdun captive (he is therefore sometimes referred to as 'Godfrey the Prisoner'). In 998 Reginar IV of Mons established himself in Mons (a region previously controlled by Godfrey of Verdun), while his brother, Lambert I of Leuven, did the same in Leuven.

In 1006 Baldwin IV of Flanders conquered Valenciennes with its castle, which had previously been governed by Count Arnulf.

In 1007 Henry II attacked Flanders and won Valenciennes back, but then he handed it out as a fiefdom to Baldwin IV. What happened to Arnulf and his family remains unclear. Herman of Ename fought in 1015 against Lambert I of Leuven, but let his daughter Ida[10] marry Reginar V of Mons, the nephew of Lambert I. When Herman of Ename retired to the monastery of St. Vanne (Verdun) he left Reginar V in control of Ename. Maybe, therefore, this marriage can be seen as an attempt to bring peace and stability to the region.

In 1020 emperor Henry II. was again in Lorraine to attack Ghent, which belonged to Baldwin IV of Flanders.

Milis and Callebaut (1990, 472 f.) mention justifiably that before this point the conflicts concerning Ename and the surrounding area were usually connected to the house of Mons and the French king, but not to the house of Flanders. They think this could be due to family connections because Mathilde of Saxony was the mother of both Arnulf II of Flanders and Herman of Ename. When the castle and the march came into possession of Reginar V of Mons, the attitude of Flanders towards Ename could have changed.

Finally Baldwin IV was able, through treachery, to destroy the castle of Ename in 1033:

[10]Franz (1940) gives her name as Mathildis, probably assuming her to be named after her mother. Herman's daughter is not named in most sources, but Van Acker (1987) argues convincingly that her name was probably Ida.

In diebus Ragineri Longicolli traditum est fraudulenter castrum, quod dicitur Eiham, et datum est Baldewino Brabato, qui castrum funditus destruxit. (Bethmann (Ed.), 1844, 399)[11]

But strangely enough Baldwin IV seems to have only gained possession of the southern part of the margraviate while the castle and settlement in Ename remained in the possession of Reginar V of Mons (Franz, 1940, 240). At least it was later in the possession of his son Herman of Mons (also known as Herman of Hainaut), who must have inherited it from his father.[12] In 1040 Herman of Mons obtained the margraviate of Valenciennes, which was probably the reason why he and Baldwin V of Flanders in 1047 traded the southern part of the margraviate of Ename and the town of Valenciennes (which went to Herman of Mons) for the northern part of the margraviate of Ename. Both gained from this exchange: Herman now held the complete margraviate of Valenciennes and the adjoining land of the margraviate of Ename while Baldwin controlled the northern part of Ename, which was close to his lands (see Fig. 7.3). As a replacement for the destroyed castle at Ename, Baldwin V built a new castle in Oudenaarde in 1051 from where he controlled the region:

Idem Balduinus apud Aldenardam castellum constituit, per quod, everso apud Eham castello, Bracbantum usque fluvium Teneram de regno Lothariensi sibi usurpavit. (Bethmann (Ed.), 1851, 320)[13]

In 1063 Baldwin V of Flanders and his wife, Adela of France, founded a monastery in Ename. The note in the *Auctarium Affligimense* is short:

1063 Hoc in anno cenobium Eihamense a Baldewino comite et Athela comitissa constructum est. (Gorissen, 1952, 115)[14]

A deed from 1063, which records Baldwin V and Adela giving Ename to Abbot Walbertus is more detailed:

Damus porro tibi et scripto confirmamus Eiham antiquum castellum sicut nos recepimus de manu Herimanni comitis, et sicut nos hactenus [te]nemus qu[iete et] solide, [et] aquam totam cum toto teloneo, et cum sex mansis de terra arabili, sicut sunt siti et dispositi uel pariter uel separatim, per[tinent t]amen ad ipsum alodium. (Gysseling and Koch, 1950, 268)[15]

Walbertus was a monk from the Abbey of St. Vaast in Arras, and returned later (before 1085) to this monastery (Batselier, 1977, 26). It is not mentioned where the other monks came from, whether they followed Walbertus from Arras or whether they came from other places.

From the source above which explicitly mentions the *antiquum castellum*, and a papal document from the 28th January 1070 (*'Gualberto uenerabili abbati monasterii sancti Saluatoris siti in episcopatu Camaracensi, loco qui dicitur Eiham, [. . .]'*)[16] which names St. Saviour's church as the abbey church, as well as from other indications (cf. Batselier (1977, 12)), Milis and Callebaut (1990, 487 f.) inferred that the monastery was first established inside the castle and later transferred to St. Saviour's church. Milis and Callebaut (1990, 488 f.) also suggest that the castle was not really *funditus destruxit* when the monastery was founded, but that some part (possibly the keep) must have been still standing for it to be used by the monks.

The deed from 1063 can be understood in another way, too, namely that the *antiquum castellum* refers to the castle and the settlement. In combination with the toll, the stretch of water and the arable land *per[tinent t]amen ad ipsum alodium* these would account for the

[11] 'At the time of Reginar V Longneck the castle, which is called Ename, was fraudulently handed over and given to Baldwin of Brabant, who destroyed the castle completely.' Gorissen (1952, 113) gives the date as 1034 (see also his comment on pages 35 f.) and remarks that the destruction of the castle took place while it was in possession of Herman of Mons, i.e. not at the time of Reginar V (Gorissen, 1952, 53) though in 1034 Reginar V was probably still alive. His presumed year of death is 1039 or later (see Van Acker (1987, 282), who refers to Vanderkindere (1902)) and therefore would have still held Ename.

[12] Maybe Baldwin IV held Ename in all but name. He conquered the land and later, by means of the following exchange of lands, could legitimise his 'possession' of the land. (Matthias Untermann, personal comment).

[13] 'The same Baldwin built a castle in Oudenaarde, for the destroyed castle in Ename, he appropriated Brabant, [part] of the Lorraine kingdom, up to the river Dender.'

[14] '1063 In this year the monastery of Ename was constructed by Count Baldwin and Countess Adela.'

[15] 'We further give you and confirm in writing the old castle of Ename, as we have received it from the hand of Count Herman, and as we hold it peacefully and strongly until now, and the complete [stretch of] water with the complete toll, and with six hides of arable land, as are situated and distributed both adjacent and separate, they belong nevertheless to just that allodium.'

[16] 'Walbertus, venerable Abbot of St. Saviour's monastery situated in the diocese of Cambrai, in the place which is called Ename, [. . .]'

whole of Ename as mentioned by the *Gesta Episcoporum Cameracensium*.

Additionally both of the contemporary written sources, which refer to the founding of the monastery, say that the monastery was 'built': 1063 *'cenobium Eihamense [...] constructum est.'* and 1064 *'coenobium [ed]ificare cupientes'*:

The following extract from the 1064 deed by Libertus, the bishop of Cambrai, in which he confirms the foundation of Ename Abbey, shows the latter quotation in context.

> *Est igitur in territorio Bracbatensi locus quem dicunt Eiham, qui iam pridem satis floruerat seculariter et m[en]ibus et colonis, sed et ecclesiarum religiosa honestate. Verum bello insurgente et inualescente, pompe omnes preterierunt, ecclesiarum [qu]idem sancta uenustas migrauit. S[ed] nondum ruina macerie. Comes praeterea Balduinus et religiosa eius coniunx Adela, diuine karitatis spiritu uisitati, in eodem loco pro pristina reuerentia ecclesiarum, ob suam suorumque perpetuam salutem, coenobium [ed]ificare cupientes, [...].* (Gysseling and Koch, 1950, 269)[17]

The monks were meant to use St. Mary's church (Milis and Callebaut, 1990, 495), which had fallen into disuse after the destruction of the castle in 1033. Where inside the castle this church was situated is not really clear, but it is assumed that it was located inside the 'great hall' (see Section 7.2.3). Therefore the complete destruction of the 'castle' (i.e. the keep), the reuse of St. Mary's church, which is possibly located in the great hall, and a new building for the monks would be in accordance with the written sources:

The initial plan could have been to give St. Saviour's church to the monks, but it is clear that during the time their living quarters were being constructed they would need to stay elsewhere. The great hall inside the castle could have been used for this purpose; it was particularly suitable as St. Mary's church was already attached to it.

[17]'There is therefore in the Brabant territory a place called Ename, which already in the past has flourished in worldly respects sufficiently both city walls and settlers, but also the religious dignity of the churches. In truth with war uprising and growing, all splendour has passed away, indeed the holy beauty of the churches is gone. Nothing but a heap of ruins. Count Baldwin and also his religious wife Adela visited by the spirit of divine charity wishing to build a monastery in this place out of the pristine reverence for the churches, for her perpetual salvation and theirs, ...'

This is indeed the version of events presented in the latest virtual reconstruction in the Museum of Ename. It is further assumed that the keep was really destroyed by Baldwin IV, and that the stones of this building were used to construct the new abbey.[18]

In 1139 abbot Snellard built a new abbey church, which replaced the old St. Saviour's church:

> *1139 Deiecta est vetusta Eihamensis ecclesia et incepta est nova a Snellardo abbate.* (Gorissen, 1952, 128)[19]

Meanwhile, on the other hand, St. Lawrence's church, being the parish church for the settlement of Ename, remained mainly unaltered. There are few written sources concerning the church, but it seems that from *c.* 1177-1182 to *c.* 1207-1224 St. Lawrence's church was used by the monks (Milis and Callebaut, 1990, 492). This could be due to building activities at St. Saviour's church, described by several sources (Milis and Callebaut, 1990, 493) or due to the damage caused by a flood, which took place around 1181.

7.1.1 Berings' collection of records

A collection of church records concerning St. Lawrence's church was compiled by Berings (1994), who relates the events and gives excerpts of the written sources in his unpublished manuscript. The sources he collected begin at the end of the 16th century. His focus lies on maintenance work and alterations to the church, which were done before the restoration in 1908–1909. For this period bills and receipts as well as Ronse deanery visitation reports were consulted, though not all of these documents have survived (Berings, 1994, 1 f.). Berings lists the events in chronological order.

It becomes clear that the church had suffered in 1566, during the Thirty Years' War; all altars are reported to have been desecrated, though the number of original altars seems to be a matter of great confusion (one record assumes that there might have been seven due to the number of relics). Also the number of conse-

[18]There is no proof that the stones from the keep were really reused for this purpose. No mortar analysis was made on the stones of the abbey, nor were remains of chipped off mortar found anywhere close to the two sites. (Daniël Pletinckx, personal comment.)

[19]'1139 The old church of Ename is pulled down and the new one is begun by Abbot Snellard.'

crated and not yet consecrated altars varies considerably in the early sources[20].

The roof is one of the main problems of the church. Between 1563 and 1623 repair works on the roof are frequently mentioned. These works came to an end when finally the wooden shingles were substituted for slate (Berings, 1994, 5 f.). The records also contain indications as to where the sanctuary of the church might be found, and indeed that there was at this time only one sanctuary: in 1593 a leaky window caused damage to it, and the sanctuary is mentioned in 1598 to be the only decorated place in the church. In 1607 the leaky roof of the tower is noted with the remark that this could lead to water coming into the sanctuary, which is situated underneath it: *'Turris multis detecta ita ut per fornicem aqua chorum intret qui sub turri est.'* (Berings, 1994, 6 f.).[21] So on the one hand this source reveals the position of the sanctuary (and therefore that of the high altar) whilst also informing us that there was a 'vault'. So far the vault is thought to have become incorporated into the tower at a later time (cf. Callebaut (1992)), but this view has been revised lately (see Section 7.2.6.4).

Berings (1994, 7) assumes that the sanctuary under the tower in the east was also the place where the altar stood before 1566 i.e. before the destruction during the Thirty Years' War. He further considers the possibility of the church having two sanctuaries at the time shortly before its destruction in 1566, but dismisses this possibility due to the images on a pilgrimage pennon (1596) (Fig. 7.17, right) and a map (Fig. 7.27, left) (1595–1596), which depict the church with an entrance to the west.[22]

Bills and receipts from 1654–1655 tell of building work in wood being done in the belfry. Beams[23] and planks for ceilings were prepared for this task as well as some ironwork undertaken.

Some decorative work took place in these years as well: as mentioned before, the St. Mary's altar was consecrated in 1655. In preparation for this event the sanctuary was painted in white and black (1654) and a picture of St. Mary was commissioned. In the same year, the provost of Ename Abbey paid for a large picture in the main sanctuary (Berings, 1994, 11).

The question which arises is, what connection existed between the abbey and parish church resulting in the provost paying for the picture? This is especially interesting in the light of an older source relating to the leaky tower roof (1607): *'Voluit decanus id [bene] corrigi: sed allegatur exiguitas bonorum fabrice et incolarum paupertas'*[24] (Berings, 1994, 7). Here the population should have paid for the repair, and there is no mention of the abbey or persons connected with it[25]. But already in 1628, the provost of the abbey had bought two estates in Ename, one to be used as house for the pastor, the other to become a school; work on the former took place during 1655 and 1656 (Berings, 1994, 11 f.). Berings interprets these as possible signs of the counter-reformation. The visitation report for the years 1657 and 1658 states that the abbot of Ename Abbey is indeed at that time the prebendary of St. Lawrence's church, which explains why he pays some of the bills. This connection between the abbey and St. Lawrence church makes sense and confirms an earlier hypothesis that the deed of 1063 (Gysseling and Koch, 1950, 268) really refers to the whole *allodium* of Ename being given to the monks. They then would not only be responsible for their own abbey and St. Mary's church in the castle, but also for St. Lawrence church.

The above mentioned visitation report also makes mention of a sacristy, but without further details concerning its whereabouts.

In 1666 the church receives a new floor, though the material is not enough to pave the whole of the church. The records make mention of 400 'fourteen inch'[26] stones in total and 200 'ten inch' stones, plus 300 further stones without mentioning their size and 30 'fourteen inch' stones in addition to ash and lime (Berings, 1994, 12). This would amount to 71 m^2 plus

[20] 1608: two unconsecrated, one partially consecrated; 1609: one consecrated; 1611: three consecrated, two unconsecrated; 1613: one consecrated, three unconsecrated; 1614: one consecrated, one unconsecrated and one partially consecrated; 1616: two consecrated, three unconsecrated; *etc.* (The number of the altars are copied and written over in the documents resulting in a muddle.) From 1619 to 1626 consistently one consecrated altar is mentioned, from 1627 onward the records speak of two consecrated altars. From the record of 1650 it becomes clear that the third, unconsecrated, altar is dedicated to St. Mary. 1655 all three altars: high altar, St. Lawrence's altar and St. Mary's altar are consecrated (Berings, 1994, 4 f.).

[21] 'The tower is very much roofless so that water could come through the vault into the choir, which is underneath the tower.'

[22] The reliability of these images has been cast into doubt because of other inconsistencies. Despite those, the entrance to the west has been assessed to be one of the more reliable parts of the drawings. See Section 7.2.7.

[23] Five beams with a total length of 29.40 m, one beam of 7.28 m, four pieces of 5.04 m, six pieces of 2.10 m and five pieces of 1.68 m. For the ceilings: 620.48 m of planks (Berings, 1994, 10).

[24] 'The dean wanted this well mended: but it was argued [with] the scarcity of on-site cash goods and the poverty of the population.'

[25] Maybe they were the persons responsible for the site cash—and that coffer is said to be empty, too.

[26] *'Veerthien duim'*, a *duim* being around 2.7 cm.

the 300 stones of unknown size[27]. As Berings (1994, 13) already mentions this is hardly enough to pave the whole church, so either old paving stones were reused or more likely (the four masons only needed 17 days for their work) only part of the floor was renewed.

Otherwise, the church was only embellished between 1661 and 1671: a new altar, a chancel, an altar rail and steps in front of the side altars were installed, walls and altars were painted and sometimes gilded. In 1673 a new window over the entrance to the church is in production, some repairs on the roof done and roughly 100 m^2 of new paving stones laid (Berings, 1994, 13). In the following year another 50 m^2 of paving stones are delivered[28]. In 1675 'the great window' is set into its place and a vault repaired. It is not clear where these works were carried out, but could it not be possible that the window is identical to the one mentioned in 1673?

In 1691 the floor around the main altar is laid anew, but even so a year later a new altar is bought, which Berings (1994, 15) assumes to be the high altar. This new altar is made from wood and put in its proper place in the same year.

An interesting document from 1696 mentions that a mason came on behalf of the provost to Ename to have a look at the *'steenen van St Caemelincx [?] oft de salye dienstich waeren tot het maeken van de vaute van (choor van) St Laureijns'*.[29] Indeed in 1699 a bill for vaulting the area above the St. Lawrence's altar is presented. Further embellishments include repair work on the window and new paintings on the walls.

It is explicitly mentioned that the St. Lawrence's altar is *'ad latus'*, i.e. in an aisle, and that the baptismal font is situated in a separate room.[30]

In 1718, 38,000 stones and 237 sacks of lime are bought for repairs to the wall surrounding the churchyard. Berings (1994, 17) relates that this is not the first time a written source refers to this wall, but indeed that it appears in the earliest visitation reports. The churchyard seems to have been only partly surrounded by a wall, the gaps being filled with hedges. Furthermore the wall was crumbling away in many places leading to the planting of more hedges. In 1718 the problem seems to have been addressed in another way: with stones and mortar.

Besides the installation of new seats only repair works are undertaken in 1722–1727.

Two receipts dating from 1740 mention that the window over the *'endeldeure'* was taken out and that a new *'groete venster boeven de hendeldiere'* was made. Already in 1673 and 1675 a great window was mentioned; in particular the source from 1673 stated that this window was to be found over the entrance. According to Berings (1994, 19) the *endeldeure* or *hendeldiere* was the door furthest away from the altar. In that case this would be the west door. In the same year another window above a door is mentioned: the one over the *'lijkdeur'*[31]. Berings mentions that there is a door in the north-eastern part of the church, which could have been used as a *lijkdeur*. He could be referring to the door in the northern wall of the eastern sanctuary, but this door leads to the stair tower (Fig. 7.15). There is a door in the eastern wall of the northern aisle, but access would be restricted by one of the altars; moreover the portal/baptistery lies behind it. Another possibility is the door mentioned in Callebaut et al. (2000)[32]. It is approximately in the middle of the northern wall of the northern aisle and leaves enough space for a window, but it is not really in the eastern part of the church.

In 1754 and 1755 a slater and a carpenter work on the roof, neither the place[33] nor the kind of work done are further specified. Berings (1994, 19) believes that the western roof could have been the one affected by these works, but both Hoffsummer (1992) and Devliegher (1969) claim that the roof in the western part is old. The dendrochronological dates suggest that the trees were felled between 1175 and 1185. What these texts do not mention is that every second rafter of these old roof trusses is a much later addition (Fig. 7.5), but probably later than the 17th century (see Sections 7.2.6). Neither can the beams of the tower be the ones in question. There is an inscription on one of them, which refers to the year 1778 (Fig. 7.6). So either the beams of the nave and the western sanc-

[27] It could be assumed that they were of the smaller size, i.e. 27×27 cm, as the bigger size of fourteen inches is explicitly referred to. This would result in further 21.87 m^2 of paving stones.

[28] Berings (1994, 14) mentions that 1678 a mason is paid for laying new stones in the aisles, though this would maybe seem either to be a bit late for the stones delivered in 1674, or the payment of the craftsman is late.

[29] '...stones of St. Caemelincx, if those are useful to make the vault for the St. Lawrence's choir.' Berings (1994, 15) thinks this means that the vault was renewed, but could it not be possible that the vault was newly constructed. The words in the bill from 1696 are *'het maeken'*, and 1699 *'ghevauteert thebben'*?

[30] This could refer to the 'portal', which was later used as the baptistery. See Section 7.2.6.5.

[31] German: *'Totenpforte'*; a door, usually in the eastern part of the church, which leads to the cemetery.

[32] http://www.ename974.org/Ndl/pagina/NuhV/000425_1.html.

[33] The terms used are *'ghevel'* and *'seije ghevels'*, i.e. front and sides.

Figure 7.5: Northern side of the roof truss over the western sanctuary. Every second rafter is a later addition and not connected to a tie beam. The tie beams themselves are truncated and connected to a trimmer. (Photograph by Joyce Wittur).

tuary are earlier (or were later replaced), or only minor works were carried out, which did not involve the major beams. In addition the aisles could have been effected. The latter are not mentioned in any of the publications concerning the roof of St. Lawrence's church[34].

Bills for the year 1762 include payments for 51,500 bricks. Receipts refer to 29 workdays for a mason and 91.5 workdays for a carpenter. Unfortunately their work is not specified. Another receipt speaks of removing the slates to work on the *'ghevel'* and later closing up the same roof *'vermaeckt van den slaeper'*[35] The definition of *slaeper* or *slaper* is ambiguous. Berings (1994, 20) refers to the Middlenederlands woordenboek by E. Verwijs and J. Verdam (1912), who define it as: 'The overlapping beam, on which the rafters lie'. But there are other definitions, too. Dirk de Vries told me that according to E. J. Haslinghuis and H. Janse, Bouwkundige termen, Leiden (1997) *slaper* can have several meanings: a slanting beam under a hipped roof (instead of a roof truss) or a horizontal beam under a valley. He added that whether *slaper* could also refer to a summer[36] is hard to define especially because the regional uses of words in Flanders could then and can still vary. A modern source (http://www.mathys.info/bouwinfo/slaper.htm (Accessed: 6th of August 2012.)) gives the meaning of *slaper* as: *'Een balk die bij o.a. een dakkapel toegepast wordt en die de gordingen die doorgezaagd zijn, op moet vangen.'*[37] This means a trimmer—similar to the ones found in the western sanctuary (Fig. 7.5) and the nave (see Sections 7.2.6.1 and 7.2.6.3). In 1767 there is another reference to *'het uijtdecken den slaeper op de neerkerke'*.[38] Berings thinks that the 'lower church' could only refer to the western part of the building, because it is furthest away from the high altar. He even goes so far as to assume that

[34]They were also not accessible for inspection.

[35]'...changed by the dormant'.

[36]German: Unterzug

[37]'A beam, which is used, for example, for a dormer and has to support the rafters, which were sawn apart.'

[38]'...the covering(?) of the *slaper* on the lower church.'

Figure 7.6: Beam in the tower roof with inscription and date: 1778. (Photograph by Daniël Pletinckx).

this is connected with the reversal of the orientation of the church. As can be deduced from the previous written sources, the high altar was in the east, under the tower. However, before the latest restoration in 1999 it was still in the western part of the church. Therefore, the orientation must have been changed. The bricks could help to find the place where the building work and maybe the roof work was done. The church is built mainly from Tournai limestone, except for the western sanctuary where the west wall is mainly built from bricks, although it still includes a western doorway (see Fig. 7.7). Also, the topmost floor of the tower was until 1907/08 constructed from bricks (Callebaut, 1992, 439).

Assuming that the bricks were meant instead for the western part of the church, the next entry probably refers to the door within that wall: in 1762 wood was delivered to build a *'hendeldeure'* for the church. Berings (1994, 20) explains that the *endeldeur* is the door furthest from the altar[39], therefore the new opening in the brick-built west wall received a new door.

Another receipt mentions the delivery of 36 windows. Ten will be necessary for the clerestory, another ten for the aisles. There were probably five window openings in the western sanctuary: one above the door, two each to the north and the south. The tower needs five windows on the ground floor: two to the north, two to the east and one to the south, and and six on the first floor: two to the north, three to the east and one to the south. This would account for all 36 windows, assuming that the entrance is really still in the west and that the window in the eastern wall of the nave has been closed up already.

No bills between 1765 and 1778 have survived (Berings, 1994, 21). But from the date on the eastern portal (1778) it is clear that at this time the orientation of the church must have changed, and the new

[39]In his interpretation this means that the door is in the eastern wall of the eastern sanctuary—the orientation of the church now being reversed. This is rather unlikely as the portal, which was installed there, bore the date 1778.

Figure 7.7: St. Lawrence's church. Two photographs showing the western wall of the western sanctuary during excavation and restoration. The bricks used for this wall are visible, also the walled up entrance in the centre. (Photographs by Daniël Pletinckx).

opening was cut into the tower. It was also the time when a new organ was fitted on the eastern wall of the nave and other refurbishments were undertaken in the tower. They resulted in the tower having four storeys instead of three, as was the case before the changes occurred.

Some receipts refer to the organ, others to wood cut for the *'docksael'*[40] (Berings, 1994, 21 f.). In 1777 and 1778 several receipts for masonry works are to be found, though the works are not specified. They are probably related to the changes on the eastern wall of the nave and the new entrance to the tower. Another receipt of interest dates from 1778 and tells of materials delivered to help move the altars[41].

These were all the major works during this period on St. Lawrence's church.

7.2 Archaeological sources

From the written sources it is known that the site of Ename consisted of a castle with a St. Mary's church and a pre-urban settlement with a St. Saviour's and a St. Lawrence's church (Fig. 7.8).

Albeit St. Lawrence's church is the focus of this chapter, the other settlement parts and St. Vedastus' church in Nederename will be mentioned shortly. So far non of the excavations have been fully published. Therefore it is necessary to draw on preliminary reports, information for the general public and personal comments by the people involved.

[40] A *doksaal* is actually a rood loft, but can also allude to an organ loft.

[41] Berings (1994, 22) believed this could refer to the side altars alone, as he thought that the reversal of the church orientation had already happened at an earlier date.

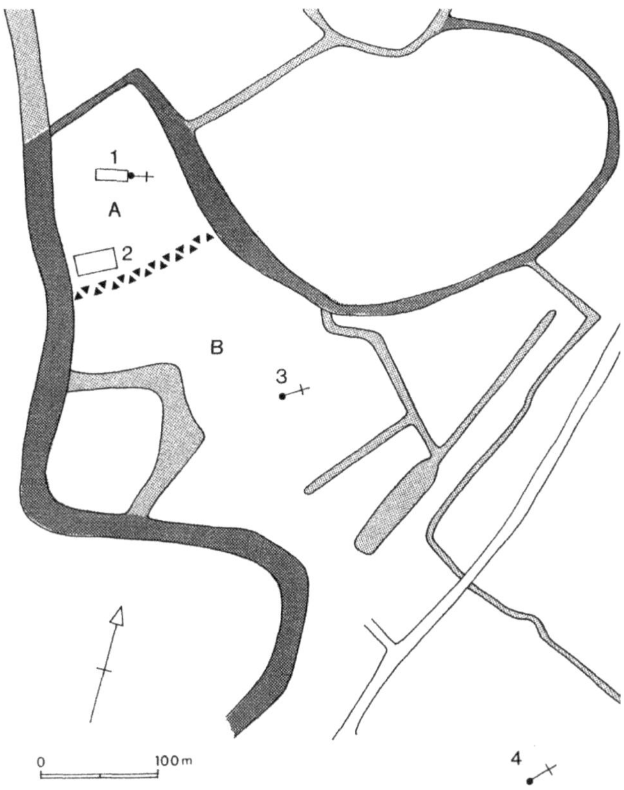

Figure 7.8: Map of Ename in early mediaeval times. A: *castrum*, 1: great hall, 2: keep, B: *portus*, 3: St. Saviour's church, 4: St. Lawrence's church. The dark grey areas denote the River Scheldt, the light grey areas are water filled ditches. The feature indicated by triangles between the castle and the settlement is another ditch (Milis and Callebaut, 1990, 461 fig. 1).

7.2.1 Early occupational traces in Ename

The archaeological excavations in Ename unearthed finds from prehistoric times (Milis and Callebaut, 1990, 462) while traces of Roman land use were discovered during the excavation in St. Lawrence's church (Callebaut et al., 2000)[42]. But it seems that there was a hiatus in the settlement activities during the Merovingian period because the next occupational layer dates to Carolingian times (Milis and Callebaut, 1990, 462). The date is based on potsherds, but no traces of buildings could be identified.

[42]http://www.ename974.org/Ndl/pagina/NuhV/000316_1.html and http://www.ename974.org/Ndl/pagina/NuhV/000726_1.html.

7.2.2 St. Vedastus' church in Nederename

St. Vedastus' church in Nederename was named as the mother church of the churches in Ename in a document dating from 1064 (see Section 7.1). It has not been excavated, though a part of the old building is still preserved. It serves as the sanctuary of the modern church, but was originally the nave. The initial building (see Fig. 7.9) consisted of a nave and a recessed rectangular sanctuary which were separated by an triumphal arch. Before the original sanctuary was broken down in 1939 (Callebaut, 1999), to make way for the new nave, measurements on the inside of the building were taken.

The old nave is 13.55 m long and 8.20 m wide. The walls are ca 73 cm thick and made from coursed, roughly hewn Tournai limestone (Fig. 7.10, top left and bottom). To the east adjoined the sanctuary, which was 7.60 m long and 5.60 m wide and had ca 70 cm thick walls (Van de Walle, 1952, 21 f.). It is unknown how many windows the sanctuary had. The northern side of the former nave still retains its original window openings, i.e. four small, high-set round-arched windows, around 97×53 cm in size on the outside (Fig. 7.10, top left). They are double-splay windows with sloped sills. The arches are constructed similarly to the ones of St. Lawrence's church in Ename. The imposts create the semblance of a horseshoe-shaped arch. The arch itself is constructed from flat stones, which are not set radially, but leave space for a wedge-shaped keystone (cf. St. Lawrence's church: 7.21 and 7.24, left). The southern side of the of the former nave is less well preserved (Fig. 7.10, bottom) because new, larger window openings were cut into the wall during the 17th (Van de Walle, 1952, 23) or 18th (Callebaut, 1999) century, but fortunately the outline of a walled-up door remains. Its visible height (in 1950) was 1.90 m, its width 1.10 m. Van de Walle (1952, 23) assumes that this door was the only entrance to the church, though he does not state on what grounds he dismisses an entrance to the West.

He also states that the brickwork tower to the west of the former nave was added in the late Gothic period, a view which is repeated by Callebaut (1999)[43]. The tower clearly displays the date of 1837 on its southern side (Fig. 7.10, top right), though there is at least one clear building joint visible in the brickwork while the socle of the tower is made from coursed, roughly hewn limestone ashlars (Fig. 7.10, bottom).

[43]He believes that it was constructed in the 15th century.

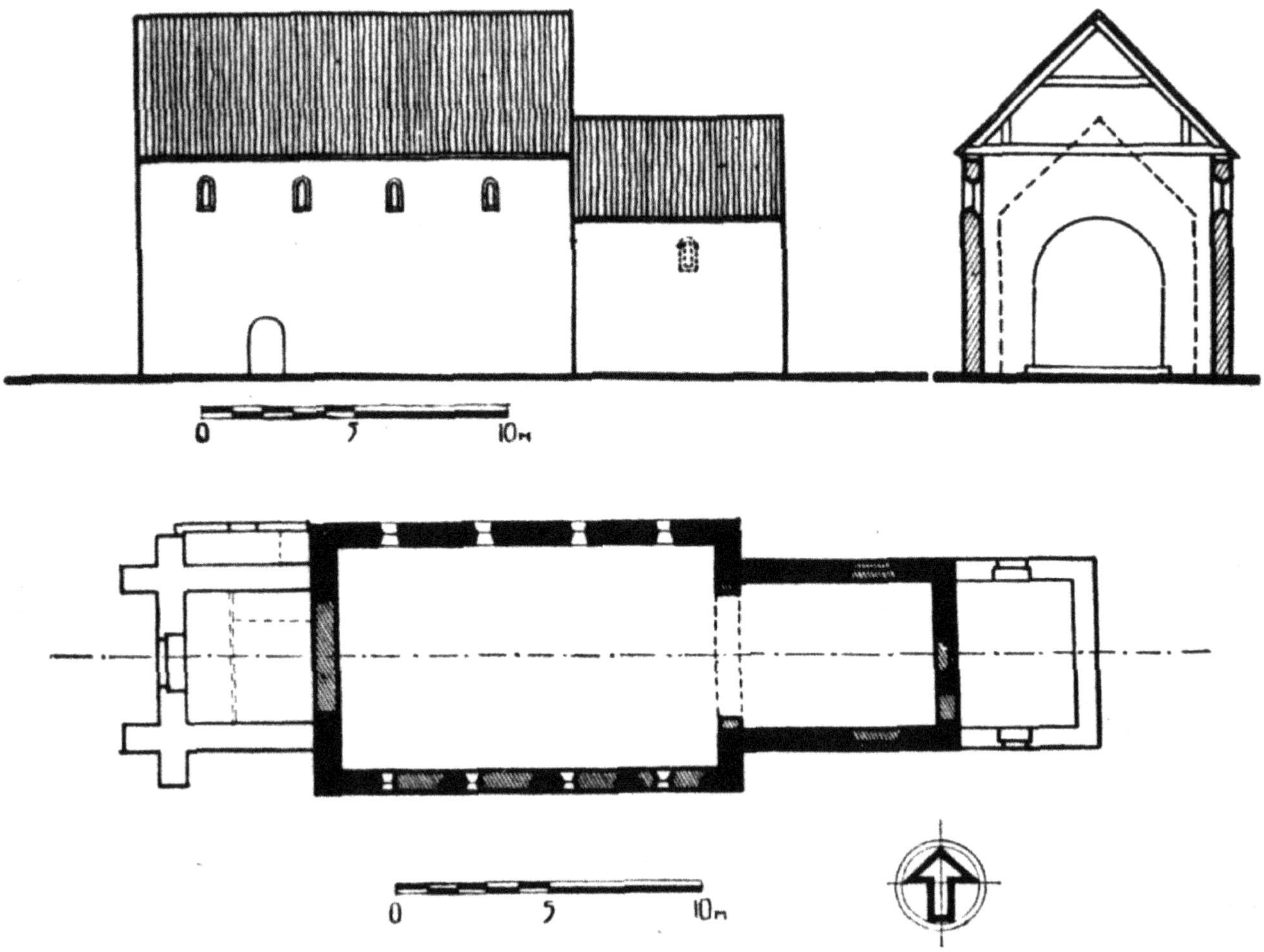

Figure 7.9: St. Vedastus' church, Nederename. Top: reconstruction of the original church, elevation and cross section. Bottom: plan of the church before 1937. The then still-standing walls of the presumably original building are indicated in black, lost pieces of the original wall are hatched, white parts indicate later additions (after Van de Walle (1952, 28)).

The vestry to the east of the demolished sanctuary was a later addition, too (tower and vestry are indicated in white in Fig. 7.9).

Whether the oldest, still-standing part of the church in Nederename belongs to the church, which existed there before the castle and settlement in Ename were founded, remains unclear. So far a 10th century date is assumed[44]. The building part in question could also belong to a later building—maybe in the same place. Only an excavation could shed light on this and whether there were even earlier predecessors.

[44]Mortar samples were taken for dating (Callebaut, 1999), but no results have been published, yet.

7.2.3 The castle

Van de Walle undertook the first archaeological excavations at the 'great hall' in 1945. In 1984 it was re-excavated by Callebaut, who also excavated the keep in 1983 (Callebaut et al., 1998, 231). The castle was situated on a peninsula in a Scheldt meander, which is coloured dark grey in Fig. 7.8. The head of the peninsula was separated by a ditch (marked with triangles), which was up to 18 m wide and more than 5.70 m deep (Milis and Callebaut, 1990, 466). In the south-western corner of the enclosed area Callebaut and van der Plaetsen (1992, 296) cut a 1.4 m wide wall running parallel to the aforementioned ditch. None of its stones remain in the excavated area, which was rather small due to a railway line crossing there. Callebaut thinks that it might be the remainder of the

Figure 7.10: St. Vedastus' church. Top left: the second window from the east of the former nave. The construction is similar to the arches of St. Lawrence's church in Ename, namely the flat haunch stones, the wedge-shaped keystone and the horseshoe shape of the arch. Top Right: the tower of St. Vedastus' church with the date 1837. Bottom: the tower possesses a socle of coursed, roughly hewn limestone ashlars and shows at least two different building stages at its southern side. The southern side of former nave lost its original window openings when the large baroque windows were introduced. Under the first window to the west the closed-up original entrance to the church is visible. (Photographs by Joyce Wittur).

wall separating the castle from the rest of the peninsula (Fig. 7.11).

A building, identified as a keep (Fig. 7.8, 2), stood near the ditch and next to the river in the southwestern corner of the enclosed area. Callebaut et al. (1998, 233) give its inside width as 27 by 10 m. This measurement must refer to the foundations as none of the walls remain. The foundations rested on two layers of wooden beams, which were well preserved, but unfortunately the wood type was unsuitable for dendrochronological examination[45]. Upon this rested a 1 m high layer of Tournai limestones set in loam, forming the basis for the foundations set in mortar. These were up to 3 m deep, between 3 and 4.4 m wide and cut through several previous occupation layers (Callebaut et al., 1998, 233). The ceramics found in the occupational layers were used between the Carolingian period and the 11th century and therefore provided little help for determining when the construction of the keep started (Callebaut et al., 1998, 232). North of the keep was a ditch, which ran parallel to the keep's north wall. It also cuts through the oldest occupational phases and was over 1.9 m deep, 9 m wide and at least 17 m long. Its fill shows in the lower part sediment mixed with a great amount of bones, potsherds and iron objects while the upper part contained loamy sand with lime- and mortar-rich strata (Callebaut et al., 1998, 233). From this description, it could be inferred that the lower part of the ditch was used as a rubbish pit while the higher strata seem to contain materials associated with building activities.

Additionally a long pit and several post holes were discovered. The post holes, being 60–70 cm wide, belonged to a wooden building, which was situated north of both the keep and the above-mentioned ditch (Fig. 7.11). Two phases were visible despite it not being possible to excavate the whole area (Callebaut and van der Plaetsen, 1992, 299; Milis and Callebaut, 1990, 467). Callebaut and van der Plaetsen (1992, 299) ask whether this could be the building, which is called *gynaeceum* in Holder-Egger (Ed.) (1888)[46].

Van de Walle excavated the great hall (Fig. 7.8, 1 and Fig. 7.12), which is thought to be a residential building. It was separated into three unequal parts: to the west a small *camera*, in the middle the *aula* and finally to the east a *capella* with an apse. Unfortunately the site was in a bad condition when it was re-excavated in 1994, because a brickyard had used the local clay in the 1940s. The stratigraphy and the foundations were severely damaged. 36 m of the building could be uncovered, but only in the western part could traces of the foundations be found. The building was 11 m wide (measured from the outside), and the lowest part of the foundations was built from Tournai limestone set in loamy sand. The upper parts were set in lime mortar and had a width of 1.9 m (Callebaut et al., 1998, 233; Callebaut and van der Plaetsen, 1992, 296). In contrast to the western part of the great hall on Van de Walle's plan (Fig. 7.12), the excavations revealed that the *camera* was a later addition to the '*aula* and *capella*'-complex (Callebaut, 1996a, 5). It is not as wide as the older building, but is slightly recessed (see first image on the left on http://www.ename974.org/Ndl/pagina/project_onderzoek.html and cf. reconstruction in Fig. 7.11). The excavation drawings by Van de Walle hint at the former existence of a rectangular ditch surrounding the great hall.

The names for the rooms are not based on any other evidence than the building remains themselves. It could be agreed upon that a room with an apse is probably a *capella*, especially if the presence of a church and canons is affirmed. However, rooms with an apse can have other uses too, and from the plan drawn by Van de Walle it appears that the wall separating the *aula* from the *capella* could be uncertain (see Fig. 7.12). The naming of the other rooms is mere speculation especially as it can be assumed that there were upper storeys with more rooms. Additionally Callebaut et al. (1998, 233) argue that the keep and the great hall were contemporary[47] and possibly the wooden building, too, which would provide the inhabitants of the castle with a great choice of rooms.

7.2.4 The pre-urban settlement

It is known from the written sources that Godfrey and Mathilde founded a settlement (Fig. 7.8, B) for which their son, Herman of Ename, later built two churches. As mentioned above, the fact that two churches (Fig. 7.8, 3 and 4) were built may be an indication that the settlement was very successful at attracting people. This also means that the settlement

[45] The beams were made from beech trees and neither the Netherlands nor Flanders had a dendrochronological curve for this kind of wood (Callebaut et al., 1998, 234 f.).

[46] The *gynaeceum* was probably located outside the castle, see comment in Section 7.1.

[47] They try to prove this point in Callebaut et al. (1998) with a series of radiocarbon dates taken from different parts of the castle. The result is that the foundation of the great hall was probably built between 950 and 1030 (p. 236, table 1) and the foundation of the keep could have been built between 955 and 1020 (p. 238, table 2). This means that the buildings could have been built simultaneously (or almost at the same time), but they could just as well have been constructed one after another with some decades in between.

Figure 7.11: Reconstruction of Ename Castle around 1020. A rampart encircles the area of the castle alongside the river Scheldt. In the direction of the pre-urban settlement it is replaced by a wall and a ditch. Inside the castle area the great hall is visible in the foreground, behind it to the right there is a wooden building followed by a ditch and finally the keep. It has been archaeologically proven that the harbour depicted on the left side of the image never existed in this place. It is assumed that it was instead on the bank to the right of the peninsula (Daniël Pletinckx, personal comment). In the background the pre-urban settlement with its two churches is depicted. (Reconstruction by Daniël Pletinckx).

was not limited to the peninsula, but must have at least bordered on St. Lawrence's church[48].

Only a few remains of this occupation have so far been excavated and these were situated next to St. Saviour's church. Callebaut, who excavated this area (Fig. 7.8, 3), identified three occupational strata (Milis and Callebaut, 1990, 474). The foundations of the abbey, which were the focus of the excavation, destroyed parts of these previous settlement phases.

The oldest stratum has a light grey colour and is probably Carolingian, which means that it predates the founding of the castle. This phase only appeared in some patches. Some post holes associated with this stratum were identified, hinting at a light settlement density (Callebaut, 1987, 213).

The other two strata give evidence of lively settlement activities. The area close to St. Saviour's church was partitioned into parcels by means of ditches, which also helped to drain the land. The parcels were unevenly shaped and were not orientated towards a road (Milis and Callebaut, 1990, 480). With the exception of the church, all the buildings were constructed from wood, and their footprints sometimes overlap, which is evidence for several sub-phases (Callebaut, 1987, 213). Post holes and hearths indicate main and subsidiary buildings (e.g. sunken-featured buildings), but no complete footprint of a main building could be excavated. The sunken-featured buildings were better preserved; some of them contained objects associated with weaving (Milis and Callebaut, 1990, 480).

[48]The fact that there were entrances in the western aisles of St. Lawrence church may indicate that occupation can be expected there (Callebaut et al., 2000, http://www.ename974.org/Ndl/pagina/NuhV/991118_1.html), but it may also be due to symmetry: the aisles had entrances to the east as well.

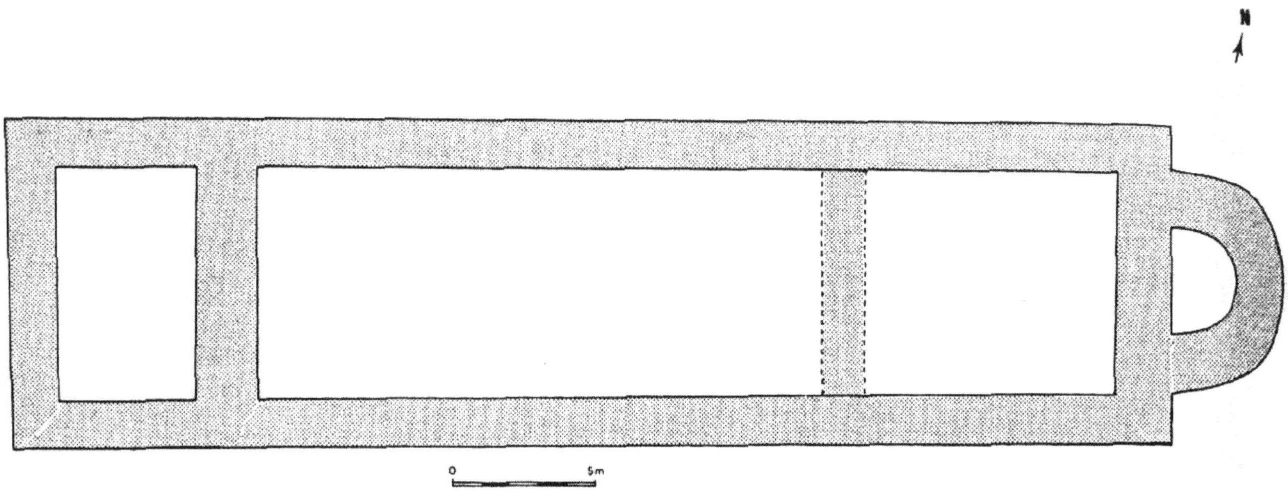

Figure 7.12: Great hall, Ename. From left to right Van de Walle referred to the rooms as *camera, aula* and *capella*. The new excavations have shown that the *camera* was a later addition and that it is indeed not as wide as the remainder of the building (after Callebaut and van der Plaetsen (1992, 298), plan by Van de Walle). (Image copyright Onroerend Erfgoed).

In the third settlement phase, which is characterised by a blackish grey soil with bits of charcoal and burnt daub, two large new ditches were dug, which did not pay heed to the older settlement structures. At least one of these ditches, which was 8.5 m wide and could be excavated over a length of 68 m, may have had a defensive function (Callebaut, 1987, 216 f.). It was part of a larger system and is depicted on historic maps (e.g. by Jan Bale, 1661-1663), and being marked in light grey in Fig. 7.8 east and south of St. Saviour's church (Fig. 7.8, 3). Van de Walle excavated part of the ditch between 1941 and 1943 and found several wooden posts at its side. Callebaut (1987, 217), too, cut a test trench through the ditch and also found some preserved wooden posts, but the excavated parts were too small to tell with certainty whether the settlement was fortified during this period.

If these ditches really belong to a defensive system they would divide the area close to the castle from that around St. Lawrence's church. The question is, why would that be desirable? Could the enclosed area be seen as a bailey?

On the other hand, the third building phase also saw the filling up of the ditches surrounding the plots of phase two and the building of new houses on top of them (Callebaut, 1987, 217).

There are several indications for crafts in the pre-urban settlement. Next to the first St. Saviour's church Van de Walle excavated a pottery kiln, while during the new excavations evidence of bronze casting was found (Callebaut et al., 2000, 303): 55 crucible sherds (belonging to at least 22 objects) and slag in addition to three patches of burnt ground, which may point to kilns, were discovered (De Groote, 2000). One crucible fragment had tiny droplets of gold on its inside. This and a touchstone with two traces of gold make the presence of a goldsmith in the pre-urban settlement probable. Trade can be demonstrated best by looking at the ceramics found in Ename. Milis and Callebaut (1990, 479) could identify wares from the Rhineland (Pingsdorf Ware) and white wares from the Meuse region while the provenance of other wares could not yet be established. Another function for the pre-urban and the surrounding rural settlements appears to be to provide the inhabitants of the castle with food. The bones found at the castle derive from young animals, which were not butchered there, but probably came from farms in close proximity (Milis and Callebaut, 1990, 480).

In two places traces of roadways were unearthed. One was constructed with branches of trees or thin tree trunks, the other was covered with flat Tournai limestones, but their date is unknown (Milis and Callebaut, 1990, 480 f.).

7.2.5 St. Saviour's church and the Abbey

From 1942 to 1946 Van de Walle excavated St. Saviour's church (Fig. 7.8, 3) of which only the foundations remain. He found the remnants of two buildings: one was the original church of the pre-urban settlement (Fig. 7.13, 1 and 2), the other

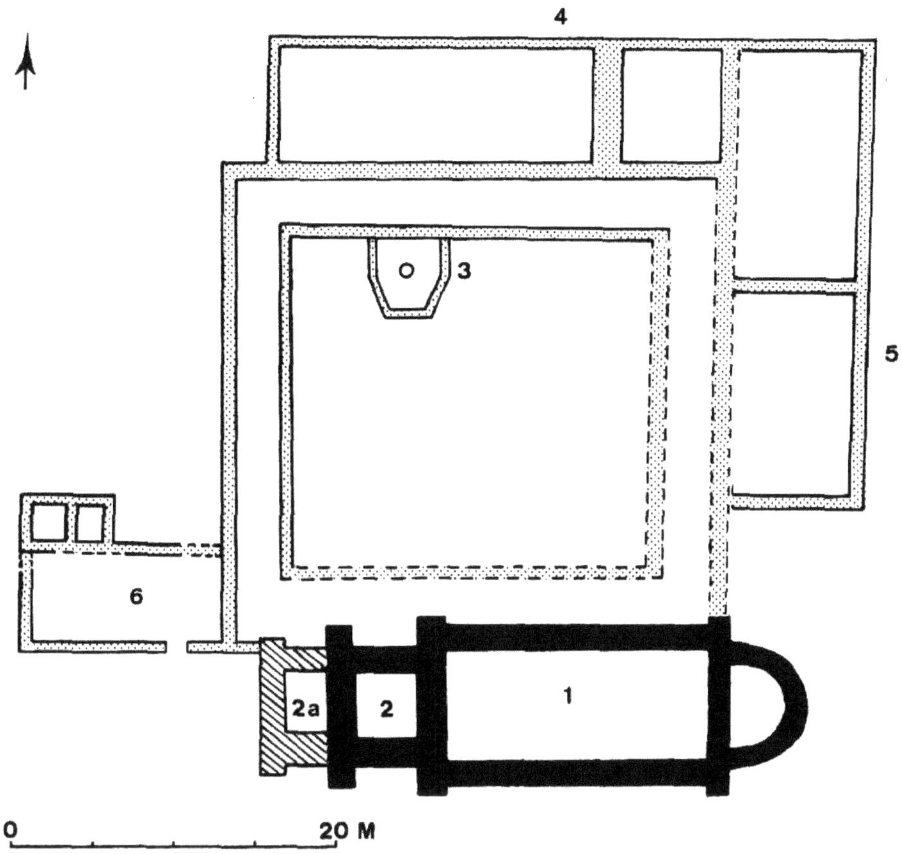

Figure 7.13: St. Saviour's church and abbey grounds before 1139. 1: St. Saviour's church, 2: western annexe, 2a: enlargement of the western annexe, 3: lavatorium, 4: refectory, 5: chapter house, 6: abbot's house with latrines (Callebaut and van der Plaetsen, 1992, 301 fig. 11). (Image copyright Onroerend Erfgoed).

the church built under Abbot Snellard (Fig. 7.14, 1), for which the work began in 1139 (Callebaut, 1985, 89). There was no trace of a wooden predecessor to the first building (Callebaut and van der Plaetsen, 1992, 300).

The first church consisted of a recessed square structure (4×4 m on the inside) to the west, a nave (6.5×16 m) and a recessed apse to the east. The western annexe was elongated at some point so that the annexe was thereafter 8×4 m wide on the inside (Callebaut, 1987, 213; Callebaut and van der Plaetsen, 1992, 300). The building was made from Tournai limestone and possessed reinforced corners[49]. Only a few stones of this building remain in the ground; the robber trenches have a width of 1.35 to 1.55 m and at the annexe 1.15 m. Only the foundations of the northern nave wall remain in place because they were reused as foundations for the northern aisle wall when Abbot Snellard built the new church (Callebaut, 1985, 92). The apse suggests that the church had its sanctuary to the east, but a western sanctuary has also been proposed (Callebaut, 1985, 93). Callebaut (1985, 93) argued at first that the annexe was probably not a tower because its foundations are even thinner than the ones of the nave. Besides that, enlarging a tower would make very little sense. He, therefore, argued for its use as sanctuary. Later, after the enlargement, this part could have been used as a choir for the monks.[50] After the church was fully excavated he revised his opinion, seeing the eastern apse[51] as the sanctuary and the western building part as an annexe (Callebaut and van der Plaetsen, 1992, 300). A cemetery belonging to the first phase of the church could not be found; only two single graves were dis-

[49]These features may have been buttresses. Similar reinforcements can be found at St. Lawrence's church, but they were only present at the foundations. Whether these reinforcements were meant to become buttresses and fell victim to a change of plan, or have always been planned not to extend above ground cannot be said (cf. Section 7.2.6).

[50]St. Lawrence's church also had two sanctuaries from the outset and would provide a good parallel for this argument.

[51]The apse is similar to the one of the great hall's *capella* (cf. Fig. 7.12).

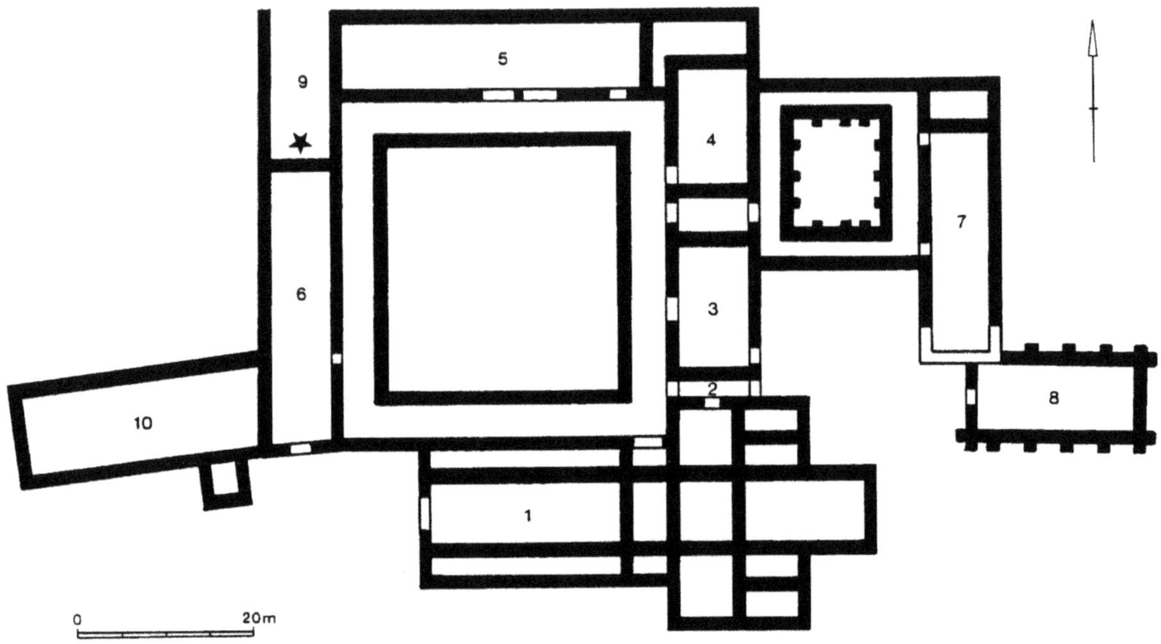

Figure 7.14: St. Saviour's church and abbey grounds around 1500. 1: St. Saviour's church, 2: library, 3: chapter house, 4: parlour?, 5: refectory, 6: guest-quarter, 7: infirmary, 8: chapel, 9: kitchen, 10: abbot's lodging (Ervynck and Neer, 1992, 420 fig. 1). (Image copyright Onroerend Erfgoed).

covered lying to the north of the annexe (Callebaut, 1987, 213). Like the other parcels in the second phase of the pre-urban settlement St. Saviour's church was surrounded by a drainage ditch. Maybe a surplus of ground water is the reason for the lack of graves.

Between 1030 and 1050 an enormous ditch was excavated around St. Saviour's church. It was 10–18 m wide and around 4 m deep (Callebaut and De Groote, 1998).

Before the monks could move from the castle to St. Saviour's church, the buildings north of the church were destroyed and the ground was raised by up to 1.5 m (Callebaut, 1985, 94). Then the annexe was enlarged and work started on the cloister. On the southern side of the northern cloister wing lay the lavatorium[52] (Fig. 7.13, 3). Inside, the foundation for the lavabo[53] was still in situ. Opposite the lavatorium is usually the refectory, where the monks took their meals (Fig. 7.13, 4). On the western side of the cloister, next to the church the archaeologists discovered a building (12×5.8 m) with a latrine attached (Callebaut, 1987, 219). The presence of the latrine was the reason for assuming that this could have been the the abbot's house because of the higher standard of comfort (Callebaut and van der Plaetsen, 1992, 305). The number of Benedictine monks living in the abbey never surpassed twelve (Batselier, 1977, 25).

Milis and Callebaut (1990, 488) state that the domestic buildings for the abbey were situated between St. Saviour's church and the ditch, which separates the castle from the rest of the peninsula. A farmstead, bakery, brewery and a butchery are mentioned as well as a mill.[54]

In 1139 Abbot Snellard started to build a new church: a basilica with a transept, from which four chapels could be reached (Fig. 7.14, 1)[55]. These chapels flanked the rectangular eastern choir. The new western wall lay just a bit to the west of the western wall of the elongated annexe belonging to the first church (Fig. 7.13, 2a). The new northern wall of the northern aisle rests on the foundations of the old northern nave wall (Fig. 7.13, 1) while the western wall of the transept appears partly to use the old eastern wall of the nave as a foundation.

At first the cloister retained its size, but to the north of the abbot's house (Fig. 7.13, 6) a new building was erected. It did not exceed the length of the west-

[52] A room to wash the hands before entering the refectory, for shaving and cutting hair.

[53] The fountain inside the lavatorium.

[54] They are not on the plans, but situated directly north of the refectory.

[55] According to Milis and Callebaut (1990, 488) the choir is flanked by two towers. From the plan (Fig. 7.14, 1) it is not quite clear where they stood. On top of the chapels?

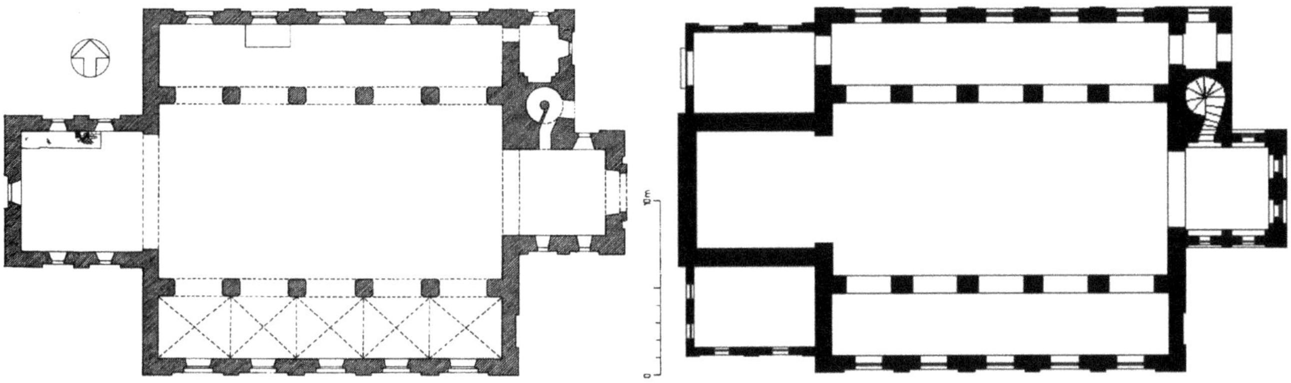

Figure 7.15: St. Lawrence's church: left: 1969 (Devliegher, 1969, 86 fig. 1); right: 2002 (Image by Daniël Pletinckx).

ern cloister wing, and it did not border on the refectory. In the second half of the 12th century the cloister was expanded to the west and the north (Fig. 7.14) (Ameels, 2000). The rooms to the east were kept but refurbished, while the buildings to the north and the west had to give way to new buildings (Callebaut, 1987, 219 f.).

A second cloister (12×10.6 m), which belonged to the infirmary (Fig. 7.14, 7), was built east of the parlour and the chapter house (Fig. 7.14, 3 and 4). This building activity may be seen in connection with the refurbishments of the eastern rooms of the main cloister because at that time a passage between the two cloister areas was established (Callebaut, 1987, 220 f.).

Between 1578 and 1582 the abbey was destroyed by the Calvinists, who brought the stones to Oudenaarde to build fortifications. This period ended with the arrival of Spanish troops. In 1596 the monks returned to Ename Abbey (Ameels et al., 2000), an event which is commemorated by a pilgrimage pennon (Fig. 7.17). At the beginning of the 17th century the rebuilding of the abbey started. The buildings west of the cloister seem to have suffered gravely and were rebuilt with bricks; also a new abbot's house was constructed[56] and the chapter house underwent refurbishments twice (Callebaut, 1987, 221, 224).

The French Revolution put an end to the abbey; it was dissolved in 1795 and the buildings were broken up (Callebaut and van der Plaetsen, 1992, 307).

7.2.6 St. Lawrence's church

For a long time it was believed that St. Lawrence's church dated from the 12th century replacing an older church which possessed no aisles (cf. Devliegher, 1969, 85). The excavations and the building research, which will be discussed, will paint a different picture. So far no final report dealing with the research on St. Lawrence's church has been published.

The orientation of the church is slightly 'off', but as in the previous chapter the cardinal directions are used for convenience. The church is a basilica where nave and aisles form a square of 19.2×19.2 m. The modern sanctuary lies in the east, i.e. the ground floor of the tower (4.8×4.8 m), where it had probably been before the altars were moved in 1778 (see Section 7.1.1). Between 1778 and 1999 the sanctuary was in the western part (6.70×6.85 m) of the church. The original design could have been a church with two sanctuaries. This can be assumed regarding the original entrances, which were at the eastern and western sides of the aisles.

A small-scale excavation took place in 1968 during the installation of a central heating system. Two small parts of the floor were opened (Fig. 7.15, left): one in the northern aisle and another in the western sanctuary. Two different flooring types were discovered. In the northern aisle the floor was once covered with bricks. 22 cm deeper than this floor a layer of burnt material was found. The floor discovered in the western sanctuary was 32 cm higher than the one in the aisle and consisted of glazed tiles (6.5×6.5 cm). The glaze was yellow or black/green, and the tiles formed a meander pattern. The border closest to the wall contained bricks and larger tiles (19×19 cm) (Devliegher, 1969, 86 f.).

Devliegher (1969, 88) had a look at the roof as well and concluded that the roof truss over the western sanctuary and the western part of the nave dated from the 12th century. This was affirmed by Hoffsummer (1992): the construction method (late 12th

[56] Fig. 7.14, 10 shows only part of the 36 m long building, which actually extended further to the west.

and early 13th century) and the dendrochronological dates (probably between 1175 and 1185) point towards the last third of the 12th century. Unfortunately both articles lack information concerning the construction date of the eastern part of the nave roof, nor are the roofs of the aisles or the tower considered[57]. From the way the eastern part of the nave roof is constructed, it could be dated to the 19th century[58]. It and the roof truss above the western sanctuary show some later changes. The tie beams were cut and stabilised by a trimmer. These are changes, which could point to wooden vaults (see Section 7.2.6.1).

After Devliegher's report, renewed research started in 1990 (Callebaut et al., 2000)[59] when cracks in the eastern tower were becoming severe and endangered its stability. The Instituut voor het Archeologisch Patrimonium in collaboration with the Bestuur voor Monumenten en Landschappen initiated the examination of the tower and small scale excavations at its foundations (Callebaut, 1996b, 4). The latter should also establish how the tower and the nave were connected (see Fig. 7.16).

The foundations at the south-eastern corner (Fig. 7.16, D on the plan, and in the photograph) consist of two ca 40 cm thick layers: the lower one is made of Tournai limestone set in sand, the upper is composed of *blauwe hardsteen*, a bluish-grey limestone, set in lime mortar. The corner is reinforced by extending the foundations. Whether these extensions were intended as the foundations of buttresses or only reinforced the foundations is unclear (Callebaut, 1992, 451). In a second trench the foundations of the eastern wall of the nave and the southern aisle were exposed. The foundation part closest to the tower (Fig. 7.16, F) is similar in construction to the foundation at the corner of the tower. Between this foundation and the adjoining one (Fig. 7.16, E) exists a building joint. Foundation E is less deep and slightly recessed in comparison to foundation F. Another foundation protrudes from foundation F at right angles, which could hint at a similar design to that found at the corner of the tower, i.e. it could be marking a corner. The wall of the nave is recessed by 30 to 40 cm in relation to the foundations, while the tower wall is only recessed by 10 cm. This led Callebaut (1992, 451 f.) to the conclusion that the foundations were initially built for a church with an eastern annexe and a nave, but without aisles (Fig. 7.16, 1). While the foundations were being laid a change of plan occurred and new foundations were added (Fig. 7.16, E). The above-grade masonry of the basilica now standing on these foundations shows no building joint near foundation F, which means that the walls of the aisles are not a later addition, but that the change in plan occurred before the walls were built. During the excavations on the inside of the church the hypothesis that the original plan comprised only a nave and (possibly) a single sanctuary and was changed in favour of a basilica with two sanctuaries could be partly confirmed. Foundation F ends and foundation E does not only proceed to the south, but also lies on top of foundation F. The foundations of the nave walls of the early plan became the foundations for the pillars of the basilica (Callebaut et al., 2000)[60]. The latest research suggests that the western sanctuary was also part of the original plan for St. Lawrence's church (Dirk Callebaut, personal comment), which would have resulted in a church quite similar to St. Martin's church in Velzeke.

Further research in 1992 was concerned with the above-grade masonry. St. Lawrence's church is mainly built from coursed Tournai limestone rubble, which is sometimes laid in a herringbone pattern to level the courses. The quoins were made from worked limestones. Some tiles were used for the arches of the window openings; the latter ones were constructed from flat stones and often a wedge-shaped stone is used as a keystone. In later times bricks were used (see Section 7.1.1). Most of the visible brickwork from the outside was removed during restorations in 1907–1908. Callebaut (1992, 439) identified some of the works done during this time by comparing two old photographs taken before and just after the church restoration. On one picture (Callebaut, 1992, 439 fig. 4), which was taken before 1907, the western wall of the western sanctuary is made of bricks. In this wall a closed-up window is still visible as well as an annexe building (the sacristy) with the same breadth as the western sanctuary, but with considerably less height. In the southern wall of the western sanctuary Callebaut identified a high square window and two smaller, closed-up window openings set high into the walls. The latter ones have arches. In the clerestory the arches of the original window openings are still visible even though they were partly closed with new masonry and smaller windows were fitted into them. On the other hand the windows in the aisles are very large

[57]For the latter an inscription on one of the beams gives the date of 1778 (Fig. 7.6).

[58]Matthias Untermann, personal comment.

[59]http://www.ename974.org/Ndl/pagina/kerk_overzicht.html.

[60]Three images at http://www.ename974.org/Ndl/pagina/NuhV/000622_1.html.

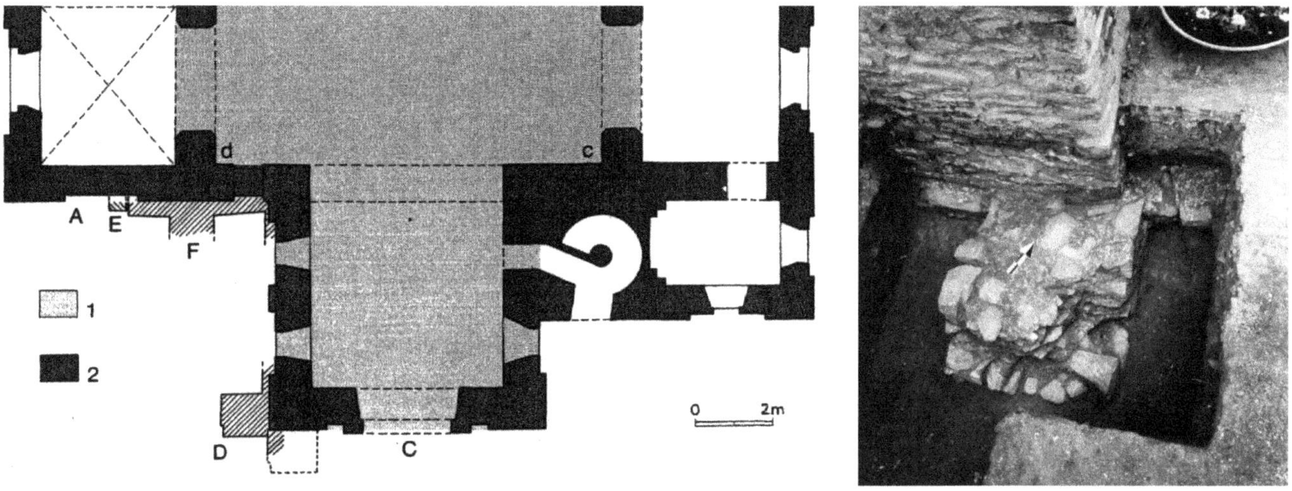

Figure 7.16: Foundation outside St. Lawrence's church. Left: east tower and nave of St. Lawrence's church. A: entrance to the southern aisle, C: entrance to the tower dating from 1778, D: reinforced corner—two extended foundations at right angles, E: foundation, which is separated by a building joint from foundation F, F: this foundation is similar to D, and it is deeper and protrudes a little in comparison to E (detail of Callebaut (1992, 442 fig. 8)). (Image copyright Onroerend Erfgoed). Right: photograph of foundation D. (Photograph by Daniël Pletinckx).

and probably destroyed the original openings there. The top storey of the tower was also constructed from bricks.

A postcard, which shows the church after the restoration, depicts the changes which were made (Callebaut, 1992, 440 fig. 5). The sacristy was pulled down and instead two annexe buildings north and south of the western sanctuary were built (see Fig. 7.15, right). The western wall of the western sanctuary appears from the outside to be constructed from limestone. A new round-arched window is set into the western wall and the walled-up windows in the side walls are open again. The window openings in the clerestory and the aisles are arched. The topmost floor of the tower, now built in limestone, appears to have a slightly darker mortar and the same goes for the topmost layers of the staircase and the the topmost layers of its northern annexe, the so-called portal. Callebaut (1992, 439) also reports that inside the church the vaults of the northern aisle and the nave were removed and replaced by flat wooden ceilings.

New research started in 1999 (Callebaut et al., 2000)[61]: the church was freed from its modern plaster on the inside so that more detailed building research could take place. Additionally, excavations inside the church were begun after the walls had been studied. This work started in the western sanctuary, progressed to the aisles, then the nave and finally to the eastern sanctuary. A restoration of the building was the final step before the church was handed back to the Ename congregation in 2002.

The subsequent sections follow the order of the excavations, but also deal with the results of the buildings research and provide links to the previously discussed sources.

7.2.6.1 The western sanctuary

The sanctuary is straight-ended and has a flat wooden ceiling (Fig. 7.17, left). There are two windows in the southern and northern wall and one window in the western wall, which was inserted during the restoration in 1907–1908. If Berings' interpretation of the *'endeldeure'* and its position are correct, then it can be assumed that at least in the 17th and 18th century a window in the western wall was present as well (cf. Section 7.1.1). This hypothesis is supported by the shape of the original window opening (Fig. 7.7): a segmental arch similar to the enlarged window openings in the aisles, and the reduced ones in the clerestory on the photograph which was taken before the restoration in 1907–08. They are supposed to have been installed in the 18th century too (Callebaut and van der Plaetsen, 1992, 439). The the sanctuary is 80 cm lower than the nave and separated from it by a 6.41 m wide triumphal arch.

As Berings already suggested and the later excavation and restoration works on the church revealed, the western sanctuary was not always used as such.

[61] http://www.ename974.org/Ndl/pagina/kerk_overzicht.html.

Figure 7.17: Western sanctuary. Left: western sanctuary just before the end of the restoration; some stonework is still visible. (Photograph by Daniël Pletinckx). Right: the representations of St. Lawrence's church on the two sides of the pilgrimage pennon. The images show the church in the year 1596, when the monks returned to Ename Abbey. On the lower image a procession around the church can be seen. (MAS Volkskundemuseum Antwerpen, AF. 06014 and AF. 06013).

The image on the right in Fig. 7.7 reveals that there were several entrances at different times. The sacristy which was pulled down in 1907–1908 must have had an entrance. It could be assumed that the two smaller doorways with the lintels could have been used as such. Then there is the '*endeldeure*', which may also be depicted on three images dating from 1595 and 1596: one on a map (Fig. 7.27, left) (Callebaut, 1992, 464 fig. 37) and the others on two sides of a pilgrimage pennon (Fig. 7.17, right). They show a door on the western side of the church. It has already been discussed (Section 7.1.1) whether in the 18th century a new western wall with a west door was installed. The latter (cf. Fig. 7.7) could have been replacing an older door in the same place, i.e. the one depicted in Fig. 7.17 on the two images, at the right-hand side.

The southern and northern wall as well as the triumphal arch appear to be original, as they are both constructed from Tournai limestone, and the same principle of construction for the arch is used as in other parts (e.g. windows) of the original building. It was discovered that there had been originally four niches each in the southern and the northern wall (Fig. 7.17). They had been closed up at a later date, maybe in the 18th century when wooden panelling was installed in front of them (Callebaut, 1992, 462 f.). The niches are 1 m wide and 0.39 m deep and topped with a horseshoe shaped arch; the wall-faces in between are 60 cm wide. Most probably three more of these arches originally decorated the western wall of the sanctuary (cf. reconstruction in Fig. 7.17, left). In Fig. 7.7, especially in the right image, the remains of the blind arcade are still visible on the left and right side of the central 'doorway', below the window sill (Callebaut et al., 2000)[62]. Underneath the baroque in-

[62] http://www.ename974.org/Ndl/pagina/NuhV/991116_2.html.

fill of stones, the niches on the southern and northern side have retained some traces of murals and several graffiti. Among them is an alphabet (Callebaut et al., 2000)[63], which probably dates to the 11th century (Van Der Meiren (Ed.), 1999). Layers of paint were conserved behind the wooden panelling, too, which covered the lower parts of the walls (cf. Section 7.1.1). These parts did not undergo the refurbishments in the 1960s, during which all the visible walls were stripped of their plaster.

The excavations in this part of the church started with three trial trenches. They revealed a, probably Gothic, floor, which is very similar to the one found in the chapter house of St. Saviour's Abbey (Callebaut, 1987, 219 f.). The floor was cast, decorated with grooves and painted, so that it resembled coloured tiles set in geometric patterns (Callebaut et al., 2000)[64]. The archaeologists discovered that the church was built on Early Mediaeval agricultural land, which had been used in Roman times, too[65]. A ciborium had been installed in the western sanctuary during the restoration in 1907–1908. Its brick foundations destroyed the original stratigraphy of the ground, and had to be removed before the excavation could proceed (Callebaut et al., 2000)[66]. In the process, the floor covering made from glazed tiles set into patterns, which had already been found by Devliegher (1969), was rediscovered in another part of the western sanctuary (Callebaut et al., 2000)[67]. The second image on another web-page (Callebaut et al., 2000)[68] shows four well-defined floor levels of the church on top of some sloping strata, which are interpreted as agricultural layers. Underneath the different floor levels three pits could be identified, one central in the east, one central in the north and another central in the south of the western sanctuary[69]. They were filled with earth, pieces of lime mortar and charcoal and can therefore be brought into connection with the building of the church. Some of the charcoal has been removed for ^{14}C analysis, but no results have been published so far. From the description on the website (Callebaut et al., 2000)[70] it is not clear what kind of pits the archaeologists were dealing with. For example, did they hold posts for scaffolding? The foundations of the 20th century ciborium, which were situated in the corners of the excavated area, may have destroyed more of these pits.

On the images in Fig. 7.7 an installation in the former opening of the central door is visible. It consists of a socle and still retains three thin brick walls: one on the back and two at the sides. It is covered with a flat object (a stone?). None of the available literature mentions it[71]. It could be the remains of an altar, which would have been situated behind the main altar after 1778, i.e. the reversal of the church orientation. Just below this installation two rectangular stones, which belong to a grave, protrude from the lower strata of the floor.

The attic above the western sanctuary can be reached via the nave attic; the wall between the nave and the western sanctuary still possesses its original door opening (1.78 m wide and 2.25 m high) (Callebaut, 1992, 450). Otherwise, most of the original gable has been lost through both the construction of a roof truss with a steeper slope, and the fire which occurred some time between 1175 and 1185. It now consists mainly of bricks. As mentioned above, the roof structure dates partly to the 12th century, but the tie-beams were cut at a later date and two trimmers were inserted to support their ends (Figures 7.5 and 7.18, top). Such a construction makes it possible to build a wooden vault so that its apex lies higher than the tie-beams. Given this fact and the source concerning the *slaeper*, which can also mean 'trimmer', it can be hypothesised that this vault was built in 1762 and may have been the cause for filling in the windows in the northern and southern wall. This vault, which had the form of a segmental arch, was removed at the same time as the vault in the nave, i.e. 1907–1908 (Callebaut, 1992, 462).

7.2.6.2 The aisles

The aisles are 3.66 m wide and have five bays. They are separated from the nave by arcades which rest on rectangular pillars. While the northern aisle has since 1907–1908 had a flat wooden ceiling, the southern aisle still possesses the 17th century ribbed vaults,

[63] http://www.ename974.org/Ndl/pagina/NuhV/990905_2.html.

[64] http://www.ename974.org/Ndl/pagina/NuhV/990905_3.html.

[65] For more information on roman remains see http://www.ename974.org/Ndl/pagina/NuhV/000316_1.html and http://www.ename974.org/Ndl/pagina/NuhV/000504_1.html.

[66] http://www.ename974.org/Ndl/pagina/NuhV/000202_2.html.

[67] http://www.ename974.org/Ndl/pagina/NuhV/000223_1.html.

[68] http://www.ename974.org/Ndl/pagina/NuhV/000225_3.html.

[69] See Callebaut et al. (2000), http://www.ename974.org/Ndl/pagina/NuhV/000302_1.html.

[70] http://www.ename974.org/Ndl/pagina/NuhV/000307_3.html and http://www.ename974.org/Ndl/pagina/NuhV/000307_3v.html.

[71] Pletinckx (personal comment) thought it might have belonged to some kind of seat.

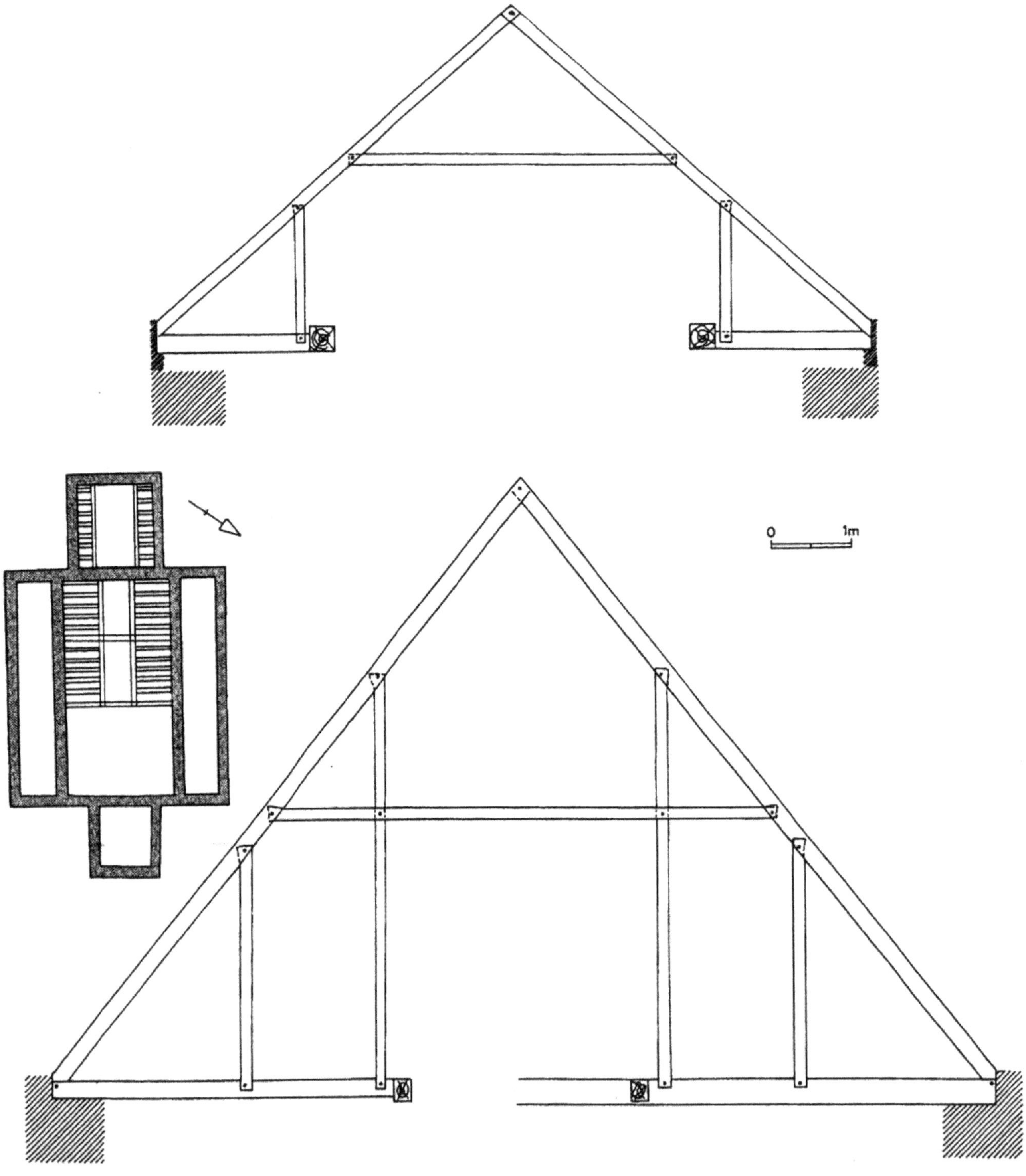

Figure 7.18: Roof truss in the nave (top) and in the western sanctuary (bottom) (Devliegher, 1969, 87 fig. 4).

which had been present in the north as well. Each bay contains a window, though their original form is unknown due to the enlargement of the window openings in the 18th century (Callebaut, 1992, 443). On the outside, the wall is textured by lesenes, which correspond with the bays. In the third bay of northern aisle a *lijkdeur* (cf. Section 7.1.1) is visible (Fig. 7.19, left). Callebaut (1992, 443) thinks that it is a later addition due to the regular stonework displayed there. The door is, nevertheless, older than the window, which cuts through its arch, indicating that the previous windows sat higher in the aisle walls at the time the door was installed.

The excavations brought new insights regarding the aisles. Callebaut (1992, 443) had noticed a door in the eastern wall of the southern aisle, which he considered to be original. He also considered the possibility of more doors in the aisles, and his assumption proved to be correct when in 1999 the side altars were removed and behind them closed-up entrances were found. Finally four entrances, one to the east and to the west in each of the aisles, were discovered (Callebaut et al., 2000)[72] and later exposed (Callebaut et al., 2000)[73] with the intention of reopening some of them. In 2000 the north-eastern entrance was reopened. The modern door, which had previously connected the aisle with the modern baptistery[74] to the east, was bricked up [75].

Finally the excavations in the aisles started (Callebaut et al., 2000)[76]. Each of the aisles was split up into three trenches. During the excavations in both aisles, the floor covering made from tiles and bricks, which Devliegher (1969) had found in the northern aisle, was rediscovered (Callebaut et al., 2000)[77]. The floors consisted of different kinds of tiles, with bricks being used for repairs. In the north-eastern trench much of this floor covering was removed when later gravepits were dug through it (Callebaut et al., 2000)[78]. As it turned out, the aisles were split up into side chapels, with differing floor coverings and with different painted decorations on the walls[79]. In the middle of the northern aisle the foundation of an altar and the remains of a wall, which separated two chapels, were found (Callebaut et al., 2000)[80]. On top of this (probably Late Mediaeval) floor, bones of small rodents and birds were unearthed. The archaeologists think that these bones were introduced into the church by birds of prey. They may have lived in the church during the Thirty Years' War, during which the church fell into disuse. The region around Oudenaarde was conquered in 1578 by the Calvinists, who also destroyed Ename Abbey. A recapture for the Catholic side by Spanish troops followed in 1582. In the end all of the altars were desecrated and it took a long time to renovate the church (cf. Section 7.1.1). The pilgrimage pennon was dated 1596 and the fact that in the same year the monks returned to the abbey is taken by Callebaut et al. (2000)[81] as a sign that the mass at St. Lawrence may have been resumed in the same year. In the southern aisle, underneath the tile floor mentioned above, a similar floor covering to the one in the western sanctuary (and the chapter house of Ename Abbey, cf. Section 7.2.6.1) was found: it consists of cast mortar, with incised lines and painting. It is therefore assumed that it belongs to the 12th century in dependence on the date of the chapter house floor (Callebaut et al., 2000)[82].

A layer of burnt ground was discovered in the whole church (Callebaut et al., 2000)[83]. This layer contained charcoal and in the southern aisle also pieces of green molten glass. The walls, too, show signs of burning. Callebaut et al. (2000) assume that this fire could have taken place just before the new roof beams were installed into the western part of the nave and the western sanctuary (see the beginning of Section 7.2.6). The predecessors of these beams could have been destroyed by the fire, which also melted the glass in the aisle windows. Charcoal samples were taken to be examined via radiocarbon dating in order to test the hypothesis for the date (Callebaut et al., 2000)[84]. In the northern aisle a lead oven was excavated; it

[72] http://www.ename974.org/Ndl/pagina/NuhV/991022_1.html, http://www.ename974.org/Ndl/pagina/NuhV/991118_1.html.

[73] http://www.ename974.org/Ndl/pagina/NuhV/000201_4.html.

[74] The room was built as a portal, but was later used as a baptistery.

[75] See photograph on Callebaut et al. (2000), http://www.ename974.org/Ndl/pagina/NuhV/000214_1.html.

[76] http://www.ename974.org/Ndl/pagina/NuhV/000221_1.html.

[77] http://www.ename974.org/Ndl/pagina/NuhV/000223_2.html.

[78] http://www.ename974.org/Ndl/pagina/NuhV/000225_1.html.

[79] Unfortunately the situation is not clear enough to tell how many chapels originally existed. Dirk Callebaut, personal comment.

[80] http://www.ename974.org/Ndl/pagina/NuhV/000229_3.html. The dividing wall between the chapels can be seen on http://www.ename974.org/Ndl/pagina/NuhV/000307_2.html. It is the first wall from the front.

[81] http://www.ename974.org/Ndl/pagina/NuhV/000229_2.html.

[82] http://www.ename974.org/Ndl/pagina/NuhV/000307_2.html.

[83] http://www.ename974.org/Ndl/pagina/NuhV/000310_1.html.

[84] http://www.ename974.org/Ndl/pagina/NuhV/000308_2.html. The results of this test are not revealed.

cut into the burnt layer and probably signifies the rebuilding of the church after the fire[85]. Samples from the materials found inside the oven were removed for further analysis (Callebaut et al., 2000)[86].

Post holes with a diameter of 40 cm were discovered (Callebaut et al., 2000)[87]. They and the corresponding putlog holes in the walls are traces of the scaffolding, which was used when the church was built (Callebaut et al., 2000)[88].

The foundations of the outer walls of the church were originally 0.9 m deep (Callebaut et al., 2000)[89]. After they had been excavated and the wooden panelling which covered the wall from the inside was removed, the *'lijkdeur'* in the northern wall could be studied. A plaster analysis will help to establish when the door was introduced and when it was finally closed up (Callebaut et al., 2000)[90]. So far it is assumed that the door was in use at the time when the pilgrimage pennon was created (i.e. 1596, see Fig. 7.17, right), because it is marked on there. The pennon also shows small windows, though the representation does not go so far as to depict one over the door. When the larger windows were inserted into the wall (i.e. in the 18th century) the door may have fallen into disuse. Another indication for the age of the door is the doorsill. It is lower than the modern floor height, but not deep enough for an original opening. Analysing the stratigraphy of the excavation will help to establish a better relative chronology for this door opening.

Callebaut et al. (2000)[91] shows that the foundations underneath the pillars belong probably, as discussed above, to the first planning phase of the church. The lowest part consists of limestone set in soil, while the upper parts were set in mortar. The lowest floor level[92] of the church lay above the foundations. The original layer of plaster covering the pillars is visible above it.

[85]The oven could have, for example, provided lead for lead cames in stained glass windows.

[86]http://www.ename974.org/Ndl/pagina/NuhV/000316_2.html and http://www.ename974.org/Ndl/pagina/NuhV/000426_2.html.

[87]http://www.ename974.org/Ndl/pagina/NuhV/000330_1.html.

[88]http://www.ename974.org/Ndl/pagina/NuhV/000411_1.html.

[89]http://www.ename974.org/Ndl/pagina/NuhV/000420_1.html.

[90]http://www.ename974.org/Ndl/pagina/NuhV/000425_1.html.

[91]http://www.ename974.org/Ndl/pagina/NuhV/000426_1.html.

[92]It is marked with red lines in the second image on http://www.ename974.org/Ndl/pagina/NuhV/000426_1.html.

7.2.6.3 The nave

The window openings in the clerestory lie within a blind arch, one in each bay. Their height is not the original one. In the 18th century the windows had been even smaller, because a vaulted roof, in the form of an segmental arch (Callebaut, 1992, 462), had been introduced into the nave. The vault hung so low that the upper parts of the windows had to be closed up (Callebaut, 1992, 443). The sills are not original, either. It appears that the roof ridges of the aisles were elevated[93], which made it necessary to raise the sills of the clerestory windows; four different phases can be discerned (Callebaut et al., 2000)[94]. The original extent of the windows can be gleaned from the original plaster layer, which encompasses the windows. At first the wall was constructed with a flat windowsill. In preparation for the insertion of the glass windows, two 'socles' on each side of the window opening were constructed, on which the window frame could come to rest (Fig. 7.19, right). These 'socles' were already at the height of the slanting window sill, which was built after the windows were inserted into the window openings[95]. These three steps belong to the initial building phase. The fourth step occurred when the sill was raised due to the higher roof ridge.

Besides the clerestory windows the nave had two window openings in its eastern wall. One was detected in 1992 (Callebaut, 1992), the other window came to light in 2000 (Callebaut et al., 2000)[96]. The windows are 3.8 m high (the same height as the original clerestory window openings) and 1 to 1.2 m wide. They have horseshoe shaped arches and were at a later point closed up with limestone. One of the reasons may lie in the stair tower which was built right in front of the northern window rendering it useless. The stair tower is a later addition[97], but it is necessary in order to reach the upper storey of the eastern sanctuary. Possible solutions to reaching the upper storey would be a wooden stair tower, a simple wooden staircase or access from the inside of the church.

After the aisles were filled up with sand for static reasons, the excavation was able to proceed in the nave.

[93]Callebaut et al. (2000), http://www.ename974.org/Ndl/pagina/NuhV/991220_2.html thought that this may have happened around 1180, when the roofs above the nave and the western sanctuary were renewed, but it became clear later that the raised roof ridge must be prior to the fire (see below).

[94]http://www.ename974.org/Ndl/pagina/NuhV/000201_2.html.

[95]Daniël Pletinckx, personal comment.

[96]http://www.ename974.org/Ndl/pagina/NuhV/000201_3.html.

[97]There are building joints marking it as such.

Figure 7.19: St. Lawrence's church: *lijkdeur* and clerestory window. Left: the *lijkdeur* in the northern aisle is a later addition. Its method of construction differs from the other door and window openings of the church. The door was cut by the window above it, when this was enlarged in the 18th century. Right: clerestory window in the northern wall. After the restoration in 2002 the 'socles' on which the original window frame rested were left visible. It has to be imagined that the socles would have disappeared behind a wedge shaped infill, which created the sloped windowsill. The distance between the socles and the modern lower end of the window panes is due to the steepening of the aisle roofs. (Photographs by Joyce Wittur).

The first old floor level which was discovered consisted of black and yellow glazed, square tiles, which form different patterns. The tiles show signs of hard wear, so that only a little of the glaze remains. Repairs to this floor were made with bricks (Callebaut et al., 2000)[98].

The burnt layer which had been found in the aisles is also present in the nave. Samples of this layer were sieved, which resulted in the discovery of charred grains, some hazelnuts and grass seeds. This shows that the church had a side use as a granary (Callebaut et al., 2000)[99]. The analysis of the grains under the microscope revealed that they consisted mainly of barley and oats with some wheat and rye. The amount and distribution of the grains indicates that the whole attic was used for the storage purposes (Bastiaens and Tency (2001) and Vera Ameels, personal comment). In order to make sure that these grains were really charred during the fire around 1075/85 they were taken to be radiocarbon dated (Ameels et al., 2000), but no results have been published so far.

In Fig. 7.15 the pillars of the nave are depicted differently in the two images. The older plan shows the situation which probably arose in the 18th century. The originally square pillars and arches were robbed of their corners and the imposts were removed. During their restoration, which started in 2001, the pillars were returned to their original form (Callebaut et al., 2000)[100]. The form of the imposts could be deduced from remains which had been preserved at the

[98]http://www.ename974.org/Ndl/pagina/NuhV/000518_1.html.

[99]http://www.ename974.org/Ndl/pagina/NuhV/000719_1.html.

[100]http://www.ename974.org/Ndl/pagina/NuhV/0109018_1.html.

pilasters at the end of the arcades, though their form was adapted slightly[101].

Building research in the attics of the aisles revealed that the outside of St. Lawrence church was originally covered in plaster. This plaster was of the same quality as that inside the church (Callebaut et al., 2000)[102]. The same studies also led at first to the opinion that the roof ridges of the eastern sanctuary, the nave and the western sanctuary were initially one[103]. This was later proved to be wrong. Instead there was a slight difference in height between the three parts of the building, whereby the nave was slightly higher than the others.

The original form of the church, with its lesenes, blind arches, the gentle slope of the roofs (around 20–25 degrees) (Callebaut et al., 2000)[104] and the different height of the roof ridges, suggests that the church was built by someone who was familiar with mediterranean-style buildings. Roman-style *tegulae* were used for St. Lawrence's but also for St. Saviour's church (De Groote, 2001). These and the gentle slope of the roof are not well suited for our region, because the wind blows the water under the *tegulae* and into the building. Additionally the different height levels of the roof (sanctuaries and nave) allow the rain water to run along the exposed gable walls of the nave and from there into the lower parts of the building.

There are also indications that originally there was no ceiling, but that the roof beams were visible from below (Callebaut et al., 2000)[105]. However, at the time of the fire which occurred at some time between 1175 and 1185, the roof-space was not open, but as the charred grains and nuts prove, having been found all over the church floor, it was used for storage. This leads to the conclusion, that the roof truss must have been modified in between times, and probably this went hand in hand with a steepening of the roof pitch to prevent rain water from entering the church[106]. The roof which was installed after the fire is now only preserved in the western part of the nave, and even there every second rafter is of a more modern date[107]. Just as in the roof over the western sanctuary the tie-beams (with two exceptions) were cut just beyond the queen struts (Fig. 7.18, bottom). Two trimmers in an east-west direction were attached to their ends, leaving a gap of approximately 3 m. This change allowed for a wooden vault, probably the same vault which was removed during the restoration in 1907–1908 (see the beginning of Section 7.2.6).

[101] Daniël Pletinckx, personal comment.

[102] http://www.ename974.org/Ndl/pagina/NuhV/020930_5.html.

[103] Daniël Pletinckx, personal comment. Some of the older reconstructions mirror this view.

[104] http://www.ename974.org/Ndl/pagina/NuhV/000201_2.html.

[105] http://www.ename974.org/Ndl/pagina/NuhV/020930_5.html. In this case a description of the findings which lead to these assumptions would have been useful.

[106] Daniël Pletinckx, personal comment.

[107] They were sawn instead of axed.

7.2.6.4 The eastern sanctuary and the tower

The situation at the tower is very complex because many changes have occurred over time. As mentioned above, the latest research comes to the conclusion that there was initially no tower, but that the roof ridge of the western sanctuary and the eastern sanctuary were slightly lower than the roof ridge of the nave.

The situation presented itself quite differently in 1992 (Fig. 7.20). Callebaut (1992, 444) gives the height of the original, still-standing, masonry of the tower as 18 m and mentions that the walls are about 1.1 m thick. The tower has blind arches, which texture each of its storeys, though their number varies. The arches of the blind arcades and the window openings are horseshoe shaped and often the imprint of the centring is preserved (Callebaut, 1992, 445). The ground floor possesses two blind arcades on each side (south, east, north), though the two blind arches to the east were in 1992 still cut through by the portal of 1778. Nevertheless, they were still partly visible from the outside. When the portal was removed in 2001 Ottonian paintings were discovered in the arches. Because of cracks in the plaster, it is hard to identify what they represent (Callebaut et al., 2000)[108]. Of the two blind arcades to the north, only one is visible from the outside, and even that is partly covered by the stair tower (Fig. 7.21). Originally each of the blind arches to the south and the east contained a window opening. The eastern blind arch on the north side also contained a window (Callebaut et al., 2000)[109], but the other one, which is now occupied by the door to the stair tower, was meant to hold a doorway from the outset.

The first storey originally had three blind arches on each side, though only the two eastern ones on the south side contained window openings, while all of the blind arches to the east had them. To the north

[108] http://www.ename974.org/Ndl/pagina/NuhV/0109018_3.html.

[109] http://www.ename974.org/Ndl/pagina/NuhV/000207_2.html.

Figure 7.20: West-east cross section of St. Lawrence's church in 1992 (Callebaut, 1992, 444 fig. 9). (Image copyright Onroerend Erfgoed).

Figure 7.21: East tower of St. Lawrence's church in 2009. Left: the tower from the east, with the stair tower and the portal to the north. Right: view from the north. (Photographs by Joyce Wittur).

only one of the blind arches is visible from the outside, the others being covered once again by the stair tower. On the inside, the middle arch had contained a window opening, but was later used for the installation of a fireplace. The other arch contains a doorway, which is original. The stair tower is a later addition to the north side of the tower, which means that originally a wooden stair tower must have existed to connect the ground floor with the first storey[110]. On the eastern wall in the central blind arch the remains of an 1.3 m high altar were found (Callebaut et al., 2000)[111]. It was made from Tournai limestone set in mortar and was covered with a layer of plaster. Only its eastern part remains in situ. The rest was removed when a vault was introduced into the tower and cut through the place where the altar formerly stood.

The second storey again has two blind arches on the southern and eastern faces. On the northern side only one blind arch, to the east, is present. The same accounts for the inside. The western part of the wall contains the door opening without a framing, blind arch. This indicates that the second storey obviously takes account of the stair tower and has no ornamentation on the outside that would collide with it. The second storey is also a little bit smaller in plan. The recess is clearly visible on the outside. These indications make it clear that this building part is a later addition, which fits the hypothesis that the church had no tower at first. The window openings inside the arches are small and square in comparison to the high round arched windows in the two lower storeys. Callebaut (1992, 447) believes that the upper layers of this storey disappeared when, at an unknown time, the third storey was demolished, so

[110] In an older version of the reconstruction of the church in 1020 (Fig. 7.22) a simple stairway instead of a stair tower is depicted, though this is unlikely. The priest who was using the altar in the first storey, his vestments and the sacred vessels would have been exposed to the elements

[111] http://www.ename974.org/Ndl/pagina/NuhV/000207_1.html.

Figure 7.22: Old reconstruction of St. Lawrence's church, which depicts an open, wooden staircase (Reconstruction by Daniël Pletinckx).

that the exact room height cannot be established any more. The doorway (1.42 m wide and 2.76 m high), which connects the second storey of the tower with the attic above the nave, is original (Callebaut, 1992, 450), though the masonry of the southern, eastern and northern walls of the second storey is not. Above the doorway, too, later masonry is visible. This has two reasons, one that the roof originally had a gentler slope, and also that the fire which occurred at some time between 1175 and 1185 destroyed part of it. The fire also left traces on this tower storey indicating that it must have been added before 1175/85.

The topmost storey is probably an even later addition. It cannot be reached via the stair tower, but only from the second storey by means of a ladder or a wooden staircase. As mentioned before, the upper storey was constructed of bricks prior to the restoration in 1907–1908, and it had a completely different appearance at that time as well (Callebaut, 1992, 439 fig. 4). Central on each wall face sat a round opening for the dials of the tower clock. During the restoration the storey was rebuilt in limestone and included two round-headed windows flanked by two blind arches on each side. The dials of the clock are mounted in front of the window openings.

The images on the pilgrimage pennon (Fig. 7.17, right) and the map (Fig. 7.27, left) mentioned earlier pose another problem concerning the tower. In these representations the tower is depicted with a round or octagonal tower top. One of the hypotheses concerning the discrepancy between the images and the state of the tower today is that the upper parts of the tower were removed and built anew during the 18th century (Callebaut et al., 2000)[112].[113]

[112] http://www.ename974.org/Ndl/pagina/NuhV/0109018_2.html.

[113] Another possibility is that the images are unreliable in this respect. They also do not show the stair tower, but an additional

In 1778 a new doorway was cut into the east side of the tower. At around the same time an organ loft was built (Fig. 7.20, B) which separated the upper part of the ground floor from the lower part. The organ loft was spanned by a vault constructed from Tournai limestone, but the part, which bordered on the nave, was built from bricks (Callebaut, 1992, 445). The question at this point is whether the vault (Fig. 7.20, C) had already existed before that time, or whether it was constructed around the same time as the organ loft. One of the sources cited by Berings (1994, 6 f.): *'Turris multis detecta ita ut per fornicem aqua chorum intret qui sub turri est.'* mentions a vault for the year 1607. Until recently it has been believed that the vault dated from the 18th century (Callebaut et al., 2000)[114], but it is now believed to be earlier[115]. However, the vault had cut the first storey in half, so that instead of two storeys (as indicated by the blind arcades in Fig. 7.20) the result were three storeys, separated by the floors B and C.

This resulted in the windows of the ground floor being split into an upper part, which shed light into the organ loft, and a lower part, which brought light to the entrance area. The room above the vault received light from the windows of the former first storey, whereupon the lower part of these windows had to be bricked up.

The weight and pressure of the vault was too heavy for the tower and cracks in its surface appeared, so it was decided in 1992 to remove the vault. In this process four holes in the north and south walls, which once held the beams of the first storey floor, were discovered in 1999 (Callebaut et al., 2000)[116]. In the same year a new concrete floor was cast at the height of these holes (Callebaut et al., 2000)[117].

The wall between the tower and the nave has so far not been mentioned. Regarding the changes inside the tower it is obvious that there must have been changes to this wall as well. On the ground floor the tower originally had a round arch opening towards the nave; the spring line of this arch was rediscovered in 1999 (Callebaut et al., 2000)[118] and the arch itself reconstructed in the following year (Callebaut et al., 2000)[119]. On top of the new concrete arch a parapet was constructed. After this it was possible to inspect the west wall in the first storey more easily.

Originally the first floor possessed three 1.3 m wide arches opening towards the nave, which were spanned by a framing arch (Fig. 7.23). The smaller arches were probably once supported by columns which have disappeared.[120] When the vault was built, its pointed arch cut into the arcade, so that the columns were removed and in their stead uneven masonry was placed between the pointed arch and their stilted ends (Callebaut, 1992, 454 fig. 22). In a later phase the arcades themselves were neatly closed up with bricks and Tournai limestone (Callebaut, 1992, 447). The recess between the wall of the nave and the huge framing arch, which had once spanned the three coupled arches was partly filled up at an even later date. The filling reached as high as the apex of the pointed arch. Right above the apex, a small socle (40 cm high and 40 cm wide) was built. Callebaut (1992, 447) assumes that it may have supported a painting. Finally, in the 18th century, just before the organ was installed, the remainder of the framing arch was bricked up. The vault had received a flooring made from bricks, which was finally higher than the sills of the first storey windows. The windows were bricked up on the inside, to match the new floor height. The central blind arch in the northern wall of the first storey was converted into a fireplace (Callebaut, 1992, 454 fig. 21). A brick staircase was built to enable entry from the original door opening to the new level [121].

The sanctuary on the ground floor, the chapel on the first floor and the entrances in the aisles could indicate that the eastern tower was planned as an 'eastwork' (i.e. a westwork in the east, e.g. at Mainz Cathedral), which would in turn suggest, that the main altar stood in the west. That the western sanctuary was the more important one is suggested by its larger size, while the rich decoration in the east may indicate that this was the main facade of the building.

The remains of an altar were found beneath the tower floor (Callebaut et al., 2000)[122]. The altar was built from Tournai limestone resting on a foundation layer of the same material. All the masonry except for the

door in the southern aisle, which has never existed.

[114]http://www.ename974.org/Ndl/pagina/NuhV/991125_1.html.

[115]Dirk Callebaut, personal comment.

[116]http://www.ename974.org/Ndl/pagina/NuhV/991125_1.html.

[117]http://www.ename974.org/Ndl/pagina/NuhV/991220_1.html.

[118]http://www.ename974.org/Ndl/pagina/NuhV/991129_1.html.

[119]http://www.ename974.org/Ndl/pagina/NuhV/000201_1.html.

[120]For a reconstruction of this part see (Callebaut et al., 2000), http://www.ename974.org/Ndl/pagina/NuhV/000202_1.html.

[121]See Callebaut et al. (2000), http://www.ename974.org/Ndl/pagina/NuhV/991116_3.html for the removal of this staircase.

[122]http://www.ename974.org/Ndl/pagina/NuhV/001004_1.html.

Figure 7.23: Restoration of the coupled arches on the first floor of the east tower. Left: the bricked up coupled arches with a framing arch spanning them, underneath the brick arch belonging to the vault, which was added at a later point. The uneven masonry, which had supported the stilted ends of the arches after the columns were taken out and the brick arch inserted, is already removed. Right: new columns were introduced, the parapet constructed at its former height. (Photographs by Daniël Pletinckx).

eastern side was plastered. On the first picture of the web-page it can be seen that several undisturbed layers lie over the stump. They date to the 17th or 18th century[123] and would therefore match what is known from the written sources (see Section 7.1.1) concerning the change of orientation of the church. The latest date for the destruction of the altar would be 1778, when the door was inserted into the tower, otherwise, it would have blocked the entrance.

7.2.6.5 The stair tower and the portal

The stair tower (Fig. 7.21) is 2.70×3 m in plan and has three narrow rectangular windows to the east and one above the portal roof to the north (Callebaut, 1992, 448). A building joint between the stair tower and the eastern sanctuary identifies it as a later addition, and it is quite probable that it had a wooden predecessor (Callebaut et al., 2000)[124]. The stair tower was built in several stages. At first only the two lower storeys of the east tower were connected by it. It must have been on the verge of closing up the lowest part of the window in the eastern wall of the nave, but could have been fitted there if the windows in the eastern wall of the nave were constructed like the ones in the clerestory (see Section 7.2.6.3), i.e. the windowsill lay higher after the window was inserted.

Another building joint between the first and the second storey can be seen on the inside of the stair tower itself: a decrease in the thickness of the walls and of the newel post of the spiral stairs. The upper part

[123] Dirk Callebaut, personal comment

[124] http://www.ename974.org/Ndl/pagina/NuhV/0109024_1.html.

of the stair tower was built at the same time as the second storey of the eastern tower because the door opening leading from the stair tower to the inside of the tower still preserves the original imprint of the centring (Callebaut, 1992, 457 fig. 26), which shows that the arch of both building parts was constructed at the same time.

East of the northern aisle and north of the stair tower lies an annexe, which Callebaut et al. (2000) refer to as a portal. It was connected to the northern aisle by one of the original entrances to the church. The lower part of the stair tower was built simultaneously with the portal. The latter has a size of 2.20×2.75 m in plan (Callebaut, 1992, 450). Just like the stair tower itself the height of the portal walls was increased during the restoration in 1907–1908, which shows in a building joint in its eastern wall. The portal possesses another building joint towards the northern aisle where the lesene of the aisle is flush with the face of the portal wall. There are two openings in this building part, a door to the east and a window to the north. Both are possibly original, though they show some traces of later interventions. The imposts of the northern arch are noticeably more pronounced than the ones of the other entrance (Fig. 7.24) and also in comparison with the other round-arched openings of the church, which seem to possess a horseshoe shape rather than to possess imposts.

As the excavations, which were performed at the outside of this building, show, an apse was added to the east of the portal at some point and, at a later date, was pulled down again (Callebaut et al., 2000)[125]. The apse is clearly not part of the first building phase because under the curving foundations lie foundations which project at right angles. Hereby the southernmost of these extensions (second image on the webpage) belonged to a buttress. Its remains can still be traced in the uprising wall (fig 7.25, 1 and 2). The masonry above disturbance 2 is continuous, linking the portal with the stair tower. Maybe these extensions to the foundations mirror the other corner foundations, e.g. the ones at the south-eastern corner of the tower (Fig. 7.16), which were 'enforced', too. It would be necessary to look at the other foundations outside the church in order to say more about the phenomenon of the enforced corners and where they appear. A grave, which is dated to the beginning of the 13th century due to its anthropomorphic form, is situated partly below the foundation of the apse, which gives a *datum post quem* for the latter[126]. At the time when the apse was attached to the annexe it must have changed its use from being a portal to a chapel. It is debatable whether the entrance to the north would have remained open after this change took place. Eventually the annexe was used as a baptistery.

Inside the portal a masonry grave was discovered (Callebaut et al., 2000)[127], which was clearly constructed at the same time as the portal itself. The grave had an anthropomorphic shape[128] and was plastered on the inside. It held the remains of a woman, whose bones were still in their anatomical position. The bones were radiocarbon dated with the astonishing result that the woman must have died much earlier than the construction of the portal.[129] This curious situation will hopefully be resolved by letting the bones be radiocarbon dated anew.[130]

7.2.6.6 Murals

The murals are not part of the computer-generated reconstructions, but some of the painted layers have influenced the research of the church. Marjan Buyle has published some of the results in Callebaut (1992, 463–465). The unpublished report by Van Dijck (1995) is more elaborate.

In the 1960s most of the old plaster was chipped off the walls; the murals survived only on those parts which lay behind the wood panelling, confessionals, altars, chancel and bricked up parts of the walls (Callebaut et al., 2000)[131]. The mural on the tympanum on the eastern wall of the nave is the most interesting. The first two layers underneath the mural consist of plaster. While the first layer is of a high quality, the second is not. The intrados of the framing arch received a floral painting on top of these layers (Fig. 7.26, bottom right). It is not clear if the tympanum was also decorated during this phase because the later layers of paint covering it were not removed.

[125] http://www.ename974.org/Ndl/pagina/NuhV/020930_1.html.

[126] I cannot follow this argumentation, as I cannot see on the photograph that the grave lies below the foundations. If it lies under any foundations, then underneath the 'enforced' corner, which must be earlier than the apse itself.

[127] http://www.ename974.org/Ndl/pagina/NuhV/001208_1.html.

[128] The grave had a cut-out for the head, which prevents the use of a coffin (cf. to the first image on the web-site). In the paragraph above it was already stated that Callebaut et al. (2000) believe anthropomorphic graves to date from the 13th century, which would be contradictory to the findings here.

[129] Daniël Pletinckx and Vera Ameels, personal comments.

[130] Dirk Callebaut, personal comment.

[131] http://www.ename974.org/Ndl/pagina/NuhV/991116_1.html.

Figure 7.24: The two openings of the so-called portal. Left: the door opening to the east. The imposts appear as slight curves, which give the arch a horseshoe shape. Right: the door opening to the north. The lower part of the opening has been closed up to make a window; the imposts are very pronounced. (Photographs by Joyce Wittur).

A third layer of plaster was applied only to the tympanum. Upon it 'Christ in Majesty' was portrayed. The painter(s) tried to copy the Byzantine style and made use of malachite and ultramarine. The latter was a very expensive pigment (Van Dijck, 1995, 15, 27). Christ is depicted in a 'quatrefoil' mandorla[132] and holds a book in his left hand. Further to the right parts of a bird are visible, though its head is missing. It could belong to a depiction of the Four Living Creatures, which are commonly part of 'Christ in Majesty'. As Van Dijck (1995, 30) pointed out, it seems as if a design was painted on this tympanum, which was actually too large for the available space. There is hardly enough space for Christ's feet and a full quatrefoil mandorla because of the central arch below (Fig. 7.26, left). The quatrefoil mandorla points to a 12th century date for the execution of the mural[133]. Van Dijck (1995, 31) mentions that experts (without naming any) were asked to date the painting and they replied that the motif appears from the beginning of the 12th century.[134] The opinion of the experts is nevertheless dismissed in favour of a date at the beginning of the 11th century, which is said to be proven by the ^{14}C analysis performed on a piece of charcoal found in the layer of plaster on which the mural was painted. The plaster is contemporaneous with the painting because the sketches for the mandorla were made while the plaster was still wet. Van Strydonck (1997, 98 ff.) uses these radiocarbon dates from Ename as an example. The question he wants to answer is whether the mural was painted shortly after the construction of the church or after

[132]The tympanum would not have granted enough space for an actual quatrefoil mandorla, but the arrangement of the lobes shows that this form was intended.

[133]It should also be considered that two layers of plaster lie underneath this painting, which could indicate that this mural does not belong to the earliest phase of the interior decoration.

[134]A date in the second half of the 12th century or later seems slightly more likely.

Figure 7.25: The eastern side of the stair tower and the so-called portal. 1: remainder of the buttress, which is still in the ground, 2: replacement masonry where the buttress has been broken out of the wall, 3: later door opening, now closed up, which led into the stair tower. (Photograph by Joyce Wittur).

1200 'when the church regained prestige'.[135] One piece of charcoal from the mortar was dated twice. The calibrated ^{14}C results lie between AD 933–1024 ($\pm 1\ \sigma$ range) and AD 960–1050 ($\pm 2\ \sigma$ range). As stated in the article this fits with the time of Herman of Ename. The old wood effect is discounted by Van Strydonck (1997). This effect occurs because the wood of the inner tree rings is older than the wood of the outer tree rings. If the tree in question was cut in 1200, then it would need to be at least 170 years old for its inner rings to provide the necessary date. The probability for this would only be 10%. Additionally such big trees would be valuable and not usually used as fuel e.g. for lime kilns (Van Strydonck, 1997, 100). It seems dangerous to place so much significance on just one sample, especially as it is not clear where the charcoal has come from. Only one of many possibilities could be that it came from the first roof of the church, for which the trees would have been felled while the church was under construction (fitting the radiocarbon dates). The tiles, which covered the roof in its first phase (Callebaut et al., 2000)[136], could also be the source of another material identified in the layer of plaster on which the 'Christ in Majesty' was painted: ground terracotta.

It is obvious that Van Dijck (1995, 31) had problems finding examples of quatrefoil mandorla from the 10th and 11th century. Her examples from Mayr-Harting (1991, 101 fig. 56, 128 fig. 77) are neither mandorlas, nor are they truly quatrefoil, but rather overlapping circles or other borders which separate parts of the illustration[137]. Due to the Byzantine style[138] of the painting she also looked to Byzantium for analo-

[135] The line of enquiry seems not a good choice, especially as the above mentioned experts favour a date in the 12th century. In the late 12th century there was enough money to repair the church after the fire (1175–1185), not only the roof, but also the expensive window panes.

[136] http://www.ename974.org/Ndl/pagina/NuhV/000201_2.html.

[137] The first image in question is the opening of St John's Gospel from the Uta Codex, Regensburg, *c.* 1020. The second is the Emperor Henry II seated in Majesty. Gospel Book, Regensburg, *c.* 1022.

[138] The image draws on Byzantine forms, although certainly no Byzantine painters were working on it; some characteristics of the

Figure 7.26: The murals of the tympanum. Left: the whole tympanum. Top right: detail of the *majestas domini*. Bottom right: detail of the floral paintings on the intrados of the framing arch. (Photographs left and top right by Daniël Pletinckx, photograph bottom right by Joyce Wittur).

gies, but could not find any which would fit the time frame set by the ^{14}C date (Van Dijck, 1995, 41). If this mural was painted soon after the church was built, how can the previous plaster and painting layers be explained? If the mural was produced before 1175–1185, then there should also be traces of the fire.

Later the 'Christ in Majesty' was covered with a layer of plaster and a following layer of whitewash, which also covers the intrados of the framing arch. Angels were painted on the whitewash (Van Dijck, 1995, 16). On the left side three angels enter the tympanum from the arch: only their wings, arms and heads are visible (Fig. 7.26, left). On the right side only two angels can be seen entering the tympanum; more of them may be destroyed. Centrally on the left side of the tympanum a larger angel is holding a cross; the right side of the mural is mostly lost. This painting has been dated to the 12th century and is claimed to either show the 'Deposition of Christ' (Callebaut et al., 2000)[139] or the '*Arma Christi*' (Van Dijck, 1995, 44)[140]. The intrados is decorated with a meander.

The following 14 layers are all monochrome. After the seventh layer the new vault was introduced, the three arches were closed up and the lower part of the framing arch was bricked up. After the 14th layer the whole framing arch was closed up (Van Dijck, 1995, 18 ff.).

7.2.7 Pictorial Sources

The photograph and the postcard used by Callebaut (1992) to show the differences before and after the restoration in 1907/08 have been mentioned before. They only show one side of the church from the outside, but still allow us to see some of the changes, which occurred during the restoration.

The images on the map (Fig. 7.27, left) and the pilgrimage pennon (Fig. 7.17, right) dating from 1595/96 are a bit more ambiguous.

Both images, as reproduced by Callebaut (1992, 464 figs. 36 and 37), show St. Lawrence's church from the same perspective (cf. Fig. 7.17, top right). The images are very similar. They depict a basilica with a western annexe and a tower to the east. The western annexe has a lower roof line than the nave, but the images vary according to how much lower it is. The annexe is shown with an entrance to the west and a window above it as well as two windows to the south. The clerestory possesses five windows, while the southern aisle has some very small windows, but the more important feature seems to be the clearly visible door. The tower to the east appears to be square in its lower part but round at the top.

production of the murals make this unlikely. For example, highlights were achieved by letting the white plaster surface shine through instead of using white paint to highlight them.

139 http://www.ename974.org/Ndl/pagina/NuhV/991116_1.html.

140 The 'Deposition of Christ' would lack a portrayal of Christ unless he was shown on the right side of the tympanum. On the other hand there should be more instruments of passion, but only the cross for an '*Arma Christi*' scene.

At the height of the clerestory it either possesses two windows (on the pilgrimage pennon) or none (on the map); the upper storey shows, on both renderings, two windows, which are higher than the roof of the nave.

There is a third image, which is on the reverse of the pilgrimage pennon and also shows the church, but this time from the north. It shows the same details as the images discussed above including a door in the northern aisle (Fig. 7.17). The tower on this image appears to be completely square, but for the roof, which seems to be round.

Callebaut (1992, 441) calls the faithfulness and the usefulness of these pictures into question, because the roof ridge and the roof edge of the western annexe is too low in comparison with the physical building. Additionally the images show a door in the southern aisle which never existed there, while the second image on the pilgrimage pennon depicting the church from the north is right to display a door in the aisle: the *lijkdeur*. Whether the tower ever had a round top is doubtful. The image is obviously no faithful representation of the church, but more a sketch, which captures the most prominent features of the building, but pays no great heed to detail. Maybe only one side was sketched (the north) and the image reversed to give a view from the south as well. This could explain the 'additional' door in the southern aisle, and the fact that the cemetery wall has exactly three entrances. Through mirroring, the misunderstanding concerning the topmost storey of the tower could have occurred. The image on the map probably copied the southern view on the pilgrimage pennon. The entrance to the cemetery to the west is missing, and there is only one window in the aisle.

A fourth image of St. Lawrence's church dating from 1596 was discovered in the Rijksarchief te Brussel in 2006 (Fig. 7.27, right). It is part of a map, which was used in a legal dispute concerning the fishing rights of Ename Abbey in the river Scheldt (Van Der Meiren and Derde, 2006). Unfortunately the image of St. Lawrence's church bears no resemblance to the physical building: it shows an edifice with a single-pitch roof, behind it a square tower and behind the tower a building with a saddleback roof. Putting the tower in the middle of the 'church arrangement' seems to be one of the characteristics of this draughtsperson.[141] It seems like a pictorial shortcut for 'church' rather than a faithful depiction.

[141] The churches of Nederename, Eine and Heurne also have their tower in front or behind two lower building parts.

7.3 The project 'Ename 974'

The Project was named 'Ename 974' because 974 is thought to be the year when the mediaeval settlement was established. Ename 974 is concerned with four places in Ename, which constitute an archaeological park: the archaeological excavations at the abbey (the castle, which lies close to it, is inaccessible due to the railway lines which run across part of the site), the nature reserve Bos t'Ename, the Provincial Archaeological Museum and St. Lawrence's church (Callebaut et al., 2000)[142]. The Provinciaal Bestuur van Oost-Vlaanderen (the Province of East-Flanders), the Instituut voor het Archeologisch Patrimonium van de Vlaamse Gemeenschap (the Institute for the Archaeological Heritage of the Flemish Community) and the city of Oudenaarde are responsible for the park (Callebaut and Sunderland, 1998, 51).

All of these sites, with the exception of the nature reserve, are places where reconstructions were used to inform the general public about the history of Ename. Once more it is necessary to consider all the reconstructions because they are interlinked and build upon each other; yet again St. Lawrence's church will be the focus.

7.3.1 General aims and ambitions

The first and foremost aim of the project is to inform the general public about the results (Ename Abbey) and also the proceedings (St. Lawrence's church) of the research undertaken at those sites. In the museum the archaeological finds are the centre of attention. The 'general public' in this case refers especially to the people living in Ename. They were informed in multiple ways: in the earliest phases of the abbey excavation with panels, regular guided tours and brochures (Callebaut, 2002, 184). Later a quarterly magazine was printed and distributed free to all households from April 1996 to May 2007. Evening programmes for adults, and school tours were organised. Two information kiosks were installed: one at the site of the abbey (1997), another close to St. Lawrence's church (1999). Information and reconstructions were displayed inside the Ename Museum (1998) and, after the restoration was finished, inside St. Lawrence's church (2002). Besides that, the webpage *Nieuws uit het Verleden*[143] (1999) provided fre-

[142] http://www.ename974.org/Ndl/pagina/project_overzicht.html.

[143] http://www.ename974.org/Ndl/pagina/NuhV/NuhV_home.html.

Figure 7.27: Two maps from 1595/96 depicting St. Lawrence's church of Ename. Left: the depiction of St. Lawrence's church on the map seems to be a copy of the pilgrimage pennon (Rijksuniversiteit Gent, map 885). Right: the church shown here bears no true resemblance to the physical St. Lawrence's church (Algemeen Rijksarchief, Grote Raad, Beroepen uit Vlaanderen. Particulieren nr. 404).

quently updated information about the excavations inside St. Lawrence's church. It included reconstructions, photographs and films, covering topics of interest to a lay audience. The plan was to involve the local community in the activities and create awareness of the history of their village, so as to let them participate in the exhibitions[144] and also to gain funding for the projects (Callebaut, 2002, 179). Open Monument days and other events also helped to appeal to the public. Besides the local focus, it was also hoped that Ename would gain a reputation as an important tourist site, thus bringing more life to the village (Callebaut, 2002, 180).

The latest insights into archaeological, historical and conservational issues were meant to be presented with 'great attention to scholarly accuracy and by means of multimedia technologies' (Pletinckx et al., 2000, 45). The reconstructions should not only provide stunning pictures, but also reflect the research which went into them. On the other hand, it was planned to look for new ways of presenting this information (Callebaut, 2002, 179), and several solutions were eventually found.

[144]For example, some older residents told stories about the past, which are available from the kiosks, others donated photographs to the museum.

The evolution of buildings and the landscape is another central feature of the reconstructions. They are seen in combination with the other multimedia applications, as effective communication tools (Pletinckx et al., 2001, 198).

7.3.2 Earlier projects

The kiosks TimeScope 2 and 3, which are concerned with St. Lawrence's church, had two predecessors, TimeFrame™/TimeScope 1 and TimeLine, with different foci (abbey and artefacts) from the later systems. Still, these earlier projects had an impact on the following systems (Pletinckx et al., 2001, 200) and upon each other. TimeFrame™ deals with the abbey, leading to know-how and models, which could be reused in TimeLine, while the sources for the abbey found their way into its follow up system, TimeTravel. TimeLine and TimeTravel already show reconstructions of the whole 9 km^2 area of Ename, including St. Lawrence's church. Accordingly the new findings made during the excavation and restoration work inside the church led to changes in the museum display. Therefore the two projects preceding TimeScope 2 are discussed here briefly.

7.3.2.1 TimeFrame™: Abbey

The first application was called TimeFrame™ or (since *c.* 2000) TimeScope 1 and designed for the abbey in 1997. The site consists of hardly more than foundations, which belong to several phases (see Section 7.2.5). During the excavation the different layers of flooring, the graves and all traces of the pre-urban settlement were removed. To let visitors experience the site and to make the cultural heritage remains more comprehensible, it was decided to establish a provincial archaeologic park (Balthazar et al., 1997), i.e. an open air museum. The features, which remained after the excavation, are hard to identify as specific structures and do not captivate the attention of visitors (Pletinckx et al., 2000, 45). One of the possibilities would have been to create a physical reconstruction, but due to the many problems attached (e.g. if the reconstruction had to be modified) it was agreed that a non-invasive approach would be best suited: therefore, an interactive computer system would provide the necessary insight (Callebaut and Sunderland, 1998).

On Behalf of the Provinciaal Bestuur van Oost-Vlaanderen a concept for TimeFrame™ was created by John Sunderland (museum designer), André De Clercq (research and development at Barco) and Daniël Pletinckx (civil engineer with a specialisation in information technology). The technical realisation was executed by IBM (Great Britain), while the Instituut voor het Archeologisch Patrimonium van de Vlaamse Gemeenschap was responsible for the archaeological data for the computer reconstructions (Callebaut et al., 2000)[145].

The reconstructions were based on historic images, excavation plans and analogies, i.e. still-standing monuments from the same period (Callebaut, 2002, 182). Starting points were the archaeological drawings. From them the proportions of the buildings were calculated. First, only the geometry of the buildings was modelled, and only later were details added. Additionally the surrounding area was considered, with not only the terrain being reconstructed, but also trees and the neighbouring premises to make the scene appear more realistic.[146] A lifelike appearance was also the aim behind adding colour and structure to the buildings. The textures for the outside of the church were created from photographs taken on site. Finally an animation was created, which shows the reconstruction from a bird eye's perspective: the observer flies around the church and then through it (Callebaut, 2002, 183). The usefulness of this kind of animation is debatable. It does give a good view of the site, but no one in the past ever experienced the site like this. A walk through instead of a fly through would have been much more appropriate because inside the church the higher vantage point holds no benefits for the user. Alongside the images, spoken explanations are available in Dutch, French and English, while Gregorian background music gives a monastic feeling (Callebaut, 2002, 183).

The TimeFrame™ system consisted of a kiosk (to keep the hardware and the visitors sheltered) a camera, a computer, two displays and a touch screen.[147] The camera was filming the foundations of the abbey upon which the reconstructions could be projected. For this, augmented reality and multimedia[148] content were employed. The augmented reality was used to clarify the relationship between the reconstruction and the physical remains on site, in order to make the latter more meaningful to the observer (Callebaut, 1997). The user is thus able to tell the different remains apart, such as were left by the people and their buildings during the evolution of the site. At the same time a camera provides life images, which are combined with the reconstructions to give the observer an air of 'authenticity', with the actual weather conditions as well as visitors walking among the ruins (Callebaut et al., 2000)[149]. What the authors meant here by 'authenticity' is certainly the authenticity of the site and the ruins, i.e. the basis on which the reconstruction is built, and not an authenticity of the reconstruction itself. The latter is displayed as semi-transparent and purplish in colour (Fig. 7.28) and accordingly not very 'authentic' to behold. While the semi-opaque reconstructions are visible on the two big screens, the visitor also has a view on the extant remains outside the kiosk (Callebaut et al., 2000)[150].

Additional information was provided in the form of spoken explanations, short features (about two minutes long) (Callebaut, 1997) and photographs (e.g. from the excavations and the finds), plans, draw-

145 http://www.ename974.org/Ndl/pagina/archeo_tijdsvenster1.html.

146 For a critique of this approach see Section 5.4.

147 The system is nowadays (2009) out of use. It had been very difficult to update the proprietary system and the hardware had to be exchanged several times (Daniël Pletinckx and Vera Ameels, personal comments).

148 The term multimedia is this context should be substituted by multicodal and multimodal. See Boxed Feature 2 on page 29.

149 http://www.ename974.org/Ndl/pagina/archeo_concept.html.

150 http://www.ename974.org/Ndl/pagina/archeo_tijdsvenster1.html.

Figure 7.28: How TimeFrame^{TM} or TimeScope 1 works: a camera inside the kiosk delivers an image of the ruins, over which the semi-transparent reconstruction is projected. This augmented reality application enables the visitor to make sense of the foundations on site and also to understand the relationship between the physical remains and the reconstruction. (Image copyright pam Ename).

ings and virtual reconstructions, which were projected on top of the images produced by the camera.[151] Animated virtual reconstructions showed the evolution of the site and its buildings. In a second step the reconstruction of a particular phase could also be explored from the inside (Callebaut and Sunderland, 1998, 54) and from the air too (Callebaut, 1997). These reconstructions are opaque models with textures and have a more 'realistic' look. When exploring the reconstructions it is also possible to view additional information which explains on what information (plans, photographs, *etc.*) the reconstruction is based (Callebaut et al., 2000)[152].

The prototype was launched on the 2nd of September 1997, and questionnaires were handed to the visitors to record their opinions and suggestions with the intention of having the system updated and fully operational in 1998. It was also planned to have several TimeFrame^{TM} systems distributed all over the archaeological site (Callebaut, 1997). The kiosks were to provide the user with several alternative routes between them and with several different interpretative programs, which would be adjusted to specific target audiences (school children, tourists and locals) (Callebaut, 2002, 183). These plans have never been realised.

7.3.2.2 TimeLine: Museum

In 1998 the Museum in Ename was opened. It contains a room, which exhibits on one side some of the archaeological finds, while the other is reserved for a huge display with a touch-screen attached to it (Fig. 7.29). The aim was to provide the visitor with the possibility of seeing the artefacts in the showcases in both their archaeological and spatial contexts via the reconstructions displayed on the screen (Pletinckx et al., 2000). The finds range from prehistory to the 19th century, but the focus lies on three main phases: the early Middle Ages (974–1025), when the castle and the pre-urban settlement were founded and flourished, the time when the abbey dominated the village (1063–1795) and finally the time when only the village remained (19th and 20th century). These phases were further divided resulting in nine

[151] Daniël Pletinckx had suggested having the reconstructions mirror the weather and daylight conditions of the physical site, to make the models more realistic. This was not implemented. Personal comment.

[152] http://www.ename974.org/Ndl/pagina/archeo_concept.html.

sub-phases. For each sub-phase a complete virtual model of the village and its surroundings was created, showing the development of the settlement and the evolution of the landscape (Callebaut and Tack, 1998). The reconstructions, which were made for TimeFrame™ with their focus on the abbey, could be reused. Otherwise the reconstructions of each phase were once again based on all the data then available: archaeological remains, texts, drawings, maps, architectural analogies, structural stability and contemporary building materials and techniques (Pletinckx et al., 2000, 47).

Daniël Pletinckx designed the system (Callebaut and Tack, 1998). The touch screen is used to interact with it, while the huge flat screen provides good visibility of the scene to a group of people. In each reconstructed time-phase the positions of the contemporary artefacts are indicated by ranging rods, which are sticking in the virtual ground (see Pletinckx et al., 2001, 199 fig. 5). The users can move through time or explore a particular virtual model, but only from a bird's eye view. By selecting artefacts or contexts they can learn more about them (Pletinckx et al., 2001, 199).

TimeLine was followed by a second application, which is called TimeTravel. It offers access to a database, which holds the source materials used in the reconstructions: excavation results, texts and historical images. It also allows the visitor to explore the virtual reconstructions by 'flying' or 'from the ground' (see Pletinckx et al. (2001, 199 fig. 6)).[153]

7.3.3 TimeScope 2: St. Lawrence's church

After the initial research on St. Lawrence's church in 1992 with its surprising results (finding a 10th century church still standing) it was decided in 1999 that the church should be systematically investigated and finally restored. The restoration was funded by the European Commission and received additional financial support from the Flemish Ministry for Binnenlandse Aangelegenheden, Ambtenarenzaken en Sport (Van Der Meiren (Ed.), 1999). It was aimed at bringing back the original two storey structure in its tower, the plastering, the interior, the floor height and the entrances (Pletinckx et al., 2001, 200).

7.3.3.1 Aims for TimeScope 2

For the planned research it was necessary to close the church to both its congregation and the public, but access and insight were to be provided in other ways, namely by a new multimedia application, which was commissioned by the Provinciebestuur van Oost-Vlanderen (Van Der Meiren and Callebaut, 1999). It would offer information about the history of the church, allow a look at the ongoing work inside it and present virtual images of how the church had looked before and what it will look like after the restoration (Pletinckx et al., 2001, 200). It would also explain the steps of the work in progress and the reasons for them. Dealing with a still-standing monument meant other methods and approaches than the ones used for TimeScope 1 were necessary (Pletinckx et al., 2001, 201).

The target audience was primarily the local population, but it was also aimed at tourists and people, who are unable to visit the site in person (Pletinckx et al., 2001, 201).

Yet again it was thought to be important that the reconstructions presented to the public should be verifiable, not necessarily by the users of the information systems, but a detailed documentation was planned in order to help explain why certain choices were made. Every time new information became available hypotheses had to be checked, the reconstructions updated and iterated until the most probable solution was found (Pletinckx et al., 2001, 202). This resulted in only one reconstruction, which was thought to be the most probable.

The information was meant to be accessible from three points: from a kiosk outside the church (Fig. 7.30), from the same huge display in the museum, which was also used for the TimeLine and TimeTravel application, and from the internet. A digital camera made photographs of the excavation and restoration work inside the church, which were broadcast to the kiosk (in a short version), the museum (in a long version) and the internet. Additionally a hand-held camera was purchased to make films of special activities. For 2000 there were plans to install a panoramic camera, which was to have been fixed to the ceiling of the nave. It was supposed to be an interactive device, which the visitors could use to

[153] TimeTravel also provides a link to another exhibit in the museum: The feast of 1000 years. This diorama shows persons during a feast, who come from different times and belong to different social classes. Each of the persons is linked to an artefact in a showcase and will tell a story including the artefact when a button is pressed. These people reappear in the TimeTravel application, so the visitor gains insight into where these persons lived or worked in Ename (Van der Donckt et al., 2002, 54).

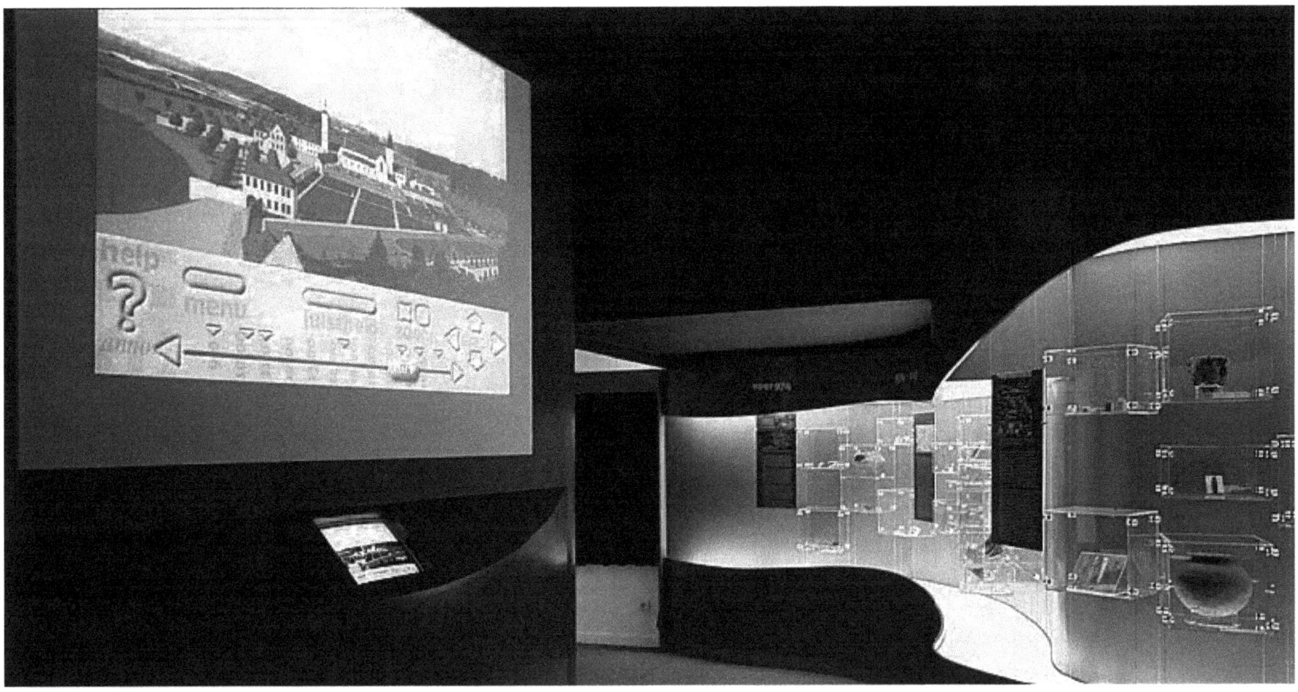

Figure 7.29: The TimeLine display inside the museum. On the right the artefacts are displayed in a showcase. On the left the reconstruction of Ename can be viewed. The visitor can find out more about the artefacts, e.g. the spot where they were unearthed, by using the touch screen underneath the display. It is also possible to explore the reconstructions and to see which artefacts are contemporary. (Image copyright pam Ename)

look around the church. (Callebaut et al., 2000; Van Der Meiren and Callebaut, 1999)[154]. Instead a webcam was installed in April 2000 because of its better resolution and because Belgian law forbids cameras in work areas. This also meant that the webcam took different photographs during the day and one of the images was selected to be visible in the museum, kiosk and on the internet. The panoramic camera was to be used on special occasions and then from a ground perspective, so that the viewer seemed to be in the midst of the archaeologists (Callebaut et al., 2000)[155].

The new TimeScope application was meant to be more than a digital logbook, and to offer access to images showing how the restoration would affect the building. This was supposed to be achieved by projecting virtual images over photographs taken on site (similar to the technology used for TimeScope 1 (Callebaut, 2002, 183)) and explaining the scientific reasons for these changes (Van Der Meiren and Callebaut, 1999).

The reconstructions, multimedia presentations and the work progress inside the church were also accessible from the internet. It was planned to offer the contents in different formats for different audiences: schoolchildren, adults, scholars (Pletinckx et al., 2000, 46).

It was hoped that the web-pages would result in an increase in visitor numbers to Ename and that a wider, especially international, audience would be informed about the project and the church (Pletinckx et al., 2000, 47).

7.3.3.2 Arriving at the model(s)

The plural in the heading indicates that not one model, but several models were made. They are not alternatives, but show different states of progress during the modelling process. In contrast to 'Progetto *Insula* del Centenario' (chapter 6) several of these older models (or stills created thereof) are available online offering an insight into the refinement of the hypotheses.

During the reconstruction of St. Lawrence's church there was close collaboration between the archaeologists and architects, who were concerned with the

[154] http://www.ename974.org/Ndl/pagina/NuhV/990907_3.html.

[155] http://www.ename974.org/Ndl/pagina/NuhV/000426_3.html.

Figure 7.30: The TimeScope 2 kiosk in the museum yard next to St. Lawrence's church. (Photograph by Daniël Pletinckx)

research, and Daniël Pletinckx, who built the 3D-models. Almost daily visits of Pletinckx to the site resulted in frequent updates of the websites and the information kiosks.[156]

The archaeological work was carried out by Dirk Callebaut, Koen De Grote, Nancy Lemay, Vera Ameels and Eva Roels and funded by the Instituut voor het Archeologisch Patrimonium van de Vlaamse Gemeenschap. Further support came from the municipality of Oudenaarde (Pletinckx et al., 2000, 48). The church was initially surveyed by architects from the Provinciaal Bestuur van Oost-Vlaanderen[157]. The documentation of the walls and the drawing of plans was produced by the architects Stefaan Browaeys (Callebaut, 1992, 436) and Zohrak Zohrabyan (Callebaut et al., 2000)[158]. The development of new technologies and standards for heritage presentation was undertaken by the Ename Center for Public Archaeology and Heritage Presentation, which was founded in 1998 (Pletinckx et al., 2000, 45) and is directed by Dirk Callebaut. Daniël Pletinckx was responsible for the New Technologies, while Neil Silberman worked on the Heritage Interpretation (Pletinckx et al., 2001, 204). Linda Van Dijck and her team were in charge of the restoration of the murals. The Afdeling Monumenten and Landschappen (Vlaamse Gemeenschap) and the Munumentenzorg en Cultuurpatrimonium (Provincie Oost-Vlaanderen) dealt with the restoration supervision, the responsible architecture office was RNR Architectenbureau (Van Der Meiren (Ed.), 1999). Several Belgian scientific institutions and universities also participated in the project (Callebaut, 2002, 181).

The documentation of the site

The historical sources, written and pictorial, as well as the findings of the archaeological work and the building research have been detailed above (see Sections 7.1 and 7.2).

After the plaster and the furnishings inside the church were removed, the walls were surveyed by Johan Van Laecke (Instituut voor her Archeologisch Patrimonium) and Zohrabyan Zohrak (Callebaut et al., 2000)[159]. As usual for archaeological excavations all features were drawn on millimetre paper and coloured. These drawings were accompanied by written descriptions and photographs (Callebaut et al., 2000)[160]. The walls were drawn stone by stone in scale 1:25 (Callebaut et al., 2000)[161].

The different layers of paint and plaster were recorded by Linda van Dijck and an inventory of the extant murals was created. This helped to decide on how to progress with the conservation (Callebaut et al., 2000)[162]. The murals, which could be detected underneath the monochrome washes and layers of plaster, were uncovered and then contact drawings of them were made on transparent film. Photographs were taken of the murals and together with the drawings were used to create 'reconstructions' of

[156] Daniël Pletinckx, personal comment.

[157] Daniël Pletinckx, personal comment.

[158] http://www.ename974.org/Ndl/pagina/NuhV/001208_2.html.

[159] http://www.ename974.org/Ndl/pagina/NuhV/000229_1.html.

[160] http://www.ename974.org/Ndl/pagina/NuhV/000307_3.html and http://www.ename974.org/Ndl/pagina/NuhV/000421_1.html.

[161] http://www.ename974.org/Ndl/pagina/NuhV/000420_1.html and http://www.ename974.org/Ndl/pagina/NuhV/001208_2.html.

[162] http://www.ename974.org/Ndl/pagina/NuhV/991203_1.html.

the paintings with a raster graphics editor on a computer (Ameels, 2005).

A photograph on the website shows that in one part of the church a geophysical survey was made though the text does not make further mention of it (Callebaut et al., 2000)[163].

The modelling

The model was based on the surveys of the Provinciaal Bestuur van Oost-Vlaanderen, which were undertaken at a very early stage. The later surveys only became available when time for the project was running out.[164] Neither are any of the murals included as part of the reconstructions. Other findings were incorporated into the model as soon as the new information became available. This was necessary to keep the public up-to-date with the latest developments and hypotheses concerning St Lawrence's church. Examples of this process are found in the last QuickTime VR[165] on Callebaut et al. (2000)[166], where the western wall of the western sanctuary is depicted with four blind arches. After the discoveries were made during the building research in this part of the church, it was decided that there was only enough space for three blind arches (Callebaut et al., 2000)[167]. One of the later reconstructions represents this view (Fig. 7.32, left). Another example involves the windows in the eastern wall of the nave, before the second window was discovered (Fig. 7.32, right): the reconstruction shows only one window to the south (Callebaut et al., 2000)[168].

The available texts do not mention anything about the modelling process itself. Daniël Pletinckx (personal comment) created the models with ArchiCAD by Graphisoft, a solid modelling software, which is mainly used by architects. There is no information on how uncertainties were dealt with. Most of the church is still standing thus reducing the necessity for dealing with this kind of problem, but several decisions inevitably had to be taken. Dirk Callebaut (personal comment) stated that analogies were used for the three windows in the west wall of the western sanctuary and they were thought to match the three blind arches in the lower part of the same wall: three arches and three window openings each time consist of a bigger arch flanked by two smaller ones. For the TimeScope 1 system, too, it was mentioned that the reconstruction had to rely on analogies, though none was named explicitly.

The reconstructions focus on the outer appearance of the church. Besides the altars there was no attempt to reconstruct the furnishings or the murals inside the church. This would have bordered on speculation in most cases anyway because, apart from the remains of the two altars in the eastern part of the church and only scarce remains of murals behind panelling or in bricked up niches, fairly little could be discovered. The late medieval period could have offered a chance for the reconstruction of the chapels in the aisles, but too little is known about their number, size and decoration.[169] Only few features hint at them: the remains of an altar, the remainder of a partitioning wall, changes in the flooring and some remains of the former decoration of the walls.

To get more insight into the reconstructions the QuickTime VR-file evolutie_kerk.mov will be considered here.[170] It is an older version of the evolution of St. Lawrence church by Daniël Pletinckx.[171] Still the reconstructions can be compared to what is known of the church (Sections 7.1 and 7.2) and therefore their data basis and how it was used can be identified. The reasoning and sources behind the reconstructions of the church have been indicated.

AD 1020 Most of the building as depicted in the model is still standing today. On the other hand several parts, which do exist today (see the model of the church in 2002), did not exist then. These are the tower from the second storey upward (Section 7.2.6.4), the stair tower and the so-called portal (Section 7.2.6.5), the annexes south and east of the western sanctuary (Section 7.2.6) and the *'lijkdeur'* (cf. Section 7.1.1 and see Section 7.2.6.2). Instead,

[163] http://www.ename974.org/Ndl/pagina/NuhV/000516_1.html, last image on the page. It can be assumed that it was for trial reasons only.

[164] Daniël Pletinckx, personal comment.

[165] See Paragraph 'Bubble worlds' in Section 10.4.1.3.

[166] http://www.ename974.org/Ndl/pagina/NuhV/990905_1.html.

[167] http://www.ename974.org/Ndl/pagina/NuhV/991116_2.html.

[168] http://www.ename974.org/Ndl/pagina/NuhV/000202_1.html, first image, and http://www.ename974.org/Ndl/pagina/NuhV/000201_3.html.

[169] Dirk Callebaut, personal comment.

[170] The file is included on the CD. To view it, a QuickTime plug-in has to be installed, it can be downloaded from http://support.apple.com/downloads/#quicktime (Accessed: 2nd of August 2012.). To manipulate the model: click, hold and drag from side to side to turn the model; click, hold and drag up and down to see different phases.

[171] The reconstruction which is now on display in the Museum in Ename is an updated version, which also shows the rest of the settlement.

Figure 7.31: New reconstruction (2006) of St. Lawrence's church without a staircase. (Reconstruction by Daniël Pletinckx)

the church had four entrances in the aisles (Section 7.2.6.2). Two of them have been reopened during the restoration. The roof was less steep than it is today and tiled with *tegulae* (Section 7.2.6.3). The window openings in the clerestory were larger (Section 7.2.6). The whole church was plastered on the outside (Section 7.2.6.3). Contrary to this older model, it is now known that the roof ridges of the nave and the two sanctuaries were not at the same height (cf. Fig. 7.31 and see Section 7.2.6.3). One of the uncertain parts of this reconstruction are the window openings in the aisles. Their size is unknown due to later changes (Section 7.2.6.2). It is also unclear how the storey above the eastern sanctuary could be reached. The entrances in the north of this building part are original so that some kind of staircase must have existed (Sections 7.2.6.3 and 7.2.6.4). An open staircase as in the reconstruction is nevertheless unlikely (as has been argued in Section 7.2.6.4). A wooden stair tower would be the more likely solution (Section 7.2.6.5). It has to be mentioned that the stair tower would need to be quite low to prevent the window in the eastern wall of the nave from being partly blocked. A later reconstruction (Fig. 7.31) shows the church without any kind of staircase. This avoids the problem of showing a possibly incorrect means of entering the upper storey, but having no stairs at all also gives the wrong impression. The western wall in the western sanctuary has been replaced with bricks at a later date (Section 7.1.1) and therefore the original window openings have been lost (Section 7.2.6.1). For the reconstruction, three windows, two smaller ones flanking a larger one, were chosen on the basis of analogies and because they mirror the interior of the church with the three blind arches.[172] The crosses which are used as acroteria seem to have no before mentioned basis.

AD 1170 The next model shows the church before the fire in 1175/85 (cf. Section 7.2.6.2). The roof has been made steeper (Sections 7.1.1, 7.2.6.1 and 7.2.6.3) so underneath it cereals and nuts could be stored (Section 7.2.6.3). The roof has a new cover made from wooden shingles. There is no other proof for this, except that no other materials (tiles or slate) have been found in the burnt layer inside the church. Now the church also has a tower with two new storeys whereby the observer is left to wonder whether they were constructed at the same time (Section 7.2.6.4). The stair tower and the portal are also built up to their final height; the previous steps are not shown in the reconstruction and would have been difficult to date, too (Section 7.2.6.5). The second storey of the tower must have existed before the fire because it

[172] Dirk Callebaut, personal comment.

has traces of burning (Section 7.2.6.4).[173] The appearance of the topmost storey of the tower is most probably designed according to analogies of other buildings from the same period because it was replaced in later times (Section 7.2.6.4) and no other sources exist. The same accounts for the pyramid roof with the cross and golden sphere on its tip. The lower part of the portal was indeed finished because traces of the fire were found in it (Callebaut et al., 2000)[174]. In the reconstruction the portal has an entrance to the east and a window to the north, and next to its entrance a buttress is visible (Section 7.2.6.5). The height of the portal and the stair tower in 1170 may still have been lower than the one shown in the reconstruction. The photograph taken after the restoration in 1907–08 shows the upper parts of both the stair tower and the portal to be freshly renewed (Section 7.2.6). Whether they were really renewed or whether they were newly built cannot be said with certainty on the given basis. Probably due to the new portal, all the other entrances to the church are shown to be closed up. This would not have been a necessary consequence. In the above cited material (Sections 7.1.1, 7.2.6, 7.2.6.1 and 7.2.6.2) nothing speaks for or against this interpretation. It also does not allow us to judge another change in the reconstruction: the window above the three windows in the western wall of the western sanctuary. This window opening still exists today, and it could have been useful for the storage purpose of the attic (e.g. for ventilation). Arguably the lower roof ridge would not have allowed for such a window, but it could be a later addition.

AD 1190 After the fire the building was changed only a little. The apse in front of the portal was added. Whether this event can be dated with any precision is questionable. At worst, it would contradict the presumed *terminus post quem* set by the 13th century grave underneath its foundations (Section 7.2.6.5). The portal window to the north has been made smaller and the portal was then probably used as a chapel (cf. Section 7.2.6.5). Now that the portal is closed a new entrance is necessary and the reconstructions show it to be in the western sanctuary. This interpretation may seem to be the necessary step following the previous reconstruction where the sole entrance to the church was through the portal. But if some of the other original entrances had still been open in the previous phase—as side entrances to the church—the new entrance in the western sanctuary would not have been necessary. It is even rather unlikely that the orientation of the church changed at such an early date, because the main altar must have initially stood in the western part of the church (Section 7.2.6.4). Additionally the western wall of the western sanctuary is built from brick (Section 7.1.1), which is used in this region from the 13th century onwards[175]. It is possible that an entrance in the western wall was created earlier and that later the wall was replaced with bricks whilst still retaining the entrance. This would be in accordance with the representations on the pilgrimage pennon and the map dated 1595/96.[176]

AD 1596 This reconstruction closely matches the drawings on the pilgrimage pennon (Fig. 7.17) and the map dating from the same year. The changes encompass the removal of the apse east of the 'portal' (Section 7.2.6.5). The latter keeps the window, but the former entrance to the east is closed up. The third storey of the tower (Section 7.2.6.4) is now shown as octagonal with a matching, high roof. The storey has a round-arched window on each of its eight sides. This reconstruction does not really match the drawings, which only show two windows from each perspective; this makes having only one window in every second wall more likely.[177] The idea that the tower top may have been octagonal may come from the way its roof is depicted in the drawings (Fig. 7.17). It is not shown in a smooth round curve suggesting a cone, but instead there seem to be facets. The reconstruction also shows the *'lijkdeur'*, but only in the north; here it is diverging from the drawings (section 7.2.6.2). The entrance to the west (Section 7.2.6.1) has been present in the previous reconstruction already, as well as the single window above it (Section 7.1.1). The attic window, on the other hand, is now missing. Another change is made according to the historic images: the gables in the west, i.e. the ones above the nave and the western sanctuary, were raised above the height of the roof truss. Only two crosses were placed on top of the church, which conforms to only one side of the pilgrimage pennon (Fig. 7.17, top right). The other side shows three crosses, while the map shows only one on top of the tower roof. It has been discussed previously whether these images are reliable sources (Section 7.2.7), and they have been found wanting. Therefore this reconstruction contains many uncertainties and some features appear only in this phase and have disappeared in the next (the octagonal tower, the

[173] Dirk Callebaut, personal comment.

[174] http://www.ename974.org/Ndl/pagina/NuhV/001208_1.html.

[175] Vera Ameels, personal comment.

[176] That these representations are not really reliable sources has been previously shown (Section 7.2.7).

[177] This has been mentioned by Daniël Pletinckxs.

raised gables[178]) while others have not been taken up in the reconstruction (the door in the southern aisle). Maybe it would have been better not to have reconstructed this phase.

AD 1655 The church has received a new slate roof cover as mentioned in the written sources (Section 7.1.1). The high altar stands in the east and there is a vault above it (Section 7.1.1 and 7.2.6.4). The changes, which the vault would have imposed on the outer appearance of the tower, are not displayed because when the reconstruction was made the vault was still thought to be an even later addition. Work on the belfry takes place, which requires ironwork, wooden beams and planks (Section 7.1.1). The reconstruction shows the tower with a square third storey (cf. Section 7.2.6.4), which possesses one window opening to each side and the dial of a clock at the front of it. This and the shape of the roof match the ones on the photograph taken before the restoration in 1907–08 (Callebaut and van der Plaetsen, 1992, 439 fig. 4). The photograph also shows that the top storey of the tower was built from bricks, a fact which could have been hidden with whitewash and/or plaster, as the reconstruction shows it. The model also shows the loss of the higher gable walls in the west.

AD 1778 The orientation of the church has been changed once more (Section 7.1.1), with the main altar now being in the west. Two new openings were cut into the tower. The main entrance to the east (Section 7.2.6.4 and cf. Section 7.1.1) and a side entrance to the north; the latter can be deduced from a building joint, which is still visible on the outside[179] of the building and also from photographs taken during the excavation, which show the doorstep and the later masonry infill. It was probably assumed to have been installed after the orientation of the church had changed. The windows on the ground floor of the tower had to be altered, too (Section 7.2.6.4), due to the installation of the organ and the organ loft (Section 7.2.6.4 and cf. Section 7.1.1). They were either split in two or partly closed up. The windows on the first floor are also shown to be smaller, though this change probably occurred earlier, when the vault was built into the tower (*terminus ante quem* 1607) because its extrados was higher than the original window sills (Section 7.2.6.4). The receipts from 1762 tell of many changes (Section 7.1.1), which are partly mirrored in the reconstruction, for example, the new windows. The vault in the nave caused the height of the window openings to be diminished and made new clerestory windows necessary (Sections 7.2.6 and 7.2.6.3). Possibly at the same time the aisles received larger windows (Sections 7.2.6 and 7.2.6.2) and the western wall of the former western sanctuary was built anew including a door opening and a window above it (Section 7.2.6.1). The enlarged aisle windows meant the *'lijkdeur'* had to be closed (Section 7.2.6.2). 36 windows were delivered (Section 7.1.1). 20 being necessary for the nave and aisles, eleven for the tower and five for the former western sanctuary. It is assumed that in the time between 1762 and 1778 the windows in the western sanctuary were all closed up (Section 7.2.6.1) and two new high rectangular window openings were cut into it. There is also the sacristy which was added in the west (Section 7.2.6). These changes are visible on the photograph, which was taken before the restoration in 1907–08, and maybe it would have been more truthful to give this as the appropriate date for this reconstruction. The *'slaper'* was here interpreted as a horizontal cover of the west gable of the nave; it might have prevented water from running along the wall and into the western sanctuary[180]. Another interpretation of the *'slaper'* has already been given (Sections 7.1.1 and 7.2.6.1).[181] The fact that the biggest part of the west wall of the western sanctuary was constructed from bricks, may have again been hidden with whitewash or plaster.

AD 1907 The reconstruction is probably based on the photographs (Sections 7.2.6 and 7.2.7), which were taken after the restoration in 1907–08, together with what was visible before the new restoration started in 1999. The outer appearance of the church was changed by substituting those parts, which had been visibly built from bricks, with Tournai limestone (in the west only, on the outside) (Section 7.2.6). The portal received a door where it had previously had a window, and a new window opening was cut into its eastern side (Section 7.2.6.5). A door was

[178]The question is, whether any proof has been found to support the existence of these features.

[179]The joint is partly visible behind the man in the first image on http://www.ename974.org/Ndl/pagina/NuhV/020930_1.html (Callebaut et al., 2000).

[180]Daniël Pletinckx, personal comment.

[181]If the *slaper* is really a trimmer and allowed for a vault, and if at the same time the number of windows delivered to the church hints at the existence of five windows in the western sanctuary, the vault could not have blocked the window openings. Another possibility is that the four windows were blocked by the vault, then four other window openings would have to be found. So far the window in the portal has not been accounted for, the two new, high rectangular window openings in the western sanctuary would also need glazing, and there is a small window opening with a segmental arch in the western blind arch in the southern wall of the western sanctuary, which is neither part of the reconstruction nor mentioned by any text.

opened up allowing entry to the stair tower from the outside, while the door opening in the northern side of the tower was closed up. The tower itself received a completely new third storey with blind arches and two window openings on each side. The upper parts of the stair tower and the portal were also renewed[182] (Section 7.2.6.5). The size of the windows in the aisles was reduced, while the windows in the clerestory could be enlarged because of the removal of the vault (cf. Sections 7.2.6 and 7.2.6.3). The sacristy was pulled down, and instead, two annexe buildings north and south of the western sanctuary were built (Section 7.2.6). These buildings could be reached via the western sanctuary. Additionally the northern annexe had an entrance to the west. The closed up window opening with the segmental arch in the west wall of the western sanctuary was partly reopened, but now contained a round-arched window (Section 7.2.6). Bargeboards were fastened to each verge.

AD 2002 This reconstruction shows the church after the latest restoration works were completed. The north door of the portal was substituted by a window, while the window to the east was transformed into a door (Fig. 7.24, cf. Section 7.2.6.5). The door opening which led directly into the stair tower was closed up (Fig. 7.25), the same as the entrance to the tower in the east (Fig. 7.21, right). The windows on the ground floor and the first floor of the tower were reopened to their full height (Callebaut et al., 2000)[183]. This was made possible by the removal of the organ loft and the vault (Callebaut et al., 2000)[184]. The western sanctuary and the west walls of the nave and aisles were whitewashed to protect them from moisture. The main entrance to the church is now in the northern annexe beside the western sanctuary.

As shown, most of the sources can be identified, though some ambiguities remain due to different interpretations of the source material or new sources, which had not previously been available. Uncertainties can be identified, e.g. the tower top in the model of 1170. Maybe a warning should be attached to the visualisation of the church in 1596 due to its not very reliable model basis. A change of date (to 'before 1907') for the 1778 reconstruction was suggested because some of its features are based solely on a photograph from this time.

7.3.3.3 The throne

Having the reconstruction prepared while the research on site is still in progress can offer unique opportunities, namely that the model influences the research. This has happened in the case of St. Lawrence's church.

During the building research inside the western sanctuary a structure was discovered in the former door opening to the west[185] (see Fig. 7.7). From the idea that the two Ename churches were built for a *Festkrönung* (see Section 7.1) Daniël Pletinckx surmised that this structure could have been part of a throne, especially as its size is close to the size of the staircase underneath the throne in Aachen Cathedral. So instead of placing an altar in the western sanctuary (Fig. 7.32, top left) he placed a staircase with a throne against the west wall (Fig. 7.32, bottom left). When he virtually seated himself on the throne, he was surprised that he could not see the altar in the upper storey of the tower (Fig. 7.32, top right), because his view was blocked by the parapet on which the pillars stood. He decided to talk to the archaeologists about his findings, and they then decided to have a look again at the parapet. As it turned out, the parapet had been lower than initially assumed and by correcting its height, a line of sight between the altar in the western sanctuary (Fig. 7.32, top left) and the altar in the tower chapel (Fig. 7.32, top right) could be established.[186]

This is one of the rare *reported* cases where the virtual reconstruction had influence on the research on site.[187]

7.3.4 News from the Past: St. Lawrence's church

Besides the kiosk close to the church and the information displayed in the museum, a third line of presenting the work inside St. Lawrence's church was pursued: the internet.

[182]This is what the reconstruction chooses to show; their height may have also been increased, but if so the stair tower and the portal would need to be smaller in all the previous reconstructions.

[183]http://www.ename974.org/Ndl/pagina/NuhV/000207_2.html and http://www.ename974.org/Ndl/pagina/NuhV/000207_1.html.

[184]http://www.ename974.org/Ndl/pagina/NuhV/991116_3.html.

[185]It is most probably the remains of an altar dating from around 1778.

[186]Daniël Pletinckx, personal comment.

[187]That making reconstruction drawings while the excavation is in progress can lead to revised strategies in the fieldwork was also mentioned by James (1997, 29 f.).

Figure 7.32: Reconstruction of the inside of St Lawrence's church. Top left: western sanctuary with altar (for the latter no foundations were found during excavation), bottom left: western sanctuary with a throne (see text for explanation), top right: view into the chapel in the first storey of the 'tower', bottom right: eastern sanctuary with chapel. (Reconstruction by Daniël Pletinckx).

The website (http://www.ename974.org/Ndl/pagina/NuhV/NuhV_home.html) was officially put into operation on the 3rd of December 1998 (Pletinckx and Servaes, 2001), though the breakthrough came only later, when TimeScope 2 was installed in 1999 and helped to promote the website: the access rate for the web-pages increased significantly from September onward. For TimeScope 2 the website *'Nieuws uit het Verleden'*[188] was modified and made accessible via the kiosk. The aim of the website was to let the public partake in the archaeological and restoration work, so letting them see, discover and think for themselves. The frequent updates to the site also showed the archaeological process: finds were made, new questions were posed and some of them resolved in the course of the work on site. This led to news items, which were linked to each other, because they followed a certain line of enquiry. Updating the websites resulted in a workload comparable to a part-time occupation. Besides texts and photographs, virtual reconstructions, film clips and QuickTime VR applications were created.

In 2001 around 2500 to 3000 visitors per month looked at the website (Pletinckx and Servaes, 2001). Around 25% came from foreign countries, 7% from the Netherlands. That the public was interested in the progress of the work is indicated by the amount of people (25%), who visited the website more than once a month (Pletinckx et al., 2001, 201 f.). Pletinckx and Servaes (2001) state that around 150 pages in Dutch and English were available. This number has been reduced to 98 pages by now (2009), which are available in Dutch only. Also, several of the fea-

[188]News from the Past.

tures mentioned in the article by Pletinckx and Servaes have been removed: for example, the Ename 974 journal as PDF, or the possibility of seeing the evolution of Ename from the year 1000 till now.

7.3.5 TimeScope 3: St. Lawrence's church

Even before the restoration of the church was finished it was already anticipated that there would be a follow up system to TimeScope 2, which would then be installed inside the church. In coherence with the Venice Charter, which forbids the destruction of later structures in order to show earlier parts, the TimeScope 3 system should make it possible to see the former appearance of the church in a virtual way (Pletinckx et al., 2001, 202) after the physical church has been mostly brought back by the restorers to how it appeared in Ottonian times.

The same people who were responsible for TimeScope 2 were also responsible for this system (see Section 7.3.3.2). Additionally it has to be mentioned that the TimeScope 3 system was developed by IBM Belgium, the interface graphics by PIMC (Belgium) and the hardware by Connecto (Belgium) (Pletinckx et al., 2003, 230 f.), while the software was developed by the Ename Expertisecentrum (Pletinckx, 2004). The new version was installed not only on the new kiosk inside St. Lawrence's church (Fig. 7.33), but also on the kiosk outside the church in the museum yard, and it went into use on the 9th of June 2004.

The aim was to present the visitors with an interactive system, which allows them to choose which information they are interested in and how much they want to know about a certain topic. The system should be usable by individuals as well as groups and should take into account that the visitors to a site have a limited attention span and cannot be expected to view a full cycle of information presentations (Pletinckx et al., 2003, 225 f.).

It was decided upon a system based on small items of information, called 'nuggets', which come in different 'flavours', e.g. history or archaeology, and can be linked to form individual 'stories'. The first prototype provided 114 nuggets (Pletinckx et al., 2003, 227). These nuggets can contain written or spoken text, a timeline, (animated) images, videos, QuickTime VR panoramas or objects (Pletinckx et al., 2004, 234). Behind this system stands a database and an XML structure, which builds HTML pages according to a predefined style sheet with the required objects. The system also makes use of Java and an Apache Tomcat server (Pletinckx et al., 2004, 235). Links between the items define their place in the narrative (Pletinckx et al., 2003, 229).

The nuggets are arranged in 'primary stories' which are circular—and therefore without a designated beginning and end—and have a particular theme (Pletinckx et al., 2003, 227). The creation of these story lines was the first step in the preparation of the system (Pletinckx et al., 2004, 235). In the prototype there were three themes or 'flavours': spatial, historical and archaeological (Fig. 7.34). Items which have more than one 'flavour' to them constitute the connecting links between the primary stories. Identifying these nuggets and linking the themes was the second step in system creation (Pletinckx et al., 2004, 235). The spatial theme starts with interactive panoramas from which a point of interest can be picked. The historical theme encompasses the history of the church while the archaeological theme is concerned with the archaeological discoveries made during the excavation. As an example of switching between themes Pletinckx et al. (2003, 227) mention the Roman field structures underneath the church: the information as to what had been found would be a nugget of the spatial flavour. The information about why the archaeologists think that it was a wheat field would provide the nugget with the archaeological flavour. Finally, the fact that the province of Flanders once was a Roman province and provided food for the Empire would be the nugget with the historical flavour. For the spatial flavour it is also necessary to create panoramas or images with hot spots, which link to the information items related to the selected places. This process was the third step in system creation (Pletinckx et al., 2004, 235).

By default, the user will follow the primary story (e.g. history) unless he or she willingly changes the flavour (e.g. to archaeology). Nuggets, which have several flavours, are indicated by displaying additional buttons on the screen linking the different story lines (Pletinckx et al., 2004, 234). A short (one time) excursion to another flavour is not regarded as the wish to change the flavour of the entire primary story, but rather as a footnote to the previous item. Only if the user stays with the new flavour will the default story line be switched (Pletinckx et al., 2003, 229).

Besides being easy to use by the visitors, the system should also be easily updatable as well as open for additions and revisions. In 2003 Pletinckx et al. (2003, 229) stated that changes to the system should always

Figure 7.33: The TimeScope 3 kiosk inside St. Lawrence's church. (Photograph by Daniël Pletinckx)

be made under the supervision of the multimedia experts and the scriptwriters. This would mean that constant technical support would be necessary, if the system were always to be kept up to date. In 2004 the opinion (or perhaps the system) had been revised,[189] in that only the local heritage staff or the experts for cultural heritage were thought to be needed for setting up and maintaining the system because the information for the nuggets has simply to be entered into a form-based user interface (Pletinckx et al., 2004, 235). The hardware required to run the system are a PC and a touch screen. Additionally, the system can be adapted easily to fit other output media (e.g. the internet) (Pletinckx et al., 2003, 229), as only the style sheet has to be adapted. It is also possible to change the links between the items to give different depths of information.

Because the kiosk is inside the church, the audio content is accessible via headphones, so as not to disturb the other visitors.[190] It was planned to make TimeScope 3 programmable, so that tour guides could compose and save their own sequence of nuggets (Pletinckx et al., 2003, 230), but only some of the features envisaged were already implemented in 2009.[191]

7.3.6 Documentation guideline

One of the outcomes of the reconstruction work was to consider how best to document the process leading up to a visualisation or visualisations.

[189] Until then the TimeScope 3 system was in use on three different sites (Ename, Tervuren and Wieringen) and the creators certainly had learned from experience.

[190] Currently the headphones are not in use, probably due to the effort to recharge, distribute and recollect them. (Daniël Pletinckx, personal comment.)

[191] Daniël Pletinckx, personal comment.

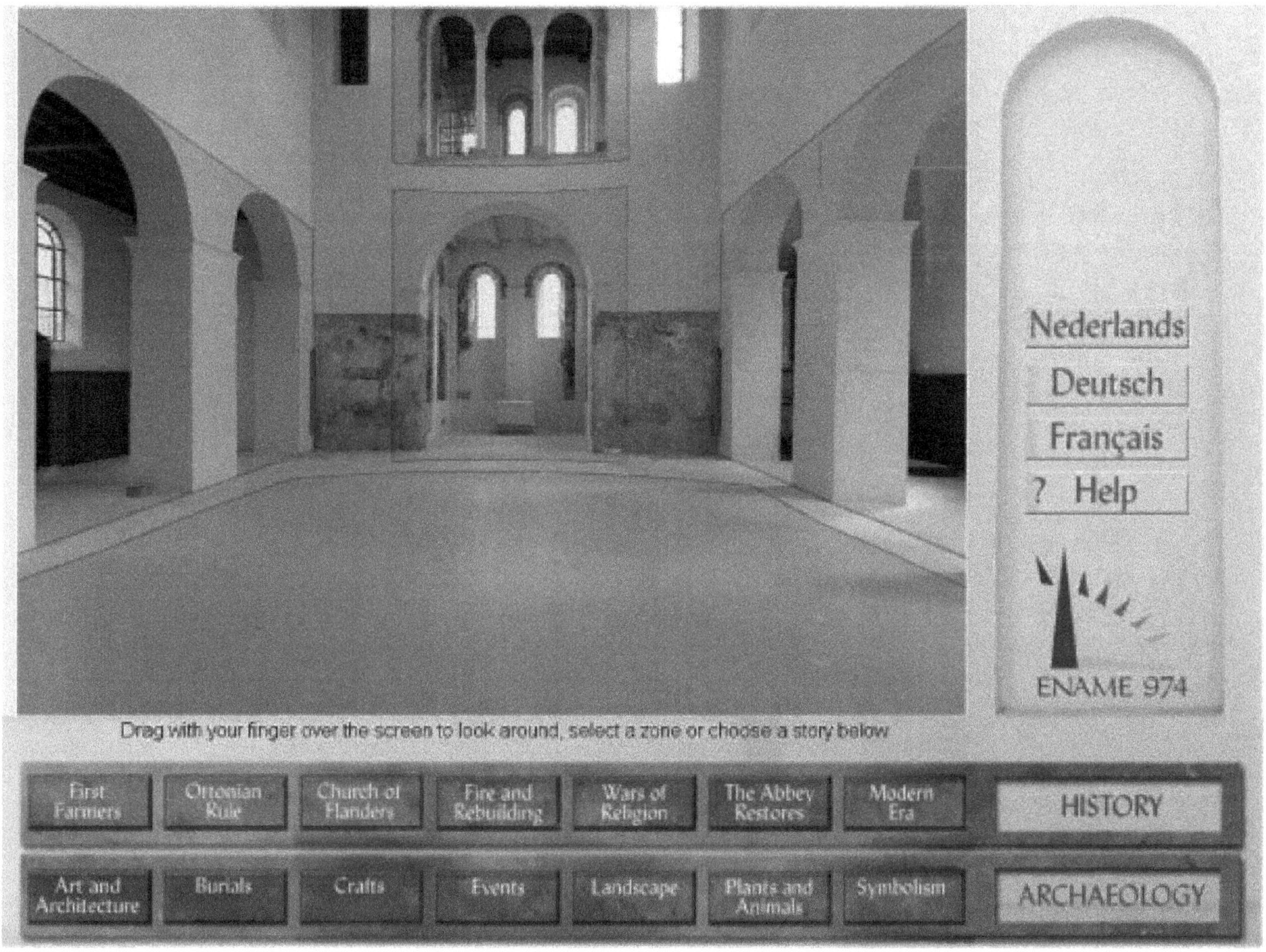

Figure 7.34: The TimeScope 3 start screen. Besides the language one of the three 'flavours' (spatial, historical and archaeological) with several starting points can be chosen. (Image by Daniël Pletinckx?)

This process was developed only after the main reconstruction work for the Ename 974 project had already been done, therefore it was not implemented for the reconstructions presented in the previous sections.[192]

A way to structure and document this process is presented by Pletinckx (2008): Interpretation Management. The aim is to record how the available sources were used in the creation of the 3D-visualisation. This is rarely done despite the fact that the interpretation of the sources is the most time consuming and complex task in this process (according to Pletinckx (2008, 4) 80% to 90% of the workload is dedicated to it). Recording can later help update the model whenever new information becomes available, because the initial line of reasoning can be retraced and changed where necessary (see also Section 5.7). Not all parts of an object can be reconstructed with a sufficient level of certainty. By documenting the decisions, especially where uncertainties are involved, the model becomes transparent for scientific scrutiny. Collecting the interpretations of different experts on sources and storing them—together with all the interpretations which are built upon them—in a structured way will help to make the model understandable and facilitate communication between disciplines (Pletinckx, 2008, 5 f.).

The methodology Pletinckx (2008, 6) proposes consists of five steps:

Creating a source database Here, all the sources (may they be texts, images, surveys, features from excavations, chemical analyses, *etc.*) considered during the visualisation process should be stored. No implicit knowledge should be used (Pletinckx, 2008, 6 f.).

Assessing the sources To judge the reliability of a source, its context, creator and the reason why it

[192] Daniël Pletinckx, personal comment.

was created should all be identified. All sources are interpretations (see chapter 9) and it is therefore important to discover possible errors, missing information, incorrect interpretations and deliberate alterations. The next step can help in detecting such possible biases in the sources. The results of the assessment should be stored as text (Pletinckx, 2008, 7 ff.).

Correlating the sources For this step sources are compared with each other in order to find correspondences, differences and inconsistencies between them (Pletinckx, 2008, 9). In particular, sources with the same content are useful to check for their consistency, disregarding whether they are of the same kind (two images) or whether they are of different kinds (image and text).[193] By this means changes through time become apparent and it is also possible to identify sources which are copies of one another (as, for example, the images on the pilgrimage pennon and the map (Pletinckx, 2008, 10 ff.), see Section 7.2.7). The correlation results should be stored in text format (Pletinckx, 2008, 13).

Creating hypothesis trees with conclusions For this step a top-down approach is suggested (Pletinckx, 2008, 13). The object which is to be visualised should be broken down into logical sub-parts. The hypothesis tree is seen as the 'formalisation of the interpretation process' necessary for the visualisation (Pletinckx, 2008, 13). Starting with the whole object, all the possible alternative interpretations are documented. Regarding the sources and the correlations, one of the interpretations is identified to be the most likely (Pletinckx, 2008, 13 f.). Then the object is broken down into smaller logical units, and these are evaluated in the same manner, *etc.* (Pletinckx, 2008, 14). The sources which support a certain interpretation are linked to the hypothesis (or sub-hypothesis) and their influence on the interpretation is documented in writing. The evaluation of the hypotheses should include pros and cons and in the end the most likely hypothesis should be pointed out, possibly with an indication of how much better it is compared to the other hypotheses (Pletinckx, 2008, 14). Unlikely hypotheses do not have to be broken down into sub-hypotheses (Pletinckx, 2008, 13). As most objects evolve over time, it is necessary to take this aspect into consideration as well. The interpretation should encompass all the evolutionary steps (Pletinckx, 2008, 13 f.).

[193]See also Section 9.1.

The reliability of a visualisation (or part of it) is connected to the reliability and the amount of the sources on which it is based (Pletinckx, 2008, 16). An indication of whether the reliability is low, medium or high is considered sufficient. A part for which only a single source or sources with low reliability exist, will be considered unreliable, while parts for which several reliable sources exist have a high reliability (Pletinckx, 2008, 16).[194] The reliability of a certain visualised part could, for example, be indicated by colour as in Fig. 7.35 (centre). Pletinckx (2008, 15) argues that unreliable parts should not be visualised for a scholarly audience, while the public needs a 'consistent image'.[195] As an example he chooses a reconstruction of Ename Abbey (Fig. 7.35). The two features indicated in red are only insufficiently backed by the sources; the entrance is supported only by a drawing and the structural analysis through a map. On the red area behind the abbey many traces of wooden buildings were found, but their size and position are unknown (Pletinckx, 2008, 17 f.). The 'scholarly visualisation' (left) does not show the wooden buildings, due to the lack of information concerning them, while the 'public visualisation' (right) does indeed show wooden buildings: on the one hand, to indicate that an abbey needs utility buildings, on the other hand, to show the continuance of utility buildings in this place, because the wooden buildings were in a later phase rebuilt in stone. Pletinckx (2008, 18) argues that the public should be told that the evidence for the buildings is non-conclusive, but rather *with* the hypothetical buildings in the visualisation than *without* them.

According to this explanation the entrance, which is also indicated in red, should consequently *not* be visible on the 'scholarly visualisation', either.

If two or more hypotheses are equally plausible, all the hypotheses in question would need sub-hypotheses and also their own visualisations. An exception can be made if the differences between the hypotheses are minimal, then a representative hypothesis, which will be visualised, can be chosen, but the other possibilities have also to be indicated (e.g. by text) (Pletinckx, 2008, 19).[196]

[194]In contrast to the Casa del Centenario project (Section 6.2.1.1) there is no list which assignes a 'level of reliability' to different source types.

[195]Cf. Section 5.10.3.

[196]This disregards the fact that images have a much stronger in-

Pletinckx (2008, 21) proposes that the uncertainty of a scene can be indicated by using other representational or rendering styles than more-or-less photorealistic three-dimensional images. He mentions an example where the rendered image of a 3D-scene was later re-drawn as a watercolour image by an artist (Pletinckx, 2008, 20 fig. 18), but he also considers non-photorealistic rendering, or colour reduction to greyscale or sepia.

Not only should be the whole evolution of a site be taken into account when creating the hypothesis trees, but it is also possible to visualise the temporal aspects of a site in order to let the viewer fully understand it (Pletinckx, 2008, 21).

Updating Updates become necessary if a new source appears or an existing source is reassessed. This will lead to new interpretations and therefore to a new hypothesis or a change in an existing hypothesis. The updated information can also influence the reliability of one or more hypotheses and, with that, the reliability of the visualisation. Pletinckx (2008, 22) argues that before the visualisation can be changed a consensus between all the experts involved in the reconstruction process has to be reached.[197] Finally, once more, the most probable hypothesis should be visualised (Pletinckx, 2008, 22).

The updating process should not only take place when new data becomes available, but also to ensure that the existing data is kept in such a condition that future use of the information is ensured, i.e. data migration.[198] These updating cycles should not be longer than two years (Pletinckx, 2008, 22).

By defining this 'Information Management' approach Pletinckx (2008, 24) aims at structuring and rationalising the interpretation process, and also at making the interpretations transparent, so that other scholars in the field can understand and contribute to the interpretations.

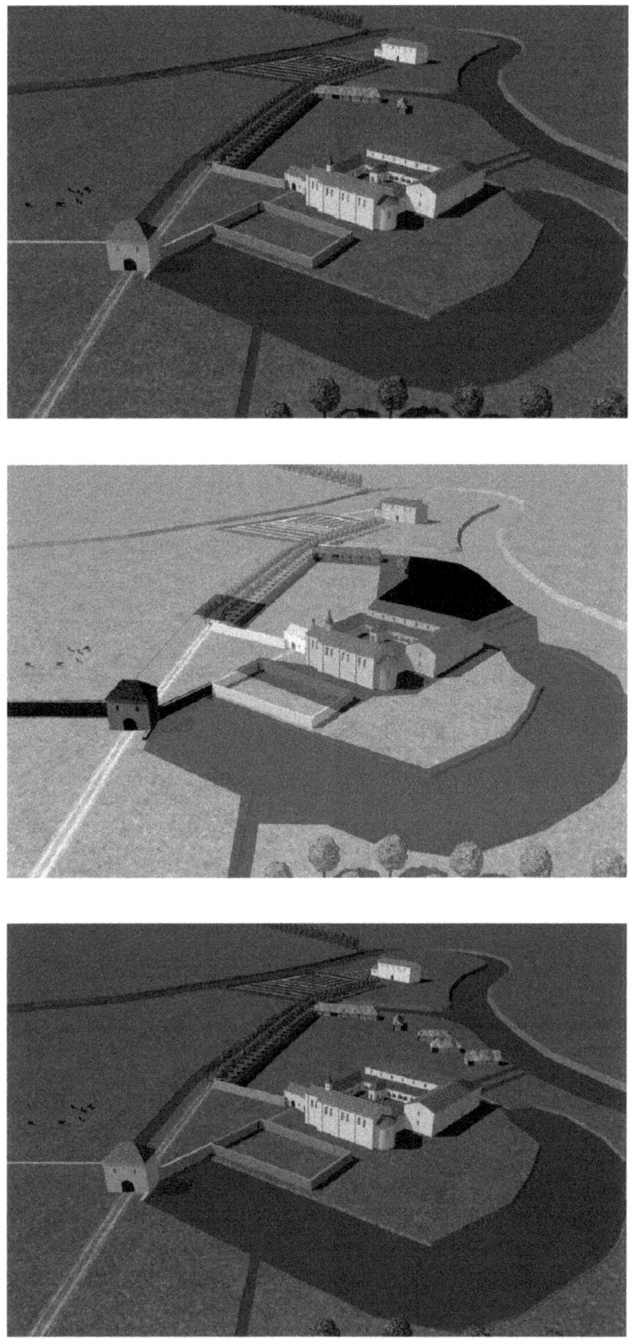

Figure 7.35: Reconstruction of Ename Abbey. Top: 'scholarly visualisation' without wooden buildings, centre: 'scholarly visualisation' with uncertainties indicated (red indicates unreliable parts, yellow a medium reliability and green reliable parts), bottom: 'public visualisation' with wooden buildings (Pletinckx, 2008, 17 f. fig. 13-15).

7.4 Have the aims been reached?

The aims for **TimeScope 2** (Section 7.3.3.1) were to inform the public and in particular the local inhabi-

fluence on the observer than mere texts, so that the visualisation will be remembered, but not the proposed alternatives.

[197]Pletinckx (2008, 22) sees this as a difficult step, not least because the experts may not be in the same place. He proposes the use of a wiki to overcome communication problems.

[198]Emulation is another possibility to keep data accessible. For detailed information see Section 10.5.

tants about the ongoing work inside the church. By providing three easily accessible information points (museum, the kiosk close to the church and the internet) and updating the information frequently, this aim has been reached. The number of visitors to the museum and the kiosk are unknown, but the websites were well frequented and the statistics (see Section 7.3.4) also hint at the frequent use of the information kiosk—otherwise the increase of website users after launching TimeScope 2 at the kiosk could not be explained. The sizeable number of website visitors from foreign countries also shows that there is an interest in this particular cultural heritage site beyond the local population.

Whether the intention, to provide the population with some insight into what their church will look like after the restoration has been finished, has been put into action, remains unsure. There are no indications on the web pages that this has happened. On the other hand visualisations were used to discuss restoration concepts with the local authorities[199], which in turn means that models of the proposed restoration result must have existed.

The steps of the work in progress were to be shown and explained. This was done by informing the public via the kiosks, the Ename 974 magazines and the website. The available information was updated frequently and did not only cover the different steps during the excavation and building research, but also the reasoning behind it and some of the interpretations attached to it. Initially it had been planned to have differing information features for different audiences (e.g. school children, adults, scholars), but this was not realised. Already during this process some of the news items of *Nieuws uit het Verleden* were grouped into story lines—a feature which returned in TimeScope 3, where it was put into use for an interactive storytelling application. The news items at the kiosks and the website were making ample use of digital photographs, webcam images, videos, QuickTime VR and 3D-models to present the information in an understandable way to the lay audience.

As anticipated webcam images were made available to the public, documenting the progress of the work inside the church. The application of the panoramic camera on special occasions amongst the archaeologists, on the other hand, cannot be verified with the available material.

The new results in research led to frequent updates of the model in order to present the most probable solution to the public. The updated model in Ename museum suggests that this process is still ongoing. Nevertheless the model is not based on the latest documentation of the site because it had not been finished when the initial project, which was concerned with informing the population about the ongoing work, came to an end. The later updates seem to be more concerned with giving in general the most probable picture of what the site may have looked like in the past instead of giving the most accurate one as regards the measurements.

The detailed documentation guidelines developed by Pletinckx (2008) were not realised for this project. It is planned to remedy this shortcoming as soon as the necessary funding arrives.[200]

Most of the aims for **TimeScope 3** (Section 7.3.5) have been reached. It is an interactive storytelling application, which requires little technical effort, and is therefore easily applicable in small cultural heritage institutions, which do not possess specialised technical staff. Besides being easy to maintain, the system is also easy to use,[201] letting the audience choose the depths of information it requires. The limited attention span of the visitors has been anticipated and a system was developed which can be started and left at any point. The contents can be viewed by groups and individuals, though only one person at a time can operate the touch screen.

Until 2009 the application was not programmable for the use of tour guides. Headphones to access the audio contents exist and work but are not always provided for the public.

7.4.1 Unexpected uses

The model of the interior of St. Lawrence's church helped to discuss the different restoration possibilities with the relevant authorities. Through it, it was possible to anticipate the overall appearance of the church.[202] It also aided the restorers in the decision-making process (Pletinckx et al., 2001, 202).

Additionally the archaeologists had been made aware of the problem concerning the broken line of sight between the altars in the western sanctuary and in the eastern first storey chapel, re the excessive height of the parapet (see Section 7.3.3.3). In this instance the model helped to point at discrepancies

[199] Daniël Pletinckx, personal comment.

[200] Daniël Pletinckx, personal comment.

[201] I had the opportunity to try it out on site.

[202] Daniël Pletinckx, personal comment.

which could later be resolved by the archaeologists, i.e. they found out that the parapet had in fact been lower and a line of sight did exist. This new finding could then be incorporated in the restoration efforts to present the correct height of the parapet. Pletinckx et al. (2001, 201) also mention that 'the VR reconstructions served to clarify the nature of the problems faced by the research team,' without going into greater detail, and also that specific building parts were examined to check hypotheses which had originally been formed while working with the model (Pletinckx et al., 2001, 202).[203]

Also interesting is also the remark by Callebaut (2002, 185) that it could be observed that visitors paid more attention to the printed information panels on site after they had watched the TimeScope 1 programme in the kiosk. It was therefore agreed that offering a mix of presentational media (traditional panels in combination with digital presentations) was an appropriate approach.

7.4.2 Appraisal

This section has to be split into several parts, as it has to deal with different aspects of the project: the reconstruction of the church (here again the model evolutie_kerk.mov will be the basis) and the two applications which partly use this model in their contents: TimeScope 2 and 3.

7.4.2.1 Reconstruction

The model of St. Lawrence's church is not restricted to one time phase only, but offers visualisations for different phases (Section 7.3.3.2). This is a very valuable asset because, as Pletinckx (2008, 13, 26) mentions, the different phases become comparable and differences between them can be evaluated critically. It is also a good method to show the public how the site evolved and that indeed only small changes took place (especially in comparison with Ename abbey, which was remodelled more substantially). The evolution of the building can also be connected to political, economical and environmental changes, to see how they affected it (Pletinckx et al., 2001, 202).

The intended goal was to create one *final* model for each time phase, i.e. the models were updated when new information led to new interpretations of the site. But images of the older model versions are available, e.g. on the website, and thus the progress of research becomes noticeable. This is one of the benefits when creating a model while the research is still ongoing. Changes which occur during the visualisation process usually remain hidden from the later users of the models. To see their evolvement and have some of the reasons for it explained (in this case by *Nieuws uit het Verleden*, i.e. TimeScope 2) is a singular event.

The fact that only one model exists for each phase also means that there are no alternative reconstructions. Pletinckx (2008, 13, 19, 22) considers this to be the ideal situation: the interpretations are discussed by the different specialists involved to arrive at what is believed to be the most probable solution. Only if the specialists are unable to agree on a most probable solution should there be alternative visualisations. Nevertheless there are cases in which alternative reconstructions are of importance: if two hypotheses are almost equally probable or if there are two (or more) competing views.[204]

This leads to the next point: the documentation. As mentioned before the project has been so far not sufficiently documented. Besides being a documentation for the people actively involved in the visualisation process, it would also be helpful for the public in order to get more insight into the usage of the sources and the inferences based upon them. This is partly achieved by *Nieuws uit het Verleden*, which gives insight into interpretation and research, but the connection directly to the model is missing.[205]

More insight into where uncertainties bias the model is also needed. For example, the reconstruction of the church in 1596 is very questionable due to its data basis (see Section 7.3.3.2). Some of the ambiguities in the models can once again be gleaned from *Nieuws uit het Verleden*, but it is necessary to follow the narratives very closely to see where changes to the reconstructions occur, and not all time phases are addressed with equal measure. The approach to make uncertainties visible by colour coding as proposed by Pletinckx (2008, 17) is not available for these models, but could have helped to remedy this shortcoming.[206]

The focus of the reconstructions lies on the outside of the church, not on its interior, which is only shown in

[203] This latter statement may refer to the above-mentioned line of sight.

[204] When following the proposed route by Pletinckx (2008, 13) then at least less probable interpretations will be partly recorded and would therefore not be lost entirely.

[205] Problems of this kind could have been tackled in the TimeScope 1 application, which offered additional information concerning the reconstructions.

[206] At least the text proves that Pletinckx (2008, 17) is aware of the problem and has considered it.

early reconstruction stages and restricted to the Ottonian building phase. It has been pointed out before that there is little evidence concerning the appearance of the interior of St. Lawrence's church, which may have led to the decision to pay more attention to the outside. Even so, there was probably some documentation on what the interior must have looked like in the later phases (e.g. the baroque times), which would have allowed a tentative reconstruction.

The final analysis of the documentation and the excavation report is in the process of being compiled by Dirk Callebaut, so that another update and improvement of the reconstructions may be expected in due course.

7.4.2.2 TimeScope 2

The plan to let the public 'participate' in the work, which is ongoing inside St. Lawrence's church, is very laudable and has helped to create awareness of the importance of the monument (Callebaut, 2002, 185), its history, the work which the archaeologists and the building researchers have carried out and certainly to create a stronger connection between the people and the heritage site.

It was not the work's progress inside the church which was deemed important, but rather *how* the work was done and *why* certain measures were taken. Therefore the focus did not lie on the 3D-visualisation, but on the interpretation of the finds and features and thereby more on the process than on the actual results. In fact the web-pages rely heavily on photographs and text, less on QTVR panoramas and 3D-reconstructions. Some activities were filmed with the hand-held camera to show special activities during the excavation (e.g. taking samples for radiocarbon dating)[207]. The process of revision is not hidden by the authors, but is included in the short narratives concerning some of the church's features: for example, the blind arches in the west wall of the western sanctuary. The first web-page which deals with this part of the church (Callebaut et al., 2000)[208] still shows a reconstruction (last QuickTime VR on the web-page) with four blind arches and presents the interpretation that the twelve niches being together in the western sanctuary may refer to the number of apostles. The hypothesis was revised after it became clear that there was indeed only one larger blind arch in the centre instead of two smaller ones, so that an image with a corrected interpretation was published (Callebaut et al., 2000)[209]. Accordingly the hypothesis concerning the apostles had been called into question. This openness makes it possible for the reader to follow the progress (and some of the dead ends) of research. This line of presentation may be more engaging for the audience than presenting mere (unchallengeable) results. Instead people are invited to think for themselves and arrive at their own judgements.

To arrive at such a presentation method it was essential to have a close collaboration between the people working on site and the people concerned with the creation of the web-pages and the modelling process. Additionally a willingness and openness on both sides to try such a cooperation was necessary and in this case worked out well.

7.4.2.3 TimeScope 3

The design of this digital storytelling application was well thought-out: considering the attention span of the visitors, and also the time constraints of the visit, which often make it impossible to see an entire conventional presentation. By having an interactive system the user can choose the information he or she is interested in and go into more depths if so desired. The application offers several starting points, which make it possible to start with a suitable topic of choice and to proceed from there. The application is easy to handle (and also easy to maintain), so that even people with little experience in technical matters can use the system.

Problems could occur if visitors tried to view the complete contents of the system (which would take several hours), or were anxious to not miss important or interesting facts about the site. In this case it would be helpful to offer a short tour with the most important facts—this should be achievable if the application were truly programmable as was planned.[210]

[207] http://www.ename974.org/Ndl/pagina/NuhV/000307_3v.html.

[208] http://www.ename974.org/Ndl/pagina/NuhV/990905_1.html.

[209] http://www.ename974.org/Ndl/pagina/NuhV/991116_2.html.

[210] The 'Feast of 1000 years' diorama in Ename museum also runs for several hours if all the features are viewed. I was told that this application was not meant to be viewed in only one session, but to offer new aspects to returning visitors (Daniël Pletinckx, personal comment). The same concept could have been followed for the TimeScope 3 application.

Chapter 8

Negotiating Avebury

The third case study is concerned with a partial reconstruction of the Avebury henge and related monuments in the region. The reconstruction has been specially designed with research questions in mind. The project set out with the aim of breaking with the conventional uses of virtual reality models in an archaeological context (i.e. representation) and, instead, of finding new ways of engaging with them. It was hoped that thus more of the research potential this technology possesses might be revealed. In order to break away from the common perception of virtual reality—which is often seen as existing in a theoretical vacuum—the reconstructions effected here are explicitly theorised in a phenomenological framework. Hypothetical Aveburys (alternative models also play an important role) were meant to be explored and then evaluated. If one of the scenarios was found to be unsatisfactory, the model was to be changed. If the whole hypothesis underlying the model was found wanting, the theory could be changed (Fig. 8.1). Avebury was selected as a case study, because it is a very important site, but nevertheless the information so far gathered is ambiguous and will in all probability remain so in some instances. It is necessary to find out more about the site and its use, but in the absence of new surveys and excavations an attempt to reinterpret the existing data with the aid of virtual reality models was made.

This chapter starts with an introduction to the site with its various monuments, commences with the history of research and then proceeds with the 'Negotiating Avebury' project description. In the monument description various references to the antiquarian record can be found. It may be advisable to consult the 'history of research'-section out of turn for additional information in that respect. Then again, ambiguities in the antiquarian record are difficult to understand if one is not acquainted with the monument itself.

The reconstruction focuses on the henge monument[1] and the two established stone avenues, i.e. the West Kennet Avenue and the Beckhampton Avenue, but there are additional contemporary or near contemporary sites which are connected with that complex.

8.1 The Avebury Region with its Monuments

The Avebury region lies in Wiltshire, southern England, on the chalk downland of the Upper Kennet Valley (Gillings et al., 2008, 1).

The late Neolithic sites in this region (Fig. 8.2) include the henge monument at Avebury and two stone avenues. One, the West Kennet Avenue, connects the southern entrance of the henge with the stone and timber circles of the Sanctuary on Overton Hill. Close to the village of West Kennet the West Kennet Avenue passes the West Kennet palisade enclosures and, closer to the henge, Falkner's Circle. The other avenue, the Beckhampton Avenue, connects the western entrance of the henge to the Longstones Cove which supersedes the Longstones Enclosure. To the south of Avebury lies Silbury Hill, an artificial mound. All these monuments are (near) contemporary and form a unique ritual landscape.[2]

[1]As some authors (Gibson, 2004, 71 f., Watson, 2004, 83 f., 89 f.) suggest the term 'henge' is not a very precise one and it has undergone a succession of re-definitions, as the sites falling into this category vary greatly in size (400 m to less than 10 m), number of entrances (1 to 4), components of their earthworks (number and order of ditches and banks) and internal settings (e.g. pits, timbers, stones). Gibson (2004, 72) remarks: 'In short, the term "henge" covers a huge range of monuments whose single consistent feature appears to have been a ditch.' Ironically enough Burl (1969, 3) states in his definition that henges must have a bank, but not necessarily a ditch.

[2]Windmill Hill which is also depicted in Fig. 8.2 is an early Neolithic causewayed enclosure.

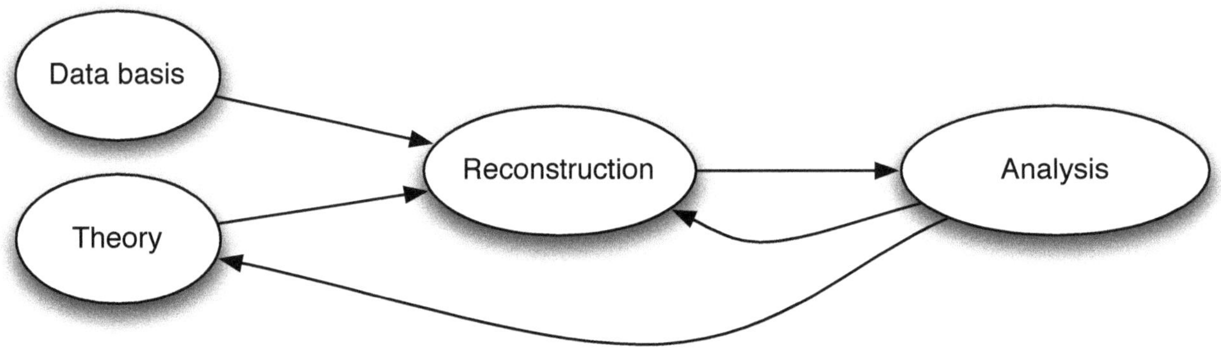

Figure 8.1: Theory Model as demonstrated in the Negotiating Avebury project.

8.1.1 Avebury

The henge at Avebury[3] (fig. 8.3) is split by its four entrances and the modern roads into four unequal sectors or quadrants.[4] Besides the roads, there are buildings, garden plots and meadows separated by fences and hedges all to be found inside the monument. Their presence significantly alters the way in which the monument can be experienced today. Already in antiquarian times they presented obstacles for survey (see Section 8.1.8 and Fig. 8.17) and continue to do so. Nevertheless modern surveys, especially resistivity and magnetometer[5] surveys, but also aerial photography of parchmarks during periods of drought, have provided a number of interesting results (Bewley et al., 1996; Ucko et al., 1991). They, in combination with the various excavations, have given a better appraisal of the site, though it still remains largely unexplored.

8.1.1.1 The bank

The henge at Avebury possesses a bank which has a diameter of about 427.7 m from crest to crest and a circumference at the crest of about 1353.3 m (Gray, 1935, 110).[6] It is roughly circular in shape (see Figs. 8.3 and 8.4) and had at the time of Harold St George Gray's survey and excavations (1908–1922) a height of *c.* 4.3–5.5 m and a width of 22.9–30.5 m (Smith, 1965, 194). The shape of the bank's crest is uneven; this could be a deliberate feature, maybe mimicking the undulating horizon of the landscape (Gillings and Pollard, 1999, 185), or due to silting effects, whereby finer material from the top of the bank was washed down and settled between larger blocks of chalk in the inner part of the bank (Smith, 1965, 194).

The four entrances to the henge are original features. The eastern entrance is the only one which has not yet been archaeologically examined, but, according to Smith (1965, 194), only the northern bank end at this entrance is still in its original state. It demonstrates that the bank ends were probably raised above the usual height of the bank—an assumption which is corroborated by the fact that the only fully excavated ditch end is broader and deeper in comparison to the ditch encountered in other places. On the other hand the geophysical survey undertaken by Ucko et al. (1991, 184 ff., plate 58) demonstrated an anomaly outside the henge which could be interpreted as an old trackway. It abuts on the bank north of the modern causeway. If this trackway could be confirmed, the assumption that the northern end of the bank is original is thrown into serious doubt.

Mutilations of the banks took place at various times: a great part of the bank north of the western entrance was levelled at some time between 1689 and 1695 (Ucko et al., 1991, 170 f.). The southern bank end at the same entrance was levelled at some point between 1721 and 1724, as witnessed by alterations in Stukeley's plans of the monument during this time (Ucko et al., 1991, 172). The western end of the south entrance was also afflicted in 1724, to let coaches pass more easily, and once more in 1762, to raise the road

[3] The henge is commonly referred to as Avebury, but this is also the name of the village which is situated east of and inside the henge. Pollard and Reynolds (2002, 202 ff.) have argued convincingly that the name Avebury originally referred to the settlement, while the variations on the name 'Waledich', which occur in historic documents, actually refer to the henge. In the present text the name Avebury refers to the henge.

[4] The entrances are not aligned to the true cardinal points of the compass, but will be referred to according to them for convenience' sake.

[5] The latter technique has been proved to obtain fewer results than the resistivity surveys (Ucko et al., 1991, 160).

[6] Dimensions given in Imperial units in the original texts were converted to the metric system.

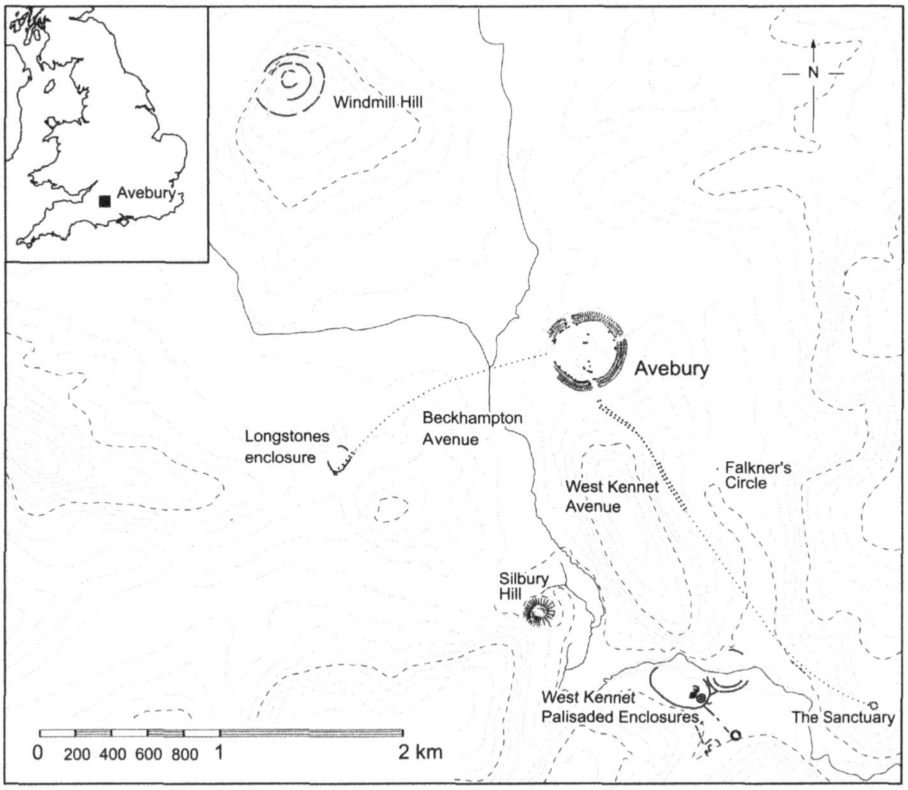

Figure 8.2: Monuments in the Avebury region, after Gillings et al. (2008, 1, fig. 1.1).

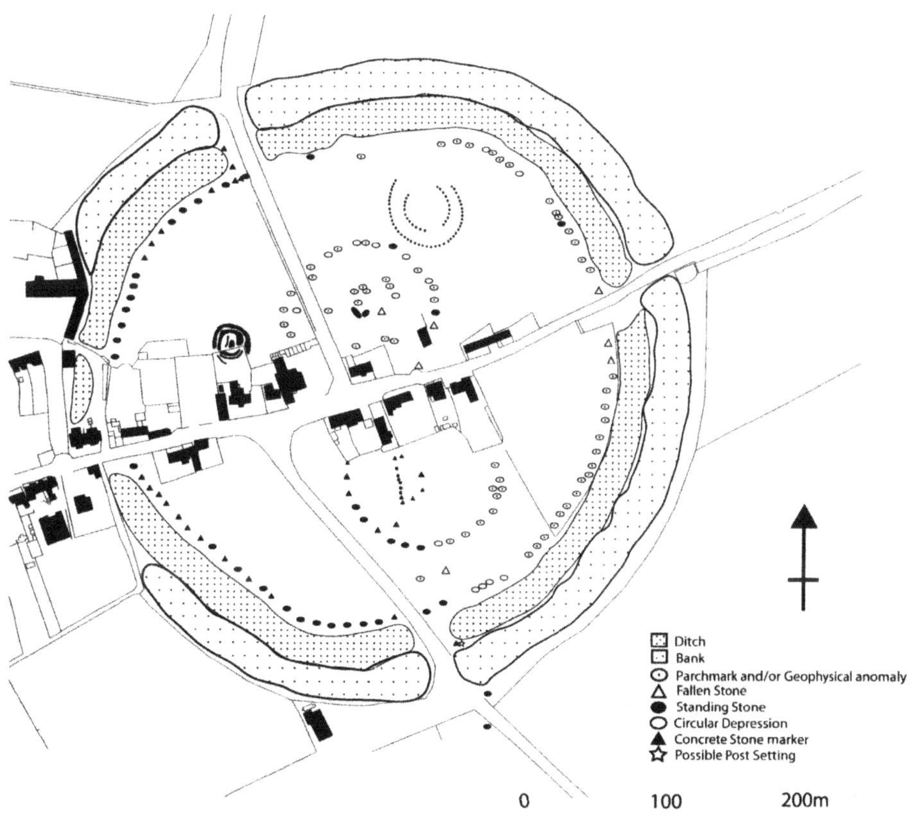

Figure 8.3: The Avebury henge, combined survey and excavation results (Gillings and Pollard, 2004, 5, fig. 1).

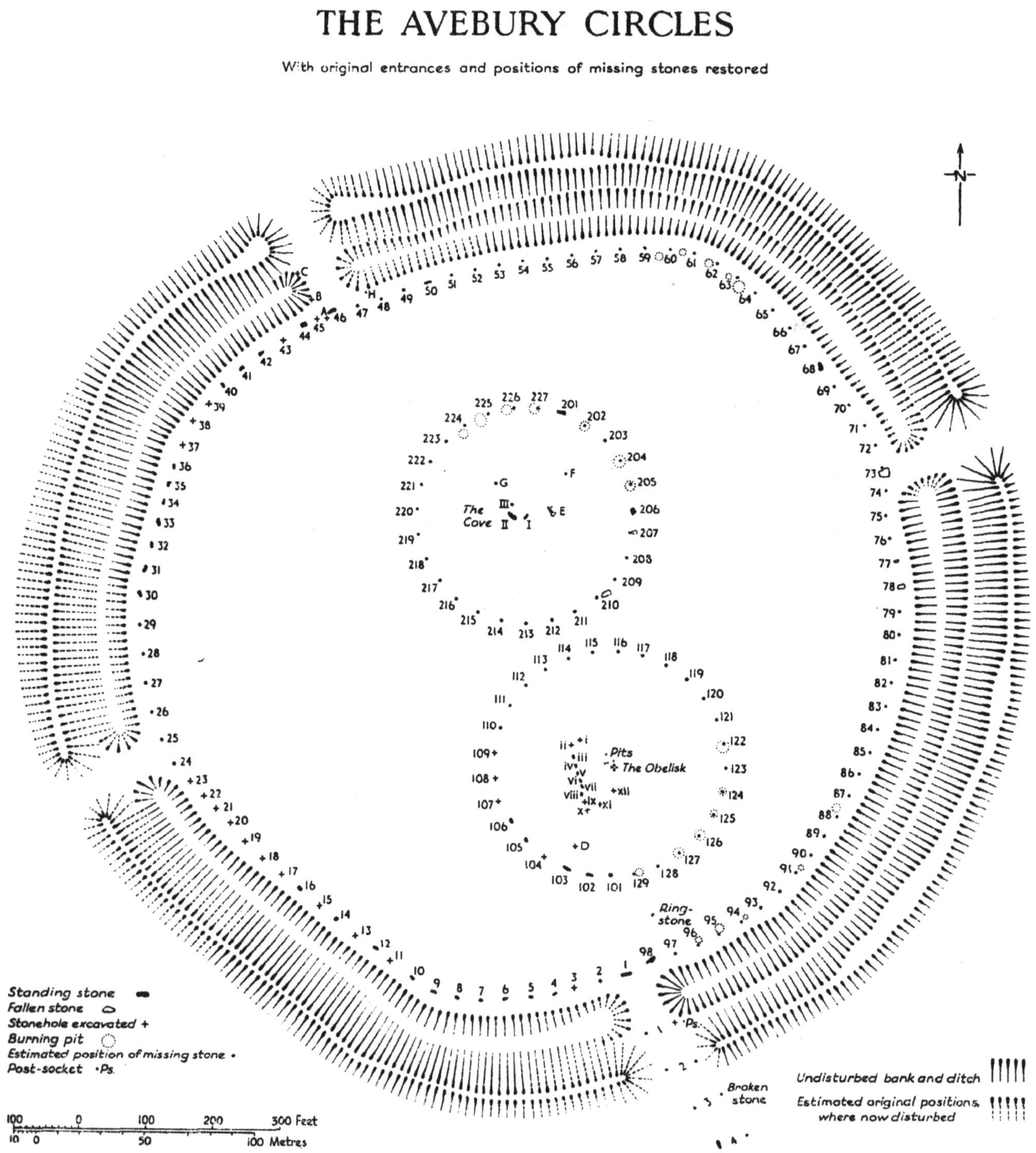

Figure 8.4: The Avebury Circles, Isobel Smith's 'tentative reconstruction' (Smith, 1965, 205, fig. 70). (By permission of Oxford University Press).

(Burl, 1979, 53).[7] Smith (1965, 182) reasons that from the survey by Philip Crocker for Sir Richard Colt Hoare it can be seen that the western bank at the south entrance was extended, probably at the beginning of the 19th century. She estimates it to end 11.3 m too far to the east.

In several trenches dug into or through the bank an 'inner rampart' or 'primary bank' was observed: e.g. in the east-south-eastern part of the south-east sector excavated by E. C. Treplin and Thomas Leslie for Sir Henry Meux (Gray, 1935, 104). Harold St George Gray himself claims not to have seen this inner bank in his excavation (Gray, 1935, 104), but Pitts and Whittle (1992, 206) on the other hand point out that on Gray's plate 42, fig. 2 a dark streak is visible which could be the soil on top of the primary bank. Gray's Cutting X lies in the southern part of the south-east sector. That the primary bank is not just a feature of this part of the monument is demonstrated by the 1969 excavation by Faith de M. Vatcher in the south-west sector, close to the western entrance (Evans, 1972, 268 and 207, fig. 95). The earthwork had been previously disturbed in this part (see above), but the primary bank was clearly visible. It had a height of *c.* 2.5 m and a width of 3 m with a core of piled turf on its inside and chalk and marl debris on top (Pitts and Whittle, 1992, 206).[8] The soil formation on top of the primary bank indicates that there was a greater gap between its construction and the construction of the later bank. It can be assumed that the primary bank was similar in circumference to the later one (Pitts and Whittle, 1992, 210).

The later bank has shown traces of vertical walling with chalk blocks in the south-west sector and some timber revetments at the western end of the northern causeway (Smith, 1965, 194 f.; Keiller, 1939, 228, 232).

8.1.1.2 The ditch

The material for the main bank derives from a subcircular ditch on the inside of the monument. It is 'dug directly into the Lower Chalk' in the north-west sector, while 'elsewhere it penetrates through the Middle Chalk' (Smith, 1965, 194). This statement will become important in the discussion concerning the phasing of the monument. The ditch is nowadays about 21.3 m wide and its bottom lies about 3.7 m below the modern surface level (Smith, 1965, 193). Gray (1935, 134) made several cuttings into the ditch and gives its dimensions as follows: its average width on the top is 9.1 m and at the flat bottom 4.6 m. Its depth varies between 5 m to 9.2 m. From this it can already be gleaned that the walls of the ditch were very steep and therefore, as Ashbee (2004) argues, prone to weathering.[9] The ditch appears to have been dug in separate sections, the remaining chalk walls between the sections being torn down at a later point (Gillings and Pollard, 2004, 9)[10].

Gray (1935) discovered that there was a causeway across the ditch at the southern entrance. His Cutting III, to the west of the modern road, did not uncover the end of the ditch, i.e. it must lie underneath the road. Cutting IX, to the east of the modern road, encompassed the ditch terminal. Right next to the causeway the ditch is especially deep (9.2 m from the brink of the ditch, 10.9 m below the level of the causeway) and also carefully squared off. In the direction away from the causeway the bottom of the ditch was found to rise considerably: 2.3 m over 7.7 m, and its bottom width diminished from 4.3 m to 2.4 m (Gray, 1935, 125 and 126, fig. 3). Further cuttings (between Cutting IX and the road) were used to examine the causeway between the two ditch ends.

During the course of the excavations in 1937, Keiller (1939, 226 f.) could establish that the northern entrance to the henge is also an original feature. A causeway was found, though only its western side, the eastern lying underneath the modern road. Similarly the western causeway was discovered in 1938 (Keiller, 1939, 232 f.). Smith (1965, 195) estimates the width of the causeways as follows: southern: about 15.2 m, western: more than 12.2 m, northern: less than 19.8 m. The width of the eastern causeway is unknown because no excavation data is available. Smith also mentions that the road level between the banks of the northern entrance is 1.4 m lower than the surface on which the banks stand, similarly the road level at the eastern entrance is 1.5 m deeper than the

[7] See also Smith (1965, 182) for the 1762 incident.

[8] The question as to where the material of this primary bank came from is discussed in (Pitts and Whittle, 1992, 210). Basically its exact source is unknown, but could possibly be revealed by further excavation.

[9] Material which collapsed into the ditch during excavation is also reported by Gray (1935, 111), though the material which caved in was the silting of the unexcavated part of the ditch. Nevertheless this could underline the assumption that the ditch had to be built in a relatively short time in order to be finished before silting could set in (Ashbee, 2004, 8).

[10] The inference that the ditch was cut in sections is made by Gillings and Pollard (2004). The referenced page in Gray (1935, 116) only states the facts, namely that the bottom of the ditch was irregular and that a small ridge of chalk crossed its bottom (Gray, 1935, plate 37). Also, the walls were very irregular 'considerable projections of hard chalk [were] left apparently for no particular purpose'.

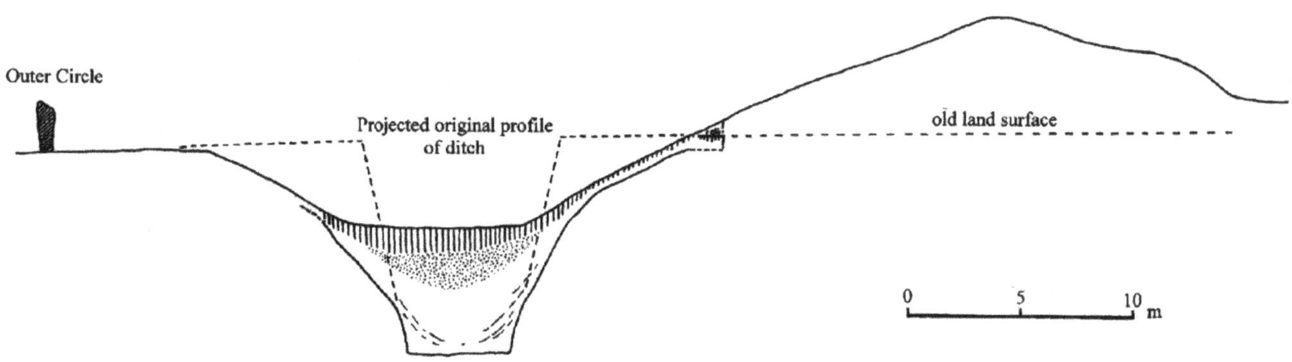

Figure 8.5: Ditch and bank and outer circle at Avebury (Pollard and Reynolds, 2002, 84, fig. 29), after Gray (1935, plate 40).

original surface. This scraping and the presumably especially high banks and deep ditches close to the causeways could have had the purpose of making the entrances appear more 'dramatic' (Smith, 1965, 194 f.).

Smith (1965, 194) also mentions a narrow berm between the bank and the ditch in the north-east sector as well as in the south-east sector, where it is reported to be 3.7 m wide. No traces of a berm could be found in the north-west and south-west sectors. That a berm must have existed between bank and ditch, as well as some level ground between the outer stone circle and the ditch, is clear from the difference between the initial average ditch width of 9.1 m and the modern ditch width of 21.3 m measured by Smith (see also Fig. 8.5).[11] This also means that the stone settings had once stood much further away from the brink of the ditch than they do today.[12]

8.1.1.3 The stone settings

Inside the monument the remains of three stone 'circles' and a number of additional stone settings can be found (fig. 8.4 and 8.3). Following the outline of the earthwork is the outer circle. Inside on a ridge lie two smaller circles, the northern and the southern circle[13]. They incorporate additional stone settings. Further stones or stone holes 'belong' to neither the southern, northern nor outer circle.

The sarsen boulders, which were used for the stone settings, are a naturally occurring material in this region, i.e. they are very hard sandstones, which were deposited during periglacial conditions (Gillings and Pollard, 2004, 32). Their form can be boulder shaped or tabular (Gillings et al., 2008, 58), corresponding to Keiller and Piggott's 'A' and 'B' type shapes respectively (Keiller and Piggott, 1936, 420). The stones seem to have been carefully selected (Gillings and Pollard, 1999, 184): some show evidence of working, like percussion or polishing marks[14] (Smith, 1965, 223), which were produced before the stones were erected. Some of them display natural cup, or cup-and-ring marks[15], hollows, folds or perforations (Smith, 1965, 223). They also vary in their surface texture, colouring, size and shapes (Pollard and Gillings, 1998, 154–158).

The average stone height in the circles lies somewhere between 3.7 m and 4 m.[16] Smith (1965, 196) observed that the stones closest to the entrances (1, 98, 46, 47) were bulkier than the stones further away. Stone 1 rises 4.2 m, stone 98 4.3 m and stone 46 4.5 m above the ground. The smallest still-standing stone, 33, has a visible height of 2.1 m. The stones of the

[11] Ashbee (2004, 7) suggests that the builders of the earthwork were well aware of what they were doing, namely building a narrow deep ditch to let it silt rapidly. In contrast, the ditch to the south of Silbury Hill (see section 8.1.6.1), was stabilised by stepped timber revetments filled with chalk rubble where it bordered against the side of the hill (Atkinson, 1970). This was presumably done to prevent landslides which could have affected the hill, too.

[12] This has not been much considered in recent arguments about the site. Most of the stones were set up from the inside of the monument, but if sufficient ground had (still) existed between ditch and stones they could have easily been set up from the other side. Could this be an indication for the phasing of the monument, that the silting process was already advanced before the stones of the outer circle were set up?

[13] To avoid confusion the term 'inner' circles will be avoided for the southern and northern circle. The confusion could occur because William Stukeley (see Section 8.1.8.1) believed that the three circles may have had inner circles. This has been proven unlikely for the outer and the southern circle, but is still being discussed for the northern circle.

[14] There is also evidence for partial removal of polishing marks, see (Gillings et al., 2008, 79 fig. 2.55).

[15] For example, see Keiller and Piggott (1936, 420).

[16] As far as this can be calculated from the table of approximate stone heights given in Smith (1965, 197).

West Kennet Avenue are smaller in size (see Section 8.1.2).

The stone holes were prepared in different ways, their depths stand in relation to the size of the stone they were meant to hold, so that the stone holes of the West Kennet Avenue were between less than 0.2 m and 0.6 m deep, while the stone holes of the Avebury circles were between 0.3 m and 0.9 m deep (Smith, 1965, 218). Smith (1965, 219) points out that a certain amount of weathering should be taken into account so that the stone holes may have been initially about 0.5 m deeper than today. The holes were prepared to receive a particular stone. Three sides of the stone hole were vertical, but the one from which the stone was to be erected was sloped. The slope could also be constructed from chalk or sarsen blocks. Sometimes 'anti-friction stakes' (their number ranging from 2 to 15) were introduced on the opposite side of the stone hole to facilitate the raising of the stone; other stakes were used to guide the stone into its correct position in the stone hole. The stakes were later removed (Smith, 1965, 219).[17]

To secure the stones in their stone holes the latter were filled with clay and sarsen boulders, rarely flint nodules or blocks of Middle or Lower Chalk being used. The clay (found in the henge and stone settings 75–78b of the West Kennet Avenue) was dark brown and thought to come from the banks of the Winterbourne. Chocolate brown clay loam was also discovered in the fills of stone holes of the Beckhampton Avenue at Longstones Field and Avebury Trusloe (Gillings et al., 2008, 75 ff., 107). Smith (1965, 221) suggests that the clay might have been used an an indicator of whether the stone had been set properly and would stand upright. Another interpretation has been offered by Gillings and Pollard (1999, 159 f.), who propose the idea that it and other 'foreign materials', like the chalk, could be part of a symbolic reference to other parts of the surrounding world and to their experienced meanings. Watson (2001, 301) suggests that the colour contrast between the dark river clay and the light coloured clay found on site may have played a role. Smith (1965, 221) believes that the Middle and Lower Chalk packing found in the South-east sector and one stone in the northern part of the West Kennet Avenue[18] must come from the ditch. Pitts and Whittle (1992, 210) point out the possibility that some of the material found at the southern circle may be reused as well, e.g. from a barrow, these were also sometimes dug into the Lower Chalk. Chalk rubble packing was observed at the Beckhampton Avenue in Longstones Field (Gillings et al., 2008, 75 ff.).

The majority of the stones rested on the bottom of the stone holes; the depressions left by them there were later used by Keiller as indicators of how the fallen or buried stones had to be reset, i.e. in his restoration attempt (Smith, 1965, 219). Gillings et al. (2008, 74, 75) state that judging the position and identifying the base of a fallen stone from the depression in the stone hole was not without ambiguity. Other stones received a foundation because they had irregular bases or were placed in a unsuitable place.[19] When set, the stone faces of the sarsen were aligned with their neighbours (Smith, 1965, 222).[20] It was also observed that stones with a twisted base were set in a way that still ensured the alignment of the upper face with the neighbouring stones (Smith, 1965, 222).

So far no datable finds from the stone holes have been recovered, indeed the complete lack of organic or other material from the settings is quite remarkable in itself (see Gillings et al. (2008, 119)). Only stone D can undoubtedly be attributed to the early Bronze Age, due to weathered beaker sherds found in a 'stake hole near the edge of Stone-hole D' (Smith, 1965, 227).[21]

The outer circle

The outer circle has a diameter of *c.* 337.7 m (Gray, 1935, 107). Many of its stones have been destroyed or buried from medieval times to the 18th century, so that the exact number of stones can only be guessed at. Smith, based on the excavations by Alexander Keiller and assuming a regular spacing of stones, estimated that there were 98 stones in the outer circle (Fig. 8.4). Her numbering scheme for these stones starts at the southern entrance to the henge and proceeds clockwise.

Keiller excavated in 1937 and 1938 (Keiller, 1939; Smith, 1965) the outer circle in the north-west and south-west quadrants. In the **north-west quadrant** (Table 8.1) four stones were still standing (32, 33, 44, 46), six had been incorporated in a wall and were broken (34, 35, 36, 40, 41, 42), two stones were buried (30,

[17] The same has been found to be true for the post holes and 'anti-friction stake' holes at the Beckhampton Avenue, see Gillings et al. (2008, 76 ff.).

[18] Stones ii–iv and xi–xii, the Ring-stone and stone 9b of the West Kennet Avenue (Smith, 1965, 247 f.).

[19] The special treatment of stone 41 and the Ring-stone is discussed below.

[20] Unaligned faces were attributed to accidental slipping (Smith, 1965, 222).

[21] For more on the problems in dating the monument complex, see Section 8.1.7.

Fate of stone	Stone number
standing	32, 33, 44, 46
fallen	34, 35, 36, 37, 40
buried	30, 31
destroyed by direct fracture	41, 42
destroyed by fire	37 (previously fallen), 38, 39, 43, 45

Table 8.1: Fate of the stones in the north-west quadrant, based on Gillings et al. (2008), supplemented with information from Smith (1965).

31), and five stones were destroyed (37, 38, 39, 43, 45).[22] Besides these 17 stones, more probably existed to the south, close to the western entrance. This area is now occupied by buildings and gardens (Fig. 8.3) and therefore the exact number of stones in this quadrant remains unknown.

In the **south-west quadrant** (Table 8.2) 22 stones or stone holes could be identified through excavation. Smith (1965, 190) mentions six buried stones (4, 5, 6, 9, 10, 16), two partly buried stones (7, 12), five buried, but later burnt stones (19, 20, 21, 22, 23) and six were destroyed by fire (3, 11, 13, 15, 17, 18). One stone had fallen (14), one was found broken in a foundation (24) and one remained standing (8).[23] In this case the number of original stone settings could be established more firmly, though some may have fallen victim to the modern roads. The stones in the south-west sector are on average 0.5 m higher than the ones in the north-west sector (Pollard and Gillings, 1998, 156).

In the **south-east quadrant** the situation looks very different. Keiller started his excavations in 1939 at the southern circle and it was planned to investigate this feature completely and then proceed with the outer circle in this quadrant. The enterprise came to a halt with the beginning of the Second World War, so that only a part of the southern circle and the position of the Ring-stone were finally excavated (Smith, 1965, 190). Information concerning the stones of the outer circle in this quadrant comes from surveys.

Two stones opposite the southern entrance are still standing (1, 98) (Figs. 8.4 and 8.3), two stones (77, 78) have fallen (Smith, 1965, 197). In 1881 under the direction of A. C. Smith and W. C. Lukis workmen were probing the ground with iron bars. Thereby 18 buried stones were discovered, 16 of them in the outer circle, the others in the northern circle. These stones were uncovered and later covered again, with wooden pegs marking their position (AAHRG, 2001, 11). Gray (1935, 103) notes that these pegs were probably lost by his time. According to Smith (1965, 184) only 15 stones in the outer circle were discovered, which coincides with the number of stones in the re-drawn plan[24] published by Gillings et al. (2008, 281 fig. 9.17). The latter also shows that probably ten of the buried stones were discovered in the south-east quadrant, while the remaining five were found in the north-east quadrant.[25] Depressions, marking possible stone holes or destruction pits,[26] were noted and mapped by several researchers (e.g. William Stukeley, A. C. Smith and W. C. Lukis, Harold St George Gray or Isobel Smith). Smith (1965), for example, marks 5 depressions in her 'tentative reconstruction' (Fig. 8.4), close to the suggested stone positions 88, 91, 94, 95 and 96. Nevertheless, the number and the position of depressions vary respectively according to the recorder. Resistivity surveys in part of the south-east quadrant were undertaken by Ucko et al. (1991, 169 fig. 58; 219 fig. 66) in 1989. Resistivity anomalies in the places of Smith's stone positions 88, 89, 90, 91, 92 and 95 were recorded.[27] The drought conditions in 1995 allowed for the documentation of parchmarks; Bewley et al. (1996, 640) mention a total of 17 parchmarks in the outer circle for the south-east and the north-east sectors. From the accompanying plan (642 f. fig. 2) it seems as if six parchmarks were recorded in the south-east sector. Additionally, the plan records parchmarks which were visible in 1990, adding three more to the count. The results of a

[22]For a plan showing the phasing of stone destruction in the north-west quadrant see Gillings et al. (2008, 334, fig. 10.24). The same authors also mention the different ways in which the stones were destroyed (Gillings et al., 2008, 296 f.).

[23]Gillings et al. (2008, 257 f., 280, fig. 9.16, 297 f.) mention how each stone was destroyed or buried.

[24]The original plan by Lukis and Smith (1885, plate V) was unavailable to me.

[25]The original plan by Smith and Lukis is unfortunately rather inaccurate in its depiction of the stone positions. See Gillings et al. (2008, 280, 290 footnote 13).

[26]Burial pits had been usually refilled carefully.

[27]The comment on p. 212 (Ucko et al., 1991), that stone position 97 (instead of stone position 95) showed on the resistivity survey must be a typing mistake. It is contradicted by fig. 66, p. 219.

Fate of stone	Stone number
standing	8
fallen	14
buried	4, 5, 6, 7, 9, 10, 12, 16, 19, 20, 21, 22, 24
destroyed by direct fracture	17, 18
destroyed by fire	3, 11, 13 (previously fallen), 15, 17, 18, 19 (previously buried), 20 (previously buried), 21 (previously buried), 22 (previously buried)
destroyed	23

Table 8.2: Fate of the stones in the south-west quadrant, based on Gillings et al. (2008), supplemented with information from Smith (1965).

geophysical survey undertaken by the National Trust in 2003 appear to be unpublished,[28] Gillings et al. (2008, 290, footnote 1) only remark on this subject that the 'buried stones [were] in exactly the positions recorded by Smith and Lukis'.[29] The same authors show in fig. 9.19 (Gillings et al., 2008, 283) 11 buried stones in the south-east quadrant and a further six in the north-east sector.[30]

The **north-east quadrant** of the outer circle has also not been excavated. Smith (1965) shows two standing stones (50, 68), a fallen one (73) and five depressions (close to the estimated stone positions 60, 61, 62 and 63) in her 'tentative reconstruction' (Fig. 8.4). A. C. Smith and W. C. Lukis mark 11 positions (Gillings et al., 2008, 281, fig. 9.17), Gray (1935, plate 29) records five, but not all in the same positions as I. Smith's.[31] Smith and Lukis mark five buried stones on their plan (Gillings et al., 2008, 281, fig. 9.17). Evidence for 10 possible stone settings comes from parchmarks in the north-east quadrant, which were visible during the drought in 1995 (Bewley et al., 1996, 642 f. fig. 2). The resistivity survey by Ucko et al. (1991, 209, 223 plate 67), which encompassed only the western part of the quadrant, i.e. Smith's positions 47–58, achieved no results due to a field boundary. Gillings et al. (2008, 283, fig. 9.19) show six buried stones in the eastern part of the quadrant.[32]

The northern circle

The outer circle encloses two smaller circles. The northern one has so far not been excavated to any great extent, only small scale excavations at individual stone settings having been carried out. Several surveys provide additional information. The northern circle has a minimal distance of about 75.9 m from the northern part of the outer circle (Gray, 1935, 110). Gray (1935, 108) gives its approximate diameter as 97.5 m, drawing a circle through stone 201, 206, 207 and slightly north of 210. The centre of this circle would lie within the Cove (see below). Later surveys suggest that the 'circle' may have a more oval shape than previously anticipated (compare Figs. 8.4 and 8.3). Resistivity anomalies for Smith's suggested stone positions 219, 220 and 221 lie around 8 m too far to the west (Ucko et al., 1991, 226 f., 223 plate 67). In the case of a 'circular' design still being envisaged, the centre of the circle would have to shift away from the Cove.

Smith (1965, 201) records two standing stones for the northern circle (201, 206) and two fallen ones (207, 210). The stones are similar in size to the ones of the outer circle. With an average distance of 11 m between the stones, Smith arrives at 27 stones in the northern circle (Fig. 8.4). But the situation is more complicated if additional survey data is taken into account. Ucko et al. (1991) make use of two resistivity surveys, i.e. in addition to their own results (223 fig. 67) they also refer to a previously unpublished survey from 1975 (224, 225 fig. 68). Their own survey extends into the north-west quadrant, where three anomalies (maybe 219, 220 and 221) are detectable. Further to the south the readings were too disturbed to give any clear readings. Similarly the resistivity survey of the northern circle in the north-eastern quadrant gave no results where former garden plots and building sites were encountered. Anomalies

[28] News items on the internet, e.g. http://news.bbc.co.uk/2/hi/uk_news/england/wiltshire/3257174.stm (Accessed: 6th of August 2012.), mention that 'at least 15 stones' were detected.

[29] This might be a slight overstatement regarding the imprecision of Smith's and Lukis' plan (see above).

[30] Is this based on the geophysical survey results?

[31] This proves yet again that it depends on the researcher which features were recorded.

[32] Based on the geophysical survey by the National Trust in 2003?

Fate of stone	Stone number
standing	101, 103
fallen	102, 104, 105, 106, 107
buried	i, ii, iii, iv, v, vi, vii, viii, ix, x, xi
destroyed by direct fracture	xii
destroyed by fire	104, 107, 108, 109, i (previously buried), ii (previously buried), ix (previously buried), x (previously buried), xi (previously buried), Obelisk, D

Table 8.3: Fate of the stones in the south-east quadrant, based on Gillings et al. (2008), supplemented with information from Smith (1965).

were detected for stone positions 222, 224, 225, 203[33], 204 and 205. Close to stone position 207 two anomalies were recorded but none for positions 223 and 202[34], suggesting that the stone intervals were not as evenly spaced as assumed by Smith (1965). The parchmarks recorded by Bewley et al. (1996, 640, 643 fig. 2) draw a similar picture. They recorded four parchmarks in the north-west sector. Three positions coincide with resistivity anomalies, and the southernmost (approximate stone position 218) lies in an area which provided no usable resistivity responses. In the north-east quadrant parchmarks of a further 11 stones were noted. They are roughly in the position of stones 222 (and another parchmark slightly south of it), 224, 225 and three parchmarks in the area of stone positions 226 and 227. 202, 203 and 205 are each marked by a parchmark, and near position 204 there are two. This fits the resistivity results rather well.

The northern circle encompasses several further settings. **The Cove** is a stone setting, which originally consisted of at least three stones (I, II, III) set perpendicular to each other, thus forming a box-like structure which is open to the east. Aubrey and Charleton were the only ones who still witnessed more than two stones *in situ*. There is some speculation about the existence of a fourth or possibly a fifth stone belonging to the original arrangement (see Ucko et al. (1991, 227) and Gillings et al. (2008, 165)). Today only Cove stone I and II remain (Fig. 8.4). Various antiquarian excavations have disturbed the ground around the Cove stones (see Section 8.1.8), as have outbuildings and 19th and 20th century rubbish dumps (Gillings et al., 2008, 160–163). Both stones are very large and stand today 4.9 m (I) and 4.4 m (II) high (a rise in ground level due to the accumulation of refuse has to be taken into account). Stone II is the biggest stone at Avebury (Gillings et al., 2008, 166) with an estimated height of 7 m and a width and depth of 4.5 m and 1.5 m. Stone II is assumed to be set deeper than 2.5 m (the excavation could not proceed beyond a depth of 2.4 m (Gillings et al., 2008, 156)). Stone I is 2.4 m wide and it is reported that stone III was similar to it in shape (Stukeley, 1743, 24).

Three small stones or a broken large stone are referred to as **stone E**. The southernmost of these stones is visible, while the others lie within the same pit (Ucko et al., 1991, 224). Gray (1935, 108 footnote 3) who had the three stones excavated, believed that they are parts of one stone. Smith (1965, 202) on the contrary thinks that they might be individual stones. Judging from their length (1.8–2.3 m (Gray, 1935, 108 footnote 3)) they are not dissimilar to the small stones i–xii discovered in the southern circle (1.3–2.4 m) some of which were also buried together in one pit (v–vii) (Smith, 1965, 201 f.).

Stone F was still standing by 1825, when it was depicted in a water-colour painting, but it is reported to be 'broken up by means of gunpowder' shortly after (Smith, 1965, 201 f.).

Stone G is mentioned by A. C. Smith as a buried stone (Smith, 1965, 201 f.); its (former) position could be confirmed by the resistivity survey (Ucko et al., 1991, 224), and it is possible that the still standing stone is depicted by Stukeley (1743) in his tables xiii and xiv. Ucko et al. (1991, 224) also argue that at least one, if not both, of the anomalies east of the stone G and III could be matched with a stone depicted in Stukeley's table xiv.

Two further possible stone positions are mentioned by Ucko et al. (1991, 226): stone S and '1964'. The approximate positions of both are marked on transparency 24 (Ucko et al., 1991). **Stone S** is depicted on Fig. 8.17, Stukeley's frontispiece (Stukeley, 1743), and

[33]No depression was marked in this place, but the resistivity anomaly was rather strong leading Ucko et al. (1991, 226) to suggest that this might be a buried stone.

[34]There is a marked depression at this site, so a former stone setting is not completely out of question.

also mentioned in his written tour[35] (Ucko et al., 1991, 279). The stone lies outside the area of the resistivity survey, but one of the parchmarks south of stone(s) E recorded by Bewley et al. (1996, 643 fig. 2) may fit the location (see also Fig. 8.3). The approximate location of the burning-pit discovered in 1964 during maintenance work (Smith, 1965, 223) was marked as **1964** on transparency 24 (Ucko et al., 1991). This position may also be matched with a parchmark south of the Cove (Bewley et al., 1996, 643 fig. 2) (Fig. 8.3).

The southern circle

The southern circle was partly excavated by Keiller in 1939. Gray (1935, 109 f.) gives its diameter as 102.7 m with a minimal distance of 40.8 m from the outer circle.

Stone positions 101–109 of the circle were excavated. From their average interval of 11 m Smith (1965, 198) calculated that there must have been originally around 29 stones in the circle (Fig. 8.4, Table 8.3). The stones of the inner circles are higher than the 'non-entrance stones' of the outer circle (Pollard and Gillings, 1998, 156 and fig. 6). Two stones are still standing (101, 103), three stones were toppled or had fallen (102, 105 and 106) and Smith assumes in a footnote (Smith, 1965, 191 footnote 1) that maybe two further stones (107, 108) had been overthrown and left lying on the ground before they were destroyed in burning pits (Smith, 1965, 191). Four stones (104, 107, 108, 109) were broken up by fire. The results of the resistivity survey by (Ucko et al., 1991, 220, 219 plate 66, transparency 21) cover Smith's stone positions 101–109 and 121–129. Modern habitation and gardens make the remainder of the area unsuitable for geophysical survey. In general the ground around the southern circle is more disturbed than that of the northern circle making the evaluation of the resistivity readings difficult. Anomalies close to the excavated stones 101–108 are observable; of more interest are the anomalies in the unexcavated part of the circle. They can be found at the stone positions 122, 123, 125, 126, 127 and 129. One anomaly lies between Smith's stone position 121 and 122, while at position 124 there is no discernible reading but a depression in the ground. Bewley et al. (1996, 640, 642 fig. 2) count ten parchmarks in the unexcavated part of the southern circle. Comparing their fig. 2 with Smith's plan (Fig. 8.4) the parchmarks appear to be close to stone positions 122, 123, 124, 125, 126, 127, 128, 129 and two additional ones lie between positions 123 and 124 (see Fig. 8.3).[36]

Like the northern circle, the southern circle has internal settings. Almost at the centre of the circle once stood the **Obelisk** (Smith, 1965, 198). The stone had fallen by Stukeley's time, but he reports it to be 6.4 m long and to have a diameter of 2.7 m (Stukeley, 1743, 24, table xvi) and thus to stand higher than the other stones of the circle. These dimensions are supported by the size of the Obelisk's burning pit, which was 2.1 m wide and 6.4 m long (Smith, 1965, 198).

Stone D stood to the north of stone 103. During the excavation its burning pit was discovered (Smith, 1965, 198). Neither the Obelisk nor stone D are visible on the resistivity survey (Ucko et al., 1991, 220, 219 plate 66) or as parchmarks (Bewley et al., 1996, 642 fig. 2).

The same accounts for the **stones i-xii**, which were discovered during excavation. Stones ii-x form a close-set linear setting to the west of the Obelisk. Stone i lies to the north-east of stone ii, while north-east of stone x lay stone xi and further away in the same direction is the stone hole of xii. Almost all of these stones, with the exception of stone xii, had been buried, some (i, ii, ix, x, xi) were later discovered and broken by fire. It is assumed that stones ix–xi were all burnt in the same burning pit (the one initially made for stone x) (Smith, 1965, 199; Gillings et al., 2008, 299). Stones v–vii were buried in one pit. Smith (1965, 199, 200 fig. 69) reports that at first a burial pit for stone v had been prepared, but in the end it was placed in one pit together with stones vi and vii, lying between the latter ones. Stone xii was broken by direct fracture (Gillings et al., 2008, 299). The stones are rather small (1.3–2.4 m (Smith, 1965, 201)) and have a reddish tinge (Gillings and Pollard, 2004, 12).[37] The excavation also uncovered two stake-holes which could have been used to lay out the alignment of stones ii–x (Smith, 1965, 199).

North of the Obelisk four **pits (A–D)** were discovered, they were filled with dark brown soil devoid of any finds or chalk (Smith, 1965, 201). Their purpose is unknown.

[35] A reconstruction of the previously unpublished and fragmentary 'written tour' has been published in Ucko et al. (1991, Appendix II).

[36] Gillings et al. (2008, 299 fig. 10.5) present a graphical comparison between parchmarks, resistivity results and depressions in that region.

[37] As before, Gillings et al. (2008, 258 f., 297–300) give information on how each stone of the southern circle and its internal settings was destroyed or buried.

Individual stone settings

Besides the outer, the southern and northern circles (both with their internal settings) some additional stone settings have been noticed. The **Ring-stone** (see Fig. 8.4) stood almost equidistant between the southern circle and the outer circle, close to the southern entrance to the henge. Stukeley (1743, 25) described it as a small stone with a hole in it[38]. The Ring-stone's stone hole was excavated in 1939 and proved to be a deep hole, which was filled with sarsen and blocks of lower chalk before the Ring-stone was set up in it (Smith, 1965, 202). This led Smith to the conclusion that the stone hole was initially prepared to receive a bigger stone and that the Ring-stone was a substitute[39]. Two attempts to break this stone by fire were made, but the base of the stone withstood the attempts (Gillings et al., 2008, 298 f.). It was recovered during the excavation (Smith, 1965, 203). The parchmarks observed in 1995 may suggest a pair to the Ring-stone (Fig. 8.3), which Bewley et al. (1996, 640, 642 fig. 2) see possibly in connection with the southern entrance to the henge (the stones seem to mark the way between stones 1 and 98 of the outer circles up to stones 101 and 129 of the southern circle. Another parchmark was discovered close to stone 98, i.e. just north of it (Bewley et al., 1996, 640, 642 fig. 2).

Stukeley (1743) had at some point assumed that the outer circle possessed an inner circle (see Fig. 8.17) to which the Ring-stone (maybe) belonged;[40] at least it is depicted thus on the supposed inner circle on most of his plans[41] (Ucko et al., 1991, 216 f.). Besides this, **a stone in the north-west quadrant** was also shown to be part of this inner circle (on plans E, S, U, V, W, X (Ucko et al., 1991)), and it is also mentioned in his written tour (Ucko et al., 1991, 273). A small cutting was made in 1937 in order to find this stone (Smith, 1965, 190), but no evidence of a stone setting was to be found. Due to the wrong scale given on plan E, Stukeley's (1743) frontispiece, the cutting was made in the wrong place (Ucko et al., 1991, 206); however, the evidence for such an inner circle is very weak.

Another stone setting, **stone H** (Figs. 8.4 and 8.17), shown on Stukeley's (1743) frontispiece (plan E) was one of the foci of attention during Piggott's excavation in 1960 (Piggott, 1964, 29).[42] It is shown standing at the end of the north-east sector ditch, close to the northern entrance. The excavation could not detect any stone setting in this place.

Stone hole A and possible stone holes **B** and **C** (Fig. 8.4) were discovered during the excavation of the north-west sector in 1937 (Keiller, 1939, 227, fig. facing p. 224). Stone hole A is situated very close to stone 46 and aligned north-south. It is impossible that two stones could have stood at the same time in each position. Keiller decided therefore that stone hole A must predate stone 46. Two further possible stone holes (B and C) were discovered at the ditch ends west of the northern entrance, one to the south, the other to the north of the ditch. Of stone hole B only the bottom existed, stone hole C was 'cut through' by the ditch[43], which was taken as evidence, that it predated the ditch. Subsequently a 'north setting' was postulated similar to the northern and southern circles and predating the earthwork (Keiller, 1939, 227). This idea was finally given up in 1960 when excavations by Piggott (1964) failed to locate any further stone settings belonging to this assumed circle in the north-east quadrant.[44] Smith (1965, 203) calls into question whether B and C were actually stone holes at all.[45]

8.1.1.4 Other features

The resistivity survey in the north-east quadrant (Ucko et al., 1991, 227, 228 plate 69) possibly discovered two concentric **timber circles** north-east of stone

[38]Stukeley (1743, 25) thought that the hole was an artificial feature, but nowadays it is rather assumed that it was a natural perforation (Smith, 1965, 202). Other stones with natural holes were found e.g. stone L6 in Longstones Field (Gillings et al., 2008, 79 f.).

[39]Stone 41 in the north-west sector of the outer circle had a similarly prepared stone hole and was thought by Smith (1965, 221) to be also a last minute substitute. In contrast Pollard and Cleal (2004, 126) believe that stone 41 and the Ring-stone were actually reset during the Bronze Age. The Bronze Age date derived from two beaker fragments found in stone hole 41. The fragments had been thought by Smith (1965, 247) to be contemporary with the stone setting (in keeping with her dating of the West Kennet Avenue stones), but she also admits that the 'precise relationship of grave [to which the sherds are thought to belong (Smith, 1965, 204)] and stone-hole must remain unknown' (Smith, 1965, 247). Further remarks: 'Extensive disturbances round the sites of Stones 41 and 42 suggested that these stones may have been overthrown and perhaps partially buried during the medieval period.' (Smith, 1965, 188) and 'An undecorated base sherd (P 348) from the disturbed stone hole of Stone 41 [...]. The original positions of beaker and skeleton are uncertain.' (Smith, 1965, 226 f.) make it hard to judge how admissible the evidence of the sherds inside the stone hole really is in any attempt to date the setting.

[40]For a full discussion of this rather complicated matter see Ucko et al. (1991, 216 f.).

[41]Plans E, O, P, Q, R, S, V, W, X.

[42]See also Smith (1965, 203) for more details.

[43]The stone hole is more likely cut through as a result of ditch weathering (Smith, 1965, 203), cf. Gillings et al. (2008, 47).

[44]Also, the geophysical surveys revealed no trace of a north setting (Ucko et al., 1991, 227).

[45]Gillings et al. (2008, 47) seem to have fewer doubts concerning their status as stone holes.

201. Their diameter is approximately 50 m and 30 m respectively (Fig. 8.3). This feature had been previously unknown and remains so far unexcavated.

Another new feature was discovered in 1995 in the north-west quadrant, west of the northern circle. Bewley et al. (1996, 639 f.) describe it as 'a double-ditched curvilinear feature [...]. There appears to be an inner feature, possibly rectangular, which may have a central pit for a post or stone.' (Fig. 8.3). The parchmarks were then inspected with magnetometry and resistivity surveys. The magnetometry survey was limited to the area of the parchmarks, but held no discernible results, because of ferrous litter in and close to the garden plots (Bewley et al., 1996, 640 f.). On the other hand the resistivity survey covered the whole quadrant. Two different probe spacings (0.5 m and 1 m) were used for the area of the parchmarks in order to obtain readings at two different 'depths'. The remainder of the quadrant was surveyed with only one probe spacing (0.5 m) and at larger intervals. Bewley et al. (1996, 641 f.) claim that the resistivity data do not show the double ditch and the central feature. On the contrary, images 1–3 in fig. 3 (Bewley et al., 1996, 644) do show double ditches with low readings and a central feature with a high reading, though not as pronounced as the parchmarks. The feature lies on slightly raised ground and could be a **barrow** and possibly pre-date the henge[46] though its date is unknown and could only be resolved by excavation.

Several other features like **field boundaries**[47], **ditches** and **trackways** were cut during excavation, appeared as parchmarks or in geophysical surveys. Gillings et al. (2008, 238–247, 245 fig. 8.9) give an overview of the boundaries in the henge. Bewley et al. (1996) note additional features which cannot all be explained as field boundaries. For example, several ditches (B–G (Bewley et al., 1996, 644 fig. 3.4)) were discovered and an area of low resistivity readings (zone F) which has so far no parallels in the other quadrants. The low readings suggest damper or deeper soil.

A **post hole** of substantial size which was discovered at the south entrance to the henge, should not be omitted (Smith, 1965, 204). It had a diameter of 1.6 m at the top and about 1 m at the bottom with a depth of 1 m (marked with 'Ps' on Fig. 8.4). The purpose of this post remains unknown.

[46]Based on the notion that there is no relation between this feature and the stone settings, which is a rather weak argument.

[47]Sometimes they incorporate the stone settings into their fabric like the outer circle in the north-west quadrant.

8.1.2 The West Kennet Avenue

The exact course of the West Kennet Avenue can so far not be established. John Aubrey (Aubrey, 1980, 37)[48], William Stukeley (Stukeley, 1743, 29)[49] and other antiquaries (see Ucko et al., 1991, 8–59) describe it as two stone rows, which connect the Avebury henge with the Sanctuary on Overton Hill (Figs. 8.2 and 8.6). Modern research makes it seem more and more unlikely that the West Kennet Avenue existed as two stone rows for all of its *c.* 2.5 km length (Fig. 8.6), and it even seems as if it may have had gaps in its course (Gillings and Pollard, 2004, 20).

8.1.2.1 The northern part of the West Kennet Avenue

The northern part of the West Kennet Avenue was excavated by Alexander Keiller and Stuart Piggott in 1934–35 and 1939. Keiller and Piggott (1936, 418) mention that when they began their excavation in 1934 all that was visible of this part of the West Kennet Avenue were three standing stones, nine fallen and a fallen stone which had been re-erected in 1912[50]. The northern part of the avenue consists of two parallel rows of stones (see Fig. 8.7). In Smith's numbering scheme each stone pair received a number beginning from the southern entrance of the henge, the stone to the east is labelled with an 'a' the stone to the west with a 'b'.[51] The excavations started in the south (stones 37a and b), where a number of stones had fallen (see Table 8.4) and thus revealed to the excavators what spacing of the stones was to be expected. In 1934 strips along the assumed stone rows were excavated in order to find the former stone positions (Smith, 1965, 185). The spacing of

[48]He describes the course of the avenue as follows: 'At the south entrance runnes a solemne Walke, [...] of stones pitched on end about seven feet high, which goes as far as Kynet (which is (at least) a measured mile from Aubury) and from Kynet it turns with a right angle eastward crossing the River, and ascends up the hill to another Monument of the same Kind (but less) [the Sanctuary] as in plate II.' (Aubrey, 1980, 37).

[49]Stukeley describes the course of the avenue in a similar way to Aubrey: 'It extends itself from this southern entrance of *Abury* town to *Overton-hill*, overhanging the village of *West-Kennet. There* [the Sanctuary] was another double circle of stones, [...]. All the way between there, and this southern entrance, which is above a mile, was set with stones on each hand, opposite to one another, and at regular distances.' Emphases by Stukeley (Stukeley, 1743, 29).

[50]According to Keiller and Piggott (1936) in a wrong position and upside down.

[51]The numbering scheme employed by Keiller and Piggott (1936) during the excavation had been different, due to the fact that excavations started from the point furthest away from the henge and progressed towards it.

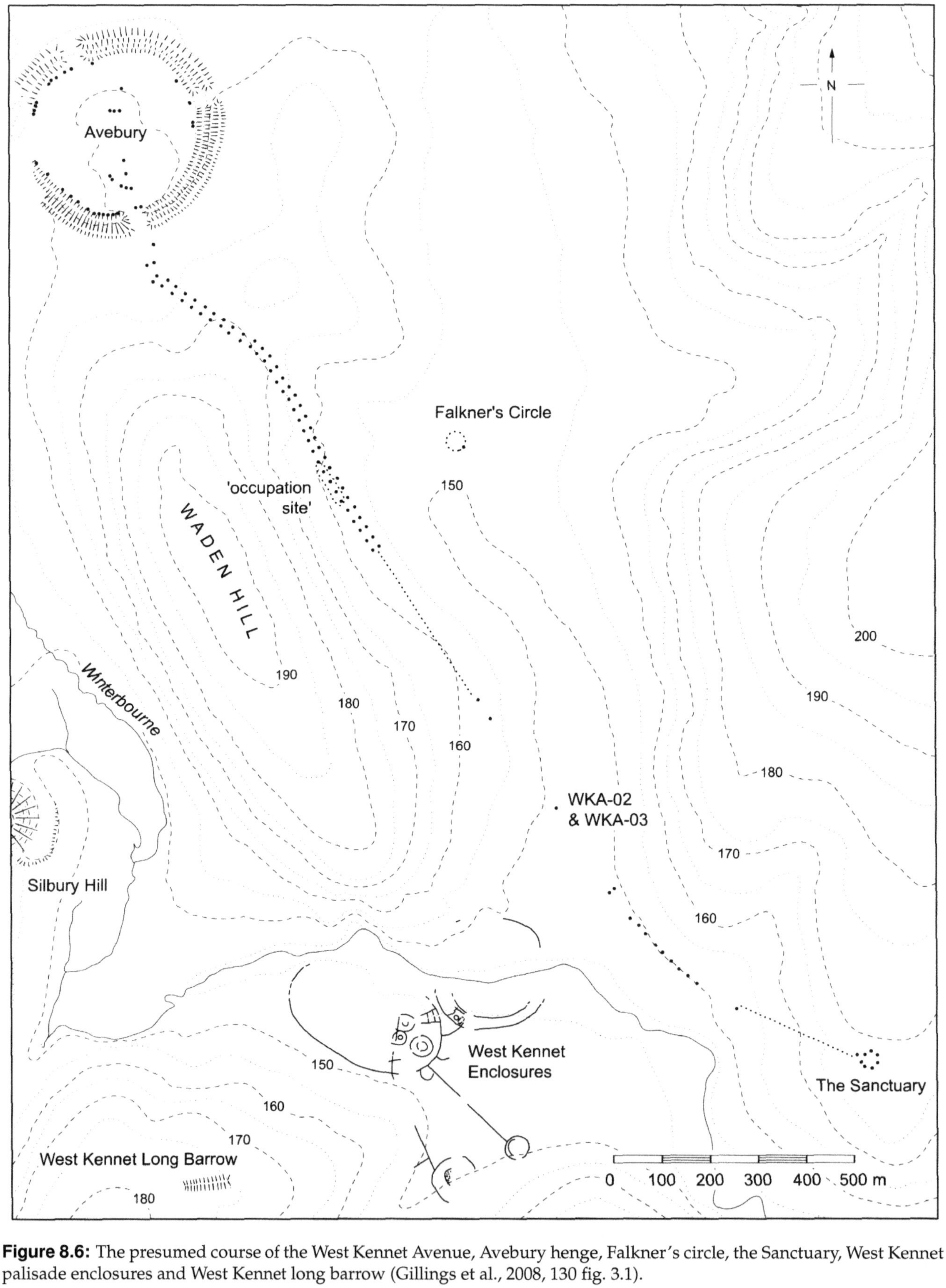

Figure 8.6: The presumed course of the West Kennet Avenue, Avebury henge, Falkner's circle, the Sanctuary, West Kennet palisade enclosures and West Kennet long barrow (Gillings et al., 2008, 130 fig. 3.1).

Fate of stone	Stone number
standing	4(b?), 21a, 33a/b
fallen	19b, 27b*, 28b*, 29b*, 30a*, 31a, 32a/b, 34a, 35a/b, 36a, 37a/b
buried	13a/b, 14a, 15a/b, 16b, 18a/b, 20a, 24b, 25a/b, 26a/b, 29a
taken away	5a, 6a/b, 8b, 9b, 10a/b, 11b, 20b, 21b, 22a/b, 27a, 28a, 31b, 36b
destroyed by direct fracture	5b, 7a/b, 8a, 9a, 11a, 12a/b, (previously buried) 20a, 22a**, (previously buried) 25a, 34b
destroyed by fire	1a**, 3a** 14b, 17a/b, 19a, 23b
not investigated	1b, 2a/b, 3b, 16a, 23a, 24a

Table 8.4: Fate of the stones in the northern third of the West Kennet Avenue, based on Gillings et al. (2008), supplemented with information from Smith (1965). Key: * so listed by Gillings et al. (2008, 300 fig. 10.6), ** so listed by Smith (1965, 185 f.).

the stones[52] was found to be quite regular so that in 1935 only small cuttings over the estimated stone positions were made (Smith, 1965, 186). Both methods leave great parts of the avenue's course unexcavated. Problems were encountered when the avenue made changes in direction, because then the avenue stones lay outside the predetermined areas of excavation. In such a case the cuts had to be widened until the position was found. Another problem was posed by allotment gardens and fowl-runs close to the entrance of the henge, which disturbed the ground (Smith, 1965, 209).[53]

Stone pairs 1–3 (together with another examination of stone pair 4) were included in the 1939 excavation of the south-east sector (Smith, 1965, 186, 208). Stone hole 1a is situated on the causeway, between the ends of bank and ditch. Due to the extended bank on the eastern side of the southern entrance (see Section 8.1.1.1), which now covers the estimated position of stone 2a, this setting was not excavated. Neither were stone settings 1b–3b because they all lie underneath the modern road[54]. Smith (1965, 208) reports that close to the bank 'lies a heap of broken fragments of sarsen, some showing sign of having been heated.' This was taken for the remains of stone 3a. The additional smaller sarsen boulders and reddish-brown clay, which were discovered together with the sarsen fragments, were construed to be the packing material from the stone hole. There is no mention of the actual stone hole being found, the fragments and additional packing being taken as proof for 'the former existence of Stone 3a on this spot.'

The average stone height (2.7–3 m)[55] is lower than the average height of the stones inside the henge circles. Keiller and Piggott (1936, 420) discovered that two distinct shapes are paired in the West Kennet Avenue, they call them types A and B. Type A stones are tall and narrow, while type B stones are rather broad and low. Both stone types can occur on either side of the avenue. The methods used for erecting and setting the stones are identical to the ones observed inside the henge (Smith, 1965, 218–222).

The somewhat irregular course of the avenue appears to consist of several straight sections. This is, as Keiller and Piggott (1936, 418) state, more apparent in plan than from the ground (Fig. 8.7). The layout becomes especially curious in the part closest to the henge. The distance between the stone pairs increases while the distance between the stones of a pair decreases; additionally, there is a pronounced dog-leg in the course of the avenue. Coming from the south the West Kennet Avenue leads to the west (stone pairs 17–6), as if bypassing the henge, but turns 85 m from the entrance (stone pair 5) abruptly to the east (Smith, 1965, 208; Thomas, 1993, 42).

[52]The average transverse spacing is 14.9 m, the longitudinal spacing 24.4 m (Smith, 1965, 206).

[53]Gillings et al. (2008, 254–256, 254 fig. 9.2, 284 fig. 9.20) give information on buried and destroyed stones (Gillings et al., 2008, 300, 300 fig. 10.6).

[54]Smith (1965, 208) mentions that Stukeley shows the sites of stones 1a–3a on the road. She probably refers to plan E, the frontispiece (Stukeley, 1743), where the positions are marked. It is reasonable to assume that Stukeley did not see the holes on the road—this would have been a hazard for the coaches—but rather that he marked positions where he thought that there may have been stones. Especially plans N, O, Q and R (Ucko et al., 1991, plates 32, 33, 35–37) show stone positions 1b–3b as possible stone positions, while plans V, W, X and E (Ucko et al., 1991, plates 14, 40–42) show them as fallen—never as a depression. Plan P (Ucko et al., 1991, plate 34) is an exception, standing West Kennet Avenue stone 4b is marked as stone 3b, positions 1b is marked as a possible stone position, while position 2b is hidden by the bank. Stone position 1a is never plotted in the place it was excavated. It is depicted in front of the centre of the ditch on all of Stukeley's draft plans and on plan E (see figures in Ucko et al. (1991, plates 14, 32–37 and 40–42)).

[55]For a list of stone heights see Smith (1965, 206).

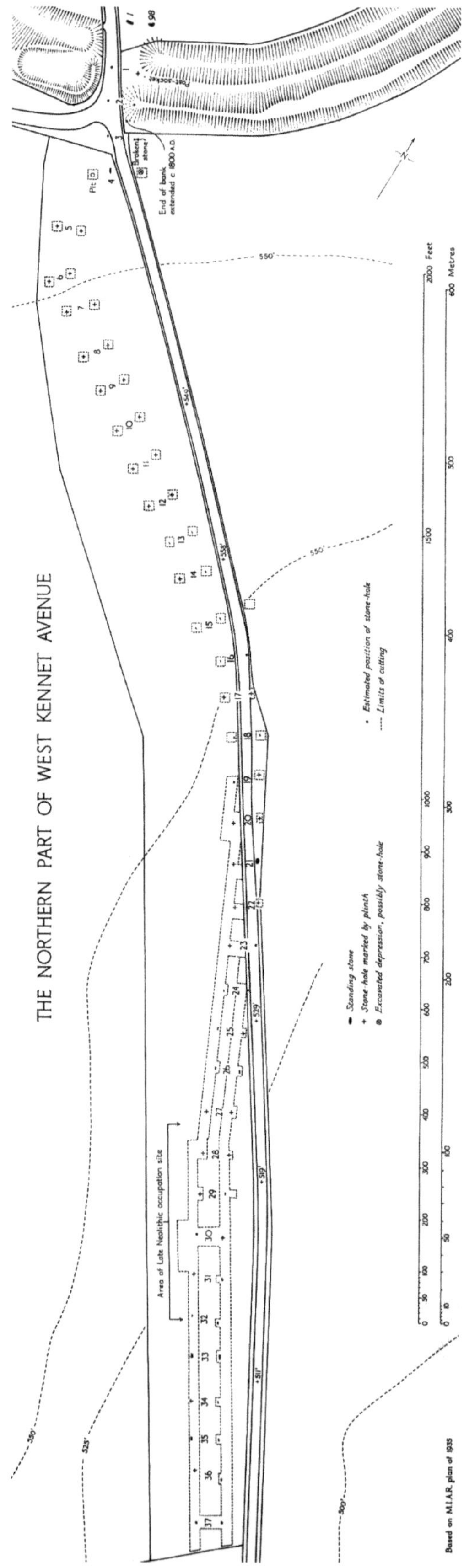

Figure 8.7: Northern part of the West Kennet Avenue (Smith, 1965, fig. 71). (By permission of Oxford University Press).

Further ambiguities arise from stone pair 4. One stone of this pair is still standing (Fig. 8.7), but the place of its pair is not so easy to identify. In 1935 Keiller believed himself to have discovered the stone hole for stone '4b' to the west of standing stone '4a' (Keiller and Piggott, 1936, fig. 1 facing p. 418; Smith, 1965, 186). He revised his opinion later and Smith (1965, 186) also assumes it to be of rather recent origin. When in 1939 the position of stone 1a was found during the excavations of the south-east quadrant a renewed search for the pair of standing stone 4 was started, this time to the east, on the other side of the road. Another depression was discovered in this place and subsequently named '4a'. Smith (1965, 186, 208) is in her publication quite confident, that the second position ('4a') is indeed the pair to the standing stone 4, whereas Keiller seemed less confident and did not, as is otherwise usual, mark the position with a concrete plinth.[56]

The marked kink or dog-leg formed by stone pairs 6–3 gave raise to a number of hypotheses, some of them were later incorporated into the 'Negotiating Avebury' project.

An arrangement similar to the dog-leg at the henge end, was documented by Aubrey (see Ucko et al., 1991, plans A and B) for the avenue's approach to the Sanctuary. The excavations at the Sanctuary in 1930 revealed that the avenue consisted of three rows of stones in this part (see Fig. 8.9). This led Gillings and Pollard (2004, 20–22) to argue that, due to the fragmentary remains at the Sanctuary, Aubrey may have felt compelled to rationalise the kink. 'This raises the tantalising possibility that a similar complexity marks the Avebury terminus of the avenue, leading to a rather confused excavation plan recovered by Keiller.' (Gillings and Pollard, 2004, 22).

Smith (1965, 209) suggests a different hypothesis. Based on Keiller's impression that the dog-leg was not a satisfactory approach to the henge,[57] she suggested that the approach may have been remodelled at a later time, arguing that the fate of stone pairs 5–12 remained unsure (no burning pits or buried stones were discovered during excavation, only the empty stone holes)[58] and that they could have been used to form a straight approach to the henge with a spacing of about 24.4 m between them. Their assumed new positions would either lie in an area unexcavated by Keiller and Piggott or situated in the allotment gardens and fowl-runs mentioned earlier.

This hypothesis is partly based on the assumption that the dog-leg is no intentional feature. Smith (1965, 247) argues this point, when discussing the date and sequence of the various features. She reasons that the dog-leg could be the result of two separate building phases: stone pairs 1–4 could have been placed after the earthwork at the henge was completed, their settings being determined by the southern causeway. Then their failure to match up with the remainder of the avenue could have only come to pass if the stone pairs beginning with pair 6 were already in place, thus antedating the henge. At the same time Smith (1965, 247) sees her own hypothesis refuted by the fact that stone hole 9b contained Middle and Lower Chalk packing stones, which in her opinion could only have come from the henge ditch.[59]

Whittle (1993, 38) argues along similar lines. The West Kennet Avenue, or a primary version of it (Pitts and Whittle, 1992, 211), could have been an early feature, especially when seen in analogy with other linear features like cursus monuments. Again the kink could indicate that it was possibly begun before the earthwork of the henge. The West Kennet Avenue would then fall into Whittle's Phase D, roughly at the same time in which the construction of the primary bank is envisaged (Phase D and the beginning of Phase E) (Whittle, 1993, 46 f.). The construction of the primary bank could also provide the material for the packing stones of stone 9b[60], if the chalk blocks were not brought from other sites (cf. Pitts and Whittle, 1992, 210).

Burl (1993, 45, 72) believes that the four stone pairs of the West Kennet Avenue which are closest to the henge were the first feature established on site and that the kink is the result of an attempt to join the avenue to the henge. He does not elaborate on whether he thinks that the avenue had been created as an independent monument in the first place.[61] Burl (1993, 72 f.) also considers the possibility of whether the

[56]That there is still some dispute concerning this stone pair is obvious from Earl and Wheatley (2002), see Section 8.2.2.2.

[57]In a letter he proposed that the whole area to the south of the causeway should be excavated (Smith, 1965, 208).

[58]Their fate seems less mysterious when regarding Table 8.4. Eight appear to have been destroyed by direct fracture while another eight stones were taken away.

[59]cf. Sections 8.1.1.2 and 8.1.1.3 concerning the providence of the chalk.

[60]As mentioned in Section 8.1.1.2 the ditch was dug directly into the Lower Chalk in the north-west sector, elsewhere into the Middle Chalk first (Smith, 1965, 194).

[61]If Burl thinks that the avenue is indeed a separate monument this would be contradictory to the general chronological sequence for the emergence of avenues in connection to stone circles which he proposes himself (see Burl, 1993, 72). In his sequence the avenue would start at the stone circle and its construction would proceed from there. There would be no reason for an 'accidental' kink affected by a join.

West Kennet Avenue was actually comprised of two avenues which were later joined at West Kennet.[62]

Gillings et al. (2008, 129), who favour the views of Thomas (1993), Barrett (1994) and Watson (2001), take the view that the dog-leg may be a deliberate feature to obscure the interior of the henge from view up to the last turn and even then stones 1 and 98 hide most of its interior (Thomas, 1993, 42). Barrett (1994, 15) sees the avenues as manifestations of pre-existing processional ways; this implies that the design is intentional. Watson (2001, 300) follows this view, seeing the final approach of the avenue as part of a choreography to heighten the sense of mystery.[63] Gillings et al. (2008, 193–196) therefore think that the primary bank may fall into the later Neolithic phase perhaps together with the Cove. Later the main ditch and bank followed 'and other architectural elements of the henge'. The construction of the avenues and the placing of stones inside the henge is thought to have taken place in the final Neolithic, with the possible continuation of activities inside the henge into the early Bronze Age.

Another indication that the West Kennet Avenue constitutes a well-choreographed monument may come from the fact that when it passes a ridge, around 300 m from the henge, taller stones have been chosen to mark this part of the course (Gillings and Pollard, 2004, 78).[64] There had been difficulties in finding the stone hole for stone 15b (Smith, 1965, 186), which occurs in this stretch of the West Kennet Avenue. It was first believed that the stone pair would flank the road, but they were discovered to lie entirely to the west of it. Stone 15b was buried and the excavators thought that they had found its stone hole to the south of the burial pit (Gillings et al., 2008, 256 fig. 9.3, 15b). Having studied Keiller's unpublished description of the stone hole, Gillings et al. (2008, 132) question this: hole 15b and its fill do not resemble the other stone holes, so that Gillings et al. (2008) suggest that it may have been a Grooved Ware pit instead and that a renewed search for the original setting needs to be undertaken.

Around 630 m from Avebury the West Kennet Avenue passes a former occupation site, which it seems to respect (see Figs. 8.6 and 8.7). A dense surface scatter of flints and pottery was found here and, in the small trenches made to find the stone holes,[65] two pits and ten holes were encountered. However, no stone hole for stone 30b could be found, though its assumed place would have been in the area with the highest find density (Smith, 1965, 210–216).[66] Gillings and Pollard (2004, 78) believe that the avenue respects the former occupation site and that therefore stone 30b never existed.

South of the excavated area several gaps were encountered in the course of the West Kennet Avenue (Fig. 8.6): one between stone pair 37 and 52b (*c.* 400 m), another between 54b and 72b (*c.* 440 m) and finally between between 83b and the Sanctuary (*c.* 280 m) (Smith, 1965, 207).

Some of these gaps have been investigated in recent times. South of the area excavated by Keiller and Piggott, Ucko et al. (1991, 188 fig. 59, 189, 192 fig. 62, 193 f.) made several surveys. The area immediately south of stone settings 37a and b (this is equivalent to stone settings 38a/b to 44a/b) was surveyed with magnetometer and resistivity probes. In a second survey the area south of setting 44a underwent a resistivity survey. Some of the stone settings of pairs 38–44 could be identified with some confidence due to the resistivity survey (see Table. 8.5).[67] The magnetometer survey yielded no anomalies, which may indicate that none of the stones was destroyed by fire *in situ* (Ucko et al., 1991, 189). These stone settings (42a–44a and 41b–44b) were also present on Crocker's surveyed plan of 1812 and it is reported by Long that three stones north of stone 52 were removed in *c.* 1823 (Smith, 1965, 207 footnote 1). When the survey was later extended to the south, stone settings 46–49 could only be conjectured (Ucko et al., 1991, 194) if the regular pattern of the avenue's stone settings was followed. Without this kind of guidance a number of alternative settings could be read from

[62] This would be in keeping with one of Aubrey's descriptions of the West Kennet Avenue (Aubrey, 1980, 50): 'West Kynet stands in the angle where the walke from Aubury hither, and that from the top of the hill, did joine.' The other description (see footnote 48) rather seems to assume that the West Kennet Avenue is a single monument.

[63] Surprisingly no-one has ever considered the opposite direction. What if the West Kennet Avenue is actually leading *away* from the henge? Which views can be gained from this perspective?

[64] As stone pairs 5–12 have not survived: could it not be possible that there is a general increase in stone size from this point towards the henge? On the other hand, looking at the stone sizes given by Smith, it could be argued that the stone pairs south of the occupation site (33–36) are rather tall (in average around 3.40 m) in comparison to stone pairs 13–16 (average 3.10 m, mostly due to tall stone pair 13).

[65] This means that there was hardly any deviation from the established excavation pattern and no effort was made to excavate the whole occupation site.

[66] Smith (1965, 212) thinks that the stone hole must have been quite shallow and that it weathered away, thus strongly assuming that there could be no divergence from the two-rowed layout of the avenue.

[67] Ucko et al. (1991, 189) remark that some of the readings were so weak that, if it had not been clear what they were looking for, they would hardly have detected them.

Fate of stone	Stone number
standing	79b, 81b
fallen	75b, 77b, 80a/b, 82b, 83b
buried	63b (buried in 1921–22)
destroyed	72b, (previously fallen) 75b, 76b, 78b
destroyed by fire	72a/b, 87
uninvestigated (visible in survey)	38a, 39a/b, 40a/b, 41a/b, 42a/b, 43a/b, 44a

Table 8.5: Fate of the stones of the West Kennet Avenue at West Kennet, based on Smith (1965), supplemented with information from Gillings et al. (2008, 135) and Ucko et al. (1991, 188–189, 192–194).

the data. Stones 52b[68] and 54b are still in position on both sides of the road, but no further stone settings could be distinguished with any confidence. Between these stones is the site where Stukeley claims that there had been a West Kennet Cove, similar to the one at the Beckhampton Avenue, i.e. Longstones Cove (see Section 8.1.4.2). It has long been discussed whether the West Kennet Cove was a figment of his imagination borne out of his desire for symmetry (Ucko et al., 1991, 94) or whether it really existed and was previously described by Aubrey as the 'Devil's Coytes'.[69] Unfortunately the site where this cove was thought to lie is nowadays underneath the road so that the notion can neither be proven nor dismissed (Ucko et al., 1991, 190–194). Within the surveyed area lay the position of stone 63b[70], which is marked on an Ordnance Survey map from 1883 (Ucko et al., 1991, 192 fig. 62). No significant readings were taken at its position.

The excavations by Gillings et al. (2008, 133, 135) were able later to identify the site of stone 63b, which is reported by Osbert Guy Stanhope Crawford to have been buried during the winter of 1921–22. No other stones were discovered even though the trench encompassed an area which should have included stone settings 63a, 64a and b.[71] This shows both the limitations of the resistivity survey in this area under the given humidity conditions and maybe the limitations of small scale excavation trenches, which are oriented on preconceived patterns of the stone settings.

The above mentioned findings, i.e. seemingly only one stone row in the geophysical survey by Ucko et al. (1991, 192 fig. 62) for stone settings 46–49 and only one excavated stone where there should have been four (Gillings et al., 2008, 139), have led to the interpretation that the avenue may have in part only consisted of one row of stones and may have had a gap, wherein a natural sarsen field was encountered (Gillings and Pollard, 2004, 20), or indeed that one should at least keep an open mind about the layout of this monument.

8.1.2.2 The avenue at West Kennet

This part of the avenue was excavated by William Edward Vincent Young in 1957–60 (Fig. 8.8). He found five stone holes (72b, 75b, 76b, 78b, 77b), two of the stones are destroyed by fire (72a/b,), four stones are fallen (77b, 80b, 82b, 83b) and two standing (79b, 81b). A 1.5 m long fragment on the other side of the road is interpreted as the remains of stone 80a (Smith, 1965, 207). The sites of stones 75b–78b now lie underneath the modern road (Smith, 1965, 187). Two more burning pits (presumably for a member of the 87th row and the 96th or 97th row each) were discovered in 1950 and 1942. One or two stones were broken up on behalf of the Commissioner for Roads in the early 19th century (these are assumed to belong to pairs 89 or 90) (Smith, 1965, 207 f.).

The avenue is linked to the outer stone ring (ring A) of the Sanctuary (see Section 8.1.3 and Fig. 8.9). This stone circle and the avenue are generally assumed to be contemporary features (Smith, 1965, 247; Pollard, 1992, 213, 217). Only three stone pairs of the West Kennet Avenue have been excavated there; they are joined to the north-west of ring A where three stones

[68](Ucko et al., 1991, 193) refer to these stones as 53b and 54b, but they would seem too far apart for being neighbours. Smith (1965, 207) sees numbers 52b and 54b in them, which appears more likely.

[69]See Footnote 76 on page 162. Aubrey (1980, 823) describes it as follows: 'Southward from Aubury, in the ploughed field—near Kynnet doe stand three huge upright stones perpendicularly, like the three stones at Aubury in fig. . they are called the Devill's Coytes.'

[70]The stone number can obviously only be an estimate.

[71]Gillings et al. (2008, 139) discuss possible reasons for not finding any other stones and whether stone 63b had been really a standing stone.

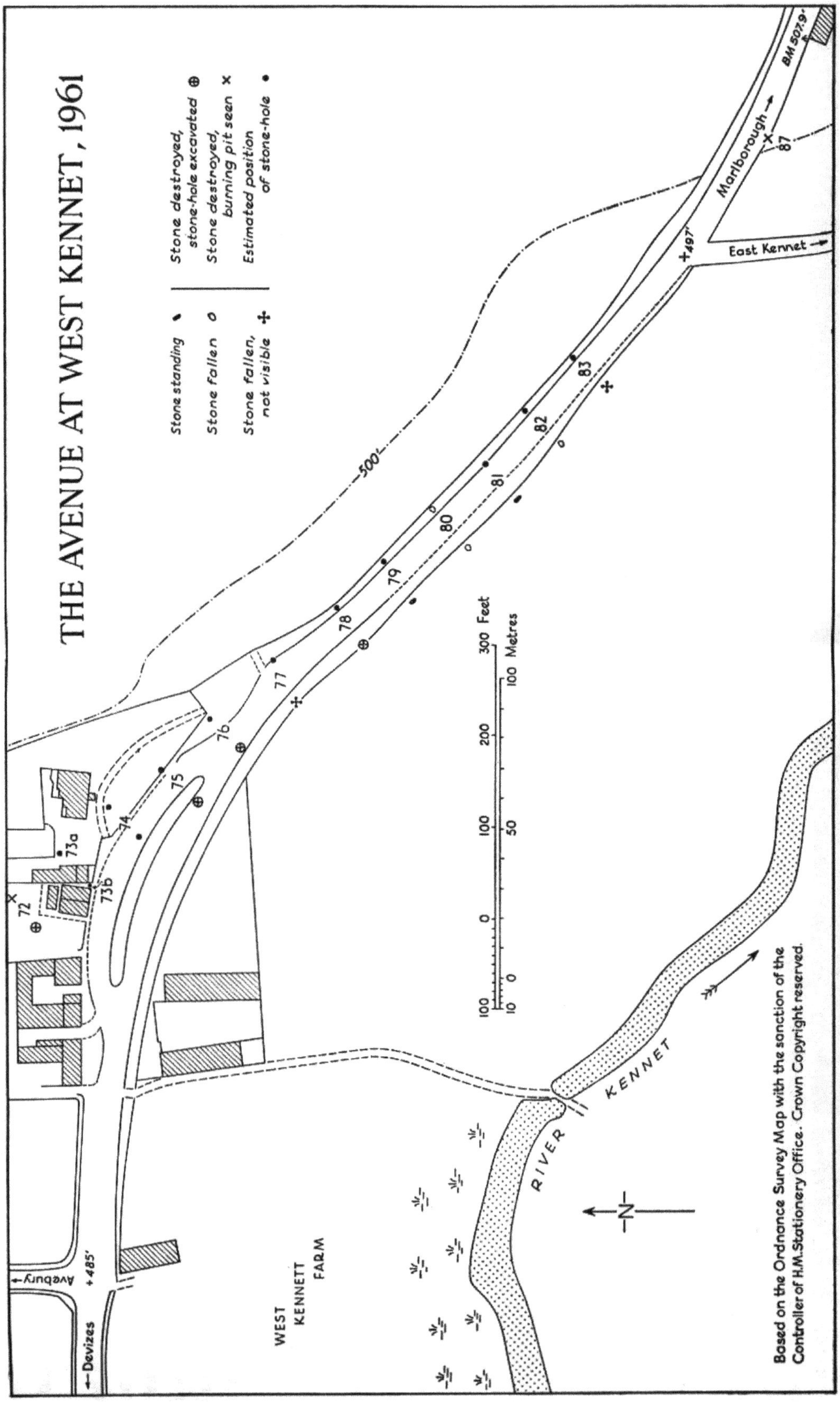

Figure 8.8: The West Kennet Avenue near West Kennet (Smith, 1965, fig. 72). (By permission of Oxford University Press).

(the outer ones aligned to the avenue stones and a central stone) are aligned radially with the Sanctuary's inner stone and timber circles (Pollard, 1992, 217, 214 fig. 1 top). The other stones of ring A have faces aligned with their neighbouring stones. The longitudinal intervals of the avenue stones are in this part rather small (5.5 m). The transverse intervals narrow from 8.5 m to 6.1 m (Smith, 1965, 208) and are, indeed, set radially to the inner timber and stone settings of the sanctuary (Pollard, 1992, 217). Stones N1 and N2 are also radially set (Pollard, 1992, 217) and in close proximity to the eastern avenue stones. Gillings and Pollard (2004, 79) suggest that the early excavation records by Maud Cunnington may hint at the possibility that the avenue was comprised of three rows of stones in this part.

8.1.2.3 Possible uses of the West Kennet Avenue

Concerning the use of the West Kennet Avenue there have also been several hypotheses. It is generally accepted that the avenue connects Avebury with the Sanctuary on Overton Hill. One hypothesis assumes that the avenue was a processional way (Thomas, 1993, 41), that it was the 'monumentalisation' of an earlier trackway or trackways (Gillings and Pollard, 2004, 81), and that it prescribed in which order monuments in the region had to be encountered (Watson, 2001, 300; Gillings and Pollard, 2004, 81). 'Perhaps, people were in some way physically playing out beliefs about their history during the act of moving along the avenue.'(Watson, 2001, 300). Usually it is assumed that processions would take place walking between the stone rows, but the geophysical data suggests that people walked both inside and alongside the rows (Ucko et al., 1991, 189 f.). The West Kennet Avenue was indeed not an 'empty' pathway, but rather in some parts littered with stones (Gillings et al., 2008, 141 f.). It has also been considered whether the avenue could therefore be meant as a pathway for the ancestors (Gillings et al., 2000, 7; Parker Pearson and Ramilisonina, 1998, 320).

8.1.3 The Sanctuary

Having mentioned the Sanctuary on several occasions in connection with the West Kennet Avenue the monument will be discussed here in short as it has no further bearing on the reconstructions discussed below. The site of the Sanctuary was called Millfield in Aubrey's time (Aubrey, 1980, 55). He recorded the stone circles (the only visible remains in his time) before the large scale destruction of the site took place[72] (Aubrey, 1980, 50 f.). It was Stukeley, who claimed that the site was called 'the Sanctuary' (Stukeley, 1743, 31). He also mentions that most of the stones had disappeared by 1723, when he first visited the site, and that the rest were removed during the winter of 1724[73]. He recorded the stone holes which were visible on the ground and the still remaining stones (Stukeley, 1743, tables 20 and 21), but obviously did not notice the dog-leg in the avenue nor the change in the intervals between the avenue stones. In 1930 the site of the Sanctuary was rediscovered and excavated under the direction of Maud Cunnington (Cunnington, 1931), who discovered that the Sanctuary had not only consisted of two concentric stone circles (rings A and C) but also of six post rings (B, C (a combination of wood and stone), D, E, F and G) plus a central post (see Fig. 8.9) (Cunnington, 1931, 306–308, 327–329). The difference in the West Kennet Avenue's approach to the Sanctuary as seen in Aubrey's and Stukeley's record was explained on the one hand by assuming that the West Kennet Avenue consisted of three stone rows at the approach to the Sanctuary: the two stone rows, which are radially aligned to stones 1 and 41 of ring A (thought to be the 'real' West Kennet Avenue) and stones N1 and N2 (radially aligned to stone 3 of ring A). The approach may have been partly dilapidated in Aubrey's time, so that he mistook the northern stone row of the 'avenue' together with stones N1 and N2 (and another stone on this line, whose position could not be detected during the excavation) for a kink in the approach. On the other hand it was argued that Stukeley saw an even more dilapidated state of the site, where the last 4 stone pairs of the avenue's approach were missing entirely (Cunnington, 1931, 306, 321, 327 plate IV).

The Sanctuary was assumed to be a two phase monument (Cunnington, 1931, 309–311) with a timber and a stone phase following in rapid succession, the notion of a roofed structure being dismissed in favour of an open monument (Cunnington, 1931, 309). Stuart Piggott on the other hand argued for four phases of construction, due to the complexity of the layout, and envisaged roofed buildings instead of open structures (Piggott, 1940). His hypotheses were in turn largely dismissed by Joshua Pollard (Pollard, 1992), who again argues for one or two construction phases, but does not dismiss the notion of a roofed structure because some of the post holes are oblique (possibly

[72]It is quite possible that the monument was already not intact in Aubrey's time.

[73]Admittedly, Stukeley is not always the most reliable source.

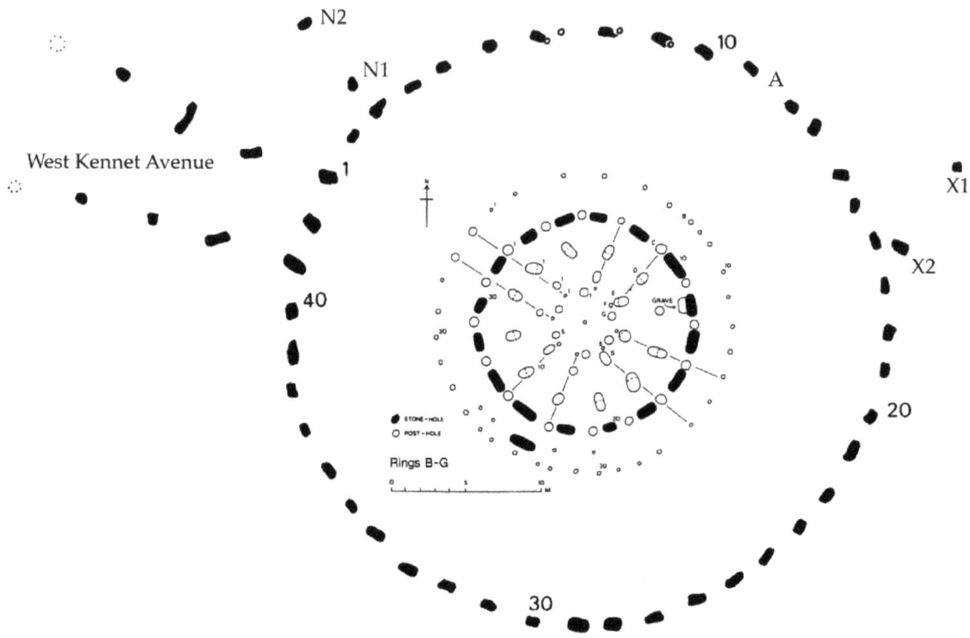

Figure 8.9: Plan of the Sanctuary on Overton Hill (after Pollard, 1992, 214 fig. 1).

due to pressure from a roof) and molluscan evidence may point to the presence of thatching (Pollard, 1992, 215 f.).

Mike Pitts, prompted by three different diaries of Maud Cunnington's foreman, W. E. V. Young—which showed discrepancies with the published excavation records by Cunnington (1931) (Pitts, 2001a, 1–3, 15–18), and his own subsequent re-excavation in part of the Sanctuary—comes to a different conclusion. Additional aid was provided by a VRML model of the site made by Jennifer Garofalini in 2000. It showed that some of the posts were too close together and that it was impossible, as an observer, to make any sense of the site (Pitts, 2001b, 238 f.). This led Pitts (2001b, 243) to the conclusion that they could not all be contemporary. Indeed, the 1999 excavation had produced evidence that there were several layers of clean chalk rubble refill in the post holes of the D, E and F rings, which had been missed by Cunnington (Pitts, 2001a, 10, 21), indicating that the posts were reset many times *before* they started to decay. Only the last post settings were left to rot *in situ* (Pitts, 2001a, 20 f.). This would mean there is a constant change in the appearance of the inner part of the site as posts were removed, holes backfilled and other posts reset in almost the same spot. This cannot be accomplished with a roofed structure. These changes in layout also indicate a controlled process of resetting and make it unnecessary for all the post holes to be occupied by a post at the same time (Pitts, 2001a, 20 f.). Rings A, B and C appear to have been more constant[74] (Pitts, 2001a, 21).

It has been generally accepted that the West Kennet Avenue is contemporary with ring A (and perhaps ring C) (Pollard, 1992, 217; Gillings and Pollard, 2004, 79), and, due to the regular positioning[75] of the post holes in rings B–G, it has been argued that these settings must belong to a uniform layout. The fact that the avenue stones and stone settings N1 and N2 are also set radially in relation to the timber settings suggests that these must have still been visible when ring A was built (Pollard, 1992, 217).

8.1.4 The Beckhampton Avenue

The existence of this second avenue has for a long time been a matter of dispute. It had been unequivocally mentioned by Stukeley (1743, 34–37), who had, however, not noticed this feature during his visits in 1719–1723, but only in his last year at Avebury in 1724 (Ucko et al., 1991, 195).[76] Its exact course and

[74] But they were also not re-excavated in 1999.

[75] A similar layout has been observed at Site IV, Mount Pleasant (Pollard, 1992, 222–224).

[76] There has been some discussion on whether Aubrey's 'Devil's Coytes' (Aubrey, 1980, 823) is the same as Stukeley's 'Longstones Cove' (this is what Stukeley (1743, 35) claims), though it seems to be rather unlikely (Ucko et al., 1991, 190). See also Section 8.1.2.1. Thomas Twining's plan in 'Avebury in Wiltshire, the Remains of a

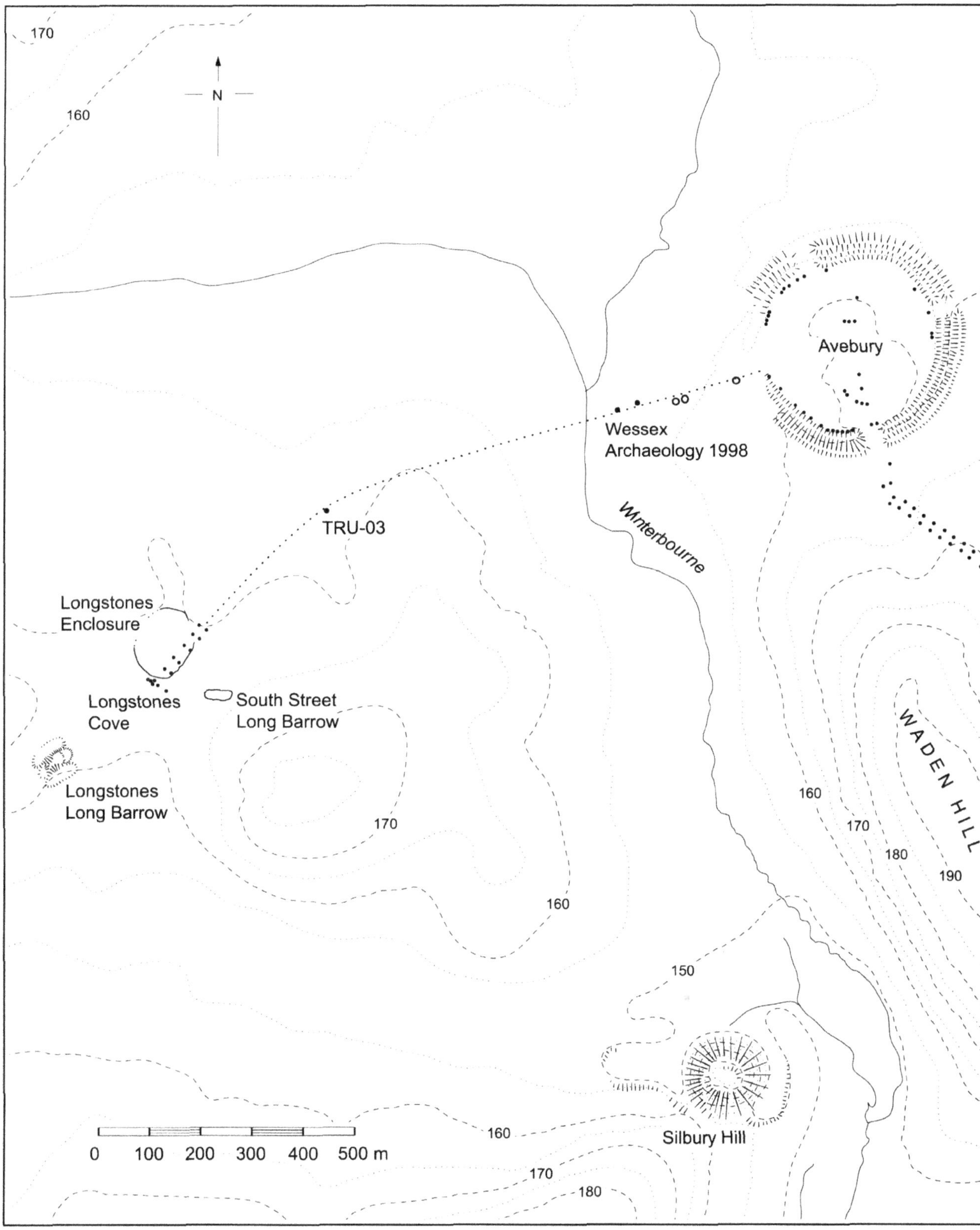

Figure 8.10: The presumed course of the Beckhampton Avenue (Gillings et al., 2008, 8 fig. 2.2).

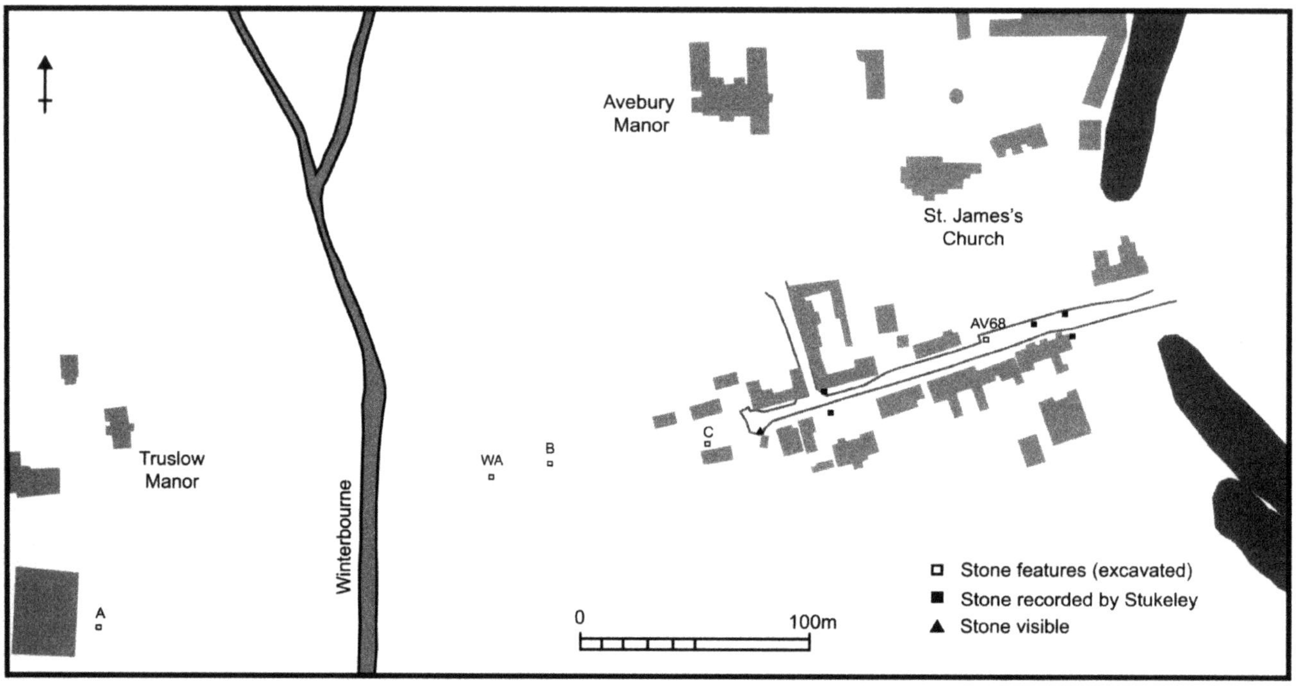

Figure 8.11: The eastern part of the Beckhampton Avenue (Gillings et al., 2008, 117 fig. 2.81).

length remains even more dubious than that of the West Kennet Avenue (Fig. 8.10). Only recently could its existence be proven by excavation (Gillings et al., 2000, 2002, 2008; Wheatley, 1999).

8.1.4.1 The eastern part of the Beckhampton Avenue

Between the western entrance to the henge and the Manor Farm at Avebury Trusloe, several observations by Faith and Lance Vatcher during maintenance work help to establish the possible stone positions for this part of the Beckhampton Avenue (Fig. 8.11) (Gillings et al., 2008, 115 f., 117 fig. 2.81). The Vatchers recorded a stone burial pit (AV68) from which the stone was later removed, one buried stone (C) and two burning pits (A, B), where, at site B, there was also a possible stone hole. Another buried stone was found between 1997–1998 (WA) and one more was visible close to the High Street (stone marked with a triangle) (Gillings et al., 2008, 116). Five additional stones were marked by Stukeley on his plans (stones marked with a square) (Gillings et al., 2008, 117 fig. 2.81). Together they all give some indication of the course and the longitudinal distances which can be expected of the avenue. The latter is in accordance with what has been learned from the excavation at Longstones Field (see Section 8.1.4.2). Maybe not completely unexpected is the discovery that the Beckhampton Avenue does not match up with the western entrance to the henge[77] (Gillings et al., 2008, 118).

To the west of Manor Farm, Avebury Trusloe, a resistivity survey was undertaken to see where and how the Beckhampton Avenue continued at this point (Gillings et al., 2008, 115, 116 fig. 2.80). The survey results left room for some interpretation. Two of the five high readings may indicate buried sarsen which lie on the estimated course of the Beckhampton Avenue. The other three anomalies (two to the north west and one to the south) could be interpreted as an elaboration of the avenue at this place. Two low resistance readings were taken for possible stone destruction pits[78]. They would also lie on the proposed course of the avenue. Excavations would be neces-

Roman Work' (1723) shows two entrances to the henge which were both connected to avenues (Ucko et al., 1991, 37 f., 38 fig. 10). This may indicate that he was observing both the West Kennet Avenue and the Beckhampton Avenue. The way in which the avenues are depicted is somewhat removed from what is now known about their courses.

[77] This phenomenon was already observed at both ends of the West Kennet Avenue, where it meets the henge and the Sanctuary, and also the linear resistivity anomaly discovered at the eastern entrance of the henge abuts at the bank instead of leading into the monument.

[78] These particular readings seem rather diffuse to me (cf. Gillings et al. (2008, 116 fig. 2.80, top)).

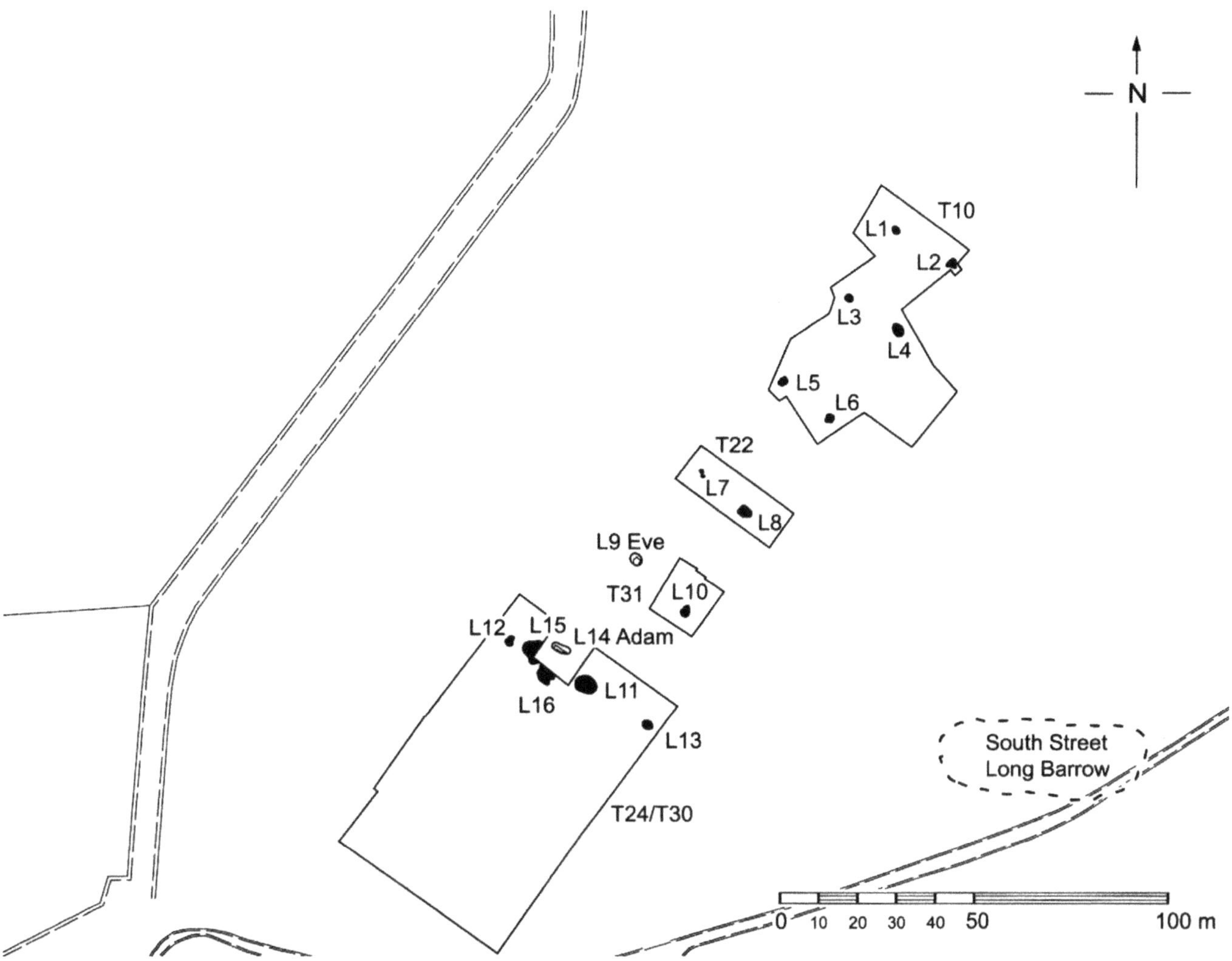

Figure 8.12: The Beckhampton Avenue at Longstones Field (Gillings et al., 2008, 63 fig. 2.36).

sary to determine the cause for these resistivity readings.

At Avebury Trusloe a geophysical survey, followed by a small scale excavation, took place (Gillings et al., 2008, 104–109). The resistivity survey was used to establish the most suitable excavation area and was scaled to include at least three stone pairs. Only two high resistance readings could be identified as possible features of which one was excavated together with the area adjoining to its south-south-east, which was thought to contain traces of its pair (Gillings et al., 2008, 105). This assumption did not hold true in the end, as neither stone, stone hole nor a destruction feature was discovered. The resistivity anomaly, on the other hand, proved to be a buried stone (Fig. 8.10, TRU-03) and part of its original stone hole (Gillings et al., 2008, 107), thus showing that the Beckhampton Avenue passed through this area.[79] Further information on the form and direction of the avenue could only be obtained by a larger excavation.

8.1.4.2 The Beckhampton Avenue at Longstones Field

The only standing stones of the Beckhampton Avenue with its terminal (Fig. 8.12), *c.* 1.3 km south-west from Avebury, are at Longstones Field and colloquially called 'Adam' and 'Eve'. Stukeley shows Adam to be part of a cove (i.e. Longstones Cove), while he claims that Eve is the last standing stone of the avenue itself (Ucko et al., 1991, 190 fig. 60, stones A and

[79]Gillings et al. (2008, 108 f.) discuss why there is no trace of the anticipated pair of the buried stone; the most probable solutions are that the narrow trench did not include the second stone setting, or that there was none. The possibility of discontinuous stone settings or the reduction of the avenue into a single row of stones in some parts has already been proposed for the West Kennet Avenue.

E; Stukeley, 1743, 35). In 1911 Adam fell and was re-erected 1912 under the guidance of Maud Cunnington (Cunnington, 1912).[80] To elucidate the situation at Longstones Field, Ucko et al. (1991, 196, 197 fig. 63) undertook in 1989 magnetometer and resistivity surveys at the proposed route of the Beckhampton Avenue. The magnetometer readings proved inconclusive. Very few negative anomalies in the resistivity readings suggest former stone positions. One was to the south of Adam, three others were over 100 m to the west of Eve (Ucko et al., 1991, 198). At the same time a ditched feature north of the Longstones was visible in the survey (Ucko et al., 1991, 199) and in 1997 as a cropmark (Gillings et al., 2000, 2). This feature and the Beckhampton Avenue were the targets of the excavations at Longstones Field by Gillings et al. (2008). Prior to excavation, in 1999, a second geophysical survey was executed focussing on the weak readings 100 m to the west of the Longstones. This time the survey yielded better results, identifying three low resistance and one high resistance anomaly (Gillings et al., 2000, 2; Gillings et al., 2008, 64, 65 fig. 2.37).

The 'oval' ditched feature[81], now called the Longstones enclosure, preceded the avenue and was deliberately backfilled probably to enable the Beckhampton Avenue to be built in this place, possibly with the intention of linking the older site to Avebury henge (Gillings et al., 2008, 57).

The excavation started in 1999 with the four stone settings from the same year visible in the geophysical survey. The excavation area was then extended to the south-west (Gillings et al., 2000, 4) and several more trenches were opened in the subsequent year. The numbering of the stone settings follows the direction of the excavation: L1 lies on the northern side of the avenue, L2, its pair, on the southern, and the numbering progresses towards the south-west (Fig. 8.12). Besides the stone which remains standing (Eve) and the re-erected one (Adam), three stones were found buried (L2, L5, L6), a fourth had been buried and was later removed (L3), another had probably been buried in a shallow pit and was later burnt (L8), six stones were destroyed by fire (L1, L4, L7, L10, L11, L16), another stone had fallen and was later burnt (L15) and two stones belonging to a previous phase were removed (L12, L13) (Gillings et al., 2008, 75–90, 225, 259–264, 303–313).

According to Gillings et al. (2008, 72, 72 table 2.18) the transverse and longitudinal intervals between the stones are similar to the ones of the West Kennet Avenue given by Smith (1965, 206). The same accounts for the stone hole sizes (Gillings et al., 2008, 73). They are between 0.2–0.5 m deep, L11, L12 and L13 (i.e. the pre-Cove setting) are between 0.6–0.8 m deep. Preparation and filling of the stone holes also matches the observations made elsewhere (cf. Gillings et al., 2008, 73 f.). The three buried avenue stones are between 2.3–3 m in height. L7 and L8 might have been taller and so is L9, Eve, which is over 3 m high. In contrast, its pair (L10) must have been rather small to judge from its stone hole (Gillings et al., 2008, 75). Therefore there is no consistent increase in stone size of the avenue towards the terminal.

The terminal of the Beckhampton Avenue shows signs of reworking (Gillings et al., 2008, 73). Besides the Longstones Cove, which had been first identified by Stukeley, two empty stone holes (L12 and L13, also called Beacock Holes) were found in 2000. These were refilled with clean chalk rubble after the stones, which they had held, were removed (Gillings et al., 2002, 254 f.). The stones were set in a line with L11, each 20 m apart from this central stone, which led the excavators to believe that these three stones belonged to a phase predating the Longstones Cove (see Fig. 8.13, phase 1).[82] The stones of this pre-Cove setting were probably bulky, which can be deduced from the depth of their stone holes. A height of 3–3.5 m is estimated (Gillings et al., 2008, 74).[83]

The Longstones Cove makes up the second phase (Fig. 8.13, phase 2). It is a rectilinear setting with splayed sides (Gillings et al., 2002, 232); L11, which is thought to belong to phase 1, remains in its position between the two avenue stone rows. The remaining Cove stones (L14, L15, L16) were centred on the northern stone row. Together the four stones form a square of *c.* 15 × 10 m (Gillings et al., 2008, 85 f.). The destruction pits and the stone holes indicate that the remaining Cove stones were of a similar size to Adam, i.e. *c.* 3.5 m (Gillings et al., 2008, 86).

[80] During the course of this work a beaker grave was discovered close to the original stone hole. Several other such secondary burials are known from the West Kennet Avenue, the Sanctuary and Avebury henge.

[81] The enclosure is radiocarbon dated to the mid-3rd millennium (this fits to the Grooved Ware sherds recovered from the ditches) rendering it contemporary with the Avebury henge (Gillings et al., 2002, 255).

[82] One could suggest that the excavators were preoccupied with symmetry here. L15 would also lie on this 'line', which is not a very exact alignment (see Gillings et al., 2002, 253 fig. 5). Furthermore there is no proof that L11 was erected at the same time as L12 and L13 and also no explanation why it should be the only stone in this setting, which was not later removed.

[83] The excavators also point out that the 40 m distance between L12 and L13 mirrors the diameter of the Sanctuary. In connection to this, the idea of standardised measuring units has been revived (Gillings et al., 2008, 126 ff.).

Fate of stone	Stone number
standing	L9 (Eve), L14 (Adam)
fallen	L15
buried	L2, L3, L5, L6, L8? (later burnt)
destroyed by fire	L1, L4, L7, L8, L10, L11, L15 (previously fallen), L16
taken away	L3 (previously buried), L12, L13

Table 8.6: Fate of the stones of the West Kennet Avenue at Longstones Field, based on Gillings et al. (2008, 75–90, 225, 259–264, 303–313).

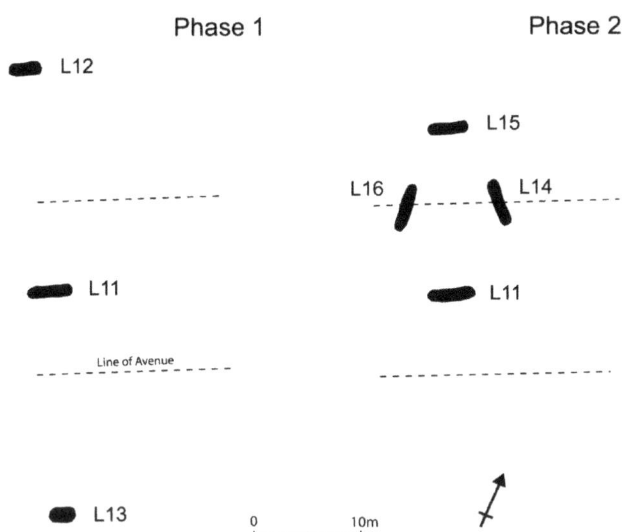

Figure 8.13: The Beckhampton Avenue terminal at Longstones Field (Gillings et al., 2008, 125 fig. 2.83).

8.1.4.3 A continuation of the Beckhampton Avenue?

Stukeley had proposed that the Beckhampton Avenue progressed after the Longstones Cove for some length; in his earlier notes he is also convinced that there must be a 'Beckhampton Temple' at its end in concordance with the Sanctuary at the end of West Kennet Avenue. He was unsure where his assumed Beckhampton temple was to be found and, in failing to produce any substantial proof for it, abandoned the idea in the end (Ucko et al., 1991, 87–89). In his later interpretation the Beckhampton Avenue formed the tail of a serpent, but still its end point, somewhere at Fox Culvert, was no better defined than before (Stukeley, 1743, 36).

The latest excavation, however, suggests that the Beckhampton Avenue did not continue after Longstones Cove. The excavated ground south-west of the Longstones Cove (see Fig. 8.12) revealed no further stone settings (Gillings et al., 2008, 73), which either indicates a gap in the course of the avenue (similar to the West Kennet Avenue) or its discontinuation (Gillings et al., 2002, 252). Some chance encounters of sarsen in the area south-west of it, (see Gillings et al., 2008, 109–115) do not constitute conclusive proof that the Beckhampton Avenue continued. The stones may belong to other settings or could have been abandoned *en route* as Gillings et al. (2008, 109) suggest, or they may well be natural sarsen.

8.1.5 Stone burial and stone-breaking

As can be learned from the descriptions above many of the stones from the henge, the avenues, the Sanctuary and the Beckhampton Avenue terminal were either removed, broken up or buried, thus altering significantly the appearance of the monuments—some of them disappearing entirely from view. The most concise research concerning these burial and destruction events is published in Gillings et al. (2008, 252–364). Several methods for stone burial and stone destruction (with a focus on burning practices) are identified and described. The scarce dating evidence was collected to give at least a rough estimate of when the events took place.

8.1.5.1 Stone burial

Stone burial seems to be the older of the two practices, but also the more long-lasting. No burial events are reported either by Aubrey, Stukeley or any of their contemporaries. On the other hand there are reports of a stone burial from the 19th century and another from 1921/1922 (Gillings et al., 2008, 253). The finds from the burial pits often comprise medieval sherds (11th to 14th century, with a focus on the 12th and 13th century (Jope, 1999, 67, 69)), which are usually abraded (Gillings et al., 2008, 267). They are thought to be the remains of manure brought into the henge, which was under the plough till the 12th or 13th century (Jope, 1999, 68 f.). They therefore do not date the burial events.

A definitive date for a stone burial is given by the discovery of the 'barber-surgeon' at stone setting 9 of the outer circle. An adult male skeleton was excavated in 1937 (Keiller, 1939, 231)[84], accompanying it were a pair of hinged scissors, and an iron implement with the remains of a wooden handle, which has been subsequently interpreted as a lancet or probe, an iron buckle and three silver coins, possibly once contained in a pouch (Smith, 1965, 178). The scissors and the iron implement led to the assumption that the man might have been a barber-surgeon[85]. The coins found with the deceased can be dated to *c.* AD 1320–1350 and rather to the earlier part of this range (Ucko et al., 1991, 178). Stone 24 of the outer circle must have been buried between Stukeley's last visit to the site (1724) and the filling of a pond at the beginning of the 19th century (oral testimony of villagers) (Gillings et al., 2008, 277, 290). Further dating evidence comes from a cattle rib found in the clean chalk backfill of stone setting L6, the radiocarbon determination suggesting a date in the 16th or 17th century[86]; a cattle tibia found in a perforation of the same stone (radiocarbon dated to the 12th–13th century) was dismissed as dating evidence[87] (Gillings et al., 2008, 278 f.).

According to this evidence Gillings et al. (2008, 279) conclude that the majority of stones must have been buried after the 12th century (due to pottery found in the burial pits) and before Stukeley's visits to Avebury (1719–1724)[88]. There is also evidence of clumping: the stones were not buried sporadically, but in groups (Gillings et al., 2008, 280). Pollard and Reynolds (2002, 246) think that this might be due to different plot owners, some burying the stones, others not. The stones were also buried in substantial numbers, i.e. *c.* 33–39% of the stones at Avebury, 23% of the West Kennet Avenue and 50% of the Beckhampton Avenue stones[89]. The burial events seem to have been episodic (adjacent stones are frequently buried in the same manner, and in the case of settings i–xi they were even buried in communal pits) and spread over *c.* 330 years (Gillings et al., 2008, 280 f.).

The hypotheses as to why the stones were buried are quite varied. At first agricultural reasons had been proposed by Stukeley, Long, Cunnington, as well as Keiller and Piggott (see Ucko et al., 1991, 178). This position changed when the 'barber-surgeon' was discovered. Keiller (1939, 231) changed his mind, because it were not (only) local farmers who had buried this stone, but the help of an outsider, the barber-surgeon, had been obtained. W. E. V. Young, Keiller's foreman, argued in his diary in 1938 that buried stones would be more hazardous to the plough because they could not be seen and avoided (Ucko et al., 1991, 179). As Pollard and Reynolds (2002, 245) point out also the majority of stones were probably buried after ploughing ceased at Avebury henge. The situation outside the henge is, of course, different, but even there not all of the stones were buried (Gillings et al., 2008, 282).[90]

That the burial practices are not simply a way to get rid of the sarsen—they could have just as easily been dragged away or toppled into the ditch if they stood in the outer circle—is perhaps best demonstrated by the deliberate refilling of the pits with clean chalk rubble. As Gillings et al. (2008, 284) put it: 'the sarsen were not so much buried as *sealed* beneath the ground' under the rammed chalk packing. This also means that at excavating the burial pit the topsoil must have been kept from mixing with the underlying chalk to allow this process (Gillings et al., 2008, 274). Another curiosity is provided by stones which did not fall correctly into their burial pit. No attempt was made to correct their position or to recut the pit even if parts of them were visible above ground (Gillings et al., 2008, 274). Another reason

[84] There is some discussion whether this man was killed by the slipping stone he intended to bury (Keiller, 1939, 231) or is indeed a murder victim concealed at a chance opportunity (Pollard and Reynolds, 2002, 244 f.). The latter seems—due to practical reasons—a bit unlikely. Digging the hole and refilling it would be labour intensive (and is usually thought to be the work of several people), and a single person hiding the corpse and refilling the pit alone might draw attention shortly after a murder incident. Both an accident and a murder could explain the presence of the coins in the man's possession. Hävernick (1975, 50) has found that several coins were present in burials with victims of violent or unexpected death, whether these persons were interred inside (usually close to to the cemetery wall) or outside a cemetery, at the gallows or in the forest and fields. It was assumed that some kind of superstition kept the living from taking the money. These burials date from the 12th to the 17th century and may present a parallel to the case of the barber-surgeon.

[85] Two other occupations were suggested: a tailor (Smith, 1965, 178) or a shepherd (cf. Gillings et al., 2008, 276).

[86] Later dates (18th–19th and 20th century) have been dismissed as unlikely, because they would have been recorded by antiquarians or later researchers.

[87] For a full discussion for the reasons of this decision see Gillings et al. (2008, 278 f.).

[88] Certainly also before Aubrey's time, who also recorded huge gaps in the stone settings.

[89] About 164 stones in total (Gillings et al., 2008, 281).

[90] There might be again the question of the individual landowners (Gillings et al., 2008, 283).

for the burial of the stones was advanced by Burl (1979, 37), who suggested that the sarsens may have been buried for religious reasons because they were seen in connection with the devil.[91] But these notions cannot be supported with any evidence (Ucko et al., 1991, 179 f.; Pollard and Reynolds, 2002, 242; Gillings et al., 2008, 286 f.). Gillings et al. (2008, 287) regard hypotheses that there were superstitious motives for the stone burials with caution for basically the same reasons. There is little that can be proven by facts either concerning why the stones were buried or why this practice stopped finally.

8.1.5.2 Stone breaking

Stone breaking took place during a much narrower time span. Stones were broken by direct fracture, using wooden wedges[92] or by fire.

The first reference to the breaking of the sarsen with the 'fire and water method' and their use for building purposes comes from Aubrey (1980, 38)[93]. Gillings et al. (2008, 292 f.) think that Aubrey may have referred to natural sarsen in the fields, especially as he never personally witnessed one of these breaking events. Aubrey thought that the houses at Avebury were built from the sarsen present in the monument, but here again it is argued that he was not an eye witness and that the material could have come from other sources outside Avebury.[94] The second description of the stone burning, this time also accompanied by a drawing and definitely referring to the stones of the monuments, comes from Stukeley (1743, 15 f.)[95]. He also mentions the high costs of burning stones, and that the material could have been obtained more cheaply from natural sarsen (colloquially called 'grey wethers'[96]) of smaller size (Stukeley, 1743, 16). The stones would then be used for building houses, but the material would be quite unsuitable for the task. Stukeley gives economical reasons for breaking the stones, claiming that it was a means of clearing the land (Stukeley, 1743, 15). The implements and materials described by Stukeley and depicted in two of his drawings match what could be archaeologically discovered (Gillings et al., 2008, 295). Several methods for breaking stones with the aid of fire could be identified, amongst others depending on whether the stone was still standing, recumbent or felled for the purpose (Gillings et al., 2008, 328–331). Straw, and tree branches were used as fuel, the latter probably from managed coppices (Gillings et al., 2008, 324). Clay pipes found in the burning pits help to date the events; they were manufactured between 1660 and 1720 (Gillings et al., 2008, 301) with the greatest bulk belonging to the ± 20 years around 1700 (Gillings et al., 2008, 338). As in the stone burials, destroyed sarsen at Avebury and the West Kennet Avenue are grouped together and neighbouring stones were often attacked with the same methods. One of the objectives seems to have been to clear the outer circle, as most of the then visible stones were destroyed and moreover some buried sarsen were indeed unearthed (Gillings et al., 2008, 339, 339 fig. 10.26). Despite the increased attention paid to the destruction-by-fire method, a great number of sarsen were either dragged away or destroyed by direct fracture (Gillings et al., 2008, 331). Stukeley blamed the stone destruction on 'farmers' in and around Avebury; at the same time he firmly closed his eyes to the dilapidations of stones and earthwork wrought by gentlemen (Gillings et al., 2004, 155). He was also convinced that the stones were cleared away for agricultural reasons. The archaeological evidence (and Stukeley's own reports) argues against this. So far only one stone hole (L15) could be proved to be deliberately backfilled. The destruction pits inside Ave-

[91] The arguments of Jope (1999, 67) run in a similar vein.

[92] So reported for stone 47? (the stone east of the road at the northern entrance) of the outer circle (Gillings et al., 2008, 292).

[93] 'Within the circumference, or Borough of this Monument, is now the Village of Aubury, which stands per crucem, as it is to be seen by scheme the Ist. The Houses are built of the Frustrum's of those huge Stones, which they invade with great Sledges for here about are no other stones to be found (except Flints). I have verbum Sacerdotis for it, that these mighty stones (as hard as marble) may be broken in what part of them you please; without any great trouble: sc Make a fire on that line of the stone, where you would have it crack; and after the stone is well heated, draw over a line with could water, & immediately give a krack with a Smiths sledge, and it will break, like the Collets at the Glass house.' (Aubrey, 1980, 38). The following passage has been crossed out by Aubrey (1980, 39): ~~'The houses in this Village are built with the Frustrum's of the Stones of this Monument/Antiquity; for hereabout are no other stones to be found, (except Flints).'~~

[94] It can be argued equally well, that his assumptions were right and his informant (a priest) on the destruction method by fire referred to stones inside Avebury. The evidence is simply not sufficient.

[95] 'The method is to dig a pit by the side of the stone, till it falls down, then to burn many loads of straw under it. They draw lines of water along it when heated, and then with smart strokes of a great sledge hammer, its prodigious bulk is divided into many lesser parts.' 'They have sometimes us'd of these stones for building houses; but say, they may have them cheaper, in more manageable pieces from the gray weathers. One of these stones will build an ordinary house; yet the stone being a kind of marble, or rather granite, is always moist and dewy in winter, which proves damp and unwholsom, and rots the furniture.' Stukeley (1743, 15 f.).

[96] The natural sarsen fields are thought to look like flocks of sheep. The term 'wether' denotes a castrated male.

bury henge were often used as dumps[97], still visible as cavities, used as holes to plant trees in, or actually fenced off. The pits outside the henge filled themselves through ploughing or sedimentation (Gillings et al., 2008, 325 f.). Gillings et al. (2004, 155–161) suggest that nonconformists might be responsible for the destruction of stones, as an act of defiance to the Anglican church, and at the same time, providing material for their chapel.[98] On the other hand Stukeley mentions no such religious reasons, and as Gillings et al. (2008, 343–350) argue, there is no strong connection between the stones of the monument and the Anglican church, which would justify these acts. Yet another hypothesis, by Edwards (2003), sees religious reasons[99] coupled with the need for infrastructure due to increased traffic, especially for road works and building materials for inns. If there was indeed an increased demand for land however, due to new settlers, why did the destruction pits remain 'unused' (Gillings et al., 2008, 351 f.)? Besides that, Stukeley mentioned that the stone was not especially suited for building purposes (Stukeley, 1743, 16).

8.1.6 Other final neolithic sites in the Avebury region

The three following sites will be described in short because they belong to the same time as the henge and the avenues and intervisibility existed between some of them. Their proximity in space and time makes them part of the same ritual landscape, though they only feature marginally in the reconstructions.[100]

8.1.6.1 Silbury Hill

This artificial mound[101] is situated *c.* 1.23 km south of Avebury and around 1 km from the West Kennet Avenue. The mound has the shape of a truncated cone and rises 37 m high (Atkinson, 1967, 259; Whittle, 1997, 7). Its top is rather flat, but could have been initially dished (Whittle, 1997, 8), with 30.5 m in diameter (Atkinson, 1967, 259). There is a pronounced terrace approximately 5 m below the flat summit (Devereux, 1991, 894), and other not so pronounced terraces are further below (Whittle, 1997, 8). The topmost terrace seems to be an original feature (Atkinson, 1970, 314)—it is surmised that the mound was built as a stepped cone whereof the steps were later infilled. The last terrace may have remained because the project was abandoned before completion (Whittle, 1997, 25 f.). The sides of the mound have a slope of *c.* 30 ° and it is surrounded by a ditch. Both the ditch and scraping of the spur, on which the mound was built, were probably employed to make the mound appear even higher. The ditch is unevenly shaped, and in its silted state about 150 m wide to the west, but 37 m to the north and east[102] (Atkinson, 1967, 259). Where it had been excavated it was over 6 m deep (Pollard and Reynolds, 2002, 120). To the south the ditch is only 27 m wide, but was found to be almost 10 m deep, with steep sides and a 14 m wide flat bottom (Atkinson, 1970, 313).[103] Probably to prevent silting and landslides, the southern side of this ditch was revetted with timbers and blocks of chalk (Atkinson, 1970, 314). It is separated by two causeways from the main ditch (Atkinson, 1967, 259), and these allow easy access to the mound. Good evidence for at least three phases exists, but construction may have paused even more frequently (Whittle, 1997, 26).[104]

Silbury Hill was built on the valley bottom, so that it cannot be seen from everywhere (Thomas, 1993, 38). Nevertheless there is intervisibility between it and the Sanctuary (Devereux, 1991, 859; Watson, 2001, 299). Additionally, the West Kennet Avenue near the Sanctuary appears to be leading towards Silbury Hill (see fig. 8.2), but then makes a turn towards the north (Thomas, 1993, 41), otherwise the line of sight between the West Kennet Avenue and Silbury Hill is broken by Waden Hill (Pollard and Reynolds, 2002, 118). It can also be seen from the Beckhampton Avenue (Gillings and Pollard, 2004, 80) and from some points of the ridge, which runs north to south through Avebury henge, and here especially from the northern circle and from the position of the Obelisk (Devereux, 1991, 895 f.; Watson, 2001, 306).

[97]Wine bottles, a wide range of ceramics and horse bones were among the finds (Gillings et al., 2008, 325 f.).

[98]This has been similarly suggested by Burl (1979, 46), but he saw in it an act of driving out 'idolatrous faiths'.

[99]Nonconformists were banned from bigger cities by the Five Mile Act, which would have made Avebury a popular location being situated just over five miles from Marlborough, Pewsey, Devizes, Calne, Wootton Bassett, Wroughton, Chiseldon, and the Ogbournes (Edwards, 2003, 44).

[100]See reconstructions by Cripps (2001), Section 8.2.2.2, and Davies (2009), Section 8.3.2.1.

[101]According to Atkinson (1967, 259) Silbury Hill is the largest ancient man-made mound in Europe.

[102]Whittle (1997, 25) mentions that the ditch is only 20 m wide in the east.

[103]In comparison: Avebury's ditch is on average at the top 9 m wide, the bottom 4.5 m wide and at its deepest excavated part almost 10 m deep.

[104]For the latest dating evidence based on radiocarbon samples and Bayesian statistics see Bayliss et al. (2007).

8.1.6.2 Falkner's circle

750 m south-east of Avebury lies Falkner's circle (Gillings et al., 2008, 142) (see Figs. 8.2 and 8.6) in clear view from the West Kennet Avenue (Gillings and Pollard, 2004, 78). It is one of several small circles which have been mentioned in antiquarian records (see Pollard and Reynolds, 2002, 110). The circle was first observed in 1840 by a Mr. Falkner, who claimed that it had once possessed 12 stones of which one was still standing, two were recumbent and the places of the other nine were only marked by hollows (Gillings et al., 2008, 142 f.). Today only the standing stone remains; it is of the same type (i.e. a grey sarsen) as the stones used for Avebury and the avenues and must have therefore been transported to the location (Gillings et al., 2008, 143). The south-eastern part of the circle was excavated in 2002 after a geophysical survey had taken place. One destruction pit and three further possible stone settings were discovered. The circle possibly had a diameter of 44 m. Gillings et al. (2008, 151) are in some doubt whether the circle was really entirely an artificial setting, or whether some parts of it were already in place due to naturally occurring sarsen in the region. The basis for this is that two of the supposed stone holes appear too shallow and too gently sloped, but stone breaking debris indicates that there must have been stones present. Although being visible from the West Kennet Avenue, this circle is not physically connected to it and, despite its diameter being similar to the Sanctuary, it is a different kind of monument since it had no timber phase and consisted of considerably fewer stones. As Gillings et al. (2008, 152) put it: 'It pales besides Avebury, Silbury Hill and the West Kennet palisades.' and may therefore just be the work of a small group of people.

8.1.6.3 West Kennet palisade enclosures

The West Kennet palisade enclosures (see Figs. 8.2 and 8.6) were first detected on an aerial photograph taken in 1950. Excavation of palisade enclosure 1 started in 1987. In 1989 palisade enclosure 2 was discovered, also by aerial photography (Whittle, 1991, 256 f.). Palisade enclosure 1 lies less than 200 m from the West Kennet Avenue, while palisade enclosure 2 is situated *c.* 700 m from Silbury Hill and almost 2 km south of Avebury[105]. Nowadays the River Kennet runs through palisade enclosure 1, but it is assumed that it was not such an active stream when the monument was first built. The monument is sub-circular in shape and possibly consisted of two concentric post rows (to the north, only one row has been discovered so far) (Whittle, 1997, 53, 57). Palisade enclosure 2 is rather oval in shape, and may have been open to the north. Three additional sub-circular or oval palisades, all with concentric inner circles, were discovered inside enclosure 2; another one lies 220 m to the south-east of palisade enclosure 2 but is connected to it via a radial ditch (Pollard and Reynolds, 2002, 115). Another ditch connects the two enclosures, while more radial ditches are situated inside enclosure 2 (Whittle, 1997, 57, 82 f.). The ditches for the post settings were around 2 m deep and included massive post settings and a backfill containing sarsen packing. The posts may have been burnt above ground, while their lower parts were allowed to rot *in situ* (Whittle, 1997, 57, 152, 157). They were probably of oak and had a diameter between 25–40 cm and were spaced 10–15 cm apart. From their insertion points it is assumed that the posts were 6–8 m high, possibly higher at the entrances to the monument[106] (Whittle, 1997, 152, 154). 2800 posts were needed for enclosure 1 and additional 1600 for enclosure 2 without any of the smaller features; this amount and the homogeneous size of the posts suggest that secondary, maybe even managed, woodland was to be found nearby (Whittle, 1997, 154, 140). Because none of the ditches were recut nor any of the posts replaced, the monument could have existed for roughly 100 years before rotting away (Whittle, 1997, 156). Nothing which could indicate a habitational or defensive use was encountered and therefore it has to be assumed that these monuments were meant for rituals (Whittle, 1997, 156).

Besides the monuments mentioned here, there were also earlier monuments (e.g. long barrows or the causewayed enclosure on Windmill Hill) which could have had an influence on how later monuments were positioned in the landscape.

8.1.7 Date and sequence of the monument complex

Pitts and Whittle (1992) first tried to establish a sequence of the monuments in the Avebury region based on radiocarbon dates. Before that there had only been discussion about the relative chronology

[105] Avebury and the West Kennet palisade enclosures are not intervisible (Parker Pearson and Ramilisonina, 1998, 319).

[106] This has been assumed due to the substantial amount of packing stones found at the entrances (Whittle, 1997, 152).

of the monuments based on their structural relationships (see e.g. Section 8.1.2.1) or on ceramics. In a later article Whittle (1993, 35) broadened the research from the confines of Avebury to the Avebury region. He divided the Neolithic Period in the Avebury area into six phases whereby the monuments so far discussed fall into phases D (3361/3109–2916/2782 cal BC), E (2916/2782–2564/2457 cal BC) and F (2564/2457–2133/1959 cal BC) (Whittle, 1993, 35; cf. Whittle, 1997, 140). Whittle argues that the West Kennet Avenue could fall into phase D based on structural reasons (see Section 8.1.2.1) and because other linear monuments would fall into this period (Whittle, 1993, 38).[107] The earthwork at Avebury is supposed to fall into phase E. Whether the stone settings are contemporary with the earthwork or a later addition is left open.[108] The radiocarbon dates suggest that Silbury Hill was built after Avebury's earthwork, maybe on the border of phases E and F. The stone settings at Avebury as well as the construction of the West Kennet palisade enclosures may have fallen into phase F (Whittle, 1993, 38 f., 46 f.). Whittle assumes that Avebury and Silbury Hill may have already fallen out of use by the end of this period.

It is also important to know that in phase D there is evidence for more scrub or woodland, after there had been more woodland clearances in phase C. Renewed clearances took place in phase E, which establishes the possibility for intervisual relationships between the above mentioned monuments. In phase F the trend for open country prevails (Whittle, 1993, 35, 39–42).

Pollard and Cleal (2004, 124–127) changed the monument sequence slightly in the light of new radiocarbon dates from Avebury. The date of the primary bank remains an enigma, basically because shards recovered from underneath the bank during the early excavations cannot be exactly localised. They may have come from underneath the primary or the main bank. This could have been built in the late 4th millennium BC or the early 3rd millennium BC (Pollard and Cleal, 2004, 124 f.) (which corresponds with Whittle's phase D). The main earthwork can be dated to the early and mid 3rd millennium BC (Pollard and Cleal, 2004, 120 table 1, 121 fig.1, 124) (falling into Whittle's phase E). It could therefore be slightly later than the Longstones enclosure and marginally earlier than Silbury Hill and the West Kennet palisade enclosures. The date of the avenues is left open, while the outer circle is thought to be contemporary or to slightly postdate the earthwork (Pollard and Cleal, 2004, 125). The radiocarbon date obtained from stone hole 41 of the outer circle shows that some activities took place in the Bronze Age (maybe also incorporating the Ring-stone and probably stone D).

Gillings et al. (2008) envisage the monument sequence yet in another, slightly different way. Structures of the Later Neolithic, *c.* 3000–2600BC (roughly equivalent with Whittle's phase F), would encompass the Longstones enclosure (2820–2660BC), the primary bank at Avebury and maybe the Cove also, with an assumed date around 3000BC (Gillings et al., 2008, 191, 193). Gillings et al. (2008, 193) also mention the main earthwork at Avebury in this context, but it is not made explicitly clear that is is meant to belong to the same phase. For the Final Neolithic, *c.* 2600–2200BC (equivalent to Whittle's phase E), the construction of the outer circle at Avebury, the avenues, the Sanctuary and Silbury Hill are all envisaged together. This is due to the Grooved Ware, which is associated with all these sites, appearing from *c.* 2800BC onward (Gillings et al., 2008, 193). The authors state, in keeping with Pollard and Cleal (2004), that the erection and replacement of stones could have continued into the Bronze Age. The construction of Silbury Hill and the West Kennet palisade enclosures is thought to fall into the same phase (Gillings et al., 2008, 194).

8.1.8 History of Research

One of the aims of the Negotiating Avebury Project was not only to build a model of what was known about Avebury at the time, but also to include models of the antiquarian records.

William Stukeley's published work on Avebury, *Abury* (Stukeley, 1743), and additional documents as presented by Ucko et al. (1991) have been variously referred to already. The same accounts for his predecessor, John Aubrey, and his *Monumenta Britannica,* which was first published in 1980 (Aubrey, 1980). Here also various additions and remarks are provided by Ucko et al. (1991).

In the following passage the antiquarian records are brought into context and some of the important facts, especially concerning the value and reliability of the records, will be outlined. The most elaborate work on this topic has been compiled by Ucko et al. (1991).

[107] That appearances may be misleading is well demonstrated by the Longstones enclosure, which is approximately 1000 years younger than other causewayed enclosures, e.g. Windmill Hill (see Section 8.1.4.2).

[108] On the other hand Gillings et al. (2008, 193) suggested that the stone settings of the Cove may predate the main earthwork at Avebury; cf. Whittle (1993, 46).

8.1.8.1 The antiquarian records

The henge monument at Avebury, was first 'discovered' by **John Aubrey** in 1649 (Ucko et al., 1991, 10 f.), though the stones and the earthwork were mentioned in much earlier records.[109] In 1663 Aubrey and **Walter Charleton** gave a presentation concerning Avebury at a Royal Society meeting, where two plans (Ucko et al., 1991, plans C and D) of the monument were shown to the audience (Ucko et al., 1991, 17–22). These plans are rather schematic, but show already several of the most prominent features of the henge. Charleton's plan, plan D[110], already describes the bank and the ditch which lies on its inside. The outer circle of stones and two inner features are depicted: one is the southern circle with the Obelisk at its centre and the other the Cove. Besides this, four entrances, each flanked by two stones at the outside of the monument, and a single road are shown (Ucko et al., 1991, 19, 129, 131 f.). The depiction is very schematic and the monument appears to be symmetric (Ucko et al., 1991, 64). Aubrey's plan, plan C, shares some of the features of Charleton's plan (the Cove, the entrance stones, the outer circle of stones and the earthwork), but also adds features (crossroads at right angles, a multitude of houses, field boundaries, the church and a stream outside the henge, as well as concentric inner stone circles) of which some are more and some less 'correct'.[111] Otherwise, he did not observe the southern circle with the Obelisk, and also omits a stone pair at the western entrance (Ucko et al., 1991, 22, 120–127). In neither plan C nor plan D is the northern circle depicted. The presentation at the Royal Society led to increased interest in the monument, resulting in a visit of King Charles II and several of his courtiers to Avebury, Silbury Hill, the Sanctuary on Overton Hill and the West Kennet Avenue in the same year (Ucko et al., 1991, 25–28).

It is quite probable that Aubrey also made his plane-table survey, plan A (Aubrey, 1980, 44 f.), in 1663 (see Fig. 8.14). Unfortunately the survey was hampered by the houses, fences, etc. inside the monument and possibly led. together with one or several measuring errors, to a distorted plan of the site (Ucko et al., 1991, 30, 108 ff.). Nevertheless the irregular outline of the bank and ditch, the stones of the outer, the northern and the southern circles, the Cove and the northern end of the West Kennet Avenue are shown, also the church, field boundaries and streets are included, but buildings from inside the henge are omitted from the design (Ucko et al., 1991, 30–32, 100, 102–108). A third plan by Aubrey, plan B (Fig. 8.15) (Aubrey, 1980, 48 f.), is based in part on his plane-table survey and depicts the relationship of Avebury, the West Kennet Avenue, the Sanctuary and Silbury Hill including some topographical reference points. One of the interesting features is how the avenue is linked to the Sanctuary: there is a kink in its approach and it narrows down close to the Sanctuary[112]. In addition the West Kennet Avenue changes course abruptly at West Kennett (Ucko et al., 1991, 32, 111, 113–116).[113] Aubrey tried to record the monument complex faithfully and was aided in his endeavour by surveying methods (which did not prevent his plan from becoming distorted) (Ucko et al., 1991, 61, 67). It also seems that he spent very little time on site (Ucko et al., 1991, 62).

Plans A and B were intended to accompany Aubrey's publication of *Templa Druidum* which he compiled around 1663–5, but he continued editing and adding to this work and to the related *Monumenta Britannica* up to the 1690s. This resulted finally in discrepancies between the text and the illustrations (Ucko et al., 1991, 32–35). Aubrey's books remained unpublished at the time, but some of his work found its way into Camden's *Britannia* (Ucko et al., 1991, 35 f.).

In 1723 **Thomas Twining** published *Avebury in Wiltshire, the Remains of a Roman Work*, which is noteworthy because its frontispiece (Fig. 8.16) shows *two* entrances to Avebury being connected to a single stone avenue (with a somewhat eccentric course), which also incorporated the Sanctuary. His record is strongly influenced by preconceptions, e.g. the avenue would mirror the shape of Britain (as the Romans thought it would look) and the design of the monument is very geometric with evenly spaced stones, right angles (where possible) and exact concentric circles (Ucko et al., 1991, 37 f., 38 fig. 10).

William Stukeley probably became aware of Avebury through Camden's *Britannia* and had, additionally, access to copied excerpts of Aubrey's *Monumenta Britannica* before he first visited the site in 1719 (Ucko et al., 1991, 39, 42–48). He visited Avebury on a yearly basis between 1719 and 1724 (Ucko et al., 1991, 48),

[109] See Ucko et al. (1991, 8 ff.). Usually the stone settings are taken as being natural—in contrast to being man-made features. Some remarks deal with the earthwork, which is frequently thought to be defensive in nature.

[110] The nomenclature established by Ucko et al. (1991) will be used here because it helps to distinguish between plans of the same author.

[111] For a full discussion see Ucko et al. (1991, chapter 3).

[112] This is also visible on his detailed plan of the Sanctuary (Aubrey, 1980, 51).

[113] A number of mistakes are also incorporated in the depiction, see Ucko et al. (1991, 111).

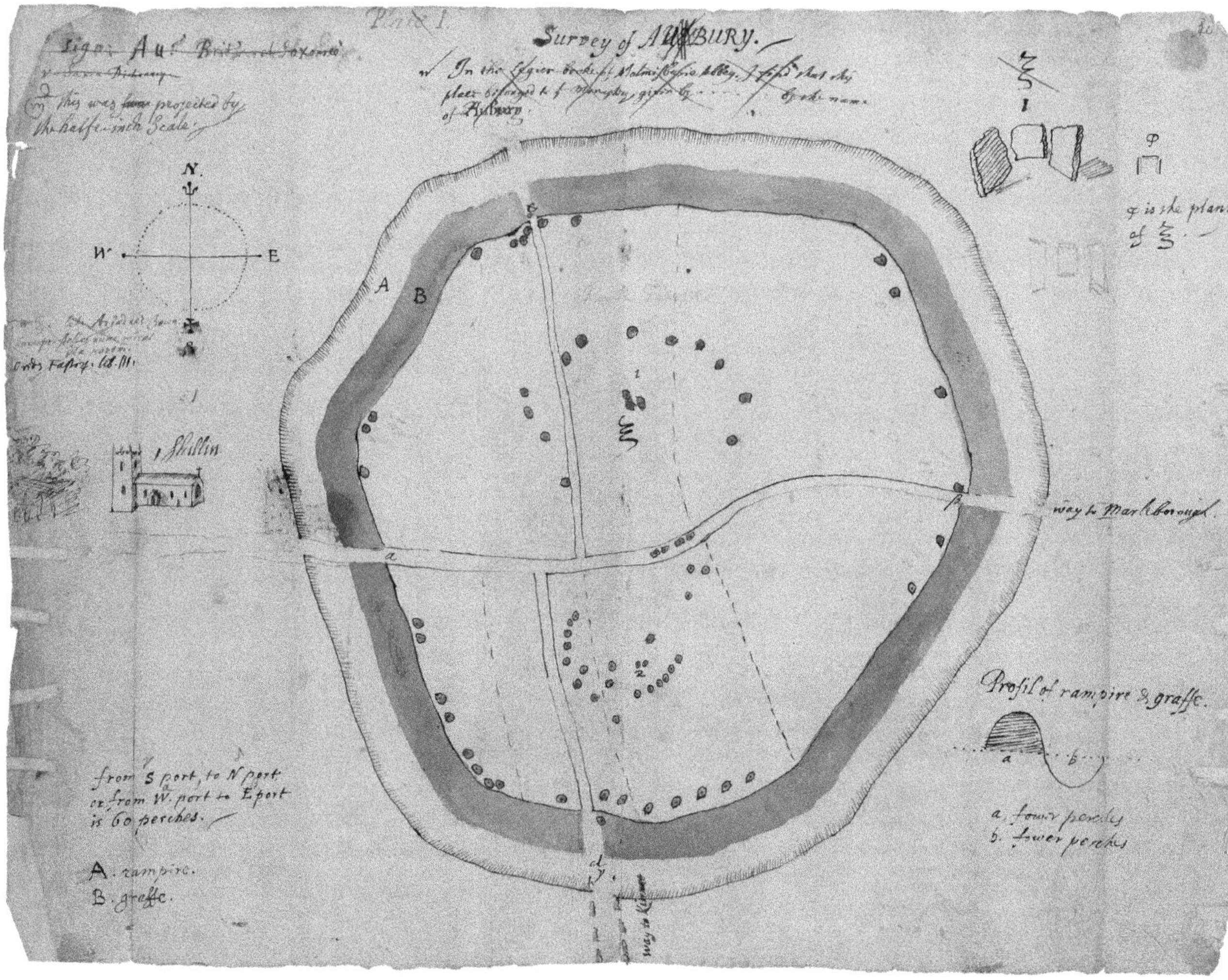

Figure 8.14: Plan A, from Aubrey's *Monumenta Britannica* (Aubrey, 1980, 44 f.), plane table survey of Avebury. (The Bodleian Libraries, The University of Oxford: MS.Top.Gen. c.24, fols. 39v–40.)

composed field notes and made drawings (also employing surveying techniques (Ucko et al., 1991, 67)), altered previous drawings (Ucko et al., 1991, 51, 133), interviewed the villagers and composed a written tour to the monument[114].

He published some of his work in *Abury*[115] (Stukeley, 1743)[116]. Besides this, several unpublished records survive, among them 11 draft plans (i.e. plans N–X, two of them incomplete) for his frontispiece to *Abury* (plan E, see Fig. 8.17)[117] (Ucko et al., 1991, 51). Piggott (1935, 26–28) tried to reconstruct the sequence of Stukeley's work and the development of his theories by means of the latter's dated drawings. Later alterations in Stukeley's drawings were assumed by Piggott to be made in the year following the date of the initial sketch, but there are several instances where this seems to be unlikely (Ucko et al., 1991, 136 f.).

None of the plans agrees entirely with another, and neither does any agree with the written tour (Ucko et al., 1991, 133, 136). As the plans cannot be brought into a definitive temporal order (leading up to plan E) and the text in *Abury* also does not correspond with plan E (Ucko et al., 1991, 132 f.), it has been argued that plan E cannot be 'a summation of all that is valid in them' (Ucko et al., 1991, 137). In addition

[114]The written tour only survives in parts and is published in Ucko et al. (1991, appendix II).

[115]The full text is available on the internet: http://www.avebury-web.co.uk/AburyWS/AburyWS.html (Accessed: 6th of August 2012.).

[116]Stukeley had previously intended to compile four books on the 'Celts': one about the Celtic people, one on Celtic religion, another on Celtic temples (among them Avebury) and a volume devoted to Stonehenge (see text quoted in Ucko et al., 1991, 52).

[117]The latter also exists in two variants (Ucko et al., 1991, 133).

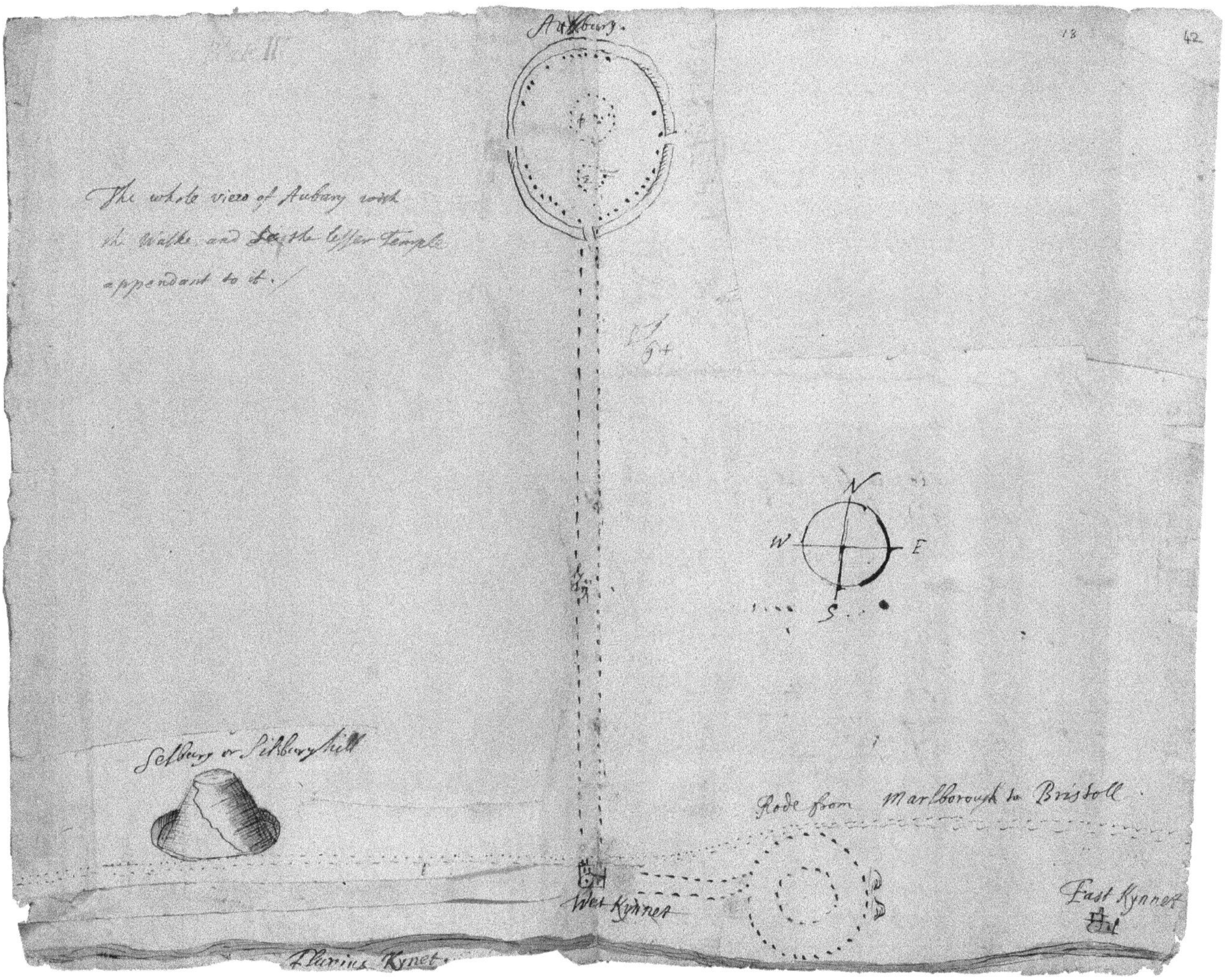

Figure 8.15: Plan B, from Aubrey's *Monumenta Britannica* (Aubrey, 1980, 48 f.), showing Avebury, the West Kennet Avenue and the Sanctuary. (The Bodleian Libraries, The University of Oxford: MS.Top.Gen. c.24, fols. 41v–42.)

to the plans and his text Stukeley also prepared several drawings, which are included as plates in *Abury* (Stukeley, 1743, tables VIII, XII–XXVIII).[118]

Stukeley states that his plans are meant to represent what he thought was 'the builders intent'—e.g. that the earthwork and outer circle were meant to be circular—and less the actual state of the monument (see quote in Ucko et al., 1991, 69). There are also inconsistencies in the positioning of the northern and southern circles and in the latter's size (Ucko et al., 1991, 152), which lead to the question concerning how much he relied on his actual survey and where he departed from it.

Some of the inconsistencies in Stukeley's record can be attributed to his preconceptions. Since 1722 Stukeley seems to have developed a 'cosmic theory', which is expressed in naming the northern and southern circle 'Lunar Temple' and 'Solar Temple' respectively (e.g. Ucko et al. (1991, plan O)), the Sanctuary is titled 'The Temple of Ertha' and a supposed monument at the end of Beckhampton Avenue is termed 'The Temple of the Infernal Regions' (Piggott, 1935, 26–28). These ideas are based on pagan religions with a growing focus on the Egyptian (Ucko et al., 1991, 74–86). At the same time he was convinced that some sort of numerical symbolism was employed by the designers of the monuments (Ucko et al., 1991, 78–81), and styled the number of stones accordingly.

[118]Table VIII shows the whole monument complex, tables XII–XVII are concerned with views of part of the henge, tables XVIII–XIX and XXII depict the West Kennet Avenue, tables XX–XXI focus on the Sanctuary, tables XXIV–XXV show explicitly the Beckhampton Avenue, but it should also be visible on table XXIII. Tables XXVI–XXVIII are dedicated to Silbury Hill, which also features prominently on table XXIII.

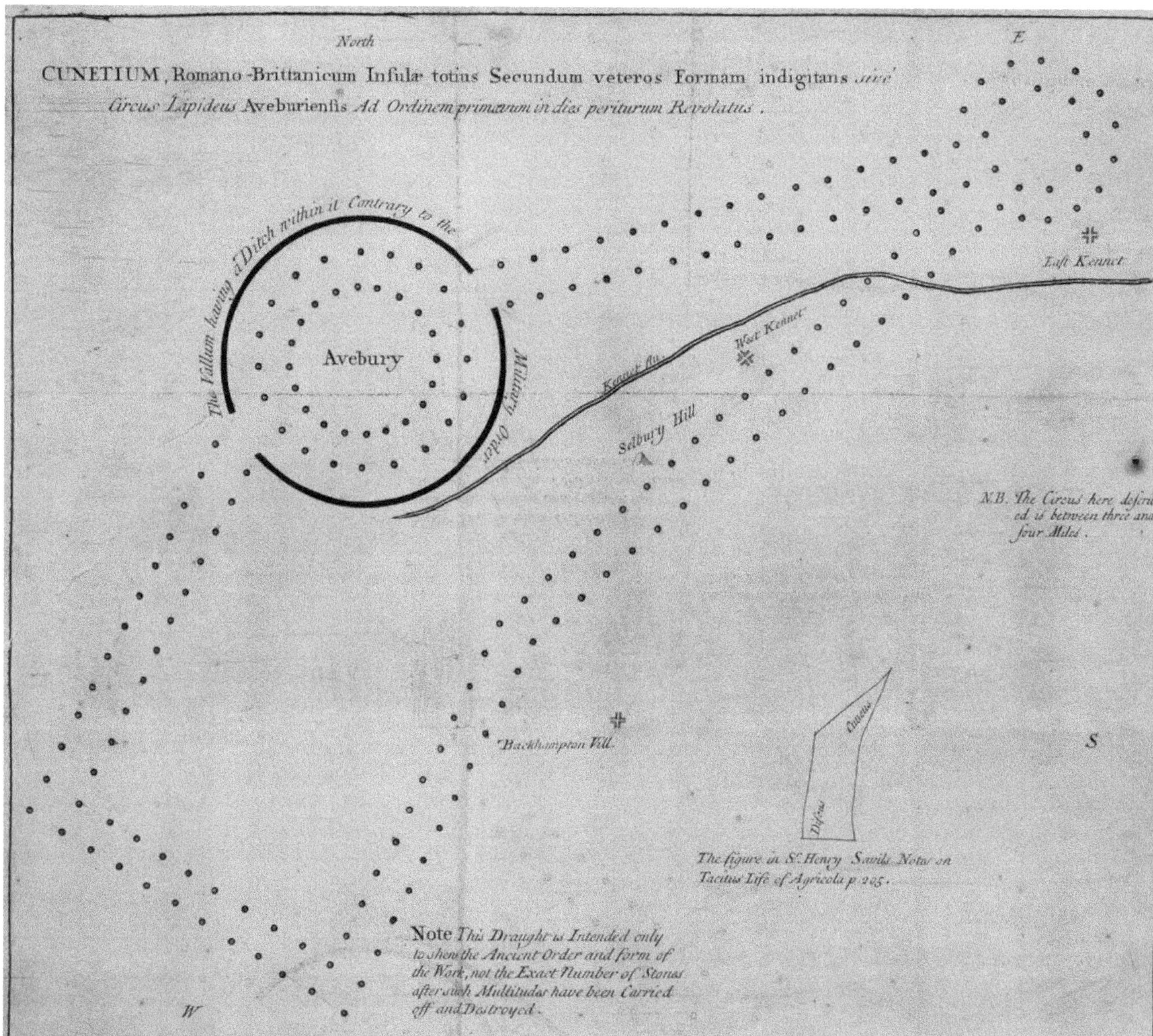

Figure 8.16: Thomas Twining's plan of Avebury, the avenues and the Sanctuary, his frontispiece to *Avebury in Wiltshire, the Remains of a Roman Work* (1723). (The Bodleian Libraries, The University of Oxford: B 7 20(9) Linc. Frontispiece.)

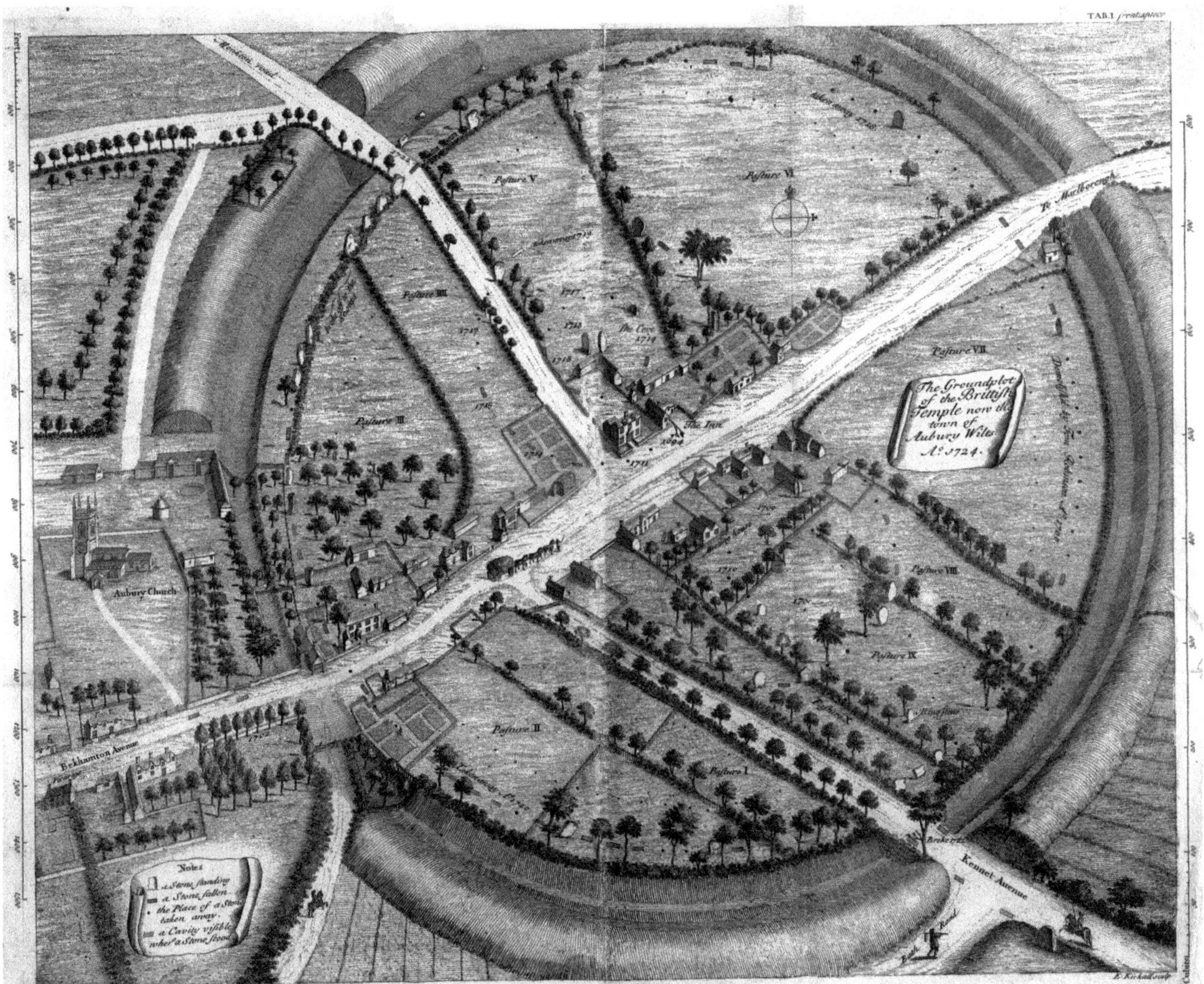

Figure 8.17: Plan E, from Stukeley's *Abury* (Stukeley, 1743, frontispiece), state E2. (The Bodleian Libraries, The University of Oxford: fol. Godw. 158 Frontispiece.)

For example, he was convinced that the northern and the southern circle each possessed an inner circle and that those were made up of 12 stones (the number of signs in the zodiac). Also, duality was important for him: each of the circles (outer, northern and southern) had an inner concentric circle, there were two avenues, and even the Sanctuary consisted of two concentric circles.[119] Besides the number symbolism he was (in 1723) convinced that the monument complex was meant to be symmetrical, even that the avenues (both with a temple at their ends) together with Avebury would be, when seen from above, a two dimensional representation of the outline of Silbury Hill (see quote in Ucko et al., 1991, 84, and 83–86, 84 fig. 21).

In 1724 he abandoned his celestial theory and deleted the titles in the plans (Piggott, 1935, 28). Ucko et al. (1991, 87–89) suggest that this was due to his failure to find the temple at the end of the Beckhampton Avenue. Thus Stukeley had to abandon his idea of a symmetrical monument complex and arrived at the hypothesis that Avebury was a 'circle' and that the avenues were meant to represent a serpent traversing through it; the Sanctuary being the head of the serpent (Ucko et al., 1991, 89). This is also the view published in *Abury* (Stukeley, 1743, esp. chapter 11), the design being based on Egyptian hieroglyphs (Stukeley, 1743, cf. table XL). To make the Sanctuary look more like a serpent's head he altered his previous plans of the monument: the concentric circles became ellipses (Ucko et al., 1991, 94). The Kennet Cove, too, was erased from the plans, while the number symbolism and some parts of the symmetrical layout of the monuments were retained (Ucko et al., 1991, 94 f.).

This shows that the antiquarian record has to be approached with some caution, but that it is still an invaluable source as, without it, knowledge of the existence and location of the Sanctuary, Falkner's circle and the Beckhampton Avenue would have been irrevocably lost.

In 1812 P. Crocker made surveys and drawings at Avebury for Sir Richard Colt Hoare's second volume of *Ancient Wilts* published in 1821. Several recordings of stone destructions followed in 1823 (Henry Browne), 1829 (Joseph Hunter), 1862 and 1878 (William Long) (Smith, 1965, 183 f.).

[119]This sometimes resulted in inconsistencies, e.g. the spacing of the avenue stones was at one point supposed to be 70 foot, but at another 'near 100 foot' to make up the desired length of the avenue with 1000 foot (Ucko et al., 1991, 79).

8.1.8.2 The earliest excavations

In 1829 Joseph Hunter reports that there had been excavations at the foot of the Cove stones, but that nothing had been observed. While in 1833 Henry Browne probably dug into the burning pit of Cove stone III (Smith, 1965, 183).

The Cove was once more the centre of attention, when in 1865 A. C. Smith, William Cunnington III and Bryan King excavated at Avebury. Four trenches were dug, three at the feet of the two remaining Cove stones (I, II) and one across the eastern opening. During their work they discovered burning debris[120] (Gillings et al., 2008, 153 f.). They also examined the ground around stone 201, stone E and two burning pits (the site of the Obelisk was one of them) (Smith, 1965, 183 f.). Additionally, several cuts into or through the bank were made. Gray (1935, 102) does give some general indications of where the cuttings were made, but the descriptions remain vague.[121]

In 1881 A. C. Smith and W. C. Lukis let workmen probe and dig for buried stones and discovered 15 of them in the outer circle (see Section 8.1.1.3); several of these had been proclaimed by Stukeley, based on oral testimony, to have been destroyed by Tom Robinson in 1700 (see Ucko et al. (1991, plan E, 211)). Two further stones were discovered at the site of the northern circle (Smith, 1965, 184 f.).

In 1894 a cutting through the bank was made for Sir Henry Meux under the supervision of E. C. Treplin and Thomas Leslie. The trench is plotted in the south-east sector on Gray's plan of the monument (Gray, 1935, plate 29). The old turf line was discovered 69 cm below the level of the modern surface. It bore signs of a fire. Additionally, an inner rampart was visible, clearly indicated by a grass surface line of about 9 cm thickness. The finds consisted of sherds of pottery, worked and unworked animal bones, flint implements and antlers, which bore the marks of having been used as tools for digging.

The later excavations are mentioned in the monument description.

[120]Smith (1965, 183) thought that it must have been the burning pit of stone III they dug into, but Gillings et al. (2008, 153 f.) think otherwise; they argue that the trench would have been in the wrong place.

[121]Smith (1965, 184) states that not all of the places can be identified. With regard to the passage in Gray (1935, 102) it seems as if the motive for the excavations was to disprove that Avebury was a graveyard so that the main interest lay in the artefacts found during the excavations.

8.2 The Negotiating Avebury project

The 3D-model was meant to be a research tool with a grounding in archaeological theory, but also a means of documenting the site.

The beginnings of the project are described by Mike Pitts (Pitts, 2001b, 178 f.). The first research proposal was written in 1996 by Mark Gillings and Joshua Pollard followed by their recording of the terrain and individual stones together with David Wheatley and Glyn Goodrick in 1997 and 1998 (Pitts, 2001b, 179; Pollard and Gillings, 1998, 149).

8.2.1 Aims and ambitions

The Negotiating Avebury project aimed at creating case studies, which reconciled archaeological theory with practical implementation (Pollard and Gillings, 1998, 148 f.). It hoped to show that virtual reality models are suitable as primary records for archaeological sites (Gillings, 2000, 59) and possess great interpretational value (Pollard and Gillings, 1998, 149). The techniques employed to create the models should be easy to use and affordable (Gillings, 2000, 59, 61), whereby the resultant models should be interactive and readily modifiable by the users (Earl and Wheatley, 2002, 8). It was planned to model several hypothetical scenarios (Goodrick and Gillings, 2000, 55) of various building and destruction phases (Pollard and Gillings, 1998, 153 f., 161). These models were then used by researchers for analysis (Gillings, 2000, 61).

The project members are aware that the Avebury monument complex is not the ideal site for interpretations due to the lack in information about the phasing and date of the individual components. But the approach proposed in the project is an attempt to arrive at novel insights on the basis of current knowledge together with non-destructive fieldwork and despite the absence of new excavations (Earl and Wheatley, 2002, 9; Pollard and Gillings, 1998, 144).

8.2.1.1 Theory and practice

One of the frequently recurring motives in the articles concerning the Negotiating Avebury project is to give 'virtual reality' a grounding in contemporary archaeological theories. Gillings (2000, 60) postulates that many of the reconstructions have been designed as 'ingenious pictures' and that thoughts concerning their further use and additional value were only secondary. In order to make good use of these models, it is necessary to consider how they relate to the actual sites they are supposed to represent (Gillings, 1999, 2002, 2005; Gillings and Goodrick, 1996; Goodrick and Gillings, 2000) and eventually to accept that perceiving them as surrogates for the actual monuments (Gillings, 1999) or otherwise detailed replicas, will lead to ever more sophisticated models, though not to applications which aid interpretation (Gillings, 2000, 60). If it can be accepted that the virtual reality models are something different, are indeed inherently wanting in detail and also lacking the authenticity of the original site (Goodrick and Gillings, 2000, 44 f.), this difference can be used to advantage (Gillings, 2005, 230 f.).

8.2.1.2 Phenomenology

The intention was to develop a **phenomenological approach** to Avebury, which took some of its motivation from John C. Barrett. Barrett (1994, 11) criticises the fact that the archaeological documentation is frequently used to compose a plan and that monuments with similar plans are then compared and grouped together. The documentation and especially the plans provide the archaeologist with an holistic view of the site, which can thus be perceived in just an instant[122]. At the same time the documentation becomes a 'frame of reference' in itself, and hence detaches the observer from the monument. In order to encounter the monument as it would have been encountered in the past, it is necessary to move in and around it, which necessarily gives the visitor only partial views of the monument and takes time to experience (Barrett, 1994, 12). It has also to be assumed that the monument was constantly changing[123], that there was no planned 'final state' and that the act of creating the monument could have been the most important part (Barrett, 1994, 13 f.).

These views are echoed in the articles by Pollard and Gillings (1998, 143 f.) and Gillings (2000, 61). Pollard and Gillings state that the monument complex

[122]Usually exploring a site takes time, as the observer has to wander around. Looking at a plan instead saves time—or as Barrett expresses it: *time is collapsed*. Barrett (1994, 12) also claims that a plan shows the site at a specific moment, but here time can be collapsed too, by showing the remains of several periods at once.

[123]This could be demonstrated for the Sanctuary. See Pitts (2001a, 20 f.) and Section 8.1.3.

was usually approached on the basis of its plan, starting with Aubrey and leading up to Keiller. Few thoughts were given to the three-dimensional aspects of the monument, with the exception of Stukeley, who had an interest in the stone settings and produced many views of them.[124] Gillings (2000, 61) stresses that while the plan focuses on a holistic understanding of the site, the two-dimensional documentation of individual monument parts, e.g. each of the stones, leads to a dissection of the monument, because the elements become separated from their context in the monument complex. Additionally, their 'bulk and texture' is lost during this process. In a phenomenological approach there should be an emphasis on 'bodily engagement with the physical world' (Gillings, 2005, 231), i.e. experiencing a place through movement—whereby a realistic perspective is maintained (Gillings, 2005, 232)—as well as other forms of activities. All this results in a fragmentary experience of the site, therefore the Negotiating Avebury Project should go further than previous studies by Thomas (1993), Barrett (1994) and Devereux (1991) (Pollard and Gillings, 1998, 144).

Thomas (1993, 41 f., 40 fig. 1.1, 41 fig. 1.2) recorded in textual form his on-site experiences of approaching the henge from the West Kennet Avenue, he also included a plan of the henge and the northern part of the West Kennet Avenue as well as a reconstruction drawing of the southern henge entrance. This kind of documentation is rather inflexible and focuses on the specific points the author wants to make. Goodrick and Gillings (2000, 48 f.) argue that documenting an on-site visit with traditional methods is rather difficult. Even a series of photographs or a video do not provide the necessary freedom to experience landscape and monuments, because the view is restricted to whatever is visible in the images.[125] Goodrick also considers the fact that usually the landscape, vegetation and the monuments have changed with time. Virtual reality models can offer a solution to this part of the problem, in providing the possibility of simulating past landscapes and exploring them, by moving inside them and being able to look around from a 'natural' perspective (Goodrick and Gillings, 2000, 52).

[124]Notably there is even less interest in the earthwork, which, as Pollard and Gillings (1998, 144) point out, was even neglected by Stukeley, who only presented it in a schematic way.

[125]See also Keen et al. (1992, 134), who refer to an audio-visual presentation of Avebury by Peter Ucko and Peter Philipps and find it wanting.

8.2.1.3 Virtual reality and geographic information system

To arrive at a three-dimensional model the monument has to be recorded accordingly. The methods used should be easy to handle (without expert knowledge), relatively quick and inexpensive in order to allow them to become routine recording methodologies with a widespread acceptance in the field (Gillings, 2000, 61; Pollard and Gillings, 1998, 149).

Nevertheless the virtual reality model should not stand on its own, but should be combined with a geographical information system (GIS) (Pollard and Gillings, 1998, 144).[126] Both approaches have their own field of application and their individual set of questions which can be addressed by them. The combination of both tools is seen by Pollard and Gillings (1998, 145) as a way to tackle complementary matters of interest and to achieve synergetic effects. The GIS provides a platform for storing, organising and combining the diverse data on the site, e.g. surveys, topography, antiquarian plans and cropmarks (Pollard and Gillings, 1998, 146). In this function it works as an administrative tool. But GIS can also be employed for inter-visibility studies (viewsheds[127]), whereby a critical approach to the technique should be chosen. Instead of the simple employment of viewsheds, thoughts regarding atmospheric phenomena (haze, mist), eyesight and the ability to distinguish a feature from its background, should all be taken into consideration through the use of 'fuzzy viewsheds' (Pollard and Gillings, 1998, 146). Thought was also given to other elements which might restrict (vegetation) or enhance (smoke trail) the visibility of a feature. Besides stationary views from or towards a certain point, the influence of movement and the changing views created by it was to be considered. Sound is also an important sensual input which can be dealt with inside a GIS.[128] Smell and texture need a more

[126]For this, detailed topographic data had to acquired as well (Pollard and Gillings, 1998, 149).

[127]In a GIS application viewsheds are calculated on the basis of a given standpoint within the landscape (usually an offset for the observer height is added). The GIS compares which points in the landscape are visible and where the topography blocks the view. Other constraints, like distance, can be taken into account (see Wheatley and Gillings (2002, 204 ff.)). The resulting viewshed, which has the form of a map, on which the visible area is marked, can be output as a binary result (visible/not visible), or fuzzy viewsheds with different degrees of probability can be employed (Wheatley and Gillings, 2002, 209 f.).

[128]Systems for traffic noise prediction and similar applications have been in existence for several years, e.g. Li et al. (2002). For an archaeological approach see Mlekuz (2004), though the project described there was not yet far developed and seems less sophisti-

complicated approach. They have to be translated into metaphors or mnemonics, i.e. indicators which trigger the memory, which can be incorporated and displayed in a GIS (Pollard and Gillings, 1998, 147). Gillings (2000, 66 f.) also suggests using the virtual reality terrain model and the surface of the stones in a similar way: as projection surfaces for physical particularities. The stones, for example, could be used as a display for polishing areas or colour trends of the stones' surface.

8.2.1.4 Lines of research

That the monument complex at Avebury underwent changes is obvious (see the monument description in Section 8.1). Dilapidations together with houses, fences and other features (including the reconstructions by Keiller) changed the way in which the monuments can be experienced today (Goodrick and Gillings, 2000, 53) and as it could be during antiquarian times. Virtual reality could help to understand 'past and present encounters' with the monuments (Pollard and Gillings, 1998, 144). For example, **visibility issues** could be investigated. Questions of visibility between parts of the monument complex (Pollard and Gillings, 1998, 144) can be addressed by a combination of virtual reality and geographical information systems (Pollard and Gillings, 1998, 146)—contrasting the results from the GIS viewsheds with the impressions gained from viewing the virtual reality model (Cripps, 2001, 1). Visibility also plays a role in trying to shed light on the manufacturing of the antiquarian records, such as the frontispiece (plan E) by Stukeley (1743) (Pollard and Gillings, 1998, 153). In the latter instance it has been assumed that Stukeley's recording may have been hampered by both the modern features and the still extant stones encountered inside the henge. The impact these features could have had on the documentation of the site could be clarified by the virtual reality model (Pollard and Gillings, 1998, 154).[129]

It is possible to move through the virtual reality models freely and get, for the most part, only a visual impression of the site (Earl and Wheatley, 2002, 7 f.). Other senses, with possibly the exception of aural stimuli, cannot be addressed so easily.[130] The lack of real *embodiment* is also unresolved, but the latter would be required for a truly phenomenological approach. An observer cannot be 'embodied' by simply looking at a two-dimensional computer screen, even if he or she is restricted to a moderate pace within the virtual model and looking from the perspective of an individual situated in the landscape (as suggested by Goodrick and Gillings (2000, 52 f.)). Much more immersive techniques would be necessary to approach a level close to embodiment. And even then the remaining sensory input (smells, the sensation of walking on uneven ground etc.) would be lacking from the experience. Nevertheless the visual examination of the model can help to tackle questions concerning ritual and its orchestration especially in connection with visibility (shutter effects, hiding and revealing parts of the monument, other monuments or the landscape, as well as ritual actions taking place in the monuments) and movement (Earl and Wheatley, 2002, 11; Pollard and Gillings, 1998, 144, 155; Thomas, 1993, 42 f.).

It is possible for the modeller (and, ideally, it should be possible for any user of the virtual reality systems, too) to alter the virtual spaces and thus create scenarios for **hypothesis testing** (Earl and Wheatley, 2002, 6 ff.)[131]. For example, questions concerning the relationship between the southern entrance of the henge and the West Kennet Avenue were tackled. Their connection is still ambiguous due to medieval and modern disturbances and despite Keiller's excavations (see also Sections 8.1.1 and 8.1.2). Also other questions concerning the role and the morphology of the component parts of the monument complex could be asked (Goodrick and Gillings, 2000, 53). Nevertheless there are clear restrictions regarding the questions that can be addressed. The models cannot be used to 'see the past through the eyes of the original people' (Goodrick and Gillings, 2000, 47) because it is impossible for the observers to rid themselves

cated than the approach taken in the traffic noise prediction example.

[129] Already in 1992 a virtual reality approach to Avebury and especially to its antiquarian record was suggested by Keen et al. (1992, 130–134), which shows that some of the fields of application discussed in the Negotiating Avebury Project have been considered previously. The authors proposed several enquiries: first, comparing Aubrey's plan A quadrant by quadrant with modern information about the monument and in this way identifying features, for which there is no other record (Keen et al., 1992, 130 f.). Second, three-dimensional comparisons of (especially Stukeley's) plans and views of the monument with views of existing stones, for which 3D-models are required (Keen et al., 1992, 132). Third, simulating obstacles encountered by the antiquarians in order to evaluate, what impact they may have had on their records (Keen et al., 1992, 132). Fourth, removing all modern features from the model of the site to gain an impression of the henge as a whole (Keen et al., 1992, 133).

[130] A topic which is carefully avoided in Earl and Wheatley (2002, 5, 7). The authors mention that neither the visual sense nor any other should be considered in isolation, according to the psychologist James J. Gibson, and that phenomenologists stress the importance of all senses. But finally the analysis is again reduced to only visual impressions.

[131] This kind of interactive environment has been implemented for a case study (Peel Gap) by Gillings and Goodrick (1996).

of their own social backgrounds (Gillings, 2005, 230, 233). It also has to be kept in mind that the models are artificial (Earl and Wheatley, 2002, 7) and the results of conscious decisions by their creators (Earl and Wheatley, 2002, 6).

This leads to problems with another phenomenological principle, i.e. the significance of a place can only be judged, if the underlying meanings and values of the people, who inhabited those places, are known (Gillings, 2005, 232). These meanings are lost to us and therefore doubt is cast upon the usefulness of virtual reality applications in the phenomenologically inspired study of past cultures (Goodrick and Gillings, 2000, 54).

Despite these doubts Goodrick and Gillings (2000, 55) propose an approach which they call 'Gummidging'. The term **Gummidging** is based on a character from Barbara Euphan Todd's books, i.e. a scarecrow called Worzel Gummidge, who could change his head for another and with that his 'set of characteristics, preconceptions and behaviours' (Goodrick and Gillings, 2000, 55). Instead of exchanging the observer's head the model is modified in such a way as to represent different theories about the use and perception of the Avebury monument complex. This results in a multiplicity of Aveburys (Pollard and Gillings, 1998, 145). In Goodrick and Gillings (2000, 55) three such models under construction for the prehistoric period are mentioned:

1. Avebury of exclusion: Access to the henge and the avenue(s) is deterred. To visualise this the stones and the earthwork could, for example, be exaggerated in size.

   This idea is probably founded on the hypotheses of Parker Pearson and Ramilisonina (1998) (Gillings and Pollard, 2004, 68). Based on ethnological analogies the suggestion made by Parker Pearson and Ramilisonina (1998, 308, 319, 320 fig. 8) is that structures made from wood (e.g. the West Kennet palisade enclosures), due to their perishable nature, are associated with the living, while durable stone monuments are connected to the ancestors (e.g. Avebury and the West Kennet Avenue). Only on special occasions were the living meeting the ancestral spirits, but otherwise did not enter the henge or the avenues (Parker Pearson and Ramilisonina, 1998, 319 footnote 11, 320). Gibson (2004, 73 f.) sees the raised banks of some henge monuments as a sign that they are supposed to indicate exclusion.[132] He also assumes that stone settings, too, could mark exclusion, but then only in a symbolic way because they do not really obstruct access or sight (nor sound). The paucity of depositions inside the henge and along the avenues, may indeed hint at little ritual activity inside the monument.[133] The wear in the chalk outside the enclosed course of the West Kennet Avenue is taken as an indication for this hypothesis. People were only allowed to walk outside the avenues, while the ancestors were walking within.[134] This interpretation would fit the fact that some of the West Kennet Avenue's course is littered with small sarsen and therefore not easy to walk upon (see also Section 8.1.2.3).

2. Avebury of inclusion: Access to the avenue(s) and the earthwork bank is given, but entrance to the henge is limited.

   It has been suggested that the bank of the henge could have served as an observation point for rituals performed inside the henge, similar to the auditorium of an amphitheatre (Gillings and Pollard, 2004, 62). The ditch would act as a barrier between the viewing platform (bank) and the inside of the henge (Gibson, 2004, 73). Despite the raised vantage point, Gillings and Pollard (2004, 62) give to consider that activities inside the monument could not have been observed clearly, and, additionally, the separation of the onlookers on the bank from the 'performers' inside the henge, would establish a social hierarchy between those who watch the rituals and those who perform them (cf. Barrett (1994, 17 f.), who sees these distinctions being acted out inside the henge, by restricting the access to the southern and northern circles and with the internal settings functioning as stages). The avenues could have been used as processional ways.

3. Shamanistic/Ancestral Avebury: Free access to the henge, but the stones are turned into anthropomorphic shapes representing the ancestors. Goodrick and Gillings (2000, 55) suggest that in this case the user may be allowed to fly.[135]

[132] At Avebury the entrances are not only marked by higher banks, but also by deeper ditches, lowered causeways and bigger stone settings in comparison to the rest of the henge.

[133] Though it could be the case that building the monument was, indeed, the ritual act performed, similar to what is assumed about the Sanctuary (Pitts, 2001a, 21; Pitts, 2001b, 245 f.).

[134] A critical appraisal of this theory is to be found in Gillings and Pollard (2004, 68 f.).

[135] Possibly to indicate states of trance, and out-of-body experiences.

Based on ethnological parallels Pollard and Gillings (1998, 158–160) argue that the building materials for the henge were probably not neutral, but that associations were already formed with certain types of material and, in particular, with individual stones. Sarsen were also used in the construction of funerary monuments in the 4th millennium BC and could have therefore obtained an association with ancestors or spirits. Another factor for deeming stones special could have been their durability which contrasted with the transitional nature of people's lives. Pollard and Gillings (1998, 159) offer the explanation that this durability could have been associated with a 'permanent sacred order' and through this with the 'ancestral realm'.[136] Burials at the foot of stones and bone fragments found in stone holes (41 of the outer circle, 5b, 25a of the West Kennet Avenue, a burial in stone hole 25b of the West Kennet Avenue) are thought by the authors—largely following Smith (1965, 209 f.)—to be contemporary with the stone settings. They could indicate a shift of rites away from the funerary monuments towards the new stone settings.[137] The stones could have been considered as representations of, or even petrified forms of ancestors (Pollard and Gillings, 1998, 159). Later burials, placed next to the stones of the West Kennet Avenue (18b, 22b and 29a (Smith, 1965, 209 f.)) and the Sanctuary (stone 12 of ring C (Pollard, 1992, 224)[138]) are usually associated with beakers (Pitts and Whittle, 1992, 210, Pollard and Cleal, 2004, 127, Whittle, 1993, 39) and could be regarded in the same context. This account could be widened to include the skeleton found by Maud Cunnington in 1912 at the Beckhampton Avenue (Cunnington, 1912). Pollard and Gillings (1998, 159, 160 fig. 7) also take the deliberate deposition of specific materials inside the henge and the West Kennet Avenue into consideration. Worked flint, riverine clay, unworked sarsen, animal bones, humic soil, Lower and Middle Chalk and other materials were used in places where they were not naturally occurring materials. Besides their obvious use, e.g. as packing materials for sarsen, Pollard and Gillings (1998, 159 f.) think that they may also have some symbolic meaning and played a part in creating an *imago mundi* inside the henge with elements (and their symbolic value) from the outside world. In Gillings and Pollard (1999, 183 ff.) the authors go even further. They suggest that some natural sarsen had already gathered specific meanings and were seen in connection with ancestral spirits, before they were carefully selected and transported to Avebury. If these stones were seen as personifications of ancestors then Avebury would be a carefully selected and choreographed assembly of them[139]. The process of gathering the sarsen/ancestors in a new space inadvertently results in a change of their spatial context. In this case the irregular crest of the bank surrounding the sarsen could be seen as an effort to recreate the undulating line of the natural landscape from which they were removed and to keep the ancestors from the realisation that they were removed from their natural context (Gillings and Pollard, 1999, 185). The 'foreign' materials, like the riverine clay, could have served the same purpose. Or the bank and ditch may have had the purpose of keeping the ancestors from wandering (Gillings and Pollard, 2004, 70). The way in which the sarsen had been positioned has been interpreted as the cosmological order of ancestor spirits (in contrast to representing their former spatial relationships) (Gillings and Pollard, 1999, 185).

From the examples above it can be seen how the authors intend to modify a virtual reconstruction of Avebury and the avenues in accordance with theories about the past uses and perceptions of the site. These modifications will allow implicit notions about

[136] See also Gillings and Pollard (1999, 183 f.).

[137] Whether any of these bone fragments is really contemporary with the stone settings is questionable. Pollard and Cleal (2004, 126) suggest that stone 41 (see section 8.1.1.3) was reset in the Bronze Age. For stone 5b, Smith (1965, 210) only states that the pelvic bone and parts of two femora, belonging to two different individuals, were found in the stone hole and that there was no grave. Stone 5b has probably been destroyed by direct fracture and therefore the stone hole may be disturbed. Concerning stone 25a, the fragment of a skull only was found close to the stone (Smith, 1965, 210) and the ground had been disturbed by the burial of the stone. Smith (1965, 209 f.) assumed that stone hole 25b and the grave next to it are contemporary due to the same packing material in the grave and the stone hole, but the site is disturbed. The later stone burial pit (both burial pit and grave lie to the east of the stone hole) destroyed most of the stone hole and the grave (Gillings et al., 2008, 255, 226 fig. 9.3). Further bones (of two more individuals) were discovered in the filling of the stone burial pit, alongside a beaker. The latter is taken by Gillings et al. (2008, 255) to be an indication for a beaker burial, i.e. a later feature in relation to the stone hole. Deposits of human remains found in the ditch at Avebury are seen by Gillings and Pollard (2004, 72 f.) as belonging to an earlier phase than the later beaker burials. Nevertheless, the authors see them as ancestral relics, too, similar to the depositions at stones 41, 5b and 25a, despite their deposition in the ditch (Gillings and Pollard, 2004, 72 f.).

[138] This grave is associated with a beaker and is taken by Pollard (1992, 224) to date around 2000 BC.

[139] Or as Gillings and Pollard (1999, 184) put it: Avebury is not built *for* the ancestors but *of* them.

the monuments to be made explicit, i.e. visible. The resultant models are then meant for exploring 'ideas of ancestor worship and control over access to the monument in late Neolithic ritual practice' (Goodrick and Gillings, 2000, 55). Whether a visual stimulus is enough to change effectively the perception of an observer is debatable.

Besides these three models, which are meant to look at Avebury from a Neolithic or early Bronze Age perspective, another model was in preparation in 2000 (Goodrick and Gillings, 2000, 55) which was also proposed for the purpose of Gummidging.

Avebury of the antiquarian imagination: A planned, geometric and exact monument, 'the Palladian ideal—symmetrical, bilaterally balanced, thoroughly Georgian' (Goodrick and Gillings, 2000, 55).

Though the model is termed 'Avebury of the antiquarian imagination' the description can only refer to Stukeley's plans of Avebury because Aubrey's plans, especially plan A (Fig. 8.14), could hardly be called 'symmetrical' (see Section 8.1.8.1). Aubrey's plans do record the irregularities of the earthwork and the stone settings even though in a overly distorted way.[140] Stukeley on the other hand sees the irregularities, but believes them to be accidental and that the earthwork and stone circles were *meant* to be circular and therefore draws them accordingly (see Fig. 8.17). His obsession with symmetry and number symbolism have also been noted before (see Section 8.1.8.1).

Indeed, prior to Goodrick and Gillings (2000, 55), Pollard and Gillings (1998, 153 f.) mentioned a *'new' Stukeley plan*, which should become the basis for a virtual model of Avebury in Stukeley's time. The plan (Pollard and Gillings, 1998, 154 fig. 4) is based on plan E and additional archaeological information.

The final model, created on the basis of the *'new' Stukeley plan*, was meant to be used for finding out how the monument might have been experienced by Stukeley and his contemporaries. Especially, the influence which field boundaries, hedges, trees and buildings had on the antiquarian record, is of interest.[141] It should also be the basis for re-interpreting the work of Stukeley and for exploring ideas about the shape and development of Avebury (Pollard and Gillings, 1998, 153 f.).

Besides the interpretational changes there are also **changes over time** which the monument underwent. For analysing the actual and proposed former appearances of the monuments, a model of the extant remains was created (Goodrick and Gillings, 2000, 53). This included modelling the stones and the earthwork and incorporating the results from excavations and surveys. Pollard and Gillings (1998, 161 endnote 2) propose that this model and the accompanying GIS can be used to explore hypotheses about the sequence of construction, for example, the primary bank in relation to the later bank (see Section 8.1.1.1); the earlier settings A, B and C in relation to the outer circle (see Section 8.1.1.3); and the unresolved sequence of earthwork and outer circle (see Section 8.1.7). Besides looking at different sequences for the monument, the model can also be used to explore changes in choreography and bodily access during the proposed phases of evolvement of the monument complex (Gillings, 2000, 67). Also, the meaning and the way the monument and/or its component parts were encountered has changed over time—from the time of its construction over the destruction phases in the Middle Ages and the (early) modern era till today—as outlined in Gillings and Pollard (1999) and Gillings and Pollard (2004).

The reconstruction focus of the Negotiating Avebury project (as far as can be learned from the available literature) lies on the henge with its stone settings[142], the northern part of the West Kennet Avenue (Earl and Wheatley, 2002) and on the (at that time still unresolved) course of the Beckhampton Avenue (Cripps, 2001).[143]

[140] Some regularity was assumed as plan A shows the bank and the ditch to be of even width and omits any signs of a berm between them. See Section 8.1.8.1.

[141] I think that the influence on Aubrey's record is as interesting as the influence on Stukeley's. Maybe the model would provide a means to make more sense of Aubrey's plan A, by identifying the possible obstacles he encountered in making his plane-table survey of the site (cf. Keen et al. (1992, 130 f.)).

[142] Four models with the purpose of Gummidging, including the 'new' Stukeley plan, are mentioned by Goodrick and Gillings (2000, 55) and Pollard and Gillings (1998, 153). A research focus on the stone settings is pointed out by Pollard and Gillings (1998, 154–158).

[143] The Sanctuary has also been reconstructed in 2000 by Jennifer Garofalini for her Master of Science degree at Southampton University (Pitts, 2001a, 6; Pitts, 2001b, 239). The model helped to create doubts concerning the contemporaneity of all the post settings recorded by the Cunningtons in 1930 (Pitts, 2001b, 243). A puzzle which could be resolved by re-excavation (Pitts, 2001a, 20 f.; Pitts, 2001b, 246). (This is in fact a parallel to the 'Throne' case in the Ename 974 project, see Section 7.3.3.3.) The Sanctuary model is not mentioned in connection with the Negotiating Avebury project in

8.2.2 Arriving at the model

The following persons were involved in the Negotiating Avebury project: Mark Gillings (University of Leicester), Joshua Pollard (University of Wales College, Newport), David Wheatley (University of Southampton), and Glyn Goodrick (University of Newcastle) (Pollard and Gillings, 1998, 161 endnote 1). Wheatley was digitally surveying topography of the West Kennet Avenue, Gillings and Goodrick were recording the geometry of the stones with the aid of digital photographs, which were later used by Goodrick to create VRML models of the stones. Pollard was looking through the documentation of Alexander Keiller and drew two-dimensional elevations of the stones[144] (Pitts, 2001b, 179). Further assistance for the field work came from David Gibson, Bob Johnson, David Salisbury and Tim Sly (Pollard and Gillings, 1998, 161 endnote 1). Also students were involved in the field work (Pitts, 2001b, 179) as well as in modelling (Cripps, 2001; Earl and Wheatley, 2002). Further help and access to data and documents were provided by: Christopher Gingell (National Trust) in the form of access to the site; Rosamund Cleal and Clare Conybeare (curators of the Alexander Keiller Museum) who granted access to the Keiller archives; and Andrew David (English Heritage) who provided digital survey data of the Avebury region[145] (Pollard and Gillings, 1998, 161 endnote 1). It is to be specially remarked upon that all the people involved actively in this project are archaeologists and that in contrast to 'Progetto Insula del Centenario' and 'Ename 974' no technicians without archaeological background were employed for modelling or setting up the GIS. This guaranteed complete archaeological control over the project, its methodologies and outcome.

Financial support came from the Universities of Leicester and Southampton and material support from McCarthy Taylor Systems (Pollard and Gillings, 1998, 161 endnote 1), but in the end the virtual reality part of the project was stopped due to funding problems (Gillings et al., 2008, 2; Pitts, 2001b, 179 f.) and with the recognition that there was not enough information on the site available (Gillings et al., 2008, 131; Pitts, 2001b, 180). Instead the team moved on to the excavating the Beckhampton enclosure and the Beckhampton Avenue (Gillings et al., 2008, 131).

the literature.

[144]Something which had not been prepared for any Avebury record before (Pitts, 2001b, 179).

[145]This is possibly the data mentioned in Burton (2000), who describes a GIS for Avebury by English Heritage.

8.2.2.1 Documentation of the site and modelling

The archaeological and antiquarian background of the site has been described in Section 8.1. Additional information is available in the form of photographs taken by Keiller during his excavations (Pollard and Gillings, 1998, 150) and also Keiller's excavation diaries provide further unpublished material on his excavation findings thus opening up different routes for re-interpretation (cf. for example, Gillings et al. (2008, chapter 8–10)).[146] Another written source is provided by O. G. S. Crawford, who worked from 1920–1927 on a book about Avebury which is based on his own observations of the site and their relation to Stukeley's record. The work was never completed, but the notes survive in the Alexander Keiller Museum (Gillings and Pollard, 2004, 177 f.; Smith, 1965, 185). Drawings of the still-standing and re-erected stones in scale 1:50 were created by Pollard (Pitts, 2001b, 179; Pollard and Gillings, 1998, 150, 151 fig. 2). A significant number of photographs showing the monument today are featured in Burl (1979). The geophysical surveys by Ucko et al. (1991) have already been mentioned (Section 8.1), similar surveys have also been conducted prior to excavation for Gillings et al. (2008, 10 f., 63–70, 103 ff., 111–115, 143 f.) in Longstones Field[147], Avebury Trusloe[148], Long Barrow Field, Beckhampton Field and Falkner's Circle. Inside the henge a geophysical survey was conducted to investigate the parchmarks mentioned in Bewley et al. (1996)[149]. Aerial photography also helped to reveal several features in the henge (Bewley et al., 1996) but also the Longstones Enclosure (Gillings et al., 2008, 10) and the West Kennet palisade enclosures (Whittle, 1997, 54).

For the 3D-model of the site,[150] an English Heritage survey dating from 1990 was used to obtain the primary survey points and the general contour information (Pollard and Gillings, 1998, 149). This survey and similar other records[151] do not provide enough detail, which the Negotiating Avebury project members

[146]No revisitation of Keiller's records has been conducted for the reconstruction presented in Earl and Wheatley (2002, 14 footnote 3).

[147]Including Longstones Enclosure, Longstones Cove and the Beckhampton Avenue.

[148]Including Manor Farm

[149]See http://sdb2.eng-h.gov.uk/visitdetails.asp?visit=736 (Accessed: 6th of August 2012.). Several other surveys in the Avebury region are also listed in this database.

[150]An early version is shown in Pollard and Gillings (1998, 150 fig. 1).

[151]Among these a detailed Royal Commission on the Historical Monuments of England (RCHME) hachure plan is mentioned (Pollard and Gillings, 1998, 149).

thought essential for the interpretation of the monument. Another problem with this existing data is that only the monument and its immediate surroundings were surveyed, so it lacks further landscape context (Pollard and Gillings, 1998, 149). Therefore, beginning in April 1997, a total-station was used to survey the ground surface in order to obtain more detailed data from inside and outside the monument complex (Gillings, 2000, 61; Pollard and Gillings, 1998, 149). From the survey points Triangulated Irregular Networks (TINs) and Digital Elevation Models (DEMs) could be calculated for use in the GIS and for the surface model of the virtual reality (Gillings, 2000, 61).[152]

The extant stones were recorded by means of a digital camera and the software package PhotoModeler[153] (Gillings, 2000, 61–68). The process for recording objects with PhotoModeler is rather straightforward: a (preferably) digital camera[154] is used to take several pictures of the same object from different angles. The number of pictures taken should correspond to the complexity of the object. Gillings (2000, 68) has found that they had in some instances not taken enough photographs to model the stone surface sufficiently. Uneven lighting of the object can also result in difficulties. The photographs are then loaded into PhotoModeler and the user is asked to identify and mark the same point on the object surface in as many photographs as possible. This exercise has to be repeated with several other points. The accuracy of the resultant model is dependant on the number of identified and marked points, the number of photographs, the accuracy with which the points can be identified and the distance between the identified points. To assure the identification of surface points, the Avebury stones were primed with tiddlywinks[155] which were fixed to the stone surface with plastic putty (Gillings, 2000, 63 f.). From the photographs and the correlated points the software can calculate the geometry of the stone which then can be textured by using the same photographs. PhotoModeler can export the textured shapes in several formats, among them VRML, which was used for the Negotiating Avebury project (Gillings, 2000, 66). The thus modelled stones replaced one by one the archetypes in a preliminary model, in which the latter each marked an 'original stone position'[156] (Goodrick and Gillings, 2000, 53). In the end only the missing and destroyed stones will still be symbolised by such archetypes.

It is not mentioned on what the shape of the archetypes is based, i.e. whether they are completely generic forms or whether their shape corresponds to information which in some cases can be extracted from the shape of their burial and/or destruction pits, or from other sources. The clearly discernible West Kennet Avenue stones in Earl and Wheatley (2002, 12 fig. 2.1, 13 fig. 2.5) resemble in shape—but appear to be smoothed—the previously buried stones 13a and 13b (see Section 8.1.2.1), whereby 13a seems to be repeated on the eastern side of the avenue and 13b on the western side respectively. The Obelisk, on the other hand, is clearly depicted as a tall, thin stone (Earl and Wheatley, 2002, 13 fig. 2.3) in concordance with Stukeley's description of the stone and also with its destruction pit (see Section 8.1.1.3).

In other cases and for special purposes the stone 'archetypes' have been additionally modelled as translucent cylinders (Earl and Wheatley, 2002, 11) to indicate their uncertain status.[157]

The degree of realism—here Goodrick and Gillings (2000, 44) argue that photos themselves are not realistic when compared to the physical world, therefore, they do not want to use the term 'photo-realism'[158]—of the virtual reality reconstruction should depend on how immersed[159] the user has to be for using the model to the best advantage. For examining the shutter effects of a stone row, for example, no great realism of the reconstruction is necessary; but where emotional responses are desired, a greater depth of immersion has to be provided (Goodrick and Gillings, 2000, 46). Judging from these comments the realism of a model which is to be used for a phenomenological analysis (especially for 'Gummidging') should be rather high, because the user needs to experience the monument in its completeness, ideally in an embodied way, to judge its effect. On the other hand it seems that the people concerned with the Negotiating Avebury project believe that a

[152] Gillings makes reference to Gillings and Goodrick (1996) where under the following URL http://intarch.ac.uk/journal/issue1/gillings/software.html (Accessed: 6th of August 2012.) information about software packages used in the Peel Gap study are mentioned. It can be assumed that several of these packages were also used for the Negotiating Avebury project, but there can be no certainty except for the software PhotoModeler.

[153] http://www.photomodeler.com (Accessed: 6th of August 2012.).

[154] The camera can be calibrated beforehand, but this is not a necessity.

[155] i.e. coloured plastic chips from the game of the same name.

[156] This original phrase would be better substituted by: 'assumed stone position', in view of unresolved ambiguities. See Section 8.1.

[157] Is this kind of representation chosen on the basis of theories by James Jerome Gibson? Earl and Wheatley (2002, 5) mention that 'Gibson drew attention to the importance of solid (*real*) surfaces over simplified transparent planes [...]'. (Emphasis in the original.)

[158] cf. Shanks (1997).

[159] The word Goodrick and Gillings (2000, 46) use is 'collapsed' which is not readily understandable in this context.

reduced degree of realism is enough for their studies.[160] This seems to constitute an apparent discrepancy, but is not seen as such by the authors.[161]

Besides modelling the extant remains of the monument complex and including uncertain features (e.g. stones which may or may not have stood in a certain place) the use of 'interpretative modelling' or 'consciously subjective modelling' is being heralded (Earl and Wheatley, 2002, 6 ff.). How this can be achieved has been described in Section 8.2.1.4 where the idea of Gummidging and other hypothesis testing methods were mentioned. Not only during the modelling process, but also during the exploratory phases, the model should remain editable by everyone (Earl and Wheatley, 2002, 8). To see the reasoning behind changes to the model, or changes in interpretation which, in turn, affect the model, the editing process has to be documented. Earl and Wheatley (2002, 8) state that 'a way for mapping and organising the modelling process' was defined and used. The text does not explain how this is achieved.[162] The result of documenting 'choices' (i.e. interpretations) and 'starting points' (the underlying hypotheses) is that the resultant model becomes reversible and reactive (Earl and Wheatley, 2002, 8).

8.2.2.2 Exploring the reconstruction model

Two descriptions of how virtual reality models and GIS were used to approach research questions in the Negotiating Avebury project were available to me (Cripps, 2001; Earl and Wheatley, 2002); while another remains sketchy (Pollard and Gillings, 1998).

The 'new' Stukeley plan

This plan has already been mentioned several times in Section 8.2.1.4. Information about the state of the site in the 18th century is taken from plan E (Fig. 8.17),[163] the frontispiece to *Abury* (Stukeley, 1743).

What had interested the project members (Mark Gillings, personal comment) was Stukeley's attempt at turning Avebury into a Georgian monument which fitted the Palladian ideal of symmetry and perfect circularity of its layout and component parts. In this way he did not simply record the monument but created an Avebury according to his taste (or the tastes of his time). His frontispiece (the only plan that was published) strongly influenced later reconstructions, e.g. Crocker's plan of Avebury (Gillings, 2000, 60 fig. 1), but also other works like Crawford's manuscript or Keiller's excavations. Gillings (personal comment) argues that this makes Avebury as much a neolithic as a 17th century monument[164]. The intent was to reproduce Stukeley's style as well as the dynamics and business in the frontispiece in an attempt to show 'the didactic value of the inauthentic' (Mark Gillings, personal comment).[165] This also explains why the focus lay on plan E, because this was the image which had the greatest influence on the later perception of the site.

Besides plan E, Stukeley's 'field drawings' were also taken into consideration and were combined with observations on site, data from excavations, cropmarks and geophysical surveys (Pollard and Gillings, 1998, 153). This information was used (as the first stage) to draw a two-dimensional plan (Pollard and Gillings, 1998, 154 fig. 4). During this exercise several imprecisions on Stukeley's part were encountered, some of which had been remarked upon by Ucko et al. (1991, 211) beforehand. Among these are the misplaced hedge between pastures IX and X, a feature also mentioned by O. G. S. Crawford in his unpublished papers (Ucko et al., 1991, 211). Other difficulties include the positioning of houses which have been drawn, like the rest of the henge in plan E, in

[160]cf. Gillings (2002, 24), who gives two examples for which a different degree of realism is necessary. High realism to let people experience the inside of a cave in flickering light and rhythmic chanting in the background. Low realism for the Peel Gap study (Gillings and Goodrick, 1996) where the user should be able to explore what can be seen from a Roman watchtower and a possible manned walkway. The height of both features could be adjusted interactively.

[161]'I still firmly believe that virtual simulacra don't have to be painstakingly detailed and achingly authentic in order to be useful'. Mark Gillings, personal comment.

[162]These are, in fact, thoughts which have been addressed in detail in Section 7.3.6. Whereby Pletinckx (2008) offers concrete methodologies to solve this problem.

[163]It has to be remarked again that Ucko et al. (1991, 137), who have studied all of Stukeley's plans (cf. Ucko et al. (1991, chapter 3)) at great length, have come to the conclusion that plan E is not 'a summation of all that is valid in them' (Section 8.1.8.1).

[164]That emancipation from Stukeley's writing has still not occurred is apparent even in the latest research concerning the Avebury monument complex—the Beckhampton Avenue would not have been re-discovered if it had not been for Stukeley's records of it (Wheatley, 1999).

[165]A reproduction of style and 'the didactic value of the inauthentic' can also be found in the creation of the 'Visualisation of Filippo Juvarra's Palace Project and Garden Plan (1705)' (http://wissensgeschichte.biblhertz.it/3d-bridge-html/index3D.html (Accessed: 6th of August 2012.)). In this project, plans of a house, which was never built, were made into a 3D model. The model was then textured with cut-outs from the original ink-drawing. They give the model a very surreal, but also aesthetic appearance—completely befitting the purpose.

a bird's-eye view.[166] This led to problems with the position of some of the stones e.g. in the northern part of the southern circle (Pollard and Gillings, 1998, 153). On the other hand Stukeley's assumption that the stone settings inside the henge were meant to be circular also led to grossly misplaced former stone settings of the northern circle in the north-west sector (cf. Section 8.1.1.3).

None of the Gummidging-models was constructed, including the Stukeley model. This had been prevented by the financial restraints which were experienced from the start (Mark Gillings, personal comment). Earl and Wheatley (2002, 11) state that their 'Entrance Model' (see Section 8.2.2.2) is based on the 'Negotiating Avebury models' by Glyn Goodrick. This probably refers to the terrain and stone models based on the extant remains, which had been produced at the beginning of the project.

The two dimensional plan was probably later used by Gillings et al. (2008, 239–247, figs. 8.4–8.10) to create plans of field boundaries inside the henge in medieval, and early modern (early 18th century and late 18th century) times[167]. For this purpose additional information concerning field boundaries was obtained from Keiller's unpublished excavation documentation (Gillings et al., 2008, 238–240).

The West Kennet Avenue at the southern entrance to the henge

As mentioned in Section 8.1.2.1, there are many ambiguities concerning the connection between the southern entrance of the henge and the West Kennet Avenue. Unsatisfactory excavation results, due to roads and garden plots as well as the lack of stones (including buried ones) and the failure of the antiquarian record to shed any more light on the situation, have led to a number of hypotheses concerning this part of the monument complex.

Earl and Wheatley (2002, 9) try to address these problems by employing a VRML model of the henge and the West Kennet Avenue. As they have proposed in the theoretical part of their article, the model was constructed in a way which allows the user to choose between alternative stone settings and to freely move through the model[168] (Earl and Wheatley, 2002, 8). The goal was to explore different scenarios and to evaluate them. It is explicitly mentioned that this area of the monument complex was selected *because* the current knowledge re the monument is inadequate and there is no real chance of new excavations resolving these matters in near future (if ever, in view of the extent of the disturbances). The model was intended to present an alternative means for arriving at new interpretations on the basis of existing data (Earl and Wheatley, 2002, 9).

As a prerequisite for this study the West Kennet Avenue has to be considered a processional way (for the living). One of the reconstructed scenarios envisages the connection between the West Kennet Avenue and the henge entrance, the dog-leg, as a deliberate feature and not as some kind of expedient which had to be employed in order to effect a connection between two separate monuments. The other is built on Smith's hypothesis, that stone pairs 5 to 12 of the West Kennet Avenue could have been removed and rearranged into a straight approach towards the henge entrance (Smith, 1965, 209). None of these stones had been destroyed by fire or buried[169], neither, it seems, has Stukeley recorded any of them[170] (Earl and Wheatley, 2002, 10).

The uncertainties concerning the pair to standing West Kennet Avenue stone 4 (which involve the question whether this stone 4 is actually stone 4a or 4b) are also taken in consideration (Earl and Wheatley, 2002, 9 f.). The question of the 'correct' pair to stone 4 results in two different dog-leg shapes, depending on whether the processional way runs to the east or the west of the standing stone.

This results in three possible approaches towards the henge, two of them include the dog-leg while another has been rearranged to avoid a kink in its course. To visualise these three alternatives, a VRML model was built. One of the VRML models constructed by Glyn Goodrick for the Negotiating Avebury Project was

[166] Plans O, P and Q (see Ucko et al., 1991, 139 fig. 33, 141 fig. 34, 142 fig. 35) do not show the houses in a perspective view and might have been better sources.

[167] Gillings et al. (2008, 246, fig. 8.10).

[168] This is still far removed from really editable models, but allows at least some degree of freedom. Otherwise, full editability could be achieved if a user had downloaded the model and changed it by altering the VRML code.

[169] The fact that none of the stone holes on the West Kennet Avenue had been refilled, contrary to the two Beacock Holes discovered at Longstones Field, does speak (besides the discovery of sarsen chips) against a removal of the stones during the final Neolithic.

[170] On Stukeley's plate depicting the approach of the West Kennet Avenue towards the henge (Stukeley, 1743, plate XVIII) no other stone but stone 4 is visible. Plates XIX and XXII (Stukeley, 1743), which show approximately the same stretch of the West Kennet Avenue from two different angles, do display several avenue stones, but it remains unclear how far the entrance of the henge is situated from the first stones in these images. In this case a comparison with the virtual reconstruction could elucidate which stones could have been recorded by Stukeley.

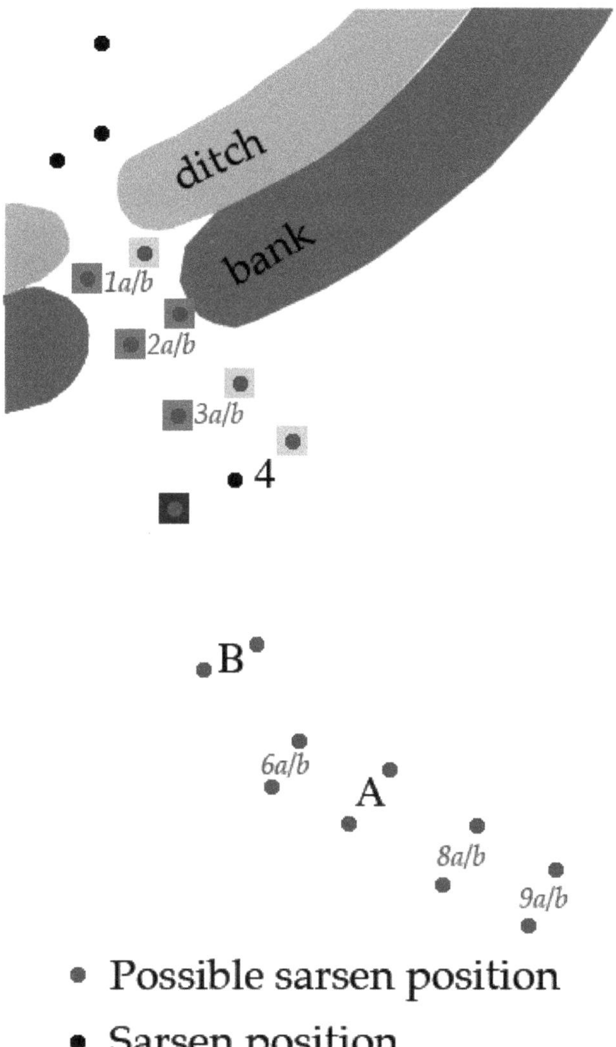

Figure 8.18: The stone settings at the southern entrance to the henge. Red: possible stone positions which were discovered by Keiller. Green: inferred stone positions which were marked by Stukeley. Blue: possible stone hole—the alternative position of standing stone 4's pair. A and B mark viewing positions in Figs. 8.20 and 8.21.

used as the starting point (Earl and Wheatley, 2002, 10 f.). This model had to be modified: at the southern entrance the height of the banks and their position were altered to represent their state shortly after construction. The remainder of the model (the remaining bank and the ditch) was left as it was (Earl and Wheatley, 2002, 11). It is unclear why the remainder of the model was not changed, as it could be a serious bias to later interpretation. The higher banks could have had, for example, a significant influence on the visibility of the Obelisk and other stones from outside the henge. It also remains uncertain whether the causeway was deepened to give a better impression of the entrance arrangement.

The user of the model is meant to be able to change its appearance (within constraints). This is facilitated by the use of Java applets. Sensitised spots in the VRML model can be clicked by the user and the model is thus toggled between different scenarios (Earl and Wheatley, 2002, 11). One of the sensitised spots is situated between stones 13a and 13b, another at the base of stone 4. Both switches affect the same change: they toggle between the straight approach to the henge or the approach which necessitates the dog-leg. A third switch changes the visualisation of stones 1a, 3a and 4a (marked in red in Fig. 8.18). These stones are seen as 'uncertain', though definitely with a different degree of uncertainty to each of them (see Section 8.1.2.1). Earl and Wheatley (2002, 11) allow for this varied degree of uncertainty by the diverse states through which the 'stones' can be cycled. Either all stones are displayed in a semi-transparent state, or only stone 1a, or stones 1a and 4a, or all three stones are visible as solids. The next click returns to the default stage: all stones are semi-transparent. Stones 1b, 2a, 2b and 3b are handled similarly (marked in green in Fig. 8.18). In their case, no real evidence exists. Therefore, as a default, they are denied a stone shape and are displayed as blue semi-transparent cylinders. This appearance can be changed into the representation of four solid stones or the positions can be left blank (Earl and Wheatley, 2002, 11).

To test the different scenarios Earl and Wheatley (2002, 11) started with the approach of the henge along the route described by Thomas (1993, 41 f.), which corresponds to the way indicated by Keiller's concrete markers. Beginning from between stone pair 13, the henge is approached virtually along the avenue. The tip of the Obelisk can be seen above the bank, while the z-feature is hidden by it (Fig. 8.19).[171] Earl and Wheatley (2002, 12) claim that the situation becomes confusing when the observer stands between stone pair 7. Taking a first glance at Figure 8.20 this is clearly the case. But here the static image is certainly adding to the confusion as it is hard to judge the distance between ourselves and the stones, whereas if on site and moving, there would be less difficulty in judging distances and the way marked out by the stones would be clearer. Additionally, there is the possibility that usage of the monuments would provide visitors with trackways which they can easily follow.[172] Earl and Wheatley (2002,

[171] What about the southern circle? It is not visible on the images, but that may be due to image quality restraints.

[172] It is remarkable that no-one before has suggested trackways

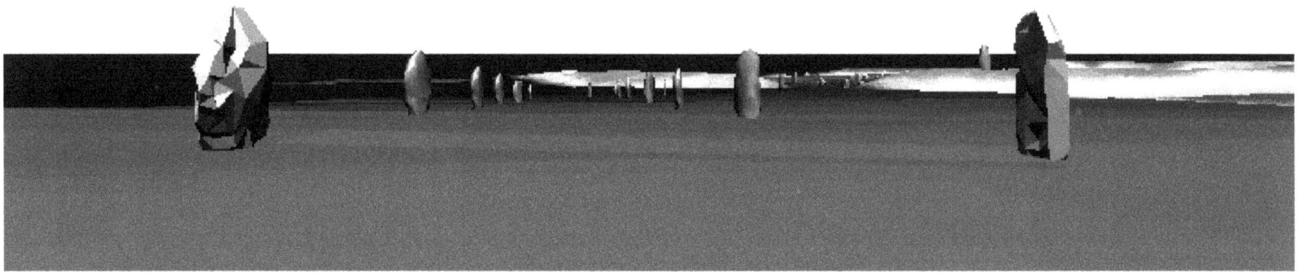

Figure 8.19: View along the West Kennet Avenue. The observer stands between stone pair 13. To the right the Obelisk is visible behind the bank of the henge (Earl and Wheatley, 2002, 12, fig. 2.1).

Figure 8.20: View towards stone pairs 6 and 5 of the West Kennet Avenue. The observer is standing between pair 7, marked with an 'A' in Fig. 8.18. At this position he or she may experience some confusion about which route to take (Earl and Wheatley, 2002, 12, fig. 2.3).

12) suggest that due to the confusion created by the architecture a visitor to Avebury might be inclined to leave the path of the avenue and take a shortcut to the entrance especially because, from the position between stone pair 7, the base of the Obelisk and the z-feature are already visible. Unfortunately this cannot be substantiated from the image in the text (Fig. 8.20). From this it becomes obvious that much depends on the exact standpoint of the observer. On the one hand a few steps forward and/or to the left would open the view on to the inner settings of the southern circle—a view which is obstructed by stone 7a. On the other hand it is possible that the 'confusion' about which route to take is in fact diminished by the same change of viewpoint.[173] If, nevertheless, the observer proceeds along the course of the avenue and arrives between stone pair 5 (marked 'B' on Fig. 8.18), the situation that opens up to him or her (Fig. 8.21) is also far from clear (Earl and Wheatley, 2002, 12). As Earl and Wheatley (2002, 12) point out, there is no good indication that a visitor to the site should proceed between standing stone 4 and the possible position of stone hole 4a, the red stone furthest to the right on Fig. 8.21. The entrance situation is approached from the side, so that no clear way can be made out and stone 4a seems to be too far away to be clearly identifiable as the pair of standing stone 4 (centre of the image). The blue cylinder marking the possible position of stone hole 4b seems to be, from this standpoint, better suited as a position for the pair of standing stone 4. The entrance to the henge and, further away, the southern circle with the Obelisk, and the northern circle with the Cove lie straight ahead[174] almost on one line (Earl and Wheatley, 2002, 12).

Earl and Wheatley (2002, 12) think that when the

as an indicator as to how the monument complex should be approached and experienced.

[173]This is a hypothesis which can only be tested on site or by accessing the virtual reality model.

[174]In the article it is also mentioned that the Ring-stone is part of the same alignment (Earl and Wheatley, 2002, 12), but it is hidden behind stone 98 of the outer circle (Earl and Wheatley, 2002, 13 caption to fig. 2.4). It should therefore not be counted among the visual cues.

Figure 8.21: View inside the henge. The observer stands between stone pair 5, indicated by a 'B' on Fig. 8.18. The coloured markings correspond with the markings on the same Figure. (After Earl and Wheatley (2002, 13, fig. 2.3); coloured markings by Joyce Wittur.)

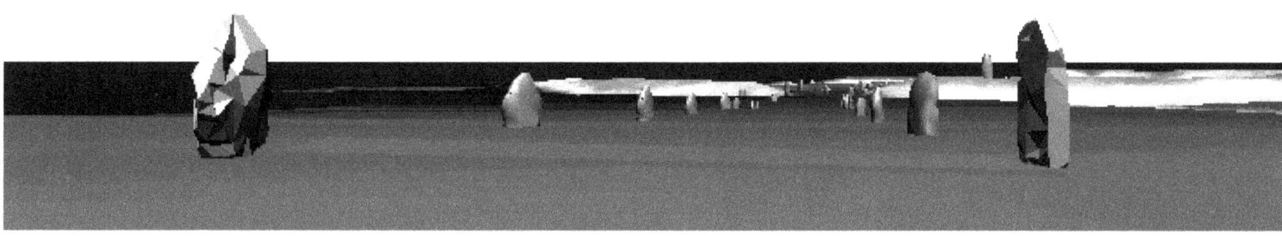

Figure 8.22: View along the West Kennet Avenue. The observer stands between stone pair 13. To the right the Obelisk is visible behind the bank of the henge. This image shows Smith's hypothetical second phase for the arrangement of stone pairs 5 to 12 (Earl and Wheatley, 2002, 13, fig. 2.5).

doubtful stone positions 1b, 2a/b, 3a/b and 4a are 'removed' in the model, the impression that the only way to proceed is between standing stone 4 and the blue marked stone position 4b is strengthened considerably.

Finally the last hypothesis, a rearrangement of the stones in prehistory, is being tested (Earl and Wheatley, 2002, 12 f.). Figure 8.22 shows the corresponding view. The route a visitor has to take is straight forward and absolutely focused onto the entrance of the henge. It is very different from what can be seen in Figure 8.19 (Earl and Wheatley, 2002, 13). In this context the approach close to the henge would have been an interesting view because obviously the 'choreography' is lost: on 'Keiller's' path it seems at first as if the visitor would bypass the henge, but at the entrance a turn is effected and the view falls directly onto both of the inner circles. This will not be the case with a straight approach. It furthermore seems to be a deliberate characteristic to avoid straight approaches towards the henge entrance: as mentioned in Sections 8.1.4.1 the Beckhampton Avenue is also not aligned with the causeway of the western henge entrance. A similar picture may be drawn by the results of the geophysical survey by Ucko et al. (1991, 184 ff.) which was conducted east of the eastern entrance to the henge (Section 8.1.1.1). Finally, the approach to the Sanctuary probably exhibits a kink (Section 8.1.3), too.

Earl and Wheatley (2002, 13) conclude that in the case of the southern henge entrance and its relation to the West Kennet Avenue a virtual reality model may, in the end, be the only way to investigate the various hypotheses because it is not unlikely that the existing disturbances of the ground will render further excavations unfruitful. They also accept that the virtual reality model will not provide new evidence which will in the end help to resolve any questions, but that it is a tool for exploration and for interpretation (Earl and Wheatley, 2002, 14). One radical, new idea offered by the authors is that there might have been a 'T junction' at the location of the dog-leg between the West Kennet Avenue, the southern henge entrance and a hypothetical path leading outside the henge in the direction of the Beckhampton Avenue.

The Beckhampton Avenue

The last example comes from the unpublished MSc dissertation of Paul Cripps (Cripps, 2001). Only the printed part was available to me, but not the accompanying CD-ROM which contains additional material, e.g. films which were rendered from the three-dimensional model (Cripps, 2001, 24 f.).

The aims of the project were to combine a GIS with an interactive three-dimensional model in order to investigate the intervisibility of the Beckhampton Avenue and other monuments in the region. This analysis was supposed to help identify 'dynamic spatial relationships', i.e. when moving along the Beckhampton Avenue which monuments are visible or not visible and in which order (Cripps, 2001, 1, 4). Cripps (2001, 10) states that thereby 'the [three-dimensional] reconstruction will be used to test hypotheses based on the results of the GIS analyses and vice versa in a reflexive manner'.[175] The similarities to the previous studies by Thomas (1993, 41 f.) and Earl and Wheatley (2002) are apparent—experiencing the (reconstruction of the) monuments by moving along an avenue—and intentional (Cripps, 2001, 11). The assumption that the avenues represent formalised routes between the monuments in the region, and with that regulate movement and vision, is central to the study (Cripps, 2001, 1). It has also been proposed that the avenues mark routes which are much earlier than most of the monument complex and that the monuments were later built in dependence on these former routes (Cripps, 2001, 5). What is suggested is that the route of the Beckhampton Avenue is based on visibility arrangements with other monuments in the region (a monumentalised sequence of encounters with other monuments). The study intends to test this hypothesis (Cripps, 2001, 12), but not only for the part described in Sections 8.1.4.1 and 8.1.4.2 but also for the continuation of the Beckhampton Avenue beyond Longstones Cove (see Section 8.1.4.3), which had been proposed by Stukeley (Cripps, 2001, 14). Cripps (2001, 15) also suggests that the three-dimensional reconstruction can be used to investigate different building phases, under the assumption that the terminal of the Beckhampton Avenue was not only succeeded by the Cove but also by the hypothetical later extension of the Beckhampton Avenue in the direction of Fox Culvert.

It was planned to link the GIS and the three-dimensional model with one another,[176] so that it becomes possible both to generate viewsheds from within the reconstruction model and respectively to generate three-dimensional views from within the GIS. It is envisaged that results from these analyses could lead to changes in the model and to renewed analyses (Cripps, 2001, 12). Several characteristics of the monuments cannot be picked out by GIS. For example, stark contrasts in colour can enhance the visibility of monuments. This is especially true for features which were cut into the chalk or constructed from this material. The glowing whiteness is easily perceptible against a darker background of soil and vegetation (Cripps, 2001, 10, 32). In contrast, the sarsen, despite their bulk, were hard to identify from a distance in the reconstruction (Cripps, 2001, 32).

Cripps (2001, 13) sees one of the main differences between his new study and previous studies of visibility relationships between the monuments (e.g. by Devereux (1991) and Wheatley (1996)) in its focus on movement.

There are some points of critique concerning this appraisal. Despite Devereux's description of places from which Silbury Hill can be seen, his approach was certainly neither static nor based on the examination of a map or on the employment of a GIS (see Devereux (1991)). From his description and the photographs it becomes evident that Devereux was exploring the landscape first-hand in order to arrive at his conclusions. The case re Wheatley's study certainly lies differently. Actually it could be argued that a GIS has the same effect as maps: i.e. detaching the observer from the monument (cf. Barrett (1994, 12)). If this is the case, what is the advantage of employing a GIS and creating viewsheds? Would not a three-dimensional model of the monuments and the landscape be sufficient to tackle the question: was the Beckhampton Avenue lined out in such a way that specific points in the landscape are visible from it? The layout was most certainly planned by exploring the landscape and not with the help of a map. So what additional information does the map based output and the calculations of a GIS provide in addition to what can be learned from the three-dimensional model? One of the possible answers to this question could be restrictions on the size of the three-dimensional model due to computational and software restraints (cf. Cripps (2001, 17 f., 37)). Points of interest, such as other monuments, could lie outside the modelled region. This can also happen to a GIS, but the spatial regions which can be managed are much larger (though this is always dependent on the resolution of the digital elevation model).

[175] It has to be kept in mind that GIS does at the moment not support truly three-dimensional data. The height value is just an attribute to the point marked by the two-dimensional (x and y) coordinates. Therefore height cannot be included into spatial calculations in the same way as the x and y coordinates can be.

[176] This link could, finally, not be developed due to time constraints and technical difficulties concerning the links to the VRML nodes (Cripps, 2001, 38).

To create the three-dimensional model the landscape and the monuments on it, different techniques were employed. The software used for modelling the monuments included AutoCAD and 3D StudioMax. The data basis for the models of the enclosures on Windmill Hill and in Longstones Field as well as the barrows ranged from digitised base plans, known measurements and accounts concerning the size and shape of the monuments, to excavation data (Cripps, 2001, 16).

For the avenue and cove stones another approach was used. Ready-made stone models from the internet were downloaded and then formed into approximately the right shape via 3D StudioMax. In this way several types of stones were created: generic stones of types A and B (see Section 8.1.2.1), stones resembling the standing stones Adam and Eve (see Section 8.1.4.2) and stones which were to represent the Beacock stones (see also Section 8.1.4.2), which were based on the shape of the Obelisk. AutoCAD was used to put the stones on their correct spatial position (concerning the x and y coordinates) inside the model (Cripps, 2001, 16 f.).

The landscape was created from the digital elevation model (DEM) used by the GIS. From this raster based DEM a vector based triangulated irregular network (TIN) was produced which was suitable for a VRML model. Several steps were necessary in this procedure, many of them concerned with simplifying the TIN without losing too much detail and also ensuring that the number of nodes did not exceed the number of nodes that VRML and the available hardware could handle (Cripps, 2001, 17 ff.). A balance between the size of the surface area and its level of detail had to be struck (Cripps, 2001, 18). Finally the landscape model was geo-referenced using AutoCAD. The surface model was then loaded into 3D StudioMax together with the previously modelled monuments in order to attain the final model. The individual parts were already in their correct position concerning their x and y coordinates, so only the height values had to be set manually for the monuments (Cripps, 2001, 17, 19). After the final touches (textures, bump maps and lighting) the model was exported as VRML, while still images and animations were rendered from within 3D StudioMax (Cripps, 2001, 20).

For generating the viewsheds the software ArcView was employed.[177] The surface models were augmented by including the prehistoric monuments and some modern features as well as the hypothetical extension to the Beckhampton Avenue (Cripps, 2001, 21). From this, binary and cumulative viewsheds were calculated.[178] The cumulative viewsheds are based on 25 individual viewsheds which were calculated from different points along the Beckhampton Avenue. The individual viewsheds were in addition used to create an animation which shows how the visible areas change when an observer is walking along the avenue (Cripps, 2001, 21).

The cumulative viewshed (see Fig. 8.23, cf. Fig. 8.2) indicates that Longstones Field, i.e. the Longstones enclosure, the terminal of the avenue and the Longstones Cove, fall into an area of low visibility, while Silbury Hill and Waden Hill are highly visible. Longstones and South Street long-barrows and Windmill Hill have only a medium visibility (Cripps, 2001, 26).

When looking at the viewsheds in a sequence, instead of the cumulative viewshed, Cripps (2001, 27) notes that several changes occur in the field of view. The walk begins at the eastern entrance of the henge. Only one image from each range in the sequence has been picked for illustration here.

- In steps 1–5 only the section closest to the henge is visible as are Silbury Hill and Waden Hill. Longstones Field and the two barrows in that area cannot be seen (Fig. 8.24).
- In steps 6–14 the field of view becomes more and more restricted, only the midsection of the avenue, including Longstones Field with South Street long barrow, remain visible. The entrance to the henge, the part of the avenue closest to it and Silbury Hill fall out of sight. Longstones long barrow and the hypothetical extension of the avenue remain invisible until, at the end of this sequence, Longstones long barrow appears (Fig. 8.25).
- During steps 15–20 the area between the henge and Longstones enclosure is obscured. The view widens to the south while it becomes restricted to the north, only the summit of Windmill Hill is still visible, but gradually also disappears from view. Silbury Hill comes into view again when Longstones long barrow is reached. The hypothetical end of the Beckhampton Avenue is still out of sight (Fig. 8.26).

[177] The part concerning the GIS will be kept short, because it does not lie within the scope of this PhD thesis.

[178] Binary viewsheds categorise points in the landscape as either visible or invisible. Cumulative viewsheds are a combination of several viewsheds, whereby degrees of visibility become apparent—from highly visible to hardly visible (cf. Wheatley and Gillings (2002, 207 ff.)).

Figure 8.23: Cumulative viewshed from 25 points along the course of the Beckhampton Avenue. Intensive red colour indicates high visibility (Cripps, 2001, 49 fig. 11).

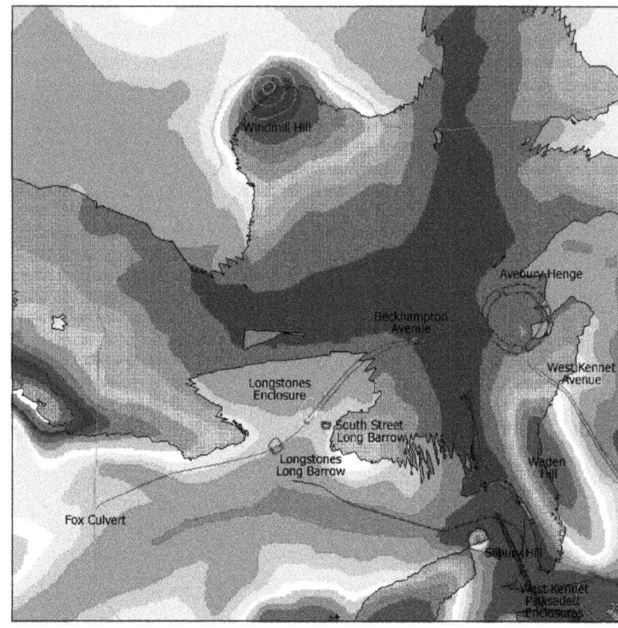

Figure 8.24: Viewshed from point 4 along the course of the Beckhampton Avenue (Cripps, 2001, 51 fig. 15).

- Between steps 21–23 the hypothetical end of the Beckhampton Avenue at Fox Culvert becomes visible. South Street and Longstones long barrow as well as Longstones Field are still in view. The henge comes back into sight, but the avenue between it and Longstones Field is invisible (Fig. 8.27).
- Silbury Hill remains in view, but Longstones Field with the two barrows and the henge disappear, only the part of the hypothetical avenue extension is now in full view (Fig. 8.28).

Different vantage points along the course of the avenue result in different views. From no point is the whole of the avenue visible, even if only the better established part (from the Longstones Field to the henge) is taken into consideration. Nor are all of the surrounding monuments (henge, Silbury Hill, Windmill Hill and Longstones Field with the two barrows) visible from one single point. This underlines the hypothesis that discrete 'visibility envelopes' are supposed to be viewed in sequence (Cripps, 2001, 27). Approaching the henge from the Beckhampton Avenue and from the West Kennet Avenue (Thomas, 1993, 42) provides a similar experience, which goes unnoticed by Cripps (2001). In both cases the entrance to the henge is hidden by a small rise in the topography (Cripps' steps 6–14) and is only revealed at a much closer stage (Cripps' steps 5–1; Thomas: 400 m from the entrance). Another parallel is presented by the hypothetical extension of the Beckhampton Avenue in relation to the West Kennet Avenue closest to the Sanctuary (Cripps, 2001, 30 f.). Both stretches curve 'around a prominence' and Silbury Hill is visible from both of their ends (Fox Culvert and the Sanctuary respectively). Also in the middle parts[179] of both avenues, the view to Silbury Hill is blocked by hills (Folly Hill and Waden Hill respectively) (Cripps, 2001, 31).

Several additional viewsheds from prominent points were created: from the western henge entrance, Longstones Cove, Longstones long barrow,[180] Silbury Hill and Windmill Hill. Only the latter two provide new information. The view from Silbury Hill includes all the monuments in the study region except Longstones Field, the avenue in this stretch and South Street long barrow (Fig 8.29, left) (Cripps, 2001, 28). One of the rendered images from the three-dimensional model shows a comparable view, i.e. from Silbury Hill towards Longstones long barrow, and highlights that the observer's view is blocked by a slight rise in the middle ground (Folly Hill) which

[179] In the case of the Beckhampton Avenue the here-mentioned 'middle part' is only a middle part if there *is* an extension. Otherwise, it would be one of the ends of the avenue.

[180] These sites are actually parts of the walk along the Beckhampton Avenue between the henge and Fox Culvert.

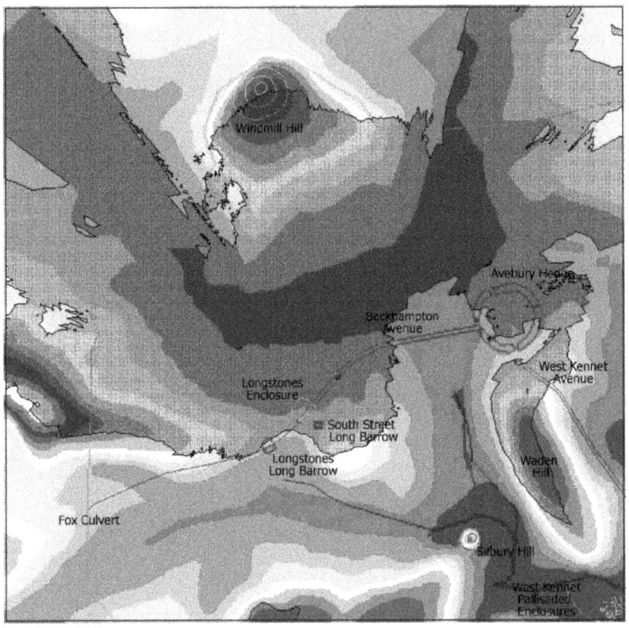

Figure 8.25: Viewshed from point 10 along the course of the Beckhampton Avenue (Cripps, 2001, 54 fig. 21).

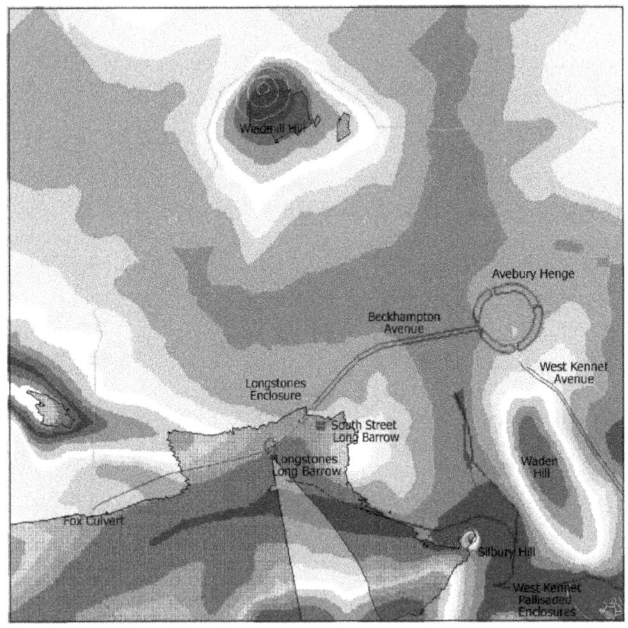

Figure 8.26: Viewshed from point 18 along the course of the Beckhampton Avenue (Cripps, 2001, 58 fig. 29).

ascends towards the right of the image, in the direction of Longstones Field (Fig. 8.29, right).

From Windmill Hill all the monuments in the study region are visible (Fig. 8.30, left), the only exception is the hypothetical extension to the Beckhampton Avenue in the direction of Fox Culvert (Cripps, 2001, 28). One single prospect from the southern bank of Windmill Hill is included in the printed part of the dissertation (Cripps, 2001, 72 fig. 51): a rendered view towards Longstones Field (Fig. 8.30, right). The monuments there (South Street long barrow, Longstones long barrow and, in this case, the Beckhampton Avenue terminal) are visible. It becomes obvious that the sarsen are not readily discernible from this distance.[181] As Cripps (2001, 28) states: the viewshed suggests a line-of-sight relationship, but under natural circumstances it may not have existed, because the features do not stand out enough to be perceived.

Apart from that, the results from the VRML model—which had been presented as animations on the CD accompanying the printed dissertation—are said (Cripps, 2001, 28) to corroborate the results from the viewsheds.[182]

[181] They can be made out against the uniform background within the model, but it is easy to imagine that they will disappear in a landscape littered with shrubs and the irregular scatter of grasses and herbs.

[182] Having no access to the CD, I cannot validate this statement.

The sequence of the monuments at Longstones Field has been visualised as well (Fig. 8.31). Cripps (2001, 28) states that it can be clearly seen how the avenue respects the earlier enclosure (Fig. 8.31, top left and right). In fact, in the rendered view this *appears* to be much more the case than in the excavation plan (Gillings et al., 2008, 14, fig. 2.6) or in Fig. 8.10. As can be seen in the latter Figure the avenue does not really enter the backfilled enclosure through its entrance (which would have been possible with only minor changes to its course), but the eastern stone row crosses the former ditch and some of the stones end up very close to it. Gillings et al. (2008, 122 f.) see in this arrangement an axial shift. Such shifts, which occurred during the 'lithicisation' of monuments, have been observed at other sites as well, among them the Sanctuary. Another shift is obvious from the reconstruction (Fig. 8.31, top right and bottom left) and the excavation data (Fig 8.13) and has been mentioned earlier: the first phase of the Beckhampton Avenue terminal is centred on the course of the avenue, while the Cove is centred instead on the western stone row (Cripps, 2001, 28). In these cases the three-dimensional reconstruction did not provide any additional information. Rather it only illustrated what had been known before from the excavation plans, and, in one instance, did 'confuse' the impression gained through excavation, i.e. when the axial shift between the enclosure and avenue phases

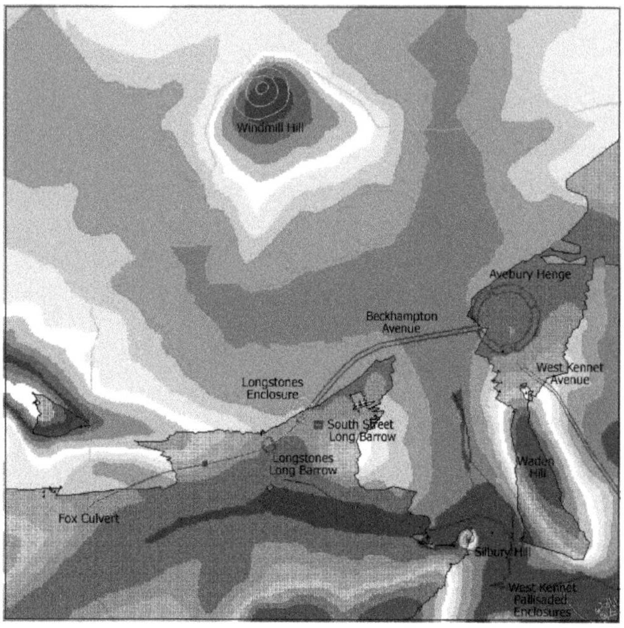

Figure 8.27: Viewshed from point 22 along the course of the Beckhampton Avenue (Cripps, 2001, 60 fig. 33).

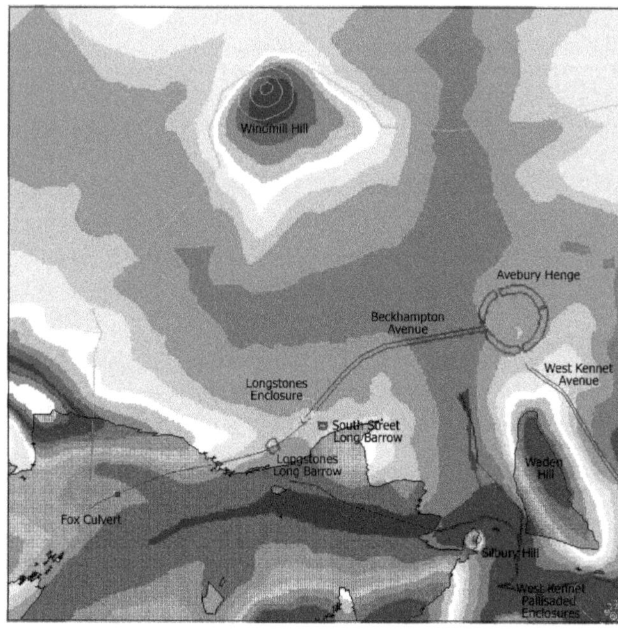

Figure 8.28: Viewshed from point 27 along the course of the Beckhampton Avenue (Cripps, 2001, 63 fig. 38).

seems to be negated in the model.[183] The last rendered image shows the hypothetical extension to the Beckhampton Avenue (Fig. 8.31, bottom right). From the image it almost seems as if there is once more a 'kink' in the avenue's course though not a very elaborate one,[184] which would be yet another example of the avenue's tendency to lead up to, but not directly into the adjoining monuments.

Cripps (2001, 28 f., 31) mentions a change in the axis between the first terminal of the Beckhampton Avenue to the Longstones Cove. He claims that the axis of the Cove could have been chosen, because it is more in alignment with Silbury Hill. The question is how this slight shift in axis could be accomplished especially because Silbury Hill is not visible from Longstones Cove and vice versa. The only solution would be via a 'ranging rod' on Folly Hill, which is blocking the view between the two monuments. The shift in alignment between the two phases of the Beckhampton Avenue terminal is also mentioned by Pollard and Reynolds (2002, 104). Of the new alignment it is said that it 'is just off-set to the west of the South Street long barrow, though anyone standing in the centre of the setting had a relatively clear view of the barrow through the sizeable gap between stones on the east side.'

There is actually no 'alignment' in the first phase of the avenue terminal because the three stones of this phase are not aligned (see Fig. 8.12 and 8.13): either the central stone departs from the axis set by the two outer stones or one of the outer stones departs from it. The Cove, in contrast, is aligned with the south-eastern Beacock hole. Disregarding the 'shift', Cripps (2001, 31) suggests that a reference to the invisible Silbury Hill could have been the main purpose of the Longstones Cove in this place, a reference which had been unnecessary if there would have been a line-of-sight.[185]

Cripps (2001, 30) concludes from his analyses of the viewsheds and the VRML model that the course of the avenue was positioned in a way which takes advantage of the topography. Monuments became visible and disappeared from view in an order. The question is whether this is a legitimate claim. Any course through the landscape would provide a series of different views. Furthermore it would require the observer to stop at intervals to take a look around because usually the gaze of someone following a marked course would be fixed either ahead or on the ground. The question of what an observer sees, looking straight ahead, is never discussed in the dissertation, in contrast to the study by Earl and Wheatley (2002). Maybe this comes from the focus on the GIS and the viewsheds created from it, with the inferior

[183] Is this due to the standpoint of the observer in the model? A trompe l'œil?

[184] Cripps (2001) does not remark on this either.

[185] Would this be also an argument for a West Kennet Cove (see Section 8.1.2.1)?

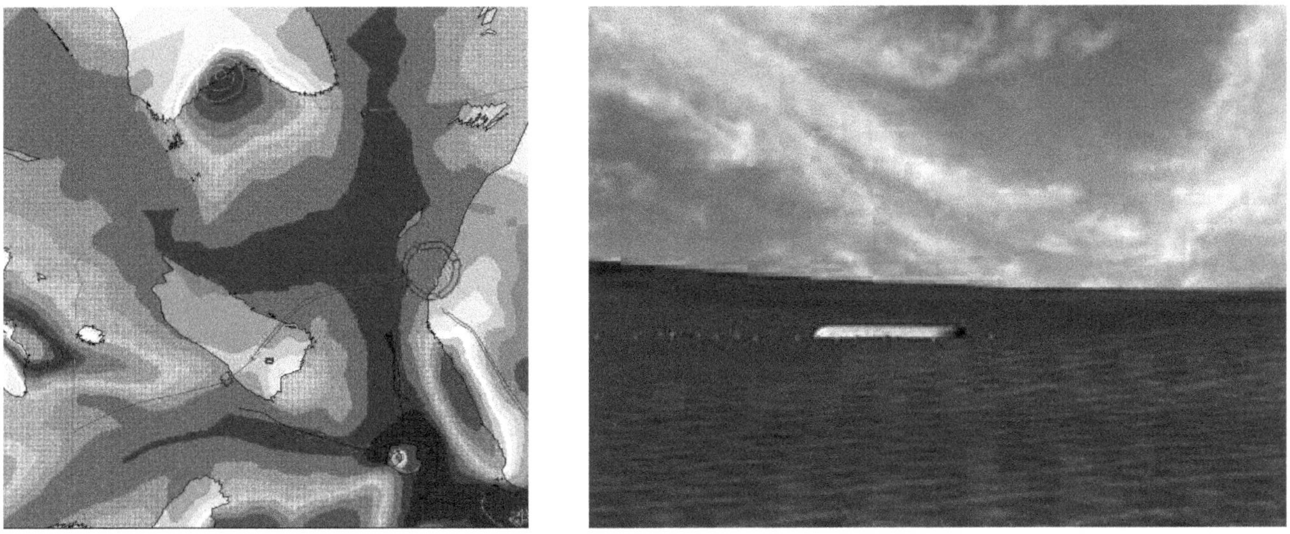

Figure 8.29: Viewshed and view from Silbury Hill. Left: viewshed from Silbury Hill (Cripps, 2001, 68 fig. 44). Right: view from Silbury Hill to Longstones long barrow (Cripps, 2001, 72 fig. 50).

Figure 8.30: Viewshed and view from the southern bank of Windmill Hill. Left: viewshed from the southern bank of Windmill Hill (Cripps, 2001, 69 fig. 45). Right: view from the southern bank of Windmill Hill towards Longstones Field (Cripps, 2001, 72 fig. 51).

role (i.e. checking the results obtained from the viewsheds) the three-dimensional model played in the research. One way to check whether the course of the Beckhampton Avenue is indeed 'special' would be to map random routes through the landscape and to compare their vistas to the ones from the Beckhampton Avenue. Is there a high/low frequency of changing views? Can many other monuments be seen, or do they stay in view for extended periods of time? What is visible straight ahead when walking along the routes? And: may there have been natural features (trees etc.) which also received attention, but we do not know about them?

Besides a comprehensive critical appraisal of the techniques used in his study Cripps (2001, 33–39) also suggests some useful additions or changes to his approach (Cripps, 2001, 31 f.):

- Fuzzy instead of binary viewsheds.
- Taking vegetation into consideration (for both the GIS and the VRML model) (cf. Section 8.1.7).
- More stress on the actual visibility of the monuments, e.g. the avenue stones,[186] in the three-dimensional model.

[186] Could this not be tested by using the restored part of the West Kennet Avenue as an example? How visible are the stones from a given distance and in contrast to the background?

Figure 8.31: Sequence of the monuments in Longstones Field. Top left: Longstones enclosure; top right: first phase of the Beckhampton Avenue terminal; bottom left: second phase of the Beckhampton Avenue terminal, the Longstones Cove; bottom right: Longstones Cove with the hypothetical extension of the Beckhampton Avenue. (Cripps, 2001, 70 fig. 46 and 47, 71 fig. 48 and 49).

8.3 Have the aims been reached?

The project stakes were set high, but only few of the anticipated analyses could be conducted.

Due to severe funding problems none of the models intended for Gummidging could be realised (Mark Gillings, personal comment). This includes the model based on the 'new' Stukeley plan, but also the models of the henge in the final Neolithic. The survey data, measurements and photographs which were collected for a present-day model of the monument were obviously used to create one[187] which was later reused in the two 'spin-offs'[188] analysed here.

[187]See Pollard and Gillings (1998, 150 fig. 1, 152 fig. 3), Gillings and Pollard (1999, 189 fig. 4) and Gillings (2000, 64–67 fig. 3–5).

[188]Namely, Earl and Wheatley (2002) and Cripps (2001).

The aims had included the following issues:

Easy and affordable creation of virtual reconstructions.
: Gillings (2000) presents a simple method for stone recording which can be used in connection with standard surveying techniques to obtain the necessary data for the creation of a virtual reality model. Though the software used for recording the stones is affordable and easy to use, creating a surface model can be much more complicated (cf. Cripps (2001, 16–20)) and requires specialist knowledge, and sometimes a multitude of software packages.

Establishing virtual reality models as primary records.
: Due to constraints regarding the previous point and lack of adoption of the proposed method-

ologies in other projects, this goal has not been reached.

Interactive and modifiable models. Through the use of VRML the models can be modified, but this requires, yet again, specialist knowledge. No procedure for sufficiently documenting the changes has been put forward, though the need for such documentation has been noted. If the model has from the start been created as an interactive model (see Earl and Wheatley (2002)), then it can easily be used and toggled between preprogrammed scenarios by anyone.

Alternatives. Already the presentation of an (interactive) model with several pre-programmed scenarios constitutes the showing of alternatives (as in Cripps (2001); Earl and Wheatley (2002)). Other scenarios had been envisaged: different phases of construction and destruction, hypothetical scenarios to help the observer 'see the past through the eyes of someone else'. These latter models came never into being, but at least ideas on how to approach this difficult subject were addressed.

Analyses and the interpretational value of virtual models. It should be demonstrated that virtual reconstructions can be more than mere 'ingenious end-products' (Gillings, 2000, 59) or painstakingly created surrogates (Pollard and Gillings, 1998, 148), indeed that they can have a place in the interpretation of the sites they have modelled. Several approaches to this goal have been proposed by the different authors (e.g. Goodrick and Gillings (2000): Gummidging, i.e. modelling different hypothetical construction phases and alternatives, and using them to test the hypothesis/explore the resultant model). Cripps (2001), Earl and Wheatley (2002) have demonstrated that models can be used to explore possible scenarios, but also that it is difficult to come to unequivocal conclusions (see esp. Earl and Wheatley (2002, 14)). It rather seems that the usefulness of virtual reality models could lie in their ability to provide new ideas or make discrepancies apparent (e.g. the 'T junction' in Earl and Wheatley (2002, 14) or the model of the Sanctuary in Pitts (2001b, 238 f.) respectively).

Bridging the gap between theory and practice. Care has been taken to build models with a phenomenological agenda. The authors have been (sometimes) aware that their approach has flaws, especially in respect of addressing *all* the senses (see Earl and Wheatley (2002, 5, 7)), but were less aware that 'real embodiment' cannot be achieved by sitting in front of a computer screen (Goodrick and Gillings, 2000, 52 f.) nor that their concept of 'Gummidging' did not really help to shed our 'male, western, academic viewing perspective' (Goodrick and Gillings, 2000, 54). At least these problems were addressed in some way instead of blanking them out completely. Some disappointment (especially after the lengthy and theoretically ambitious introduction) arose from the 'common sense' approach to the interpretation of the entrance model in Earl and Wheatley (2002) which fell far short of my expectations. Expressions like 'it makes little architectural sense' (Earl and Wheatley, 2002, 12) immediately raise the question how we, from our perspective, can judge what made sense to the people in the final Neolithic *or* whether it has to make sense at all. The same goes for the statement 'makes, superficially, a far more believable arrangement' (Earl and Wheatley, 2002, 13). In both cases the authors were not aware that they are not presenting very convincing arguments.

The models and GIS in this project were intended for research and were therefore meant to be used mainly by archaeologists. This could also explain why the need for additional information was not felt, because the people interacting with the model were experts—i.e. were aware of the sometimes scant and unequivocal evidence, with all the consequences this would have on the model and its interpretation. The use of transparent cylinders in the model by Earl and Wheatley (2002) is probably owed to the desire to make the uncertain status of the stone positions immediately *perceivable*[189] so that it can be taken into account during interpretation—in contrast to a desire to inform the uninitiated about their uncertain status. The documentation of the model could also have included additional information, but nothing definitive is mentioned in this respect (Earl and Wheatley, 2002, 8).

Additionally, all the people involved in the project were archaeologists, which probably resulted in fewer communication problems than in the projects described in Chapters 6 and 7. In both of the previously described projects 'helping with communication' was one of the major unexpected bonuses. No additional unexpected uses for the Negotiating Avebury project are mentioned in any of the texts.

[189] Based on Gibson, see Earl and Wheatley (2002, 5).

Several of the texts give references to webpages, but none of them can be accessed any longer.[190] The only model which can still be accessed via the internet is a VRML model of one of the stones from the outer circle[191].

What makes this project stand out is the multitude of fresh ideas on how virtual reality could make an impact on archaeological research. It provides critical reflection of previous fields and methods of applications and the wish to arrive at practical, affordable and novel approaches—in order to find the true potential of the technique and to move beyond low level interpretation[192] (Earl and Wheatley, 2002, 6; Gillings, 2000, 59). All this can be achieved without the necessity of too much (superfluous) detail or photo-realism, but by emphasising that the model is an artificial environment (Earl and Wheatley, 2002, 7; (Gillings, 2002)).

8.3.1 Useful additions

Useful additions are hard to suggest, because so little of the project could be put into practice. It would have been interesting to see how the interpretation of the 'Gummidging'-models would have been carried out and what effect they would have had on the observer. Even a study with a lay audience may have heralded interesting results: e.g. do consciously subjective models get across the impressions they were meant to convey? For example: do overlarge banks make an observer feel excluded? Or do people feel prompted to climb the bank to look beyond?

Also, the Stukeley model could have yielded interesting insights, and furthermore could have been used as the basis for a Gummidging exercise, because there are two very different views on the monument from Stukeley's time (Gillings and Pollard, 1999, 188):

First, the already mentioned antiquarian view, where the sarsens were little more than pieces in a wider concept. Their positions, much less the stones themselves, were recorded simply to find out the underlying construction rules for the monument.[193] Second, the villagers broke up the stones and mutilated banks and ditches, sometimes for economic or practical reasons. They saw them as obstacles for coaches (and for agriculture?) and, the stones especially, as building material, but there may have been other connotations related to superstition or religious dispute (see Section 8.1.5.2). That those two views on the monument were in opposition to each other is obvious, and could have been the basis for a '18th century villager's Avebury' model.

8.3.2 Beyond Negotiating Avebury

As mentioned above, the Negotiating Avebury project never progressed very far. Nevertheless it seems necessary to mention briefly two further articles which stand in connection to Avebury or the project.

8.3.2.1 Digital Avebury: New 'Avenues' of Research

First will be an article by Davies (2009), which has a very similar approach to Avebury as Cripps (2001). Davies combines a GIS and a VRML model of the Avebury region in a phenomenological approach to the monument complex. His point of departure is the hypothesis presented by Parker Pearson and Ramilisonina (1998) that part of the monument complex (i.e. the wooden structures) belong to the 'realm of the living', while the stone monuments belong to the 'realm of the dead/ancestors' (Davies, 2009, section 1).

From an assumed approach to the monument complex from the south, he investigates the views which can be had from south of the West Kennet palisade enclosures and from inside enclosure 2. Davies (2009, section 3.2) claims that the visitor is shepherded by the monuments and some higher ground to enter the enclosure 2 (Fig. 8.32), from where Silbury Hill (which is also associated with the 'realm of the living') can be seen but none of the barrows (which belong to the 'realm of the dead'). When leaving enclosure 2 the same way as one came in, then the only choice the architecture and landscape would offer (Fig. 8.33), would be either to leave the Avebury area

[190] Mark Gillings confirmed in a personal comment that the models were not available any more.

[191] Stone 4: http://www2.le.ac.uk/departments/archaeology/people/gillings/documents/avebury-stone-4-revisited (Accessed: 6th of August 2012.).

[192] cf. Reilly (1992, 170), who proposed that solid modelling will only have impact on low level theory with possible effects on middle level theory.

[193] From what can be judged from his recordings, it seems to have been Stukeley's aim to arrive in some way at a reconstruction of the monument as intended by its builders (or what he believed to be the intention of the builders).

Figure 8.32: View towards the West Kennet palisade enclosures from the south: Silbury Hill can be clearly seen in the background, Davies (2009, section 3.2) claims that the architecture shepherds the visitor towards enclosure 2 (Davies, 2009, fig. 3).

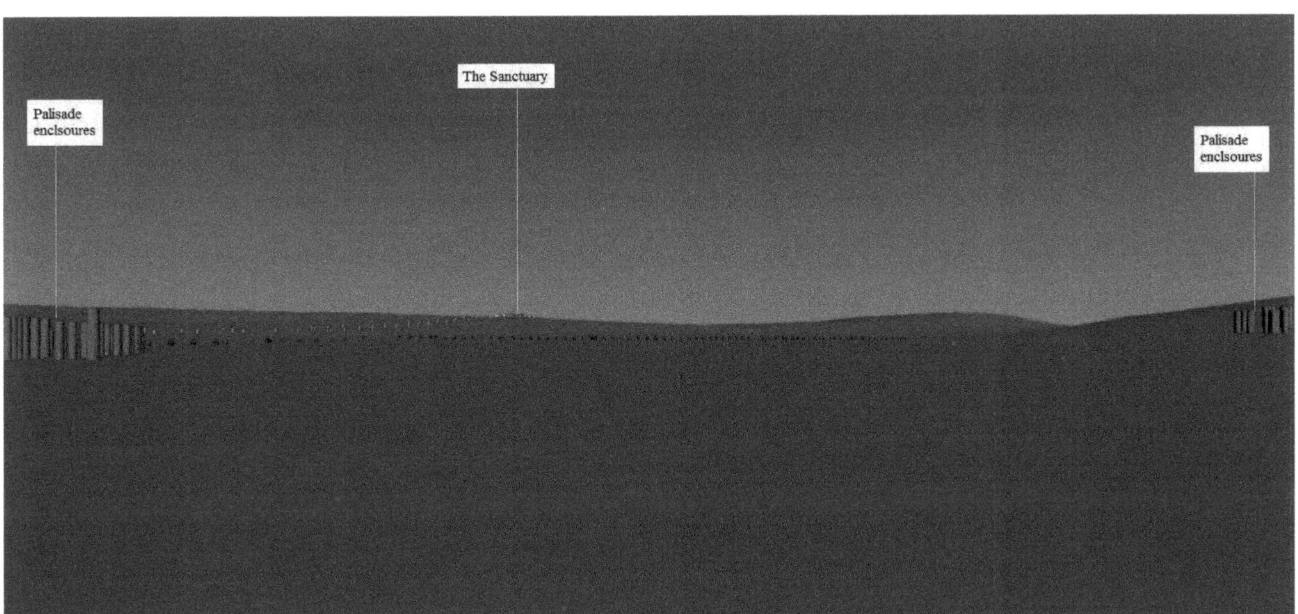

Figure 8.33: View towards the Sanctuary from the West Kennet palisade enclosures: here Davies (2009, section 3.2) claims that the architecture shepherds the visitor towards the Sanctuary (Davies, 2009, fig. 4).

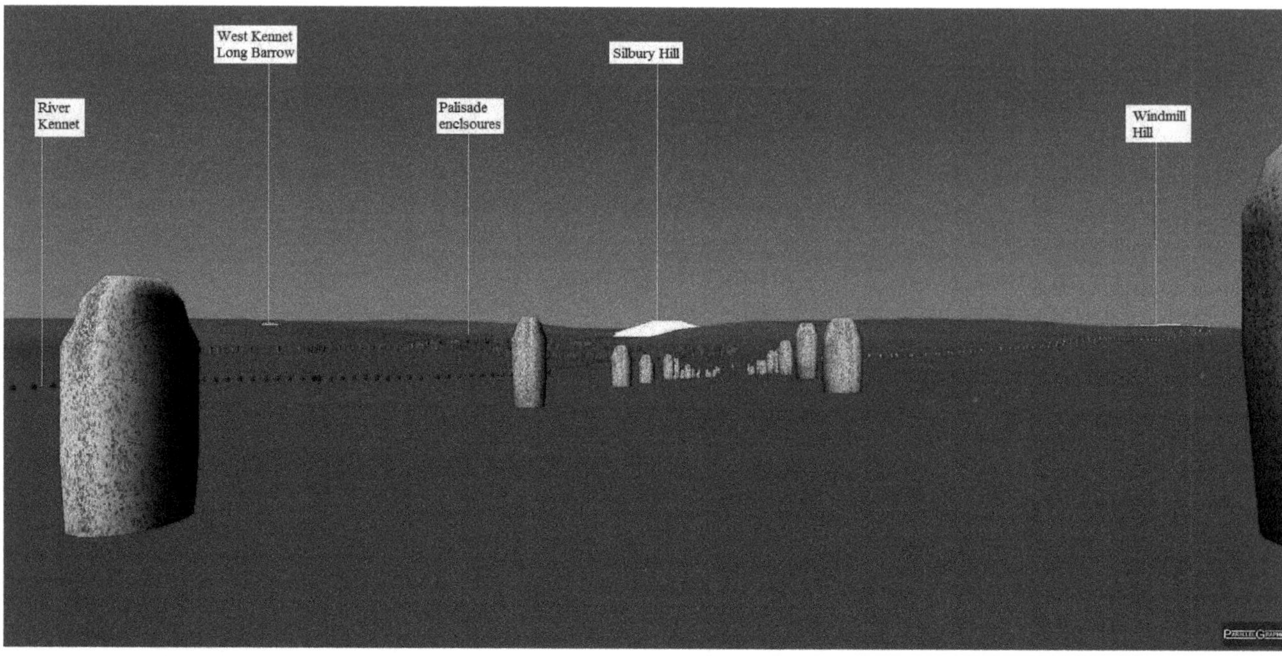

Figure 8.34: View from the Sanctuary along the West Kennet Avenue: some of the marked monuments (e.g. Windmill Hill or West Kennet long barrow) can hardly be distinguished. (Davies, 2009, fig. 6).

completely or to proceed to the Sanctuary (Davies, 2009, section 3.2).[194]

From the Sanctuary the route is clearly laid out. Davies (2009, figs. 5–13, appendix 2) provides viewsheds from different points along the West Kennet Avenue, still images from the VRML model and, additionally, an animation created from the VRML model, which shows the progress along the avenue. He argues that the beginning of the West Kennet Avenue is dominated by the West Kennet palisade enclosures and Silbury Hill (realm of the living), but also that West Kennet and East Kennet long barrows are visible (Davies, 2009, section 3.3). From the Sanctuary most of the avenue's course can be seen in advance; it seems to lead in the direction of Windmill Hill (realm of the dead)—Avebury henge being out of view (Davies, 2009, section 3.3). Sight lines which are suggested to exist by the GIS can be made out on the VRML still images as well, but in the model the ground has a uniform shade of green and often the monuments (e.g. West Kennet long barrow (Davies, 2009, figs. 5 and 6)) can only be made out as tiny slivers of white chalk (Fig. 8.34). It is possible to wonder whether the monuments may not be hidden from view by the chance appearance of a few long tufts of grass (cf. Cripps (2001, 32)).

At the point when Falkner's circle can be seen from the avenue Davies (2009, section 3.4) suggests that it could appear as if Falkner's circle might actually be the henge and that the avenue would therefore be leading up to it.[195] Before the henge comes eventually into view, the avenue swerves to the left, which is a very confusing experience, similar to the confusion encountered at the dog-leg in front of the henge entrance (cf. Earl and Wheatley (2002, 12, 12 fig 2.2)). This may be in part due to the speed with which the progress along the avenue is animated.[196] Similarly to the West Kennet palisade enclosures where Silbury Hill was always prominent in the distance, here Windmill Hill has a central posi-

[194]Some of these arguments seem not very plausible, especially that the higher ground would prevent people from crossing it. But then it is necessary to walk up the rise to the Sanctuary, even though the West Kennet Avenue is very close. Why not take the shortcut? Why not walk along the West Kennet Avenue towards the Sanctuary? The stone settings are definitely not close enough to prevent anyone from crossing or entering the avenue.

[195]There is a lot to say against that. People in the Neolithic were probably very used to judging distances and the size of things, so there is no chance of mistaking a small stone circle for a huge henge, as Davies (2009, section 3.4) suggests. Both are very different in appearance as well; the banks of the henge are hiding the stones inside, while there is no bank around Falkner's circle. If some delegation was meant to make a side excursion to visit the circle (as proposed by Davies (2009, section 3.4)), then there is no architectural indication of it. Finally, while watching the animation (Davies, 2009, appendix 2), I was never under the impression that the avenue was leading up to the circle.

[196]On the other hand the animation is already rather long, a less speedy approach would have other drawbacks.

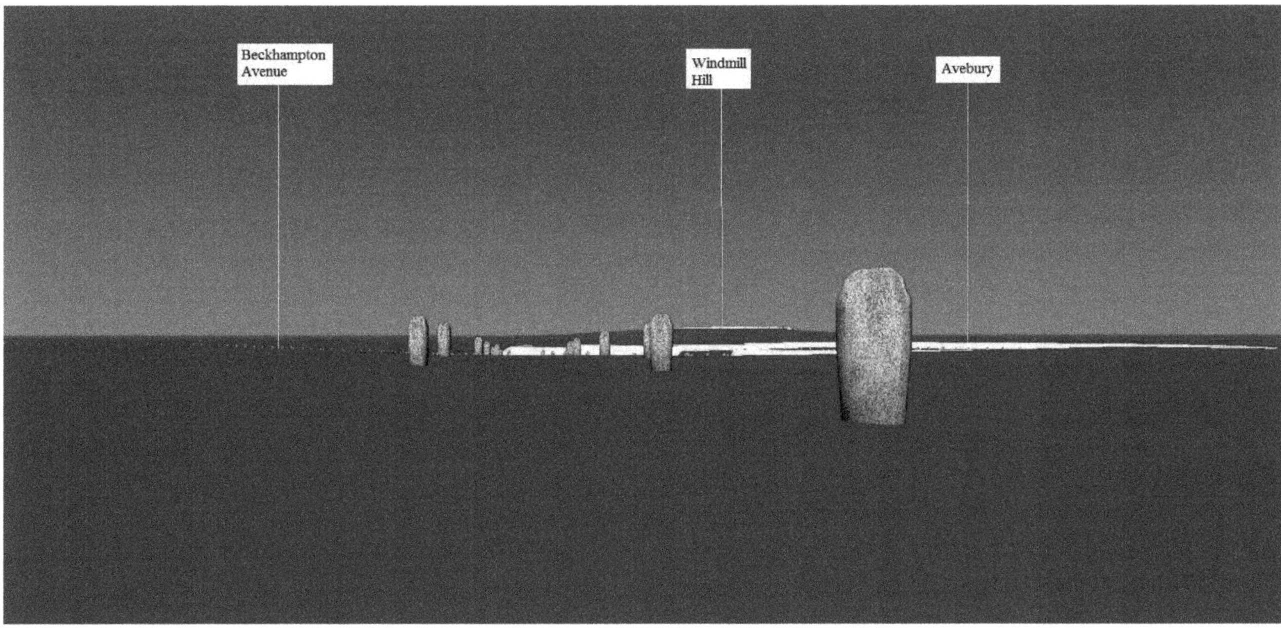

Figure 8.35: View along the West Kennet Avenue towards Avebury: the henge stands out due to its white banks. Beckhampton Avenue can hardly be seen in the background. (Davies, 2009, fig. 12).

tion on the horizon as soon as one crosses the occupation site (see Section 8.1.2.1), showing that Avebury and the West Kennet Avenue belong to the realm of the dead (Davies, 2009, section 3.5). Before entering the henge the Beckhampton Avenue also comes into view (Fig. 8.35)—again the question is whether it could have been seen on site, because the stones are only tiny specks against the uniform green background (Davies, 2009, fig. 12). Echoing Earl and Wheatley (2002, 14), Davies mentions that at one point it appears as if the West Kennet Avenue was going to join up with the Beckhampton Avenue, bypassing the henge.

The interior of the henge is little considered and no images were rendered of it (Davies, 2009, section 3.5).[197] The animation of the walk along the West Kennet Avenue stops several stone pairs before the henge is entered. Davies (2009, section 3.5) describes that after passing the banks the interior layout of the henge seems less confusing and that the architecture indicates that people should proceed to the open area to the west of the southern and northern circles.[198]

The next animation (Davies, 2009, appendix 2) shows the walk from the henge towards the Longstones Cove, two rendered stills providing additional information. At the beginning of the animation the camera pans to the south, where Silbury Hill can be seen. Davies (2009, section 3.6) mentions that West Kennet long barrow would also be visible, but in the animation the view towards the point where it should be standing is blocked by a sarsen (cf. Fig. 8.6). Instead South Street long barrow is visible as a sliver of white in the far distance (cf. viewsheds by Cripps (2001, 50 fig. 12, 65 fig. 41)). To the north (this is not shown in the animation but in a rendered still image: Fig. 8.36 (Davies, 2009, fig. 14)) the view shows Windmill Hill flanked by Horslip and Millbarrow long barrows (Davies, 2009, section 3.6). Some time before the Winterbourne is reached, South Street long barrow disappears from sight. Instead, to the north-west another barrow becomes visible (Horslip long barrow?). After crossing the Winterbourne the Beckhampton Avenue swerves to the south and the barrow is lost from direct sight and for a very long time no other monument enters it. Folly Hill then blocks the view to the south; instead, South Street long barrow and Longstones long barrow with the Longstones Cove appear (Davies, 2009, section 3.7, fig. 15). Nowhere else is a visitor in such close proximity to barrows and Windmill Hill, which leads Davies (2009, section 3.7) to the assumption that here the encounter with the ancestors has its focus.[199]

[197] This may be due to the fact that the interior of the henge was not fully modelled, for example, the ditches are missing (Davies, 2009, appendix 1).

[198] If the ring stone really had a partner it would rather appear as if the West Kennet Avenue was proceeding towards the southern circle and not away from it.

[199] He is also quoting Burl, who sees in the coves a transitional stage between chambered tombs—by imitating their chambers in the open air, (which would fit with the ancestral theme)—and

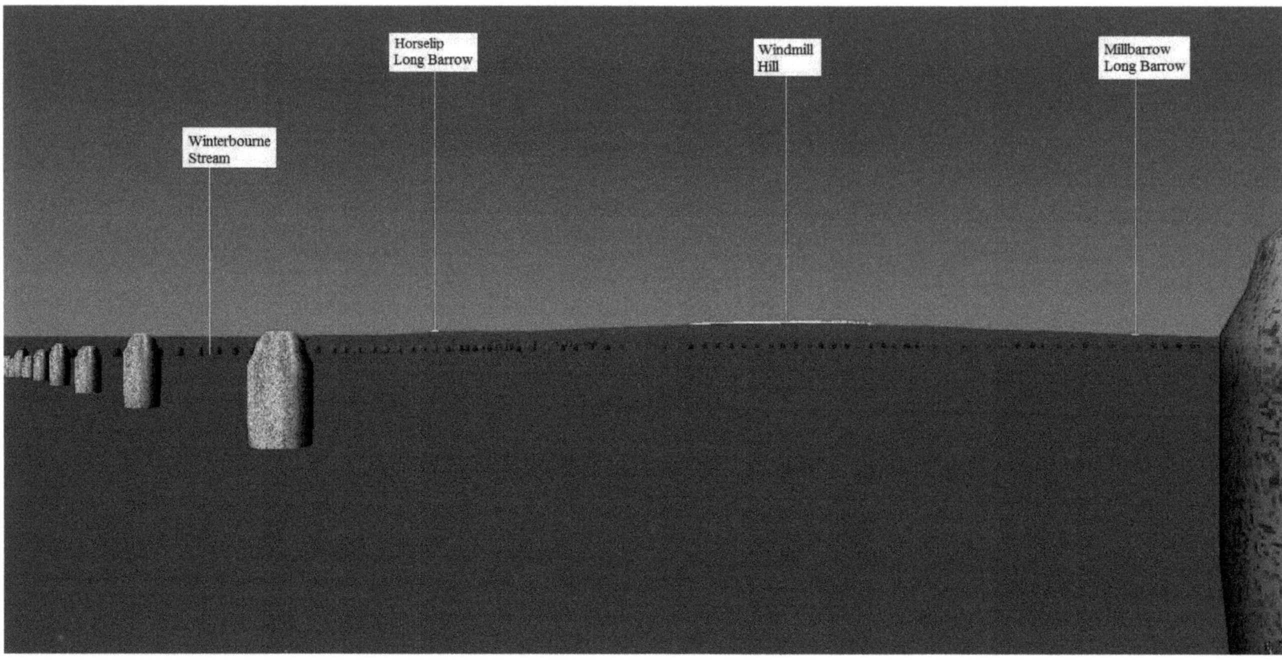

Figure 8.36: View from the western henge entrance to the north: Windmill Hill flanked by Horslip and Millbarrow long barrows (Davies, 2009, fig. 14).

One of Davies' (2009, section 3.8) discoveries during the modelling process was previously made quite similarly by Parker Pearson and Ramilisonina (1998, 319) (Fig. 8.37): the constellation of structure 4, enclosure 2 and the radial connecting both (West Kennet palisade enclosures), in connection with West Kennet long barrow to the west and Silbury Hill in the background, can be seen in parallel with the constellation of the Sanctuary, Avebury and the West Kennet Avenue connecting both, with the Longstones Cove to the west connected by the Beckhampton Avenue, and Windmill Hill in the background. Davies (2009, section 3.8) suggests another parallel constellation consisting of the Sanctuary, enclosure 1 of the West Kennet palisade enclosures and the river Kennet almost connecting both, with Silbury Hill in the background. This last variant is not so convincing in comparison.

In his conclusion Davies (2009, section 4.1) suggests that the Avebury landscape has to be divided into three zones (the realms of the living, the dead and deep time) instead of only two zones (the realms of the living and the dead) as suggested by Parker Pearson and Ramilisonina (1998, 319 ff.)[200]. The discussion as to whether the monuments were designed for the ritual practices of a few people (with a greater number of bystanders), be they some kind of elite or simply persons who are undergoing transitional rites, is raised once more. Just as the question whether the site was in use all the year round or only seasonally is posed again. Davies (2009, section 4.1) argues that the landscape was grazed and that this would not be likely if it was seen as a sacred place all the year round. Grazing would have had to stop for ceremonies.[201] All in all the approach by Davies (2009) combines imaginative elements concerning ritual and myths, which could have formed around the monuments, and computer applications (VRML models and GIS) in order to support Parker Pearson's, Ramilisonina's and his ideas about the site. It also gave him new ideas while exploring the model (Davies, 2009, section 4.2).

8.3.2.2 Beyond phenomenology

So far all the projects which have been presented here have had a phenomenological approach as their basis. The authors saw this as way to move beyond the creation of ingenious pictures (e.g. Gillings (2000, 60) or Davies (2009, section 4.2)) and to design models which had a research value. The question why ever

stone circles (Burl, 2000, 32 f.).

[200]Though the Sanctuary with its stone and timber settings (if they were contemporary) could have formed a transitional area between the living and the dead (Parker Pearson and Ramilisonina, 1998, 320).

[201]I do not see why this would be necessary, or why the presence of animals would 'desecrate' the monuments.

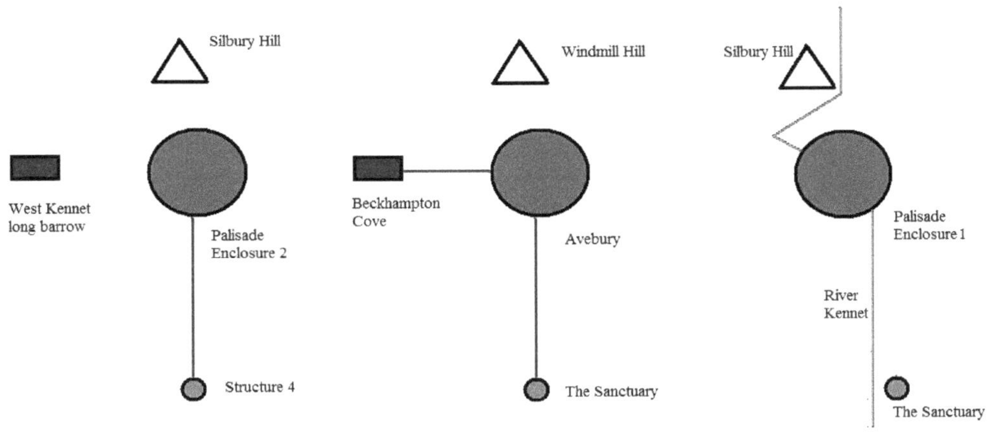

Figure 8.37: Comparison of monument layouts in the Avebury region. (Davies, 2009, fig. 16).

more 'ingenious pictures' without any added value were produced, has been answered in the way they are perceived: namely as surrogates (Gillings, 2002, 18). Instead of trying to create ever more detailed reproductions of what is there, or presumably was there, and to make them look as photo-realistic as possible, new useful applications for virtual reality models have to be found.

One of the solutions was to embed virtual reality models into an existing theoretical framework, i.e. into a phenomenological approach (Gillings, 2005, 231). That the models can be of some use in this field has been demonstrated in the preceding project descriptions, but it is also apparent that there are drawbacks which cannot be resolved. True embodiment and immersion is so far not really achievable. Only visual relationships (and maybe sound) can be incorporated into the models, while all the other senses cannot be addressed. The greatest problem lies within the observer him- or herself. We cannot escape our own social and cultural background (Gillings, 2005, 233) and we have no way of knowing what values and meanings were associated with places or material culture in the past (Gillings, 2005, 232 f.). Additionally, substituting one framework (i.e. surrogates) for another (i.e. phenomenology) is perhaps not the best solution for finding the potential of virtual reality applications in archaeology (Gillings, 2005, 233). In order to broaden the context in which the virtual models can be seen, Gillings (2005, 234)[202] suggests that they could be associated with the concept of 'mimesis'. In his preferred definition, mimesis encompasses the concept of imitation coupled with fiction and creativity (Gillings, 2005, 235). Gillings (2005, 234) hopes that seeing the virtual reality models in this manner will help us to arrive at 'more playful, creative applications', where 'emotive responses and questions' provoked by the model are the important factors. Unfortunately no concrete fields of application are mentioned. Evoking emotive responses is a problematic aim. The means of manipulating emotional responses are manifold. A landscape in the mist looks mysterious, maybe oppressive—the same landscape in full sunshine looks friendly and inviting. The same could be said for different lighting scenarios of interiors. Already some reconstruction companies have specialised in creating 'atmospheric reconstructions' (Grellert and Svenshon, 2010, 189), ingenious pictures which are meant to elicit defined emotional responses. This is apparently not the aim Marc Gillings had in mind, but it is the obvious looming pitfall. The second aim is better suited for the application of virtual reality models. That they have the potential to stimulate new ideas or raise questions has not only been proven by Earl and Wheatley (2002) (Section 8.2.2.2) and via the model of the Sanctuary by Jennifer Garofalini (Section 8.1.3) but also by the Ename 974 project in Chapter 7.

[202] See also Gillings (2002, 27 f.).

Chapter 9

Pre-Modelling

A short overview of the steps which must be considered in the creation of a virtual model has already been provided in Section 5.7, Fig. 5.4. The review lists four steps: *sources*, *inferences*, *visualisation decisions* and *modelling*. Following Pletinckx (2008, 6) the first steps in pre-modelling (i.e. belonging to step *sources*) should be source collection[1], source assessment and source correlation. From there the interpretation process should progress into the general *inference* phase where interpretations based on the sources are made (Wittur and Krömker, 2009, 89). These first two steps in Fig. 5.4 are discussed in this chapter. Chapter 10 will proceed with the *visualisation decisions* and their representation (i.e. *modelling*).

At the beginning of this chapter different kinds of data are introduced, which are frequently encountered when a reconstruction is to be created. They constitute the data basis (Section 9.1) and must be therefore subject to source criticism to establish their suitability as the foundation for further inferences (Section 9.2). The interpretation of the sources is a very individual process for which there is no standardised approach. The interpretation is dependent on the theoretical and personal preferences of the researcher, the questions asked, the available material and other factors. Hence no attempt on presenting an 'interpretational guideline' or a 'classification of interpretational approaches' is made here. Instead, Section 9.2 provides some general information regarding this topic. Alternatively, the three case studies (Chapters 6, 7 and 8) can be reviewed for some insight on different approaches.

9.1 Data basis

Different modelling aims (see Section 5.5.1) will result in different kinds of models. According to the aim and the available information about the site or artefact, their data basis will vary. Not all of the available information about a site or object will find its way into the reconstruction (see Section 5.7), nevertheless, sources which have been considered at some point should be included in the source collection (Beacham et al., 2009, section 3.1)[2]. Surveys, excavation data, building research on standing remains, but also historical documents and analogies may be used for the reconstruction.

The following sections introduce the various types of data, which can constitute the basis for a model. They show that these data cannot be used without prior examination of their content and background (i.e. source assessment (Pletinckx, 2008, 6)), nor that any kind of representation can be taken to be 'more or less value- or theory-laden than another' (Bateman, 2000, section 2).

9.1.1 Data

When encountering the 'real world' we appropriate it through interpretation/theory[3] (Lock, 2003, 3); this in turn means that the real world exists separate from our observations, but the observations (i.e. the data) are theory-dependent (Hodder and Hutson, 2003, 17).

The basis for the models are therefore 'theoretical objects', which are 'cultural products' (Lock, 2003, 3) of the time of their creation[4]. This should be kept in mind when in the following different kinds of data, which are frequently used for reconstructions, are considered.

[1] No implicit knowledge is allowed (Pletinckx, 2008, 6).

[2] http://www.londoncharter.org/principles/research-sources.html

[3] Calling a stone a stone is already such an interpretation.

[4] It has to be noted that usually not only recent excavation results are the basis for the models, but also historical text, drawings, *etc.* which were created at other points in time.

Surveys, excavation and building research records are, from an interpretational standpoint, very close to the primary data, i.e. they contain observations and interpretations. In many cases people involved in the fieldwork are also part of the reconstruction team. At the same time it is frequently the case that older archaeological records are also employed (see case studies in Chapters 6 and 8); these are usually treated with the same scrutiny as are **historical sources**[5]. If no direct evidence remains, then **analogies** can help to reconstruct missing parts (e.g. in Chapter 6).

9.1.2 Surveys, excavation and building research records

In this section it will be demonstrated that fieldwork cannot be documented without recourse to interpretation. The first subsection deals with measurements, which are a part of most surveying and recording methods, and moves on to discuss recording methods which are frequently employed to document and publish fieldwork results.

9.1.2.1 Measurements

Surveying and recording techniques, but also scientific analyses (e.g. laser scanning, georadar, photogrammetry, chemical analyses or dendrochronology) all rely on measurements. Measuring is often considered to be objective and, therefore, so are the methods and records associated with it (e.g. Riedel and Bauer, 2008, 2), but decisions have to be taken even for measurements and their recording (Lock, 2003, 5). The choice of methodology (and therefore of the measuring device) is also theory dependent (Hodder and Hutson, 2003, 18). Likewise interpretations are indispensable (Shanks and Hodder, 1995, 8) in order to make sense of the recorded phenomena.

It is necessary to know how the different measuring methods are working, what their output represents and which additional steps have to be taken so that the data can be readily used.

For example, models which were constructed from 3D point clouds (e.g. from a laser scanner) or image data (e.g. photogrammetry)[6] are often seen as surrogates[7] (e.g. Campana and Remondino, 2008, 41). Among the number of decisions which have to be taken for the recording are: when and where to record and at what resolution, using which method and recording device. Often several records have to be merged into one and there are several post-processing stages before the measured data has been transferred into a suitable format for further use. Pietroni (2005, 243 ff.), for example, describes the necessary steps for generating a 3D-model for documentation purposes starting from the point clouds generated by the laser scanner up to the textured 3D-model. Merging the point clouds, noise reduction, deleting redundant points, sampling points, smoothing the surfaces and further thinning of points are all mentioned before a triangulated mesh is produced. Before this stage there were only points in a 3D-space, whereas the mesh is creating a surface (see Section 10.2.1.1). After generating the mesh, several more steps of model simplification are carried out. Finally, textures can be applied (see Section 10.3.4), which may also need prior adjustments. From this example it becomes obvious that there are many subjective decisions which have to be made.

As the devices record 'everything' (up to a certain degree) there is no differentiation between important or unimportant features (cf. Pringle and Moulding, 1997, 22 f.). Identification of the important features, i.e. interpretation, should occur in the field (and not in front of the virtual records), because it may not only depend on visual, but also on other properties, e.g. texture. Documentation models (see Section 5.5.4) enhanced by information from observations and interpretation in the field are most certainly a very useful asset for archaeological research. But then, as Lambers and Remondino (2008, 31) mention, the budget and the time for training people to use the technology may outweigh the benefits of the 3D-documentation.

The accuracy of the measurements is a frequently mentioned concern. Johnson (2008, 4 on CD), for example, argues that models should be as accurate as possible, going so far as to suggest that peer reviewers should go out into the field and re-measure the site to ascertain that the measurements in the model are correct. Besides this very unlikely scenario, the question remains whether these accurate models are really necessary for the use envisaged (Pringle and Moulding, 1997, 22). Most communication and even a great number of research projects may simply not need this level of detail (cf. Section 5.5 for different model applications). Very detailed models are labour- and computationally-intensive and thus expensive to produce, not to mention the later main-

[5] Historical sources were used in all three case studies (Chapters 6, 7 and 8).

[6] For a discussion and explanation of various methods on different scales (regional, local, and object scale) see Lambers and Remondino (2008).

[7] For a definition of this term see Section 5.3.1.

Excursus 4: Definition of primary data, observation and interpretation

Archaeology and related disciplines deal with *primary data*[a], i.e. the external world in the form of physical remains (walls, features, finds), but as soon as an attempt to define or describe them is made, the realm of *interpretation* is entered. Besides the categories *primary data* and *interpretation* a third and intermediate category called *observation* is useful for the Argumentation Networks (see Section 5.7). *Observations* are interpretations (Thomas, 2000b, 4), but they are based immediately on the *primary data*. The term *interpretation* will be used for all other conclusions drawn from observations, other source material and interpretations.

[a]There are several other terms being used: e.g. 'empirical reality' (Lock, 2003, 2) or 'the real world' (Hodder and Hutson, 2003, 17).

tenance and data preservation efforts to keep the recorded data in a readable format (see Section 10.5). The intended use of the model should determine whether this effort is worthwhile[8].

9.1.2.2 Fieldwork

Decisions, which have an influence on the later analyses, have to be taken in the field.[9] Also, different recording and sampling techniques[10] may influence the data (Reilly, 1991, 134) upon which later interpretational steps are built (see also Section 5.5.4.1).[11]

Excavations destroy their subject matter. The excavation record and the recovered finds constitute the only tangible output of the enterprise. Some analyses (e.g. radiocarbon dating) and surveying methods (e.g. coring) require the taking of samples, which affect the substance of the site or object. Non-destructive techniques (e.g. georadar) and research methods (e.g. building research) exist, too (Andersson et al., 2007, 30–43). Their results will be presented as text and pictures.

Both drawings and written descriptions are interpretatively influenced (cf. Bateman, 2000, section 2), and even photographs taken on site provide only a partial and selective view of the scene (Shanks, 1997, 76, 83). What finally finds its way onto the pages of the printed report is a linear argumentation which guides the reader—by presenting what the author presumes to be the relevant data—to a final conclusion.[12] Pictures and text are combined and related to each other in order to give meanings to both,[13] their purpose is to underline the argument and they are selected or 'redefined' accordingly (Hodder, 2000, 94). The views presented in the report have been acquired by observing the evidence in the field, but they also have their roots (unconsciously) in the personality, experiences, social background, field of research, ideology, *etc.* of the author (Bateman, 2000, section 2).

In the absence of excavation data (see e.g. Section 8.1.1 concerning the Negotiating Avebury project) survey results can provide useful supplementary information. Though once again decisions on whether and to what extent survey results are included in the model have to be taken. Their results are often quite ambiguous and open to interpretation (cf. the initial survey results for Longstones field (Gillings et al., 2000, 2)). Likewise, the outcome of scientific analyses can provide additional data, as in the case of the Casa del Centenario project (Section 6.3.1).

Nevertheless, it is also possible that evidence is overlooked *accidentally* or left out *intentionally*, because the connection to the problem at hand is not apparent, or it is judged to be unimportant. Surveys and scientific analyses can seem too expensive or unnecessary, so that some kinds of information will be unavailable. Similar finds, structures or features may not be known to the archaeologist so that certain connections cannot be made (cf. Andersson et al., 2007, 30).

[8]http://www.londoncharter.org/principles/aims-and-methods.html

[9]For example: where does one feature end and another one begin (Shanks and Hodder, 1995, 8)?

[10]For instance: will every pot sherd be recorded with its coordinates or is it sufficient to record all sherds from the same feature as bulk finds?

[11]Computer models have also been used to compare or let students 'try out' different sampling strategies in a simulation. See Section 5.5.4.2.

[12]To counterargue relativism, Shanks and Hodder (1995, 18 f.) reason that logical coherence, founded on the trustworthiness of the excavation team, with the meaningful combination of diagrams, photographs and text, creates 'objectivity'. Only by excavating a similar site can this 'objectivity' be shaken. This is contradicted by the fact that even the same excavator can hold different opinions about the features discovered (see Section 8.1.2.1, stone pair 4).

[13]This process of collage and montage is described by Shanks (1997, p. 83 f.).

9.1.3 Historical documents

Besides, physical remains, historic sources—written, drawn, photographed or otherwise—provide useful information. All three case studies (Chapters 6, 7 and 8) draw on them. Like excavation reports historical documents are not *primary data*, but interpretations.

The available information can be divided according to its encoding[14] as follows, though often a mixture of different encodings will be encountered when dealing with historical sources:

Text: written, spoken

Pictures: images, figures[15] and photographs[16]

Numbers: measurements, amounts, *etc.*

The problems which arise from using historical sources are similar to other sources. The only distinctive features are a greater distance in time between the creation of the records and their planned use for the reconstruction, and the fact that we are in most cases not the intended recipients of the information (Moreland, 2001, 26; Ditmar-Trauth, 2006, 5; cf. Goetz, 2006, 257).

The historical authors, draughtspersons, *etc.* had intentions which might be unknown to us. The documents they created had to fulfil a certain purpose (cf. Shanks and Hodder, 1995, 8): they could be simply a reminder to the authors themselves, like a diary (see Section 8.1.3, diary by W. E. V. Young), or documents of commerce such as, for instance, a collection of receipts (Section 7.1.1); they may have been produced for clients in serving their needs, for example, an image to accompany a text (e.g. Walter-Karydi, 1998, 341 f.), or they may have been created for a contemporary audience, for example, an account of events (e.g. Pliny the Younger's account of the eruption of Mount Vesuvius). To determine how reliable we believe the sources to be, it is necessary to research the creators of the documents, their methods of work and also to deduce their intentions (Bateman, 2000, section 2). As mentioned before neither we nor the creators of the documents can escape our own background, so that our view of the past is made from the vantage point of the present (Goetz, 2006, 257).[17]

[14]See Boxed Feature 2 on page 29.

[15]See Boxed Feature 5 on page 212.

[16]Photographs differ in their mode of production from other forms of images.

[17]For an in depth discussion on how different philosophers deal with the problem of 'understanding' the past, see Johnsen and Olsen (2000).

Some points which have to be considered (see Goetz, 2006, 254), differ regarding media and encodings. Others are quite universal:

- Who was the creator (Goetz, 2006, 253 f.)? Related questions concern his or her personal, cultural and professional background, comparisons of other works by the same person and preferred methods of work; also the opinions of contemporaries can be informative (cf. Ucko et al., 1991, chapter 1).
- When was the document created (Goetz, 2006, 253 f.)? Not only the exact date, but also the zeitgeist, the state of research, ideologies and preferred methods of the time can give clues about the background of the documents (cf. Ucko et al., 1991, chapter 2). It should be considered whether the source is reflecting a contemporary state (Carver, 1986, 117 f.) or whether it was created retrospectively. Through the passing of years the memory of the original events or things may be distorted, or the information may have been handed down and may have been changed in the process [18].
- Why (Goetz, 2006, 253 f.; Moreland, 2001, 31) and for whom (cf. Walter-Karydi, 1998, 342) was the document made? And what could this mean for the contents? The information passed on can be overly positive or negative, certain persons may be favoured or disfavoured, pieces of information may be omitted or emphasised (Carver, 1986, 465 f., 468). Other sources relating to the same subject matter may present differing aspects and can therefore act as a means for comparison (Andrén, 1998, 164, 171).
- Is the source in question an original or a copy (Goetz, 2006, 253)? In case of a copy, it could be compared to the original (if it still exists), differences between the two could constitute intentional alterations or unintentional mistakes (Carver, 1986, 120). Further insight about the reception and use of a document may be presented by annotations or corrections (see e.g. Fig. 9.1).
- Many sources do not stand alone, but are included in a context, e.g. in a book or a collection. What was their purpose in appearing in this assemblage—in contrast to being viewed alone (cf. Moreland, 2001, 28)? How were they presented and who was the intended audience?

[18]Besides this, even contemporary persons may experience and remember events in different ways (cf. Goetz, 2006, 258 f.).

> And what impact were they meant to have (cf. Goetz, 2006, 257)?[19]

Answers to these questions will often be subjective (Andersson et al., 2007, 30). Especially when trying to determine the reliability of a source, researchers will arrive at different opinions due to differences in perspective (Andrén, 1998, 146), which can have a great impact on the interpretations based upon these documents. Some sources may even be rejected due to doubts about their contents (Pletinckx, 2008, 8 f.).

Carver (2002, 487) points out that not only different kinds of sources will require different approaches[20], but also the opposite: for different research questions different sources will be consulted.[21] Equally, the kind of information a source type holds can vary: 'a bus timetable has a different kind of message to impart than the Bible, and [...] a cross-slab enjoys comparable differences with a midden heap, even if both were made at the same time by the same person.' (Carver, 2002, 486).[22] The inherent difference is perceived to lie between formalised (e.g. a furnished grave) and non-formalised (e.g. a midden heap) sources (Carver, 2002, 487).

The above mentioned questions were also addressed in the three case studies:

In the Casa del Centenario project (Chapter 6) the focus lay on the 19th century excavation records, the selectively preserved finds as well as on drawings, paintings and photographs which were created during or shortly after the excavation (Capoferro Cencetti et al., 2001, 8, 10, 12). The way these heterogeneous sources were used and approached is described in Sections 6.1 and 6.3.1.

For the interpretation of the function, history and appearance of the settlement, castle and churches at Ename (Chapter 7) historical written sources and, for later periods, image sources were studied, evaluated and correlated among each other, and with modern building research and excavation results. For a detailed account see Sections 7.1, 7.1.1 (written sources), 7.2 (excavation and building research), 7.2.7 (historical images) and 7.2.6 (old photographs ~1907).

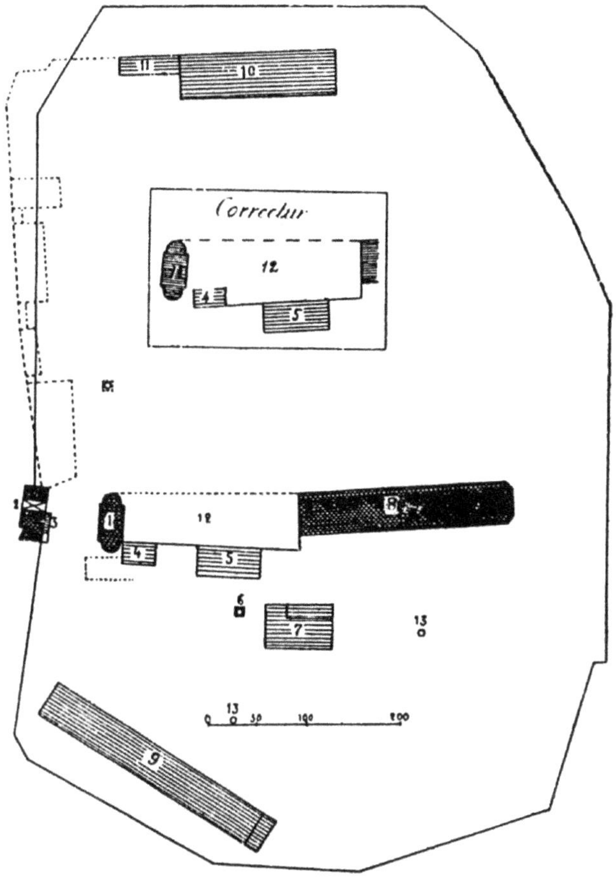

Figure 9.1: Plan of Lorsch Abbey, first half of the 18th century, with corrections. The latter indicate that the plan has been used and compared either with the buildings on site or with other plans. (Copy published by Adamy, 1891, 9 fig. 7).

The Negotiating Avebury project (Chapter 8) could draw on antiquarian records (text and pictures) as well as old excavation records. Ucko et al. (1991) provide an excellent example of how a comprehensive, critical evaluation of antiquarian records should be treated, and also how valuable the compiled information on the personal background of the antiquaries, and the circumstances under which the records were made, can be for assessing the 'reliability' of the documents. It has also been shown that the meaning and the perception of the monuments changed over time (Gillings and Pollard, 1999).[23] The antiquarian record and the earliest excavations at Avebury are described in Section 8.1.8, while the results of the excavations

[19]Cf. Moreland (2001, 31): 'documents are not neutral epistles' [...] 'they were, in fact, active in the production, negotiation and transformation of social relations.'

[20]It is also assumed that the medium/encoding which has been chosen 'to convey a message' tells something about the people who created it (Carver, 2002, 487).

[21]Their value varies in relation to the research question (Goetz, 2006, 259).

[22]See also Goetz (2006, 258 f.) on how written sources can complement, explain, place different emphases or contradict each other.

[23]The same can be demonstrated if the context of artefacts is changed (Moreland, 2001, 28).

after 1900 are included in the monument description in Section 8.1.

Specific source types have special problems attached to them. These will be addressed in the following sections.

9.1.3.1 Texts

The relation (superiority/inferiority) between textual sources and archaeological remains (cf. Andrén (1998); Moreland (2001); Scholkmann (2003)) will not be discussed here in detail. It will be postulated instead that the different source types should be taken into consideration on equal terms (see also Carver, 2002, 473).

Using text sources in an attempt to establish form is usually a difficult task, due to the ambiguities inherent in the textual medium (Andrén, 1998, 148). It is often necessary to consider supplementary information (physical artefacts or pictures (Andrén, 1998, 161)) in order to visualise what is described in them (Andrén, 1998, 148 f.).

Andrén (1998, 157–175) sees three main application areas for combining text with archaeology, though several of them can also be used to describe the relationship between other source types:

1. **correspondence** between archaeological record and text:

    (a) **classification** means to match object categories in texts with categories in the archaeological record. For this, both categorisations should be established independently from one another and then the matching process should be undertaken. This can be problematic especially if object categories in the text are based on function rather than on form (Andrén, 1998, 157 f.).[24] Jaritz (1996, 12 f.) stresses that the comparison of different source types should not be practised for its own sake, but in order to open up new contexts and information for source interpretation.

    (b) **identification** is based on classification and aims at identifying individual sites[25], people, events, *etc.* (Andrén, 1998, 162).[26]

    (c) **correlation** is also based on classification and describes the search for similar structures and patterns in both source categories. As an example Andrén (1998, 164) mentions the comparison of tax rolls with building activities.[27] Where the sources do not corroborate each other, point 3: *contrast*, is achieved.

2. **association** refers to identifying the physical context of texts and to exploring the implications for both the place in, or the object on which the text was found, and the text itself (Andrén, 1998, 168 f.).[28]

3. **contrast** is looking for differences between written statements and material culture. It can be used to 'stress complexity' or to show that 'material conditions are out of phase with social norms and ideology' (Andrén, 1998, 171). It can also indicate that a source may be unreliable.

With the exception of the category *contrast* Andrén (1998) assumes in his list of application areas that the sources used are thus without faults or bias. But mistakes and biases can be manifold (Carver, 2002, 468)[29] starting from simple spelling mistakes (West, 1973, 23 f.), to mistakes made because the writer is writing in a foreign language (West, 1973, 27) or ambiguities can be introduced due to a lack of naming conventions.

There is also ample room for mistakes during copying. Words or lines can be accidentally left out (West, 1973, 24 f.), which can happen through a lapse in concentration and/or if the copyist actually does not understand the text he or she is copying (West, 1973, 19).

[24] An example would be assigning functions and denotations for rooms in a Pompeian house (Section 6.1).

[25] For instance rediscovering and identifying the site of the Sanctuary in Avebury (Section 8.1.3).

[26] Andrén (1998, 162 f.) warns that identification can lead to arguing in circles and that it should not be considered an end in itself, for in this way it does not produce any new insights.

[27] Another example is provided in (Section 7.1.1) where the number of new windows in a receipt is matched with the number of openings in the church.

[28] This applies, for example, to inscriptions, or rooms whose purpose could be identified through the texts which have been found in them (Andrén, 1998, 168 f.).

[29] Carver (2002, 468) refers to Foucault (2009, 6 f.): 'Of course, it is obvious enough that ever since a discipline such as history has existed, documents have been used, questioned, and have given rise to questions; scholars have asked not only what these documents meant, but also whether they were telling the truth, and by what right they could claim to be doing so, whether they were sincere or deliberately misleading, well informed or ignorant, authentic or tampered with.'

It can also happen that the copyist is actively changing the text (West, 1973, 16 f.) in an attempt to clarify meaning by changing (from the copyist's point of view) 'old fashioned' language to a more 'modern' one (West, 1973, 18 f., 22) as well as adding or deleting passages[30]. More ambiguities, changed meanings and mistakes can occur when a text is translated from one language into another (Theuerkauf, 1991, 45 f.).

Very similar reasons for alterations in images have been identified by Carver (1986, 120).

9.1.3.2 Pictures

Pictures, i.e. images and figures (see Boxed Feature 5 on page 212), have different purposes. Images can occupy a spectrum from artistry to technical drawing (cf. Gaskell, 1991, 168 f.), whereby the intention and abilities of the person creating the image can significantly increase or hamper their usefulness for archaeological research (Röber, 2010, 109; cf. Pringle and Moulding, 1997, 22 and Walter-Karydi, 1998). Other important factors can be contemporary design principles and the patron's demands (Röber, 2010, 109).

As Carver (1986) points out in his analytical comparison of the Harley Psalter[31] and the Utrecht Psalter[32] (of which the former is a copy): historical illustration in manuscripts cannot be taken at face value.

He proposes that the way an artefact is depicted in an illustration does not necessarily reflect the form of the artefact used at the time the illustration was made (Carver, 1986, 117 f.). A contemporary artefact would only be depicted if either the illustrator is updating an image while copying an exemplar, or in the case where there is no exemplar for a specific object or scene which could be used as an inspiration so the illustrator has to invent it (Carver, 1986, 118).

In the case of digressions from the original image in the copy, Carver (1986, 120) envisages five possible scenarios from which the form in the illustration may have derived:

1. This is what the contemporary object looked like.
2. This is the fashionable way of presenting the object.
3. This is the traditional way of presenting the object.
4. This is the antique way of presenting the object.
5. This is a doodle.

If, through comparisons with other illustrations, possibilities 2, 3 and 4 can be eliminated, scenario 1 becomes more likely and could possibly be substantiated by comparisons with other images or by archaeological artefacts (Carver, 1986, 120). By this, it is implied that illustrative conventions have a longer lifespan than the artefact forms they depict. Whether an illustrator was unfamiliar with an object or its use can be indicated by mistakes in the depiction of the object in question (Carver, 1986, 120).

What Carver describes is a special case. Felgenhauer-Schmiedt (1993, 99) points out that the realism of the images varies with time, i.e. becomes more prominent in the Later Middle Ages,[33] but they may still be dependent on older image traditions, or can even anticipate future developments.[34] But the date of the image is not the only factor which has to be taken into consideration. The kind of object which is rendered in the image also seems to have an impact (Felgenhauer-Schmiedt, 1993, 99). Items of clothing seem to be well represented according to the fashion of the time, while, for example, ceramics do not always represent recognisable shapes and do not cover the whole spectrum of archaeologically retrieved objects (Felgenhauer-Schmiedt, 1993, 100 f.).

In many contexts images provide rather a 'typical' object form instead of an 'authentic portrait' of the object in question, thus rendering them universally understandable for their target audience (Jaritz, 1996, 10). Objects in images can also be used as staffage or in an attempt to make, especially in the topographic contexts, a view more comprehensible by presenting a recognisable structure (Röber, 2010, 101). A good example for this kind of image is provided by the 'fourth image' (Fig. 7.27, right) of St. Lawrence's church dating from 1596 (Pletinckx, 2008, 11 fig. 6) mentioned in Section 7.2.7. These 'archetypal' representations can include additional information as, for example, some kind of (e.g. positive or negative) connotation, which is meant to provoke a specific response (e.g. situating) from the observer (Jaritz, 1996,

[30]The latter two changes are especially interesting as they indicate changes of conception, interests, identities and possessions, which have occurred over time (Pohl, 2005, 351). See also West (1973, 18, 22)

[31]British Library Harley MS 603

[32]Utrecht, Universiteitsbibliotheek, MS Bibl. Rhenotraiectinae I Nr 32

[33]See also Ditmar-Trauth (2006, 5), who notes that objects in painted scenes correspond to physical objects, but does not approach the subject on more critical terms.

[34]Traditional wooden dishes are more and more replaced by metal dishes in the archaeological record from the 16th century onward, but images already depict a high quota of metal dishes in the 15th century (Felgenhauer-Schmiedt, 1993, 102).

Excursus 5: Perception of pictures

Pictures fall into two categories: *images* and *figures*. In general they are viewed with regard to the observer's previous knowledge and his or her objective (Schnotz, 2002, 70). Their content is often underestimated, as viewers believe that they can take in the gist with only a short glance, thus relevant detail can be missed (Schnotz, 2002, 65, 73; Weidenmann, 2002a, 85 f., 89).

Images show the world the way the artist has perceived or imagined it (Weidenmann, 2002a, 83). Examples could be as varied as a photograph, a watercolour painting, a comic strip or an archaeological stone-by-stone drawing of a wall. These pictures can be understood using the cognition patterns used for everyday life (Schnotz, 2002, 71). In contrast to text there is no particular order in which a picture is viewed, but the artist may introduce directive signs like arrows, enhance details, direct the focus by using boxes or labelling. Also, cultural specifics like the reading direction will influence how a picture is viewed (Schnotz, 2002, 71).[a] These visual clues are helpful for images which were designed to demonstrate objects or details, while realistic images are less useful in this context (Weidenmann, 2002a, 89 f.). If, however, the observer is to be 'situated', i.e. presented with a cognitive setting, elaborate images are most suitable. Unfortunately they are prone to becoming outdated, can be culturally dependent and tempt the viewer to look for faults (Weidenmann, 2002a, 91).

Figures or diagrams, for example, maps and statistical graphs, cannot be understood with everyday-life viewing habits (Schnotz, 2002, 66); they have to be 'read' in a different way, which has to be learned like other cultural techniques (Schnotz, 2002, 72). Familiarity with this kind of information representation has consequences not only for the comprehensibility of the figures, but children, juveniles and adults also have different levels of visual literacy (Weidenmann, 2002a, 84). When designing diagrams it is necessary to adhere to certain principles of design (see Schnotz, 2002, 74 f.) which enhance the readability. It should be made sure that the favoured type of diagram is suitable for the purpose, and moreover it is preferable to use well known diagram types (Schnotz, 2002, 77).

[a]The research of Rosenberg et al. (2008, 128) indicates that not only cultural but also subjective differences in the viewing habits exist.

10). These images are useful for context analysis and the provision of informational background, but not for reconstruction purposes.

Most interesting for reconstructions are the instances where *specific* objects are depicted, which do represent some, if not all,[35] aspects of their subject (Röber, 2010, 108 f.; cf. Jaritz, 1996, 11), e.g. the first three drawings of St. Lawrence's church mentioned in Section 7.2.7 (see also Figs. 7.17, right and 7.27, left). The same example also illustrates that the concept of contrast (Andrén, 1998, 171) can be applied to images, too. A comparison between the image on the pilgrimage pennon and the still-standing church, revealed that the former depicts a door where there has never been one (Section 7.1.1). Nevertheless, the context of the image, its purpose and meaning have to be taken into consideration[36] too, in order to make the best interpretational use of the source material (Jaritz, 1996, 12 f.; Pletinckx, 2008, 9; Röber, 2010, 109).

Technical drawings, i.e. drafts, maps and plans[37], constitute another frequently encountered image type when dealing with legacy excavation data. One of the most pressing questions in this regard concerns the accuracy of the images, the relation of measured points and free-hand drawing, the drawing scale and the amount of detail depicted. Also methods like photogrammetry should be scrutinised to establish whether the resulting plans have been drawn correctly and were checked against the original surfaces. In Fig. 9.2 different images of the same area have been selected from the documentation of Lorsch Abbey church and compiled for comparison to illustrate this point.

Three areas have been marked in the three drawings (Fig. 9.2, left, left centre and right centre) to show some of the differences:

[35]Röber (2010, 109) sums up his case study with the conclusion that none of the representations of the Petershausen Abbey in Constance depicts the original buildings absolutely faithfully. Rather, they include a varying degree of similarity or dissimilarity to the physical reality.

[36]Röber (2010, 109) mentions that the daughtspersons personal familiarity with the locality is not a guarantee of correctness of the depiction.

[37]See, for example, the discussion of the antiquarian plans by Aubrey and Stukeley in Section 8.1.8.1.

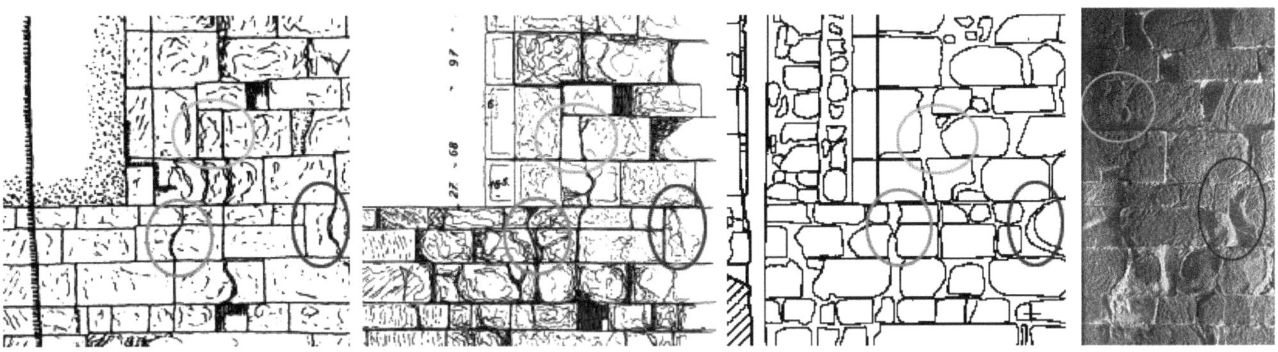

Figure 9.2: Comparison of drawings, Lorsch Abbey. Left: detail of plan 14 from Behn (1934b) depicting the western nave wall of Lorsch Abbey church from the east. Left centre: the same area taken from plan 16 (Behn, 1934b), this drawing is meant to show more detail. Right centre: once more the same area on a photogram from 1992 (Verwaltung Staatliche Schlösser und Gärten Hessen). Right: a photograph, which depicts only two of the features taken in the 1920s or 1930s (Photographic documentation for Friedrich Behn, image 7538. Original negative: Landesamt für Denkmalpflege Hessen).

Green: The first drawing shows a small stone, the second one a broken part of a big stone, the last one a gap.

Red: The drawing on the left suggests two stones in the top layer and a broken stone below, the second drawing shows two broken stones, the last one shows four separate stones.

Blue: The first drawing shows a notch on the right stone and a break underneath the notch, the next drawing has no notch and no break, while the right one shows a rounded stone.

Even though the drawings published by Behn (1934b) look much more convincing than the photograms, they have in some cases been proved to be faulty (Platz, 2007a, 12 f.). When the drawings are compared with the photograph of approximately the same area (Fig. 9.2, right) which was taken in the 1920s or 1930s, it can be seen that the far left drawing appears to be the most exact regarding the two instances visible on the photograph.

Figures which are based on statistical material should be probed to see whether the analyses used and the way the data is presented are appropriate (Shennan, 1997, 5 ff., 11, 21).

In times of digital editing methods, a lot of visible and invisible changes can be made to photographs, but in the past also editing (e.g. excision of figures, cropping and toning (Gaskell, 1991, 187)) was possible, so that photographs cannot always be taken at face value (Gaskell, 1991, 187 f.). But even without active 'manipulation', photographs are open to diverse interpretations, depending on the context in which they are embedded, i.e. other pictures, text and captions (Shanks, 1997, 81; Gaskell, 1991, 187; Brehm et al., 1998, 68 f.).[38] This also applies to photographs taken for excavation documentation because these images are always arranged and cleaned up (Shanks, 1997, 83). Therefore, some consideration should always go into what the photographs show and what they do not show, when they were taken and for what purpose.

9.1.3.3 Numbers

When dealing with measurements and quantities it is important to know how they were obtained. Especially, the accuracy of measurements and potential sources of error depend highly on the methods, which were used to take them.

For example, during 1920s and 1930s excavations at Lorsch Abbey, a compass was used to measure the extent of the trenches and all the features within them. A mistake which led to radial discrepancies between the initially recorded measurements of the trenches and features and their true orientation, was discovered during re-excavation (Platz, 2007a, 12, 13 fig. 11).

A considerable number of examples are also present in the Negotiating Avebury project concerning the antiquarian records (Section 8.1.8.1). It shall suffice to refer to the plane table survey by Aubrey where some kind of measuring mistake wreaked havoc on the resultant plan (Fig. 8.14). Or the instances where Stukeley's preconceived number symbolism let him assign

[38] Some very illustrative examples of how images can lie were presented in the exhibition 'Bilder, die lügen' in which the whole spectrum of image manipulation is presented (Brehm et al., 1998).

the 'right' amount of stones to certain features, e.g. to the Avebury circles and the West Kennet Avenue (see Section 8.1.8.1).

Old measurement units can also confuse the situation, especially if they are not indicated. In addition, they can vary locally.

9.1.4 Analogies

Usually some parts of a reconstruction cannot be based just on the evidence at hand, so further information is needed, if for instance part of a mural painting is missing (see, for example, Capoferro Cencetti et al., 2001, 22; Coralini, 2007, 30) or only the foundations of a building remain (see Kirchner and Jablonka, 2001, 238). In this case, similar objects, features or monuments (i.e. analogies) are used to give ideas about how the missing parts may have looked or the objects may have been used.

In the reconstruction of a building, the analogy can lie very close (i.e. the neighbouring house) or in comparisons with buildings in the same region, of slightly different date; sometimes other media (images or text) may be necessary, and also ethnological studies can provide information (Kirchner and Jablonka, 2001, 238). Particularly 'closeness in space and form' (Andrén, 1998, 156) are considered indicators for good analogies.

Questions of appearance are not the only reason why archaeologists are looking for analogies; dating or inferring the usage of artefacts are also important questions under consideration (Fehring, 2000, 36 ff., 196). This requires some kind of connection between those objects which are thought to be 'similar'. Parallels in appearance may not be enough to make this connection, especially if one or both of the objects only survive in small parts. Further links between objects can be constituted by contemporaneity, the objects' creation was initiated by the same person, carried out by the same craftsperson, or one object may have acted as a model for the other. Additional reasons are also possible.

What all the possible sources for a reconstruction have in common is that a scrutinising look is advisable. As mentioned before the results of this critical appraisal will differ from person to person, which consequently means that different reconstruction hypotheses will arise.

9.2 From the basis to the reconstruction: Inferences

Interpretation is very different for each site or object. This becomes apparent when looking at the three case studies in Chapters 6, 7 and 8. The aims of the projects, the questions asked, the available sources and the data incorporated in them—they all lead to very different approaches and outcomes. It is hardly possible to define a strict outline for a course of interpretative action (Theuerkauf, 1991, 35), nor is interpretation a linear process. It can lead in different directions which are determined by the initial research question (Theuerkauf, 1991, 34 f.), the personality of the researcher (Johnsen and Olsen, 2000, 111) and the available sources. The deliberation about what is significant and what is not, as well as the choice of material, also have an influence on the direction the interpretation will take (Shanks and Hodder, 1995, 7).

In the beginning there is a phase of source collection and source assessment (Section 9.1), whereof the latter is already an interpretation (Pletinckx, 2008, 6). Source assessment is related to source correlation, i.e. the sources are not only judged on their own, but have to be set in relation with each other (Pletinckx, 2008, 9-13). All sources should be taken into consideration, and one should be wary if only one (pictorial) source (for a particular feature) exists (Röber, 2010, 109). The next step is to find out what the different sources contribute to the all-over reconstruction. For this they have to be set into relation with each other which will eventually lead to new questions and again new sources will have to be consulted.

There are different levels of complexity in the interpretational process which in turn have an influence on the level of certainty of the reconstructed parts. 'Simple' interpretations are commonly based on observations (see Boxed Feature 4). For instance a foundation trench will probably have held a foundation, or a posthole indicates the former existence of a post (Klein et al., 1999, 23). Such clear cut interpretations can usually be made with considerable confidence, but the further that—in a spatial and with that in an interpretational respect—the reconstruction ventures from the extant remains, the less confidence can be ascribed to it (Hermon and Niccolucci, 2004, 2). Another basis for 'simple' interpretations belong to the realm of preconceptions based on generally observed features[39], e.g. a building needs an entrance and usu-

[39] Hermon and Niccolucci (2004, 2) call this 'common sense'.

ally possesses a roof.[40] The 'simplicity' of this kind of interpretation does not mean that some of these interpretations may turn out more complex in the visualisation than in their first conception: the questions of 'Where was the entrance and which shape did it take?' or 'Which building height and kind of roof structure has to be assumed?' can be very difficult to answer.[41] To proceed in this and other complex interpretational efforts, larger interpretational leaps, together with additional sources from further afield, will have to be employed, which will lead to less certain results.

In order to illustrate exemplarily[42] how observations in the field and written sources are used as the basis for interpretation and how different interpretations of the same source can occur, a modified and extended version of the Argumentation Network from Fig. 5.3 is presented in Fig. 9.3. The line of reasoning visualised here is concerned with the rebuilding activities at Lorsch Abbey after the great fire in 1090:

The argumentation starts with a look at the walls of the clerestory (1) and the arcades (2) noticing differences in the building technique and the material used for the purpose (3 and 4). Also, a difference in the preservation of the stone material is noted, i.e. the stones of the clerestory are burnt, while the stones of the arcades are not (6). The burn marks are thought to have been caused by the fire in 1090 (5), which is mentioned in the Lorsch Codex (9)[43]. These observations led Platz (2007b, 39) to the conclusion that the (old and partly rebuilt) clerestory was underpinned with new arcades after the fire (10). As the rebuilt arcades include the holes for the roof over the aisles, those roofs must have been rebuilt, too (4 and 10 lead to 7). The Lorsch Codex (9) provides another written source concerning the rebuilding events (15), whereby Platz (2005, 213) only picks the references to the arches, windows and the coffered ceiling (13) from the listed building activities mentioned in 15. The written source supports the observations concerning the new arcades and the coffered ceiling, but also indicates that new windows were installed. These are identified by Platz (2005, 213) as the windows in the clerestory (8 and 11). Forster (2005, 218) (16) on the other hand comes to a slightly different conclusion (14) concerning the contents of the written source (15) in the Lorsch Codex. He supports the interpretation concerning the arcades and the coffered ceiling[44] (this time including the leaden roofs), but sees the windows as belonging to the aisles—not to the clerestory[45]. He also indicates in his translation of the Latin source (15) that *'absidibus'* could refer to the 'aisles'. This presents a good example of the ambiguities which are inherent to written sources: *absidibus* can either mean apses or aisles, which clearly makes a great difference for a reconstruction.

Also, the correlation of sources (e.g. the burn marks and the fire mentioned in the Codex) and different opinions concerning these correlations (e.g. whether *'fenestris'* relates to the windows in the clerestory or in the aisles) is demonstrated. That the interpretational process is not linear is shown by the recourse to previous observations[46] and also by the referral to so far unused source material (15) as soon as it becomes relevant for the argument (13).

Additionally, the *selection* of information which is seen as supportive can be shown, for instance, by the choice of building parts mentioned in the written source (15) which find their way into the argumentation (13), but also by the statement that 'the research by Forster (2005) points into the same direction' Platz (2005, 213).

The application of Argumentation Networks makes it possible to include a level of confidence for the visualised interpretations, to let other researchers re-evaluate the arguments presented, to aid the development of alternative hypotheses and to document at which points these alternatives diverge from already existing arguments. Fig. 9.4 constitutes an example.

The left side of Fig. 9.4 shows a detail of Fig. 9.3. A different researcher re-evaluating the initial hypothesis may find that some observations are missing from the argumentation, for example, that the clerestory is not only constructed from roughly hewn stones (3), but that also some ashlar blocks are present. The observation, though not wrong, may be deemed slightly lacking and is therefore marked with a light grey background, to indicate slight uncertainty. In

[40] This kind of preconception can also include the observation that decorations in a certain culture are often symmetrical, which can in turn aid the reconstruction of lost parts. See for instance Section 6.2.1.1, Level 2, Item 2.

[41] Cf. Eiteljorg (2000, section: Reconstructions or Fantasies?) who describes the efforts of a student struggling with a reconstruction of a house on the sole basis of a plan. See also Kanter (2000, 50 ff.) concerning the difficulties of reconstructing a Kiva roof.

[42] This example does not reflect the latest research results. New research on the Lorsch Abbey church is currently (2011) conducted by a team from the Technische Universität München under the supervision of Manfred Schuller.

[43] Metadata concerning the Lorsch Codex is stored in 12.

[44] In Fig. 9.3 this is indicated by the arrows from 14 which point onto the green line between 10 and 13 and onto the yellow line between 7 and 13.

[45] This is depicted by the arrow from 14 marked 'not supported'.

[46] The interpretation 10 and previously used observation 4 lead to a new interpretation (7).

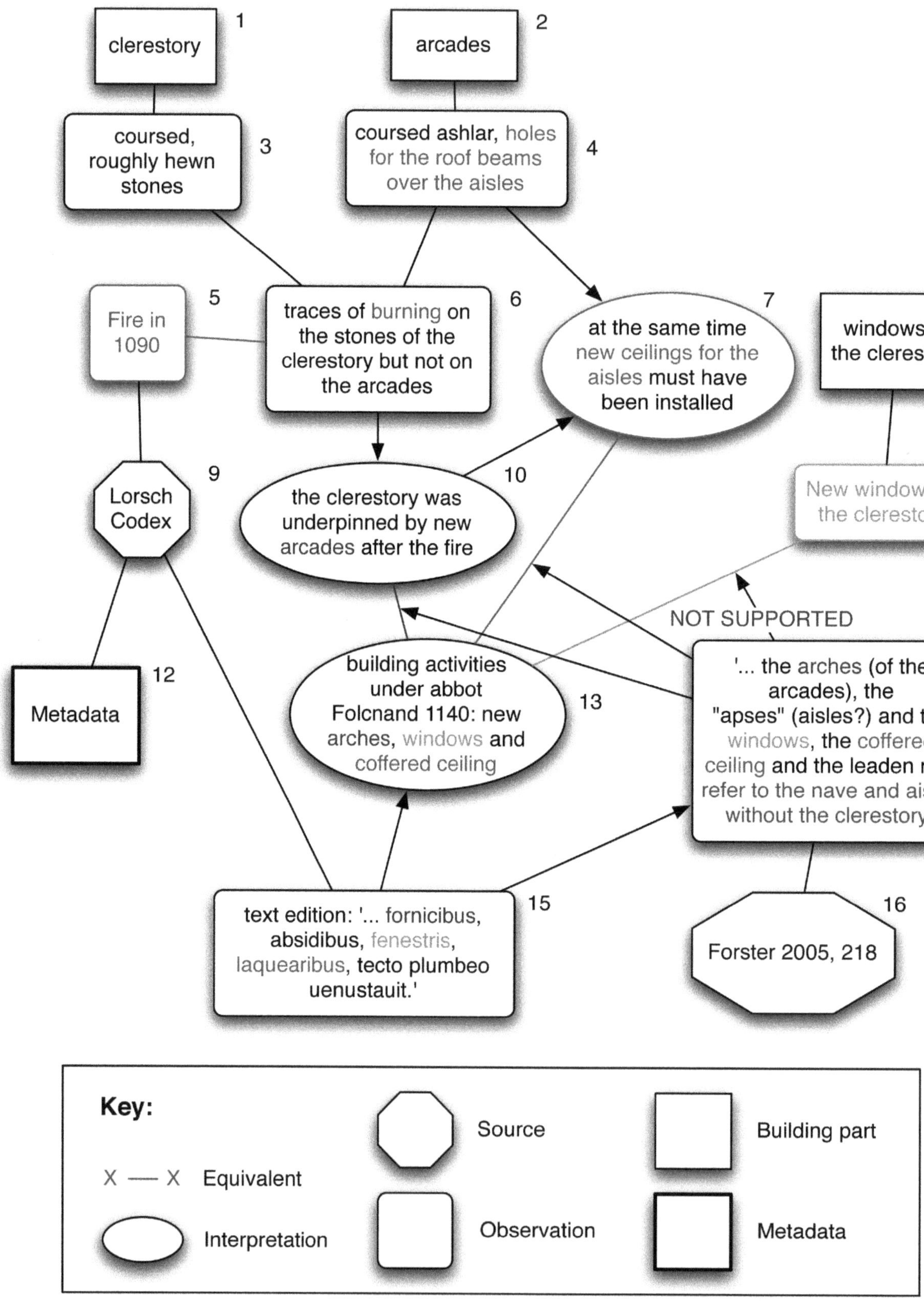

Figure 9.3: Argumentation Network with different interpretations concerning the rebuilding activities at Lorsch Abbey church after the fire in 1090. (Shortened argument taken from Platz (2005, 2007b) and Forster (2005).) (Drawing by Joyce Wittur)

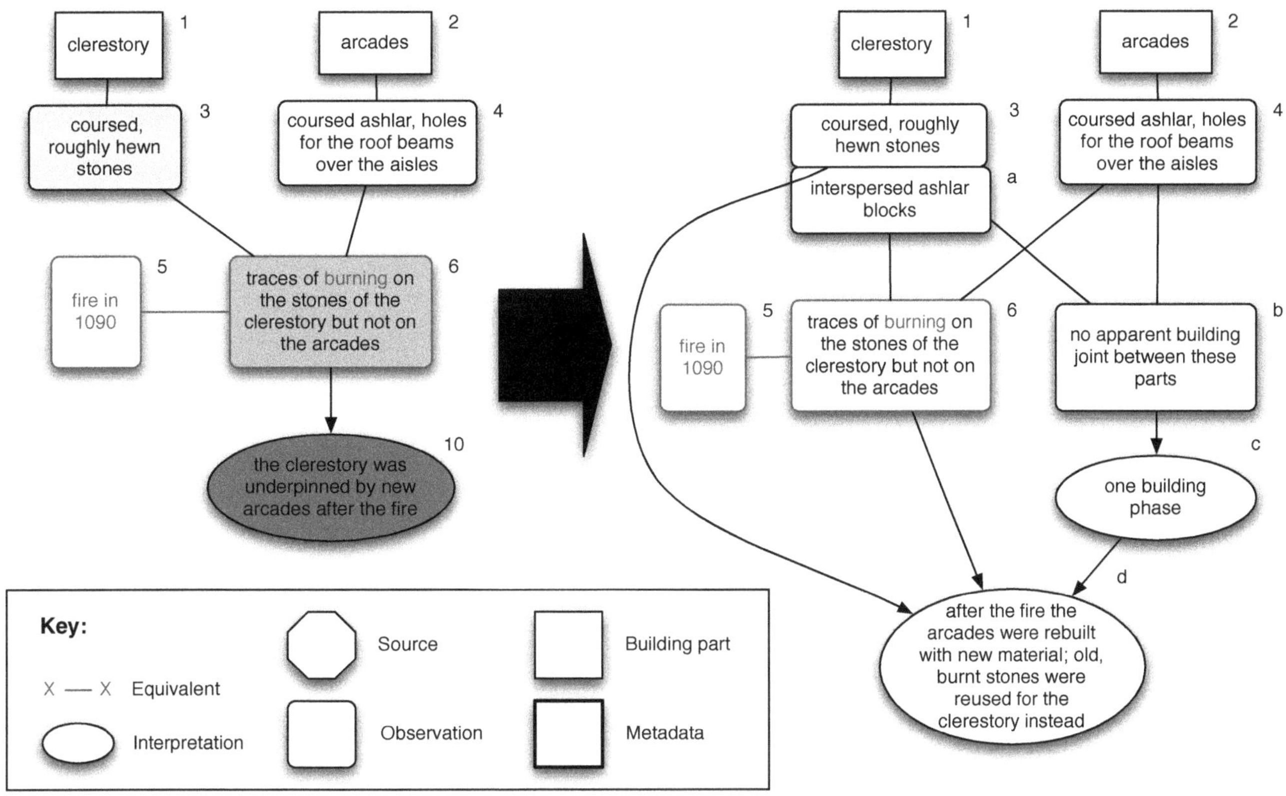

Figure 9.4: Two different hypotheses concerning the rebuilding activities at Lorsch Abbey church after the fire in 1090. Left: the re-evaluated initial hypothesis; right: the altered new hypothesis. (Shortened argument taken from Platz (2005, 2007b).) (Drawing by Joyce Wittur)

the next step of the argumentation the masonry of the arcades (4) and the clerestory (3) are compared (6). Again an observation is missing: there is no apparent building joint between these two building parts. Yet again the initial observation is deemed to be correct, but the omission concerning the building joint may be considered more serious by the re-evaluating researcher leading to a higher degree of uncertainty and therefore a deeper shade of grey in the diagram (6). The inference (10) drawn from this observation (6) is now considered extremely unreliable, because it leaves out important information, hence the dark grey colour.

The levels of confidence indicated here can also be mirrored in the virtual model.

The re-evaluating researcher could now draw up a competing hypothesis (Fig. 9.4, right) based on his/her own observations: observation 3 is supplemented to include the ashlar blocks (a). The new observation concerning the missing building joint (b) was also added. From this new information converging inferences are drawn, namely that the arcades and the clerestory belong to one building phase (c) instead of two, and that materials of different quality were used in the rebuilding process (d). These considerations could, in turn, lead to further lines of inquiry so far unexplored.

The ultimate aim for the interpretation is to make sense (Thomas, 2000b, 4), by looking for a hypothesis that incorporates most of the data (Hodder, 2000, 93 f.)[47]. It should be obvious that the latter is a utopian endeavour because we cannot know the past. As Shanks and Hodder (1995, 10, 29) point out, this relates to the concept of under- and overdetermination. The (archaeological, historical, *etc.*) data are not sufficient to account for 100% of a hypothesis (i.e. underdetermination), which means that certainty of a degree achieved in the natural sciences can never be accomplished in the humanities (Shanks and Hodder, 1995, 10). At the same time some surplus of data

[47] Hodder (2000, 94) warns against possible distortions of the data in order to make it fit the hypothesis, but distortions are always present through the act of selecting and judging information (see Shanks and Hodder, 1995, 7), the researchers' preconceptions and the questions which are posed (Thomas, 2000b, 3). Therefore it could be argued that 'distortions' are a natural part of interpretation.

which cannot be incorporated into the hypothesis also exists (i.e. overdetermination), which results in uncertainty too (Shanks and Hodder, 1995, 29). This uncertainty in the reconstructions should be made apparent in the modelling phase (Chapter 10; see also Section 5.9).

Keeping track of the interpretation process[48] (Forte, 2000, 252) can help to identify uncertainties and also to assign reliability values to (parts of) the hypotheses (Chadwick, 2004, 22; Pletinckx, 2008, 16–18).

In most cases the result of an archaeological interpretation process will not be a fully-fledged model, but a mental image (Hermon and Niccolucci, 2004, 2) or a verbal description of what the site might have looked like, possibly with several alternatives and uncertain parts. Such a mental image or verbal description will retain many ambiguities (cf. Fig. 5.4, Inferences) which have to be dealt with at the visualisation stages (Section 10.1 and 10.4).

[48]The concepts of the 'hypothesis tree' (Pletinckx, 2008, 13–16) and 'Argumentation Networks' (Wittur and Krömker, 2009, 89 f.) were introduced in Section 5.7, Figs. 5.2, 5.3 and 5.4.

Chapter 10

Modelling

While the previous chapter (Chapter 9) dealt with the first two steps of model creation (i.e. *sources* and *inferences*)[1], this chapter covers the last two steps, which are *visualisation decisions* and *modelling* (Wittur and Krömker, 2009, 89, 90 fig. 7), the latter has been termed *representation* in this chapter.[2]

The *inferences* (see Section 9.2) which were made on the basis of the data and which retain a high degree of vagueness have to be rendered more precisely in the *visualisation decisisions* step (Section 10.1). To illustrate this point, it is possible to draw upon the example in Fig. 5.4: it is not enough to say that two towers with a lower middle part adjoined to the west of the church—even though this may be about all the sources let us know. For the model it is necessary to assign a height to the towers, and to decide whether they had windows, if they were coated with plaster, what kind of roof has to be assumed and so on. To achieve this, it may become necessary to look once more for additional sources or to draw on further analogies. Additionally, it could be decided that alternative reconstructions should be used to present different hypotheses.

For the *representation* stage (Section 10.4) further decisions have to be made and additional material has to be accumulated. This time the decisions encompass the rendering method, whichever is best suited for the purpose, including the question whether a photorealistic or non-photorealistic approach should be chosen. The form of presentation (image, animation or interactive model) also puts constraints on suitable approaches to create the necessary output. Additional material, e.g. in the form of textures may be necessary and ways to convey the intended message have to be found (indicating uncertainty, displaying additional data and so on). The underlying applied ethical issues and ways to address them have been covered in detail in Chapter 5.

The section concerning the visualisation decisions (Section 10.1) has been supplemented with technical information on how shape can be modelled (Section 10.2), while the section dealing with the representation (Section 10.4) is preceded by a section about different rendering and mapping techniques (Section 10.3). It will clarify the diverse technical constraints which are placed upon the representation.

The chapter concludes with a section on data preservation (Section 10.5), a topic frequently neglected during the planning stages. Often no resources are allocated to preserve and maintain the virtual models so that sooner or later the information they convey as well as the soft- and hardware become outdated and unusable. Hence valuable research, work and resources are irrevocably lost.

10.1 Visualisation decisions

After the interpretational part (Section 9.2) the abstract ideas which resulted from the inference stage have to be transformed into three-dimensional shapes.

Starting points are records of extant remains and archaeological features (Section 10.1.1) upon which inferred model parts have to build up (Section 10.1.2) (cf. Hermon and Niccolucci, 2004, 2). The latter constitute the more problematic part of the visualisation, because their appearance has to be established with the use of of the available sources, analogies and not least imagination (Hermon, 2008, 39). The amount of available information and the certainty with which

[1] These steps were introduced in Section 5.7, Fig. 5.4.

[2] The term *representation* has been chosen to better differentiate between, first, actual shape modelling (Section 10.2), second, rendering and mapping techniques which also constitute a part of the modelling process (Section 10.3) and, third, decisions which have to be taken concerning representational aspects (Section 10.4).

the shape of the missing parts can be deduced has influence on the reliability of the reconstruction—this degree of confidence should be recognisable in the finished reconstruction (see Section 5.9.1). Also, a distinction between extant and inferred parts should be made (Hermon, 2008, 40). These are generally representational issues but they can sometimes have an impact on the geometric model: for example, the decision to mark uncertain stone positions with transparent cylinders instead of stone-like shapes (Earl and Wheatley, 2002, 11) in the Negotiating Avebury project (Section 8.2.2.1), or to present a reduced level of detail for unknown (or unimportant) areas (Freudenberg et al., 2001, 98; Maschek et al., 2010, 454 fig. 6). In such cases the decisions concerning shape and representation in the (rendered) model have to go hand in hand.

10.1.1 Extant remains

The documentation of extant remains and archaeological features provides a relatively 'secure' basis for the reconstruction process. Depending on the format of the documentation (two-dimensional, e.g. plans and sections,[3] or three-dimensional, e.g. laser scans and total-station surveys[4]) the modelling process has to proceed on different paths (see Section 10.2.1). While three-dimensional data are—more frequently than not—already in digital format, two-dimensional information (maps, plans, section and elevation drawings, *etc.*) is often analogue. These non-digital documents have to be digitised first (Grellert, 2007, 176); different processes are mentioned in Section 10.2.1.3.

10.1.2 Inferred model parts

While the extant remains may provide data relating to the topography and spatial disposition[5], as well as in some instances contribute clues as to the size[6] and shape (if, for example, an imprint survives) of features, any missing information about the appearance and form of lost parts must be derived in other ways. They usually involve the use of texts, images, analogies and educated guesses (Hermon and Niccolucci, 2004, 2).

How such vague knowledge which is provided by the extant remains and other sources can be used and supplemented with additional data to arrive at a geometric model of the missing parts will be illustrated by presenting three examples, one from each of the three case studies:

In the **Casa del Centenario** project holes in the columns in the peristyle led to the deduction that there had once been railings (Coralini, 2007, 31). Nothing was known of their original form nor of their material so that in these matters analogies had to be consulted (Section 6.3.2, Fig. 6.11). It was proposed that these railings had been made from wood in a style known from murals as well as wooden remains in Herculaneum. Hence the railings were modelled in a way which imitated a weathered surface.[7]

Building research in **Ename** showed (Section 7.2.6.4) that there were probably two columns in the chapel above the eastern sanctuary, which were at a later date removed and replaced by uneven masonry (Callebaut, 1992, 447). This in turn meant that nothing of their shape or material was known, but they had to be represented in the virtual (Callebaut et al., 2000)[8] and later in the physical reconstruction of the church (Fig. 7.23). The virtual reconstruction shows two small columns with cushion capitals of indefinite material, while the physical reconstruction mostly avoids any reminiscence of historical styles and materials by using two green columns with 'capitals' which *could* remind the observer of echinus and abacus of the Doric order.

A different approach was taken in the **Negotiating Avebury** project. Based on research results and antiquarian documents the henge and the avenue(s) were populated with stone archetypes (Section 8.2.2.1), which were later replaced by models of the extant stones (Goodrick and Gillings, 2000, 53). Where no extant stones could be taken as exemplars the archetypes remained in place. Upon what the archetypes are based, remains unknown. At least the Obelisk appears to be modelled according to Stukeley's description. In other cases the stones are represented

[3]For example, plans and elevation drawings of the St. Lawrence's church were used as the basis for the 3D-model in the Ename 974 project (Section 7.3.3.2).

[4]For instance a total-station survey was used to document the topography of Avebury henge (Section 8.2.2.1).

[5]A stone hole indicates the position of a standing stone, a foundation the former existence of a wall, *etc.*

[6]The position of the window in the eastern nave wall of the St. Lawrence's church gives an indication that the adjoining stair tower could initially not have exceeded in height above its window sill (Section 7.2.6.5).

[7]Hodgson (2001, 11) warns against the urge to make things look old even though they had been new at the time or to otherwise imprint our modern taste upon things (e.g. depict classic statues without their painted surface).

[8]http://www.ename974.org/Ndl/pagina/NuhV/000202_1.html

by translucent cylinders to indicate uncertainty (Earl and Wheatley, 2002, 11).

All these examples show that uncertainty issues in the source material were the reason for the decision making processes described above. They become necessary because thoughts or text can remain vague in relation to details, but for a visualisation decisions have to be taken inadvertently (cf. James, 1997, 26).

10.1.2.1 Multiple models

If diachronic development or several alternatives are to be presented, they also have to be defined as shapes.[9]

In the case of a multi-phase monument it is necessary to reconstruct one phase at a time and to select only the relevant parts from the source collection.[10] The separate models can then be combined. Pletinckx (2008, 21), for instance, recommends a QuickTime VR application (see Section 10.4.1.3) for this purpose. This method has also been used to create the model of St. Lawrence's church (evolutie_kerk.mov)[11]. It is also possible to present each phase as a separate model (Herrmann, 2010, 199) or to provide the user with a timeline whose manipulation lets parts of the model fade in and out according to their probability of existence at the chosen point in time (Zuk et al., 2005, 101, 102 fig. 2).

Alternatives can also be incorporated into one model, for example, the Canterbury Roman theatre (Roberts and Ryan, 1997) where alternatives can be created interactively by manipulating the VRML model via sliders. Another possibility is to present the alternatives side by side, so that the diverging models can be compared by the user (Pletinckx, 2008, 19 f., 19 fig. 17).[12]

How these shape decisions (as well as the shape information gained from extant remains) are converted into virtual 3D-shapes is discussed in the next section.

10.2 Modelling shape

Actually *shape* (this section) and *texture mapping and rendering* (Section 10.3) cannot truly be considered independently. Even the simplest model has to be rendered in some way in order to be graphically represented on the computer screen.[13] Also concerning the (archaeological) decision making process it is sometimes not possible to make a distinct division (see Section 10.1).

Still, the creation of shape precedes any sophisticated rendering process, which will usually only be employed after the shapes, the lights and the viewing position in the scene have been defined (Watt, 2000, 142).[14]

The focus of this section lies in introducing modelling strategies on a *basic* level[15] in order to raise awareness concerning dependencies between the documentation of features, their modelling and, in Section 10.3, different rendering approaches. For this purpose I refer primarily to the book by Watt (2000), which is a standard reference in the 3D computer graphics field and encompasses most modelling and rendering methods. It was felt that such an introduction would benefit readers who are unfamiliar with the complex principles behind the modelling process and do not wish to confront mathematical formulae.

Besides the technical descriptions, possible archaeological application areas will be pointed out. In some cases, references to the case studies or other archaeological projects will be presented, in other cases general considerations must suffice. The reason for the latter lies in the fact that details concerning modelling and rendering techniques are seldom part of the published descriptions of archaeological projects.

[9] Another possibility are superficial changes, e.g. new murals painted on the wall or alternative colour schemes discussed. They can be dealt with on a purely representational level (Section 10.4) by changing the textures.

[10] If, for example, a Gothic church is to be reconstructed it is necessary to select the Gothic and possibly the pre-Gothic parts of the building, but to ignore any post-Gothic additions and changes.

[11] The file is included on the CD. To view it, a QuickTime plug-in has to be installed, it can be downloaded from http://support.apple.com/downloads/#quicktime (Accessed: 2nd of August 2012.).

[12] Another interesting example, though only in hand-drawn format, is provided by Ludwig (2004, 28 fig. 4–6, 8–9 and 11) who depicts six different interpretations concerning the annexe buildings of the Einhards-Basilica at Michelstadt-Steinbach.

[13] There are other output formats, which do not incorporate rendering, for instance output in binary or text format (e.g. VRML or X3D can be output in the latter way). This does, however, not aid the mental visualisation of the 3D-scene in the user's mind.

[14] See also Section 10.3.1.

[15] For deeper insight into the underlying mathematical principles see, for example, Watt (2000) or Foley et al. (1996).

10.2.1 Wireframe, surface and solid models

These three types of models encompass the possibilities in which three-dimensional data can be represented. They are each here characterised in short.

Wireframe models or wireframe rendering describe objects as 'see-through' shapes of which only the edges are visible (cf. Fig. 6.9). They are quick to render and are therefore used when the model has to be manipulated in real time (O'Rourke, 2003, 83; Watt, 2000, 182). They are frequently considered as a model class on their own, but they are actually a 'reduced' rendition of surface (and sometimes solid) models, which will be explained in Section 10.2.1.1.

Surface models describe the surfaces of objects, but the latter remain hollow shells (O'Rourke, 2003, 29). This group includes polygon mesh models and bi-cubic parametric patches (Watt, 2000, 30 f.) (see Section 10.2.1.1). Polygon mesh models only approximate to the surface of the object they are representing (Watt, 2000, 123) and do not contain information about the enclosed space (Lock, 2003, 152). Bi-cubic parametric patches also approximate to the surface, but are so well defined that their volume can be calculated (Lock, 2003, 152; cf. Barceló (2000, 16)).[16]

Solid models additionally model the 'interior' of the object (O'Rourke, 2003, 29), popular representations are Constructive Solid Geometry (see Section 10.2.1.3) and voxels (see Section 10.2.1.1) (Watt, 2000, 31).

The choice of model type (wireframe, surface or solid) does not only depend on the wishes of the modeller, but also on the nature of the object, the computer graphics technique which is used and on the modelling application (Watt, 2000, 29), as well as on the already existing data, e.g. plans and elevations, 3D laser scans, X-ray computed tomography scans, photographs, *etc.* They will determine which modelling methods are suitable.

There are various methods for creating shape (Watt, 2000, 28):

- Shape information can be taken from physical objects, for example, automatically with a laser scanner or manually measured with a total-station (both methods were employed in the Casa del Centenario project (Capoferro Cencetti et al., 2001, 5, 24)).
- Models can be created from photographs (as in the Negotiating Avebury project (Gillings, 2000, 61–68)) and videos (e.g. in the MURALE project (Pollefeys et al., 2000, 74 f.)).
- Shape can also be modelled through a mathematical description. This is the manner in which most reconstructed parts are built, because it is a common way to create models from plans and sections (Grellert, 2007, 176 f.). For instance the reconstruction of St. Lawrence's church in Ename (see Chapter 7) has been constructed with these methods.

As the method with which the shape information was created often determines which kind of model will result from the data—and existing data is usually the starting point for any reconstruction activity—the following subsections will be ordered according to the methods which can be employed for the generation of virtual models.

10.2.1.1 Shape information from physical objects

Automated (laser scanning) or semi-automated measuring approaches (e.g. with a total-station) are very popular for archaeological applications; as examples, both the laser scans of the floors in the Casa del Centenario (Section 6.3.1) and the surface survey in the Negotiating Avebury project (Section 8.2.2.1) can be named.[17]

If the starting point is a laser scan or measurements taken with a total-station, the data consists of points in space, which have to be connected with neighbouring points to form a network (see Section 9.1.2.1): the points become *vertices* and the connecting lines are *edges* (cf. Watt, 2000, 35). In a **wireframe** representation only the network of edges is rendered (cf. left side of Fig. 6.9). Often such a network or mesh is made up of triangles (cf. Fig. 10.14), but other shapes, like quadrangles[18], *etc.*, are possible. The edges can be seen as the boundaries of a surface, i.e. a *polygon*, with a vertex at each corner (cf. Watt, 2000, 35). These polygons make up the **surface** of a surface model.

[16] Lock (2003, 152) and Wood and Chapman (1992, 123) actually count boundary-representations among the solid modelling methods.

[17] The two examples referred to here are both on local scale. Lambers and Remondino (2008, 27 ff., 30 f.) also mention examples on regional and objects scale.

[18] See e.g. left side of Fig. 6.9

Figure 10.1: Photo-modelling: various points in different parts of the pictures have to be identified. The points in the left picture correspond to the ones in the centre. They can then be marked in ImageModeler, from which the program calculates the camera positions for the shots. (Photographic documentation for Friedrich Behn, images 7497 and 7501. Original negatives: Landesamt für Denkmalpflege Hessen). The image to the right shows how the software can create a textured 3D-surface model from the images. (Modelling by Maxi Platz and Susanne Krömker).

A surface model can also consist of *bi-cubic parametric patches* instead of polygons. A bi-cubic parametric patch (Fig. 10.17) has four corner points which are connected by four edges, whereby the latter are not straight lines but (cubic) curves[19]. In consequence the patch itself is a curved surface (Watt, 2000, 66). Bi-cubic parametric patches allow for exact, smooth and curved surfaces (Watt, 2000, 69), which can be created from either 3D-scans or manually measured points (Watt, 2000, 115, 117–121) whereby the patches approximate the shape determined by the point cloud. Its characteristics make this surface model type appropriate, for example, for modelling archaeological ceramic vessels.[20]

What has been illustrated here is that measured point data will usually result in a surface model (Watt, 2000, 38 f.), but this it not always the case:

From X-ray computed tomography (as only one example) three-dimensional solid models based on voxels can be generated (Watt, 2000, 370 f.). Voxels are the three-dimensional equivalent to pixels, i.e. small cubes[21] which fill a three-dimensional space. Each voxel has to be labelled according to whether it belongs to an object or empty space, which results in a high memory consumption (Watt, 2000, 51). To create a voxel volume the space between the individual two-dimensional X-ray planes, which are recorded by the tomographic imaging system, has to be interpolated to calculate the X-ray absorption coefficient for each resultant voxel (Watt, 2000, 371, 372 fig. 13.2).

X-ray computed tomography has been used, for example, in studying mummies (Attardi et al., 2000, 79) and in the study of different archaeological vessels (Jungblut et al., forthcoming).

10.2.1.2 Shape information from photographs or video images

Surface models (or—being reduced to their edges—wireframe models) can be created in other ways too, for example, using photo-modelling techniques (cf. Gillings (2000, 61–68) and Watt (2000, 465 f.)). This can be illustrated with an example from Lorsch Abbey (Fig 10.1): two legacy photographs taken in the 1920s or 1930s, which show the same excavation location—outside the south western corner of the church—from different perspectives at different times, can be used as the starting point (Fig 10.1 left and centre). These images can be loaded into Autodesk ImageModeler[22] where their relative position in relation to each other can be computed by giving the program different views (at least two, preferably more) of the same object. The user has to mark corresponding points in at least two of the photographs. Points from different regions of the photographs and the object should be selected. This makes it possible to vectorize the pictures in '3D'. The vector-lines can then be saved and imported into AutoCAD, 3ds Max,

[19] A cubic curve can be understood as lying within a cube: the start and the end point are fixed to opposing vertices. The curve progresses in the direction of the x-axis and enters the end point from the direction of the z-axis. Two further vertices of the cube are acting as control points for the curve's shape. In order to change the curve the cube is deformed thus changing the position of the control points as well.

[20] See, for example, http://i3dea.asu.edu/data/docs_pubs/using_geometric_modeling_for_archiving_and_searching_3D_archaeological_vessels.pdf (Accessed: 7th of August 2012.) where vessels have been laser scanned and then converted into parametric patches (sections 3 and 4).

[21] This is true for voxels generated by X-ray computed tomography, but other voxel shapes are possible (Watt, 2000, 371).

[22] http://usa.autodesk.com/adsk/servlet/pc/index?id=11390028&siteID=123112 (Accessed: 7th of August 2012.).

etc. and a texture, which is generated from the photographs, can be applied (Fig 10.1 right).

In a similar way 3D-models can be created from video data or successive images (Cosmas et al., 2001, 300). For the MURALE project Pollefeys et al. (2000, 252) developed a fully automated approach. The sequence of images allows automatic matching algorithms to identify similar (automatically selected) points in consecutive images (Pollefeys et al., 2000, 254 f.). Starting from the first two images in the sequence the position of the camera is calculated and from this, through triangulation, the geometry of the object can be computed (Pollefeys et al., 2000, 255). Additional views are added to the calculation and the reconstruction is thus refined (Pollefeys et al., 2000, 255 f.) and can eventually be textured (Pollefeys et al., 2000, 251).

The photogrammetric modelling was also used to derive parametric lines and surfaces (i.e. bi-cubic parametric patch surfaces) (Cosmas et al., 2001, 300).

10.2.1.3 Mathematically generated shapes

When a reconstruction model is built according to plan and section drawings—no matter whether they represent architectural drawings (Grellert, 2007, 176 f.; see also Section 7.3.3.2 about the St. Lawrence's church), excavation documentation (Lieberwirth, 2008b, 68) or images of finds (Lock, 2003, 156, 156 fig. 4.11a, 157 figs 4.11b and c)—they are usually created as mathematically generated shapes. Reconstructed parts are also normally created by using this method.

Two-dimensional drawings can be scanned and imported into the modelling program where they are used as a backdrop for a digital drawing of the represented object(s) (Lock, 2003, 156, 156 fig. 4.11a; Grellert, 2007, 176). They can also be input using a CAD digitiser tablet, or measurements can be entered manually. The *digitising* process in a vector-based drawing program can either make use of straight lines or employ cubic curves[23].

These two-dimensional drawings are then used to produce three-dimensional shapes, whereby mathematical generation can result in surface or solid models (Watt, 2000, 39 f.). Common approaches are:

revolving/lathing: A vertical cross-section, for example, of a Roman bowl or a column, is revolved around a central axis to create the whole bowl or column (Watt, 2000, 39 f., 107 f.; Barceló, 2000, 11; O'Rourke, 2003, 66). For an example result see Fig. 10.2, left.

extrusion: A horizontal cross-section is extended along a perpendicular path into space (Barceló, 2000, 11; O'Rourke, 2003, 66 f.). An example is given in Fig. 10.2, left centre. Another example constitutes some of the walls in the Casa del Centenario project which were modelled with this technique to represent the wall and different layers of plaster on it (see Fig. 6.8, lower centre).

sweeping: Sweeping is a more sophisticated form of extrusion. A horizontal cross section and a path are still the essential requirements, though the path does not have to be perpendicular, but can be curved in three-dimensions. Additionally, the horizontal cross-section can change size, shape and rotation (Watt, 2000, 39 f., 108 f.; Barceló, 2000, 11; O'Rourke, 2003, 68). This allows for the generation of more complex objects, e.g. cushion capitals or volutes. Two simple sample shapes created with this technique are depicted in Fig. 10.2, right centre and right.

lofting: In lofting two or more polylines[24] with the same number of vertices are joined together (Barceló, 2000, 11). In this manner a two-dimensional contour map can be transformed into a surface model (O'Rourke, 2003, 68 f.): a scanned map can be loaded into a vector based drawing program in which the individual isolines are digitised as polylines. In the modelling software the isolines (i.e. polylines) are moved to their correct height on the z-axis and the lofting process will produce a surface model (see Fig. 10.3, left).

boundary surfaces: Two or four lines act as boundaries for a surface, the computer calculates a surface which lies between the boundary lines (O'Rourke, 2003, 69). The examples in Fig. 10.3 show two identical boundary lines (left centre), two different boundary lines (right centre) and four boundary lines (right) each with their respective generated surface.[25] Interpolated strata boundaries from legacy excavation section drawings could be created in this way.

[23]This means: Bézier curves, B-spline curves, Rational curves (resulting e.g. in Non-uniform Rational B-splines (NURBS)). For an in-depth discussion of this very complex matter see Watt (2000, 66–121) and Foley et al. (1996, 478–528).

[24]A polyline consists of a number of connected line segments with a vertex at each segment's end.

[25]This is similar to a bi-cubic parametric patch.

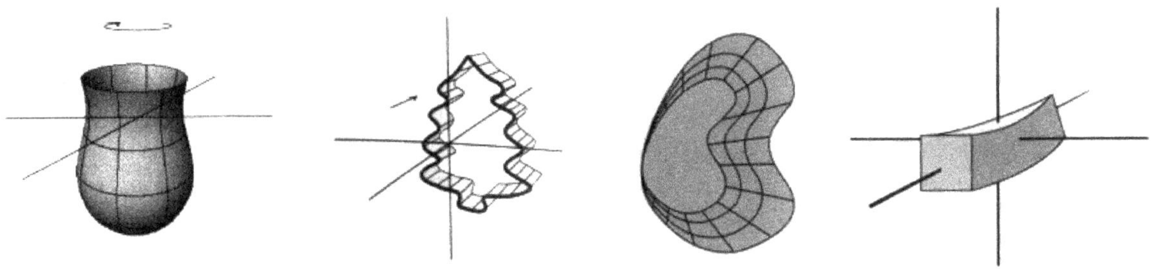

Figure 10.2: Lathing, extrusion and sweeping. Left: lathing; left centre: extrusion; right centre: sweeping, whereby the size of the cross section is increased; right: sweeping along a curved path (O'Rourke, 2003, 66 fig. 2-61b, 67 fig. 2-62a, 68 figs. 2-64a and b). (Image copyright Michael O'Rourke, 2003).

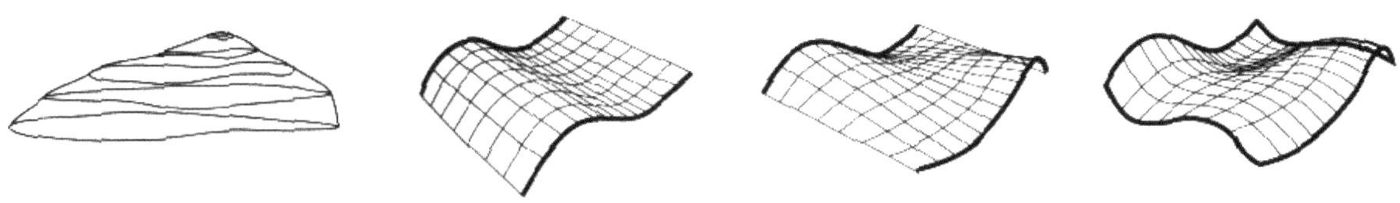

Figure 10.3: Lofting and boundary surfaces. Left: lofting; left centre: boundary surface; right centre: boundary surface; right: boundary surface (O'Rourke, 2003, 69 figs. 2-66b, 2-67a–c). (Image copyright Michael O'Rourke, 2003).

> By going one step further Lieberwirth (2008b) created voxel models from a legacy excavation documentation (see also Section 5.5.4.1).

Model parts which have been constructed in this manner are frequently combined with other modelled parts and then copied and pasted. Koob (2000, 76) gives the example of a combination of base, shaft and capital which are 'grouped' to form a column. The columns, in turn, can be replicated, arranged and grouped to form clustered columns, and so forth. This procedure has been employed excessively in the stone-by-stone reconstruction of Furness Abbey (Delooze and Wood, 1991, 143).[26]

As has been mentioned above, lathing, extrusion and sweeping can be used to create surface and solid objects. The difference between a surface and a solid object becomes apparent when the object is cut. It can be envisaged by imagining the result of a cut through a solid brick (i.e. two solid pieces) and a cut through a paper carton (i.e. two hollow pieces). These differences become important if, for example, a doorway is to be cut into a solid or surface wall (Ryan, 1996, 97, 98 fig. 2). In contrast to surface models, solid models can also be used to calculate volume, mass, *etc.* and therefore permit a number of analyses which cannot be performed with surface models (for a detailed discussion see Section 5.5.7.1).

Constructive Solid Geometry (CSG)

CSG is a special way to create solid models. Starting points in this case are geometric *primitives* (spheres, cubes, cones, cylinders, *etc.*) which are combined via Boolean set operators (union, substraction and intersection, as shown in Fig. 10.4) and linear transformations (i.e. scaling) (Ryan, 1996, 97, 99, 99 fig. 3; Watt, 2000, 46 f.). What makes CSG special is also the way in which the shape information is stored, i.e. as a tree structure in which the leaves are the names of the primitives and their dimensions, while the nodes contain the operator and the spatial relationship of the combined objects. This approach requires little storage space for the models and makes them very precise and easy to alter (Watt, 2000, 49). Objects created with CSG can also be replicated, arranged and grouped to accelerate the modelling process (Barceló, 2000, 16). Already some of the earliest archaeological reconstructions, namely the Old Minster of Winchester in 1985 and a 'temple' at Navan in Amagh (Ireland)[27], have been created using CSG (Reilly, 1992, 152, 155).

Modern modelling software can often handle surface and solid objects in one environment and also offers a variety of ways to manipulate them or convert the objects from one model type into another.

[26] For a critique of this approach, when it is applied to extant (in contrast to reconstructed) parts, see Section 5.5.7.1.

[27] The exact date of creation is not mentioned, but related publication dates (Reilly, 1992, 155) suggest an origin in ca. 1989.

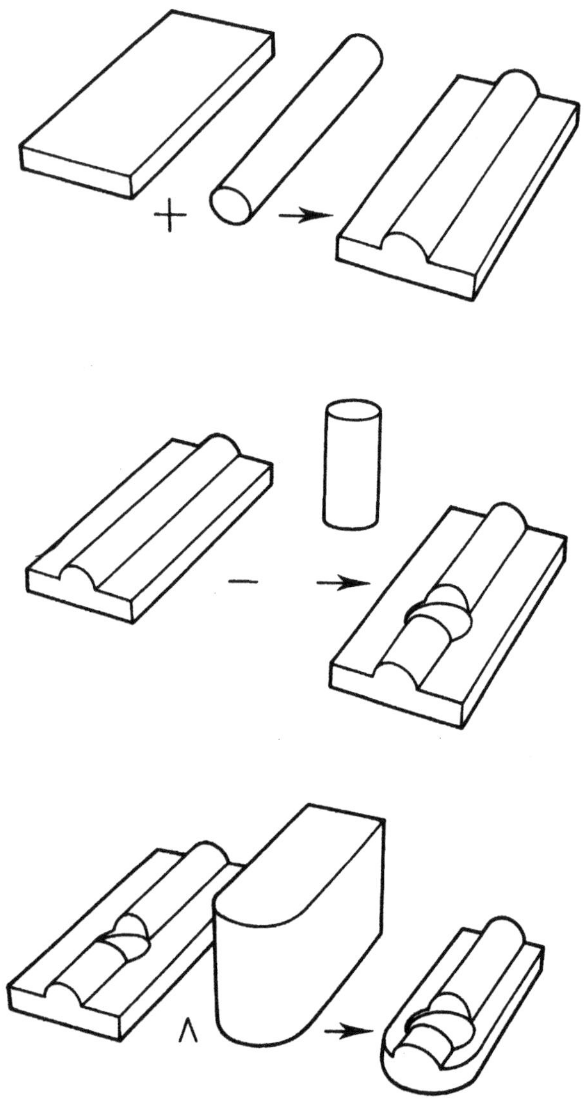

Figure 10.4: Constructive Solid Geometry (CSG) construction via Boolean operations: top: union; centre: substraction; bottom: intersection (Watt, 2000, 48 figs. 2.12a–c).

10.3 Rendering and Mapping

Rendering methods depend on which modelling methods were employed (see Section 10.2.1): polygon meshes, bi-cubic parametric surfaces, CSG and voxel models require different approaches (Watt, 2000, 123). Several approaches to rendering polygon meshes exist (Section 10.3.1), some of which provide quick results (Section 10.3.1.1). This is one of the reasons why bi-cubic parametric patches, CSG[28] and voxel representations are often converted into polygon meshes for rendering (Watt, 2000, 123 f.).

[28]The CSG representation has first to be converted into voxels, before it can be transformed into a polygon mesh (Watt, 2000, 124).

Bi-cubic parametric patches can also be rendered directly from the mathematical description of the patches (Watt, 2000, 125) (Section 10.3.2) while CSG descriptions can be rendered directly via ray tracing (Watt, 2000, 138) (Section 10.3.1.2). Voxels can be rendered directly via ray tracing or in a volume rendering approach (Watt, 2000, 140) (Section 10.3.3).

Besides the mere 'objects' in the scene many other factors play a role in rendering: the view point of the observer, light sources in the scene, the field of view, showing only surfaces which 'can be seen', shading, reflections, transparencies, textures, rasterisation of the scene for display, *etc.* (Watt, 2000, 142, 205, 223 ff.).

Again only a basic description of some common approaches is attempted in order to give the reader an overview of various methods, their strengths and weaknesses as well as their effect on the final representation of the models and, not least, of the complexity of generating visual output.

First rendering methods for polygon meshes (Section 10.3.1), ranging from some simple and quick rendering local illumination approaches over global illumination concepts to non-photo-realistic rendering, are introduced. This is followed by a description of how bi-cubic parametric patches and voxel models can be rendered without transforming them to polygon surfaces first (Sections 10.3.2 and 10.3.3). Then different mapping techniques are presented (Section 10.3.4) which are either employed to make the model look more interesting or realistic, or to circumvent the need to employ computationally expensive rendering methods for realistic looking lighting or reflections. Finally the creation of geometric shadows is addressed (Section 10.3.6) which are also used to reduce rendering time.

10.3.1 Rendering methods for polygon meshes

Polygon meshes are the most frequently generated model type, therefore this kind of model will be considered first.

In order to display a somewhat realistic looking scene, it is necessary to take several steps (Watt, 2000, 142):

1. The objects have to be defined (Watt, 2000, 142).[29] This has been described in Section 10.2.

[29]It may be wise to check whether the surface normals, which

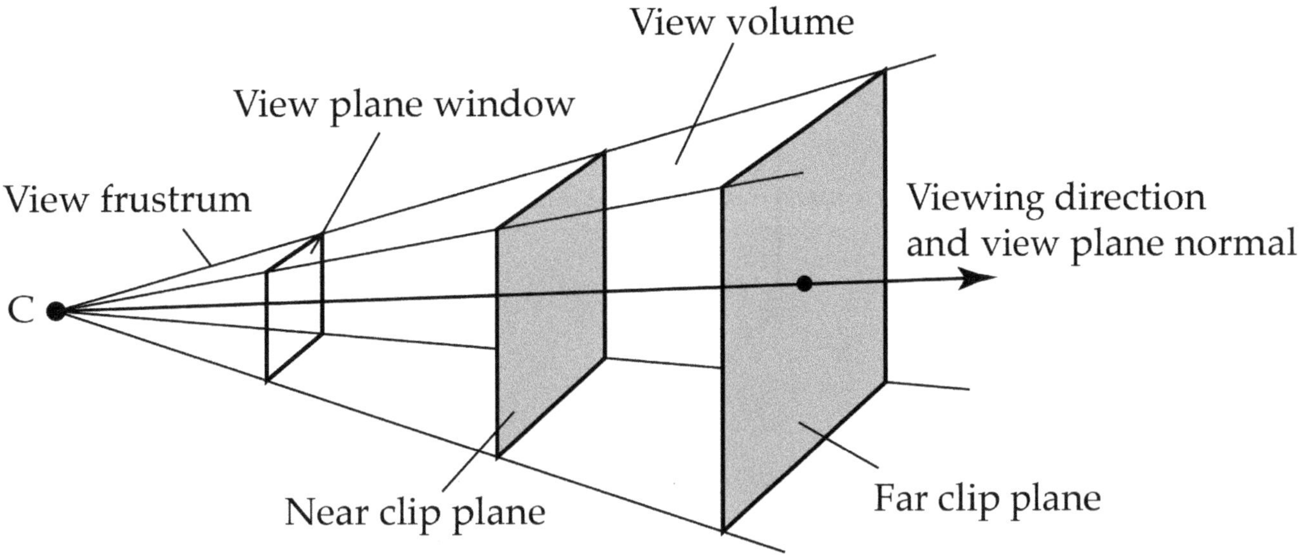

Figure 10.5: Clip planes and view volume: the view plane window can be considered as the surface of the computer display. The near clip plane (which may coincide with the view plane window) makes sure that objects (too) close to the viewer—marked as C—are not displayed. The far clip plane does the same for objects far away from the observer, thus reducing computational effort. The view volume indicates the area of the visible, virtual scene (Watt, 2000, 148 fig. 5.6).

2. A scene using the different objects has to be composed. Objects can be textured (see Section 10.3.4) and coloured. Light source types (point light (e.g. the sun) (O'Rourke, 2003, 95; Watt, 2000, 142), spot light (O'Rourke, 2003, 96), area light (O'Rourke, 2003, 97) or an ambient light (O'Rourke, 2003, 97 f.)), their position and colour as well as their spread (for spot lights) and falloff all have to be defined (O'Rourke, 2003, 95 f.). Additionally, the position and viewing direction of the observer has to be established (Watt, 2000, 143 f.).

3. Knowing from 'where' in virtual space the observer sees the scene, thus the view volume can be established (Fig. 10.5) (Watt, 2000, 144 f.). The surface of the computer display acts as a two-dimensional window, the *view plane window*, into the virtual scene (Watt, 2000, 149).[30] Different from a real window, *clipping planes* (also called *clip planes*) can be employed: the near clipping plane can be identical with the view plane window, but can also be set to a further distance (Watt, 2000, 147). Everything between this plane and the observer will not be rendered—this can e.g. enable an observer to 'look through a wall' if the near clipping plane lies beyond it. The far clipping plane cancels out every object that lies beyond it.[31] These objects have not to be considered in the rendering process. What lies between the near and far clipping planes is the *view volume* (Watt, 2000, 147 f.). Clipping planes cannot only help to see through walls, but can also be useful in the exploration of three-dimensional data: Reilly and Shennan (1989, 161) state that the model of the stratigraphy of the St. Veit-Klinglberg excavation was visually analysed by setting the depth of the near clipping plane to different values. For such explorative purposes clipping planes can also be defined arbitrarily—they do not have to be parallel to the view plane window.

Culling or back-face elimination (Watt, 2000, 147) is one of the steps which are necessary to make (wireframe) objects appear to be 'solid'. In Figs. 10.6 and 10.7 the effect of culling can be seen: in both figures the left images show the object without culling. All polygons and their edges are visible, which looks as if they were constructed as wireframes. Culling removes the invisible polygons from view for which the position of the observer has to be known as well as the direction in which the polygons face (this is indicated by the surface normal, see Fig. 10.6,

indicate where the outside of the model lies, are all pointing in the right direction. Otherwise an object may not be visible due to culling (see point 3 below and cf. Fig. 10.6); additionally, light models and texturing will provide wrong results.

[30] This is not exactly true (Watt, 2000, 149), but it provides a good mental image.

[31] This is somewhat related to our experience in the physical world, where we are also unable to see into the infinite distance, but here it is due to atmospheric effects.

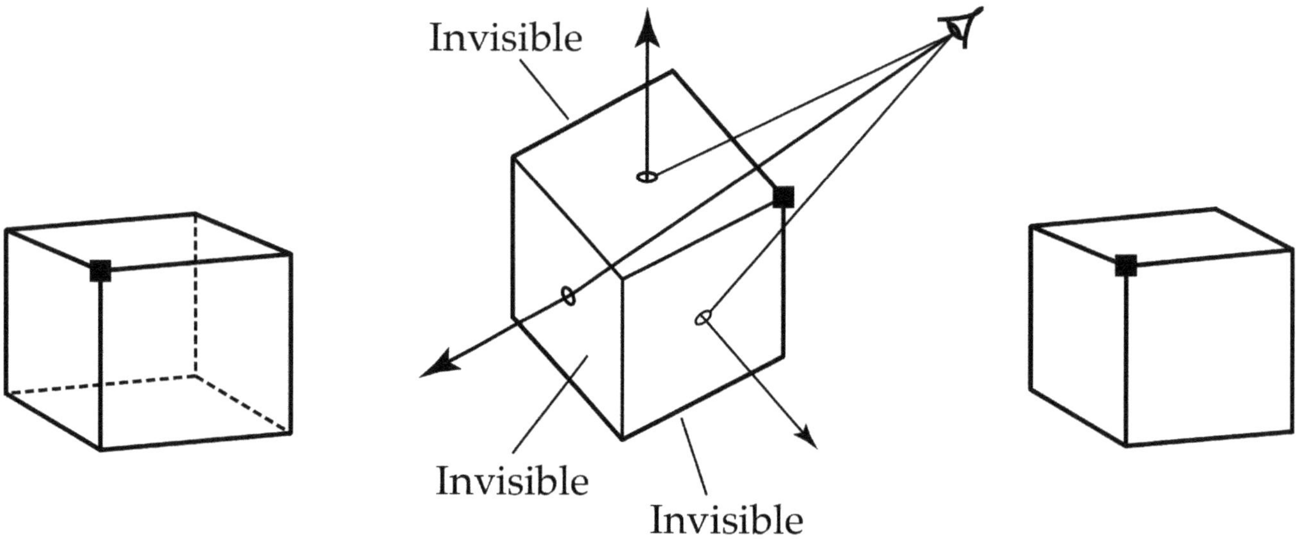

Figure 10.6: Culling: left: all polygons can be seen. Centre: using the surface normals of the polygons it can be established which surfaces a viewer can see and which are invisible. The invisible polygons will be hidden from view. Right: finished object with back face elimination (after Watt, 2000, 148 fig. 5.5).

centre where the surface normal is visualised as an arrow which stands perpendicular to the polygon's surface). When the back faces have been eliminated the observer has the impression that he or she is seeing a solid cube (Fig. 10.6, right) (Watt, 2000, 147).

4. As is demonstrated in Fig. 10.7, centre, it is not sufficient to remove whole invisible polygons. Instead, *hidden surfaces* have also to be *removed* in order to arrive at Fig. 10.7, right. Most frequently a z-buffer algorithm is employed for this task (Watt, 2000, 189).

   Before proceeding to the z-buffer algorithm, *rasterisation* has to be considered, that is the way in which the vector-based polygons are mapped on the pixel-based screen or view plane (Watt, 2000, 183). First it has to be defined which pixels are crossed by the edges of the polygon and then the interior (i.e. polygon) surface has to be filled (Watt, 2000, 183–185).

   To find out whether a polygon is (partly) hidden by another polygon the *z-buffer* algorithm is used (Watt, 2000, 189): the view plane can be seen as a xy-plane and the distance of a polygon from the observer would be on the z-axis. The z-buffer algorithm establishes for each pixel (i.e. xy-coordinate) on the screen the closest point on the z-axis in the scene. Each time a new polygon has to be loaded into the scene, the z-values for all of its surface points are compared with the z-values of the so-far closest point for this pixel. If the distance is greater, the point on the polygon is discarded, if it is closer, it will replace the previous point (Watt, 2000, 189). In the end for each pixel on the screen only the polygon point with the smallest distance from the observer will be drawn (Ryan, 1996, 100).

   Shading also belongs to this last step (Watt, 2000, 171). Several different shading techniques exist; due to their complexity they will be dealt with in separate sections. Section 10.3.1.1 explains Gouraud and Phong shading, Section 10.3.1.2 introduces ray tracing, radiosity and photon mapping and Section 10.3.1.3 explains the use of g-buffers, defines different edge types and describes Gooch- and cel-shading as two examples for non-photo-realistic shading types.

10.3.1.1 Local illumination techniques

Two simple but very popular shading techniques, Gouraud[32] and Phong[33] shading, will be described here. The images these techniques produce are not very realistic, but they are computationally inexpensive, which bestows unfailing popularity upon them (Watt, 2000, 171). Phong shading adds some refinements to Gouraud shading (Watt, 2000, 171 f.), so the latter will be described here first.

[32]Named after its inventor, Henri Gouraud (O'Rourke, 2003, 111).

[33]Named after Phong Bui Tuong (O'Rourke, 2003, 112), who developed it in 1975 (Strothotte and Schlechtweg, 2002, 253).

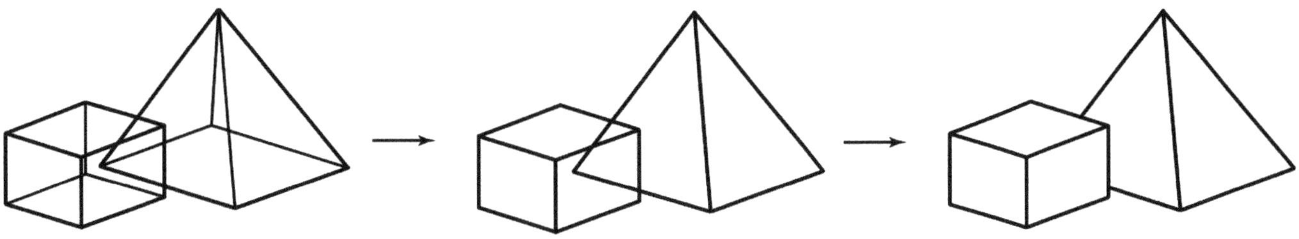

Figure 10.7: Culling and hidden surface removal: left: all polygons can be seen; centre: complete invisible polygons are removed, also called culling; right: hidden surface removal, where polygons obscure parts of other polygons (Watt, 2000, 147 fig. 5.4).

Both shadings deal with two different 'representations': one is the three-dimensional virtual object, which receives (virtual) light from a light source and reflects this light. The other is the two-dimensional shaded image of this object which we see on the computer screen (Watt, 2000, 172).

Gouraud shading

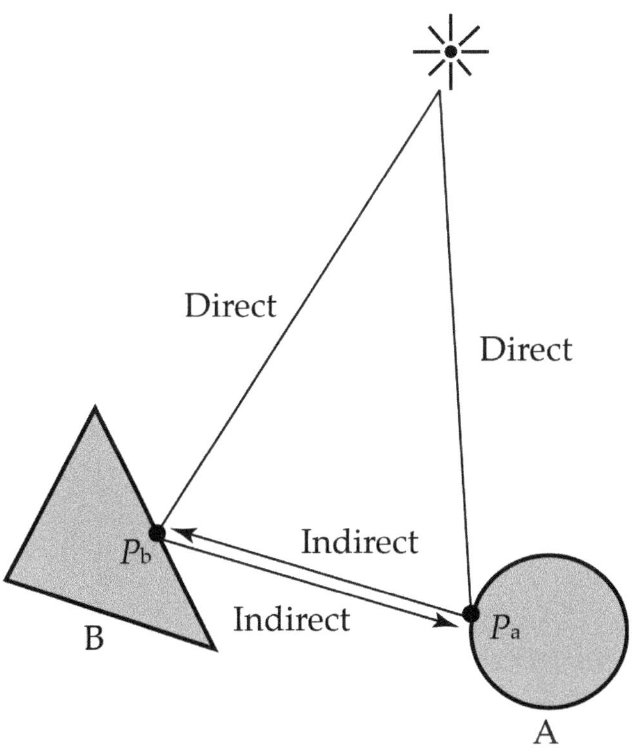

Figure 10.8: Local and global reflection models. A local reflection model is only concerned with direct illumination on points P_b and P_a, while a global reflection model (e.g. radiosity) also takes indirect illumination between objects A and B at both points into consideration (after Watt, 2000, 173 fig. 6.6).

In order to deal with these two above-mentioned 'representations' and the conversion of one into the other, two steps are necessary. First, a *local reflection model* has to be established, which is used to calculate (for each vertex) the reflection of light on the object's surface 'in 3D' (Watt, 2000, 172). Second, *shading algorithms* map the shape of the polygons onto the 2D-screen (i.e. rasterisation) and the values calculated for the vertices are used for a linear interpolation of the light intensity for the pixels onto which the polygon is mapped (Watt, 2000, 172, 179). Taking the vertices as starting points for the interpolation process lets the shading of the object's surface appear smooth (O'Rourke, 2003, 111 fig. 3-49),[34] but at the edges of the object the outlines of the polygons may leave a jagged edge (O'Rourke, 2003, 111).

The local reflection model is rather reduced (Fig. 10.8). It only considers direct illumination coming from a light source (cf. Fig 10.8), but ignores any indirect, reflected light from other objects in the scene. Neither are shadows calculated, which means that an object will not throw a shadow on any other object (Watt, 2000, 173).

As for reflections, three types are usually considered in computer graphics (Fig. 10.9): perfect specular reflection[35], imperfect specular reflection (both for glossy objects) and perfect diffuse reflection (for matte objects) (Watt, 2000, 173; O'Rourke, 2003, 111).

The *object colour* has influence on the diffuse reflection coefficients, while the *colour of the light source* has influence on the specular highlights[36] (Watt, 2000, 178).

The reflection model takes the viewing direction and the position of the light source in relation to the orien-

[34] This stands in contrast to flat/faceted shading, where the reflection at the polygon's surface normal is calculated and used to shade the whole polygon. The separate polygons are easy to identify due to their varying shade (O'Rourke, 2003, 109 f., 110 fig. 3-46). An example of flat shading provides Fig. 7.32, note, for example, the facets on the triumphal arch.

[35] In reality perfect specular reflection does not exist, but the model is used for ray tracing (see Section 10.3.1.2) (Watt, 2000, 174).

[36] A highlight is in fact just a reflection of the light source on the object.

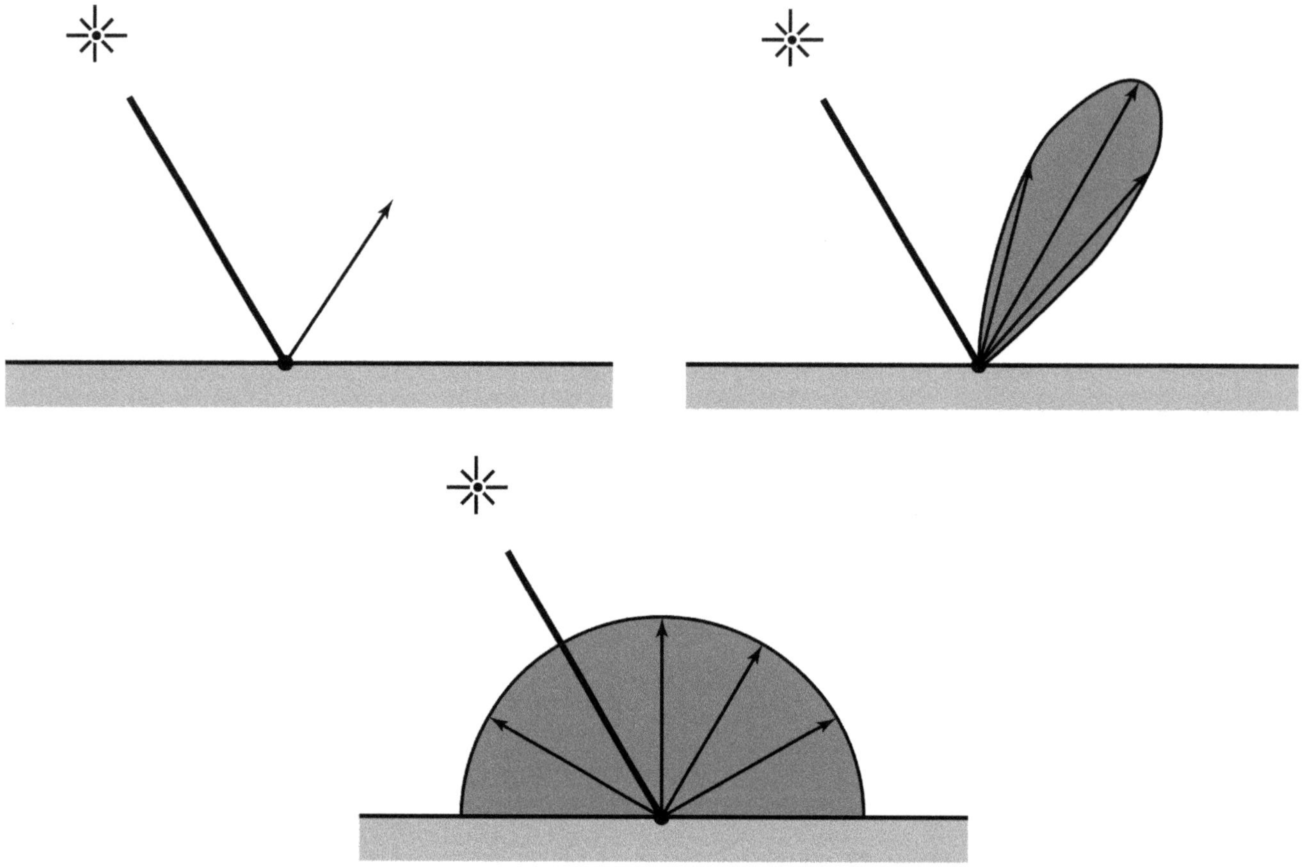

Figure 10.9: Reflections: top left: perfect specular reflection; top right: imperfect specular reflection; bottom: perfect diffuse reflection (after Watt, 2000, 174 fig. 6.7).

tation of the vertex (Gouraud shading), surface (flat shading), or pixel normal (Phong shading) all into consideration (Watt, 2000, 175). Due to the law of reflection[37] (Jarosz et al., 2008, 24), highlights from specular reflection can only be seen when the viewer faces in the same direction as the reflection (Watt, 2000, 176). But as Gouraud shading only calculates the light intensity at the vertices, while all other values are interpolated (Watt, 2000, 179 f.), highlights can only occur at a vertex, and not on the polygon surface.[38] Phong shading has found a solution for this problem.

Phong shading

Instead of only calculating a light intensity value for each mapped pixel (as is done in Gouraud shading) Phong shading also calculates a normal for each pixel (Watt, 2000, 181; O'Rourke, 2003, 111); if the normal has the right orientation to indicate that incoming light would be reflected directly at the viewer, a highlight will appear disregarding whether the pixel coincides with a vertex, edge or polygon surface in the object model.

The highlights in Phong shading are very concentrated, but some materials produce rather imperfect specular reflections, which Phong shading cannot model appropriately (O'Rourke, 2003, 112).[39]

Phong shading not only uses diffuse and specular components, but also ambient light (Watt, 2000, 174) which acts as a substitute for indirect light. In reality indirect light (Fig. 10.8) originates from diffuse reflections occurring on other object surfaces in the scene (Jarosz et al., 2008, 22).

Due to their shortcomings (there are no shadows, only direct illumination is considered and the lighting of the objects is unrealistic), both shading models

[37]The angle of reflection equals the angle of incidence.

[38]If the highlight should only fall onto a polygon surface it would not be visible at all (Watt, 2000, 181).

[39]Blinn shading can be used for this purpose (O'Rourke, 2003, 113).

have been regarded as inferior by several archaeologists (Ryan, 1996, 100; Lucet, 2000, 90; Lock, 2003, 153). These considerations are only relevant if the reconstruction strives for 'realism', which is not always the case (see Section 5.10.3).

If, indeed, photo-realistic output is sought, global illumination concepts are better suited for the purpose.

10.3.1.2 Global illumination

Global illumination concepts do not only consider the light coming directly from the light source, but also reflected light. Thus they are able to calculate shadows, transparencies and reflections (Watt, 2000, 275 f.). Three different global illumination concepts will be introduced: *ray tracing*, *radiosity* and *photon mapping*.

Ray tracing

Ray tracing is excellent for rendering specular surfaces; it can also deal with transparency, but produces only hard shadows (Ryan, 1996, 100)[40]. It cannot, however, deal with diffuse reflection and the associated colour bleeding[41] (Ryan, 1996, 100). The reasons for this will become apparent from the description of the rendering method below. Even despite these drawbacks[42], ray tracing has been applied to some archaeological reconstructions early on: Reilly (1992, 151) states that it was used on the model of Bath's Roman baths in order to create shadows. Ryan (1996, 100, 101 fig. 4) mentions that ray tracing was used to create views of the Longmarket Roman building in Canterbury.

Whitted ray tracing (introduced by Turner Whitted in 1980 (Jarosz et al., 2008, 24)) only simulates a subset of global interaction, namely only perfect specular interaction[43] (Watt, 2000, 276). Starting point for following the rays is not the light source, but the observer's eye (Watt, 2000, 343).[44] The rays are traced until the energy of the ray has dropped below a pre-defined threshold (i.e. it has been absorbed by the environment), a pre-defined recursive depth[45] has been reached or the ray hits a diffuse surface (Watt, 2000, 343 f.). It is necessary to calculate for each ray whether its path is intersected by an object in the scene (Fig. 10.10). If it hits a specular opaque surface two rays will be spawned. One is the reflection of the initial ray and will be followed in the next recursion, the other connects the intersection point with the light source in the scene. In the case of the ray hitting a (semi-)transparent specular object, three rays will be spawned: a reflected ray and a ray which connects the point to the light source—just as before—but also a refracted ray[46], which also has to be followed in the next recursion (Watt, 2000, 284). As the rays are traced backwards in the scene, the 'true colour' of any point in the scene can only be determined if it is known from which objects the ray was reflected previously, because they determine the colour and the light intensity of the ray (Jarosz et al., 2008, 24). Objects with a mirror surface will reflect objects from which rays have been reflected onto them (Watt, 2000, 286). Consequently a high recursive depth will result in a 'realistic' image, but also in high computational cost.

If the ray which connects the intersection point with the light source is itself intersected by an opaque object, the intersection point lies in shadow (Watt, 2000, 284; Jarosz et al., 2008, 23). If the intersecting object is not opaque but semi-transparent the colour component introduced by this object has to be noted (Jarosz et al., 2008, 23). In case there are several light sources in the scene each of them has to be tested for occlusion in this way (Watt, 2000, 345; cf. Jarosz et al., 2008, 38). This testing results in shadows with hard edges only (Watt, 2000, 342).

It would be theoretically possible to use ray tracing for diffuse surfaces as well, but every encounter with a diffuse surface would spawn a multitude of hemispherically reflected rays (Fig. 10.9, c) which all have to be followed in the next recursion (Watt, 2000, 285). This would make the all-over rendering process computationally *extremely* expensive (Watt, 2000, 344).

As only specular objects which have been intersected by rays become visible, diffuse objects are invisible. In order to make these objects discernible, an ambient constant (for example, a Phong reflection model) has to be added (Watt, 2000, 306, 342, 540).

[40] To remedy the hard shadows Ryan (1996, 100) and Lock (2003, 153) recommend the use of several light sources, but also caution that this will result in additional rendering time.

[41] Coloured reflected light from other objects in the scene (Ryan, 1996, 100).

[42] Especially, the predominance of diffuse objects in archaeological scenes makes the application of a rendering algorithm for specular objects seem inappropriate.

[43] The interaction between glossy surfaces.

[44] This also means that every time a new viewer position is chosen the scene has to be re-rendered completely.

[45] The first calculation cycle follows the rays from the eye until they intersect with an object. The resulting reflected (and refracted) rays are traced in the next, recursive, calculation cycle.

[46] The angle of refraction is determined by the materials involved (Jarosz et al., 2008, 26).

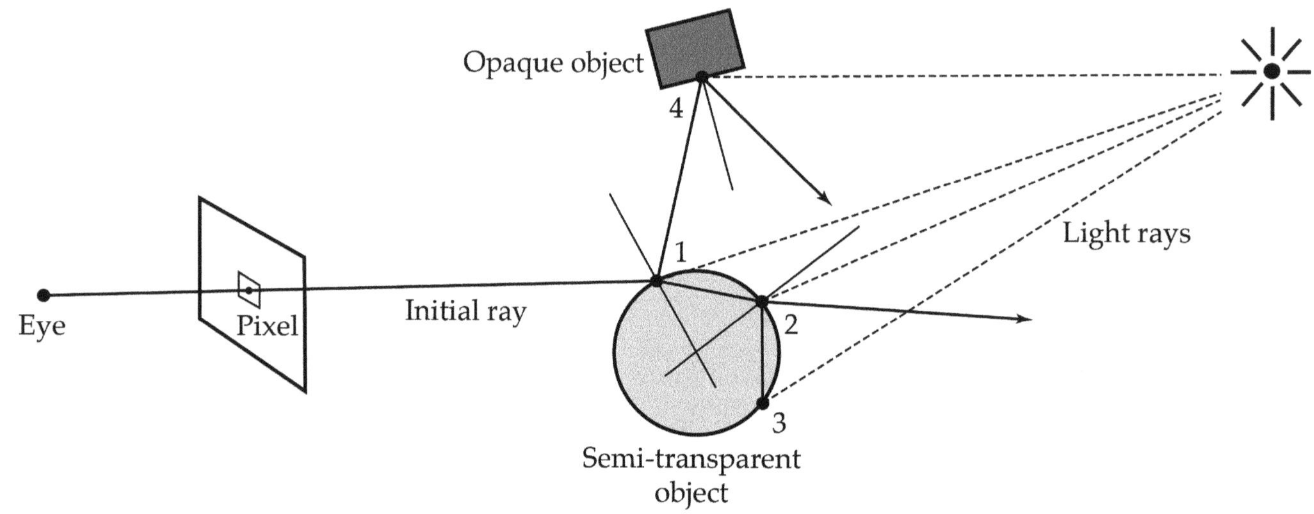

Figure 10.10: Whitted ray tracing, introduced by Turner Whitted in 1980: rays which hit the eye of the observer are traced backward through the view plane pixel (on which they will map) into the 3D-scene. When the ray hits a semi-transparent specular surface in the scene (1) three rays will be spawned: one connects the interception point with the light source, another lies in the direction of the refraction (towards point 2) and the last in the direction of the reflection (towards point 4). In the next recursion the newly spawned rays will be traced in the scene. The light ray connects point 1 with the light source; if the light ray were intersected by an opaque object, then the intersection point 1 would lie in shadow. The refracted ray is reflected and refracted once more (2) leading to three further rays which have to be traced. The reflected ray is reflected by an opaque object (4) thus only two rays have to be traced in the next recursion, and so forth (Watt, 2000, 284 fig. 10.5).

Ray tracing can also be used to render CSG and voxel models directly (Watt, 2000, 138-141).

Radiosity

Radiosity can calculate the correct colour of objects in a scene because it takes the colour of the reflected light into consideration. Also coloured, soft shadows[47] result from the process (Ryan, 1996, 100). A scene which has been rendered using radiosity is view-independent (Lock, 2003, 153). Its major drawbacks are that it is computationally very expensive (Ryan, 1996, 100) and that it cannot handle specular or transparent objects (Lucet, 2000, 90).

In an archaeological context, radiosity has been, for instance, used in an illumination case study on the Red Temple of Cacaxtla (Lucet, 2000, 91–95). Another example is provided by a study which is concerned with the natural illumination conditions in the Fugoppe Cave (Japan) with the aim of determining whether the cave art could have been painted in natural light, or whether artificial illumination would have been necessary (Masuda et al., 2004, 1, 6).

Radiosity is concerned with the relationship between diffuse surfaces (Watt, 2000, 306). For this the surfaces of the objects have to be subdivided into patches, whereupon a balance between the computational effort and the visual smoothness of the objects has to be struck (Watt, 2000, 286 f., 306).[48] In addition an energy conservation equation is necessary, which deals with the amount of reflected light between the patches (cf. Fig 10.8), whereof the distance and the orientation of the patches in relation to each other have to be taken into consideration (Watt, 2000, 306 f.).

Patches may not only reflect light, but also emit light (i.e. if they are light sources) (Watt, 2000, 307). Starting from the light emitting patches light is brought into the scene. Patches with an appropriate orientation[49] and distance to the light sources will receive light from them. The amount of received light (and colour (Lucet, 2000, 90)) is stored for each patch (Watt, 2000, 286). In the next calculation cycle, the light which has been received by the patches is shot back[50] into the scene until a high percentage of the initial light energy has been distributed around the

[47] They possess an umbra and penumbra.

[48] Large patches will result in a blocky appearance, small patches will result in a high computational cost.

[49] Perfect diffuse reflection is considered to reflect light in all the directions of a hemisphere. All patches which lie in this reflection hemisphere may receive light from it.

[50] In fact the starting point is the patch with the highest unshot amount of energy (Watt, 2000, 318).

scene, whereby patches can receive light in several stages of the calculation process (e.g. light from the initial light source, but additionally reflected light from other patches) (Watt, 2000, 286 f.). Patches which receive little or no light from other patches will stay in darkness, thus shadows are calculated automatically (Watt, 2000, 307). A radiosity calculation is view-independent (Lucet, 2000, 90), so that the finished model can be viewed from any point, but this makes it necessary to employ a separate hidden surface calculation (Watt, 2000, 307).

Radiosity is unable to deal with specular surfaces (Watt, 2000, 287) so that highlights or reflections on mirror-like surfaces do not occur. Neither can transparent objects be modelled (Lucet, 2000, 90).

Photon mapping

When comparing ray-tracing and radiosity it appears that they have contrary properties because one is only concerned with reflection between specular surfaces and the other only deals with diffuse reflection. It would be desirable to incorporate the benefits of both methods into one global illumination concept. Photon mapping, developed by Henrik Wann Jensen, provides such a concept. It can deal with both specular (in the form of ray tracing) and diffuse reflections, handle transparencies and particle systems (smoke, dust, mist, *etc.*), calculate caustics (i.e. bundled reflected or refracted light) and coloured shadows while providing a view-independent lighting calculation. To my knowledge there is no archaeological project which has used photon mapping, but it can be assumed that any scene which incorporates several of the traits named above can be rendered photo-realistically with this method—at a high computational cost.

In Photon mapping, photons—emitted from a light source—are traced through the scene[51] the aim is to calculate indirect illumination, so that the light intensity and colour of the photons is stored on diffuse surfaces (Jarosz et al., 2008, 47). The direction into which the photons are emitted is dependent on the type of light source (Jarosz et al., 2008, 47 f.). Its power and whether it is the only light source in the scene, are also of importance (Jarosz et al., 2008, 48 f.).

There are three possible scenarios for a photon which hits an object: it can be reflected, refracted or absorbed. Depending on the surface characteristics different probability values are assigned for the photon's fate. This has been called 'Russian roulette' (Jarosz et al., 2008, 51). The drawback of the Russian roulette method is that it only provides a sample and no exact value for reflection and refraction (Jarosz et al., 2008, 53). On the other hand no new photons are spawned during the process, so that the number of traced photons is equivalent to the number of emitted photons (Jarosz et al., 2008, 53).

[51] This is almost the reverse of ray tracing (Jarosz et al., 2008, 50).

When tracing the photons, only their interaction with diffuse surfaces is stored in a photon map. For specular surfaces, and for the calculation of direct illumination, ray tracing is employed (Jarosz et al., 2008, 53).

Caustic effects which occur through reflection and refraction can be refined (and then stored in a caustics photon map) by sending photons specifically in the direction of the object which creates the caustic (Jarosz et al., 2008, 58).

When the photon maps have been created, the preparations for the rendering process can take place. The information stored in the photon map, which constitutes only a sample, has to be integrated to find radiance values for the points in the scene which have not been sampled (Jarosz et al., 2008, 58, 60–63). The photon map is view-independent (Jarosz et al., 2008, 67) like radiosity. The real rendering process can take different approaches. Jarosz et al. (2008, 67 f.) mention two: one which computes the radiosity values at the vertices of a mesh by using the photon map and another based on ray tracing.

10.3.1.3 Non-photorealistic rendering (NPR)

NPR describes rendering methods which are neither necessarily bound by the laws of physics, e.g. concerning illumination (Strothotte and Schlechtweg, 2002, 206), nor does the rendered output aspire to be comparable to a photograph. Such an approach can enhance the comprehensibility of the represented scenes (see also Section 5.10.3)—as it is the case with archaeological illustrations (Pringle and Moulding, 1997, 22)—by emphasising the important features and reducing or omitting unimportant detail (Strothotte and Schlechtweg, 2002, 208). NPR renditions can be generated on the basis of two-dimensional images which encode three-dimensional properties of the scene (i.e. G-buffers) which in turn can be combined with two-dimensional 'photo-realistic' images in order to enhance the latter (Strothotte and Schlechtweg, 2002, 183 ff., 191). It is also possible to generate non-photorealistic output directly from the three-dimensional scene (Strothotte and Schlechtweg, 2002, 203).

Figure 10.11: Different G-buffers: top left: ID-buffer; top centre: normal-buffer; top right: lighting-buffer. Bottom: z-buffer (left) with the extrema of its first (centre) and second derivative (right) used for edge detection (Veryovka, 1999, 73 fig. 5.11, 75 fig. 5.12, 76 fig. 5.13).

All the approaches which are introduced in this section are applicable to polygon mesh models because they are the most common model type (Strothotte and Schlechtweg, 2002, 203) and, if necessary, both bi-cubic parametric surfaces as well as solid models can be transformed into them (Watt, 2000, 123 f.). The NPR approaches presented here constitute a selection of methods which result in comprehensible rendered images using different representational means: i.e. enhancing contour lines, hatching, Gooch- and cel-shading. Other rendering methods (e.g. stippling) can either be created in similar ways or (e.g. artistic shaders, like watercolour painting) have little influence on the comprehensibility of the result, or indeed may even hamper it.

In general it has to be said that to increase the comprehensibility of the rendered images it is necessary to have access to additional information beyond the geometric information (i.e. shape) and the graphical information (i.e. surface properties, colour, transparency, smoothness *etc.*) which are sufficient for photo-realistic renderings (Strothotte and Schlechtweg, 2002, 206 f.). This information can come from within the model (e.g. the relationship of subjects, sub-objects or dependencies between objects in the scene) or from outside the model (e.g. information stored in a database)[52] (Strothotte and Schlechtweg, 2002, 207). It can be used to *select* those *objects, features or information*[53] which will be represented as well as the *level of detail*. It can also be employed to decide which *representational style*[54] and *degree of abstraction* should be used and whether features should be *highlighted or obscured* (Strothotte and Schlechtweg, 2002, 205–208). Indeed, these are all questions which also have to be considered when a hand-drawn reconstruction is to be created (see Hodgson, 2001, 5).

G-buffers and their application in NPR

G-buffers (geometric buffers) contain information about the geometry of the scene (Strothotte and Schlechtweg, 2002, 185). They can be stored for later use in three-dimensional operations or can be rendered as two-dimensional images. Fig. 10.11 shows different examples of G-buffers:

ID-buffer (top left) depicts every object in the scene in a specific colour (or greyscale), this helps to

[52] In an archaeological context, this information could be anything that is relevant in the research context, from find densities in an excavation to specific decorative motives on pottery.

[53] Informational content can be incorporated by adding annotations to the objects in the scene.

[54] For example, whether hatching, stippling, or shading should be employed and in which manner.

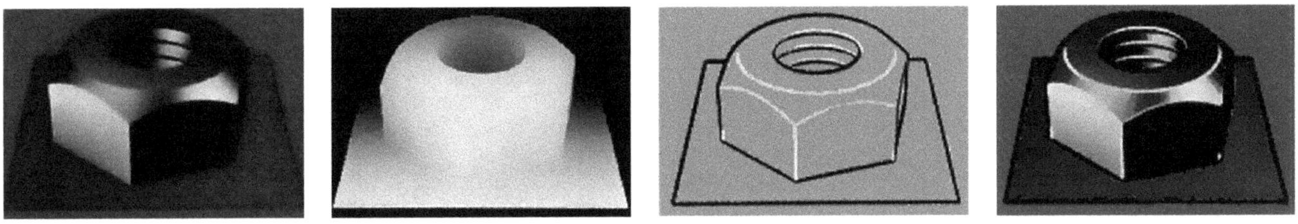

Figure 10.12: Nut with NPR edges: left: photo-realistic rendering, left centre: z-buffer, right centre: result of the edge detection, right: photo-realistic rendition combined with the edges (Saito and Takahashi, 1990, 202 fig. 7 a–d).

identify spatial relationships in the scene (cf. Strothotte and Schlechtweg, 2002, 191).[55]

normal-buffer or n-buffer (top centre) encodes the direction of the surface normal. The x, y and z coordinates are usually encoded as the red, green and blue channel in the RGB colour model (Strothotte and Schlechtweg, 2002, 192). This information can help with applying hatch lines and is used for edge detection.

lighting-buffer or shadow-buffer (top right). For each light source in the scene, or all light sources together, the pattern of light (i.e. the lighting-buffer) and shadow (i.e. the shadow-buffer) is computed (Strothotte and Schlechtweg, 2002, 192). These buffers are not implicitly identical. Lighting will include shaded areas on the objects but not necessarily shadows which one object casts onto another. This is discussed in more detail in Sections 10.3.4.2 and 10.3.6.

z-buffer (bottom left) encodes the distance from the observer (Strothotte and Schlechtweg, 2002, 191) (see Section 10.3.1 for more details).

first derivative of the z-buffer image (bottom centre). A differential operator[56] can be employed to find edges (Strothotte and Schlechtweg, 2002, 188): in this case those parts of the image exhibiting an abrupt change from light to dark pixels and vice versa.

second derivative of the z-buffer image (bottom right) can be used to find internal edges (Strothotte and Schlechtweg, 2002, 188), i.e. edges which mark the centre of a smooth light transition.

A number of other G-buffers are conceivable, e.g. a material-buffer which encodes different materials or material properties in the scene or buffers which contain user-supplied values (Strothotte and Schlechtweg, 2002, 192).

To illustrate some of the effects which can be created by using G-buffers, two examples will be presented. One depicts how added lines at the edges of an object can enhance perception. The other illustrates an approach towards generating shading via hatch lines, which result in a line-drawing appearance of the object.

Fig 10.12 illustrates how a photo-realistic rendering of a nut (left) can be enhanced by adding edge information (Strothotte and Schlechtweg, 2002, 188). The edges were detected by, first, generating a z-buffer image (left centre) and calculating from it the first and second derivative,[57] which are illustrated as black and white lines respectively in Fig 10.12, right centre. These lines were combined with the photo-realistic rendering of the object in order to clarify its shape (Strothotte and Schlechtweg, 2002, 188).

Some of the parts in the reconstruction of the church of Nola are represented in this manner (Lehmann and Haarlammert, 2010, 172 fig. 6, 175 fig. 8), e.g. those parts of the columns which still exist (see Fig. 5.6, bottom).

In the second example (Fig. 10.13) the contour lines were retrieved via the z-buffer and lighted areas are derived from a lighting-buffer, so that the hatching only appears on areas which lie in the shade. The n-buffer[58] is used to create the (cross) hatching which

[55]Object IDs are usually assigned when the object is created in the 3D-scene. There are other possibilities, too, for instance labels can be assigned during a segmentation process, which is carried out on scanned data (Susanne Krömker, personal comment). She refers to the scanned data from clay tablets, see http://www.iwr.uni-heidelberg.de/groups/ngg/projects.php?L=0 (Accessed: 7th of August 2012.).

[56]For example, a Sobel operator.

[57]Cf. the bottom three images of Fig. 10.11.

[58]In the actual example by Strothotte and Schlechtweg (2002, 189) the object is not a polygon mesh but a bi-cubic parametric surface (see Section 10.2.1). Nevertheless this approach should work equally well for polygons because they also possess object coordinates (u, v, w), which are not dependent on the scene coordinates (x, y, z). For an example of the relation of two diverging coordinate systems see Fig. 10.19, where the object coordinate system (u, v, w) and the scene coordinate system (x, y, z) have a different ro-

Figure 10.13: Hatching of a torus: left: hatching along the poloidal direction, centre: hatching along the toroidal direction, right: a combination of both hatching directions results in cross hatching. To all images a lighting-buffer was applied to remove the hatching on the illuminated surfaces (Saito and Takahashi, 1990, 202 fig. 9).

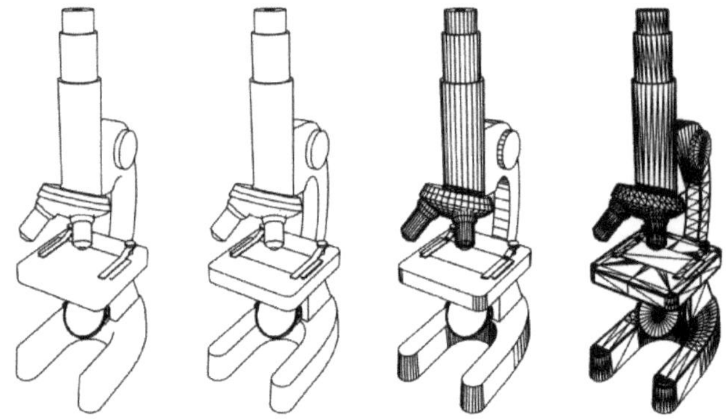

Figure 10.14: Edge Types: left: contour edges, left centre: sharp edges, right centre: smooth edges, right: triangulation edges (Strothotte and Schlechtweg, 2002, 215 fig. 7.2).

follows the geometry of the object (Strothotte and Schlechtweg, 2002, 189). All the polygons where the surface normals have an identical value on the y-axis can be selected and connected with a line (Fig. 10.13, left) as can be all the points with an identical value on the x-axis (Fig. 10.13, centre). Both hatchings can be combined to form a cross hatch (Fig. 10.13, right) (cf. Strothotte and Schlechtweg, 2002, 189).

This creates images which are very similar to traditional archaeological black and white find illustrations (see, for example, Griffiths et al. (1990, 32 fig. 13a)).

Edges

Edge detection on the basis of a two-dimensional z-buffer image is just one way to arrive at edges or contour lines. Contour lines and edges can also be derived directly from the 3D-geometry. Fig. 10.14 gives examples of different edge types. *Contour edges* (Fig. 10.14, left) separate an object from the background (Strothotte and Schlechtweg, 2002, 214) or other objects.[59] These edges can be found by employing backface culling (see Section 10.3.1) so that all 'invisible' polygons are removed. Then all the edges which border on only one polygon are selected—these should be the edges where the second polygon, which shared them, has been removed by culling (Strothotte and Schlechtweg, 2002, 211, 216), i.e. they should lie on the border between the visible and the invisible geometry and thus define the outline of the object.

Sharp edges (Fig. 10.14, left centre) indicate discontinuities between the object's faces (Strothotte and Schlechtweg, 2002, 214). They can be found by comparing the surface normals of two adjacent polygons. A threshold for the angle (e.g. every angle below 100 degrees) between the two normals has to be defined in order to separate sharp edges (Strothotte and Schlechtweg, 2002, 216). Similarly *smooth edges* (Fig. 10.14, right centre), which indicate curvatures are defined by another threshold (e.g. every angle above 150 degrees) (Strothotte and Schlechtweg, 2002, 214, 216). *Triangulation edges* (Fig. 10.14, right) show the whole polygon mesh for the 'visible' polygons (Strothotte and Schlechtweg, 2002, 215).

The augmented reality reconstruction of St. Saviour's church in Ename displays sharp edges as purple lines which help the visitor comprehend the semi-transparent rendition of the church (Fig. 7.28). A similar result is presented in Freudenberg et al. (2001, 78 fig. 2). The use of edges for the visualisation of complemented parts in a reconstruction can be seen in Fig. 5.6, top right and bottom.

Finding edges is an important prerequisite not only for non-photorealistic renderings in a line drawing style, but is also incorporated into shaded renderings like Gooch- and cel-shading.

Gooch-shading

Gooch-shading is named after Amy and Bruce Gooch (Gooch et al., 1998) who developed it. It is based on conventions for hand-drawn technical illustrations

tation. The object can be transformed freely within the scene, but will retain its own coordinate system, which is bound up with the geometry of the object.

[59] The left image in Fig. 10.14 is not the best of examples. For instance at the foot of the microscope a line part clearly does not denote a contour edge while one edge of the microscope slide is missing. Additionally, it has to be assumed that the eyepiece (ocular) and other parts of the microscope consist of several separate objects.

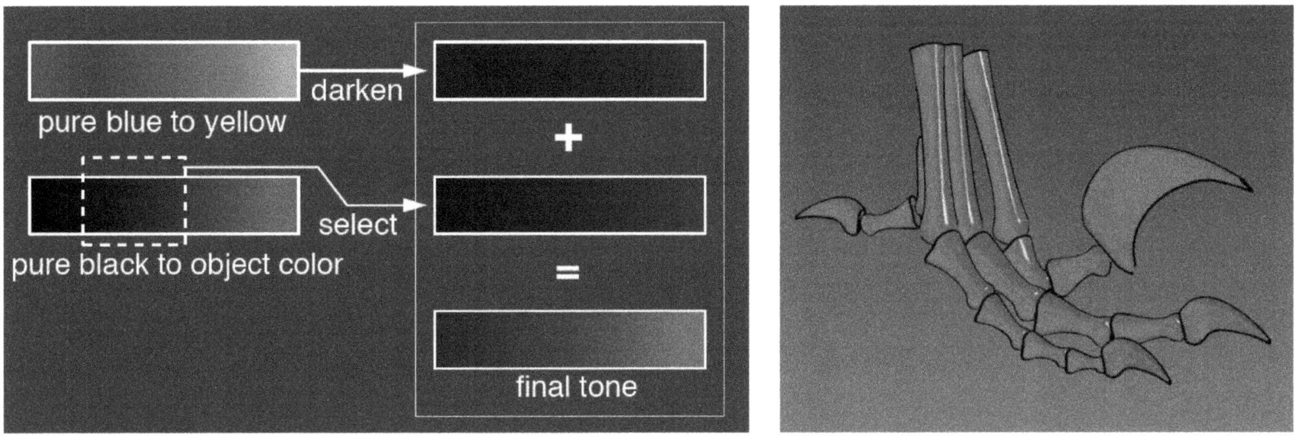

Figure 10.15: Gooch-shading. Left: colour ranges from cool blue to warm yellow to indicate the direction of the surface normal (top), from dark red to red to indicate the object colour and its shading (centre) and the combination of both ranges, i.e. from purple, over red, to orange (bottom). Right: model of a claw to which this shading has been applied. Highlights are indicated in white, edges are drawn in black (Gooch et al., 1998, 449 fig. 2, 451 fig. 6).

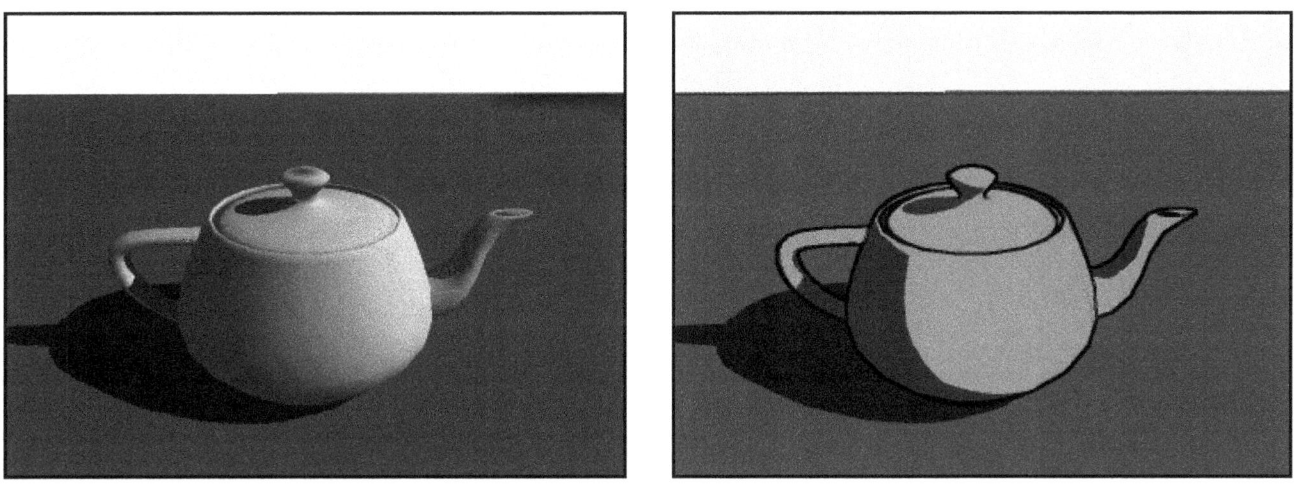

Figure 10.16: Cel-shading. Left: photo-realistic rendered teapot. Right: teapot in cel-shading (Decaudin, 1996, 3 fig. 1).

(Gooch et al., 1998, 447), namely that edge lines are drawn in black; matte objects are shaded in greyscale, but avoiding black and white extremes; warmth and coolness of colours is used to indicate the surface normal[60]; highlights are provided by a single light source; shadows are not drawn; metal objects are depicted as anisotropic (Gooch et al., 1998, 448).

In Gooch-shading pure black and white are reserved for outlines and highlights respectively (Gooch et al., 1998, 448). Shading and object colour are restricted to a narrow bandwidth which does not border on pure black or white (Fig. 10.15 left, centre range). To the object and shading colour scale a cool-to-warm colour scale is added (Gooch et al., 1998, 449); in the example provided by Gooch et al. (1998, 449) (Fig. 10.15 left, top range) it ranges from blue (cool) to yellow (warm).[61] This colour range indicates the direction of the normal in the rendered image (cf. Gooch et al., 1998, 448). The result from combining both colour ranges for a red object is a range from purple, over red, to orange (Fig. 10.15 left, bottom range).

Applying this shading principle makes slight changes in the surface orientation visible, even in shadowy areas (see Fig. 10.15, right). This makes this shading seem especially appropriate for rendering laser-scanned finds with subtle surface detail.[62]

[60] I am doubtful whether this is actually frequently the case.

[61] It should be noted that yellow is not *per se* warm and blue not *per se* cool as Gooch et al. (1998, 449) claim. All primary, secondary and tertiary colours exist in warm and cool variants.

[62] Despite its benefits regarding the comprehensibility of the rendered output, there is to my knowledge no archaeological project

Cel-shading or (car-)toon-shading

Cel-[63], cartoon- or toon-shading produces cartoon-like output from 3D-models (Fig. 10.16). This can be achieved by using two-dimensional G-buffer images (z-buffer and normal-buffer) to extract edges (Decaudin, 1996, 5 f.). The shading is generated by using a modified Phong shading, in which the diffuse colour has been copied to the ambient colour, the diffuse colour itself has been changed to black and the specular highlights are retained (Decaudin, 1996, 6). Shades and shadows are calculated using a shadow-buffer (Decaudin, 1996, 7 f.). There are several other ways to create the same effect.[64] For example, O'Rourke (2003, 260) describes an approach where the surface normal is used to find the outline of the model, which is then drawn in black, while the different shades are the result of posterisation of the usually smooth shade gradients.

Freudenberg et al. (2001, 98 fig. 3) give an example of a cel-shaded reconstruction of the building excavated on the Domplatz in Magdeburg.

10.3.2 Rendering of bi-cubic parametric surfaces

As has been mentioned before bi-cubic parametric surfaces can be transformed into polygon meshes before rendering, but the polygons can only achieve an approximation of the patch surface. Therefore it may be desirable to render the bi-cubic parametric surfaces directly in order to lose none of their mathematical precision.[65]

For polygon meshes, the vertices can be used to find the maximum and minimum extent of the polygon; the edges of the polygon are straight and the polygon's surface is flat. In contrast, bi-cubic patches exhibit none of these properties (Watt, 2000, 125 f.). It is, for example, possible that the silhouette edge of the patch intersects (from the viewer's perspective) with the boundary curve of the patch (see Fig. 10.17). In order to arrive at a surface representation a scan plane is projected through the patches (XZ scan plane in Fig. 10.17). The intersections with the edges and the patch surfaces are calculated by solving equations,

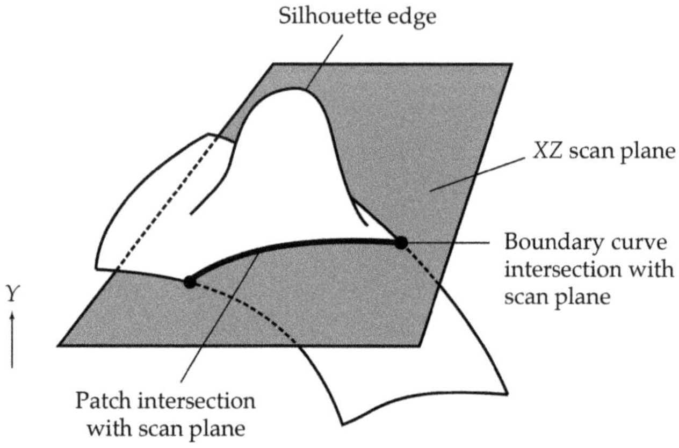

Figure 10.17: Scan line algorithm for patches (after Watt, 2000, 127 fig. 4.1).

then the scan plane is moved along the y-axis and the calculations are repeated (Watt, 2000, 126 f.). In this way x and z values of surface points (whereby only the points closest to the observer are registered) for each scan plane position on the y-axis, result in a surface representation (Watt, 2000, 127 f.).

10.3.3 Volume rendering

Volume rendering is a method to visualise voxel models. As has been described in Sections 10.2.1.1 and 10.2.1.3 voxel models can be the output of X-ray computed tomography scans, or can be mathematically generated like the excavation reconstructions by Nigro et al. (2002) and Lieberwirth (2008b).

As a space subdivision method, the voxels will represent different media (Watt, 2000, 373), i.e. in a medical example this could be skin, muscles and bones, or in an archaeological example, different strata in an excavation.[66]

There are different viewing options for this kind of model. On the one hand it might be desirable to view only objects with a specific property (e.g. all bronze finds or all strata containing charcoal); voxels possessing this specified property can be selected and represented in a specific way, for example, by setting their opacity values to 100% while all other voxels are

which used Gooch-shading.

[63]Short for 'celluloid'—the material onto which hand-drawn animations were painted in the past.

[64]One of them is described on http://en.wikipedia.org/wiki/Cel-shaded_animation. (Accessed: 18th of March 2011.)

[65]Possible applications for bi-cubic parametric surfaces have been suggested in Sections 10.2.1.1, 10.2.1.2 and 10.2.1.3.

[66]Problems occur if it has to be assumed that a voxel contains not only one material but a mix of materials and it is necessary to interpolate this mix (Watt, 2000, 376 f.). The example presented by Watt (2000, 378 f.) for classifying materials in CT scans is not readily applicable to archaeological data not dealing with human remains. Human CT scan results possess a natural order of layers (air → fat → soft tissue → bone), while archaeological data holds a much higher number of possibilities for neighbouring materials.

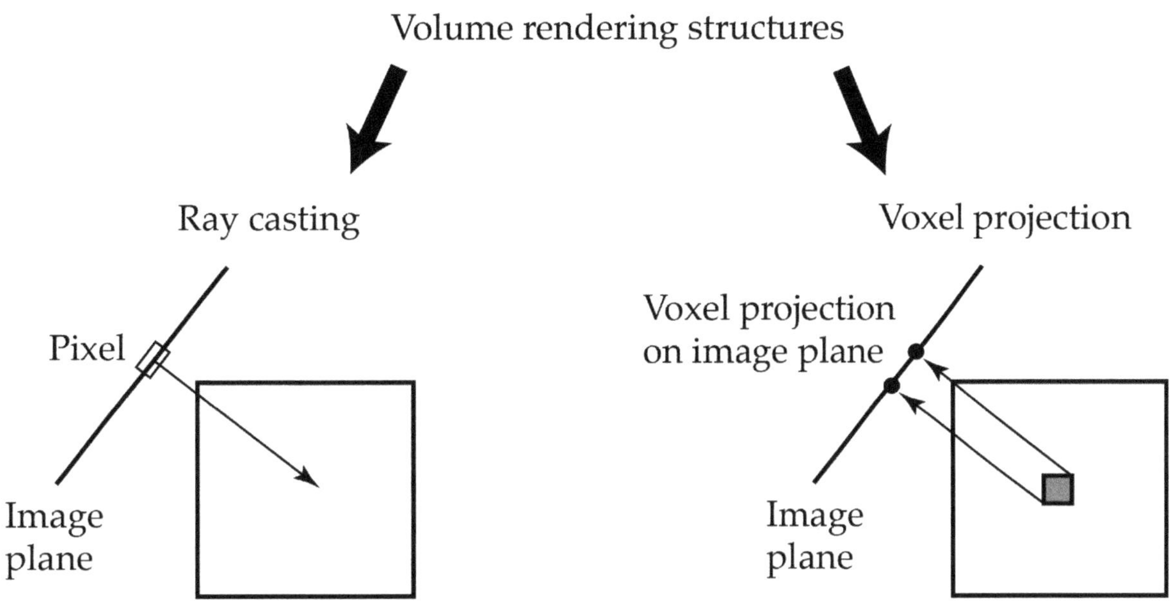

Figure 10.18: Volume rendering structures: ray casting or voxel projection. Ray casting: from each pixel at the image plane a ray is emitted and the voxel(s) it hits are recorded (see also Figs. 10.19 and 10.20). Voxel projection: the extent of the voxel is mapped onto (usually several) pixels on the image plane, this is called a *footprint* (after Watt, 2000, 385 fig. 13.12).

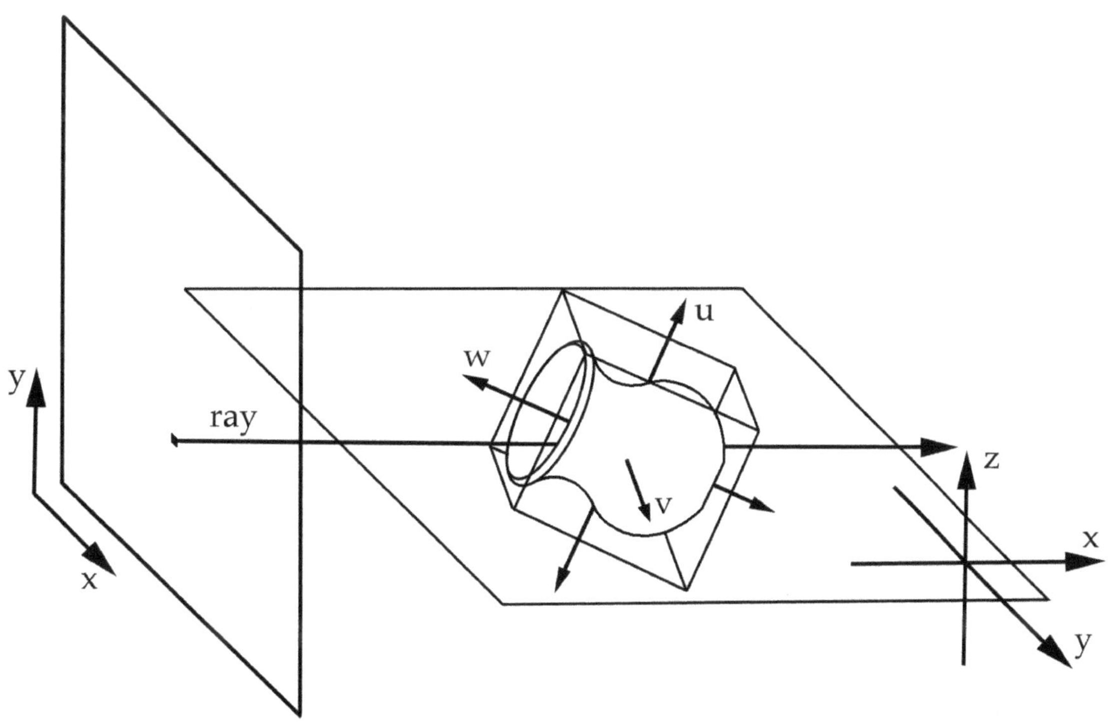

Figure 10.19: Ray casting: from each pixel on the image plane (x, y) a ray is sent into model space where it encounters the volume data and passes through it. The rotation of the volume data (u, v, w) will often not correspond to the orientation of the image plane. The values of the voxels which the ray encounters (cf. Fig. 10.20) are sampled. (Drawing by Joyce Wittur).

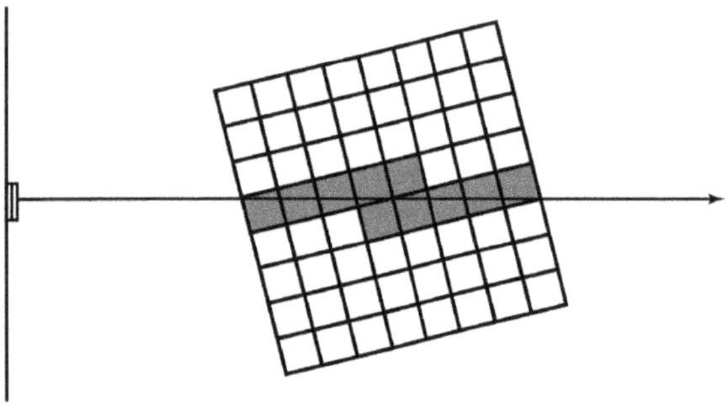

Figure 10.20: Ray Casting and resampling. The values of the voxels which have been traversed by a specific ray (grey) are sampled and have to be resampled according to the ray's traversal length through them. In this specific case the first to the fourth voxel on the left and right all contribute an equal amount of colour and opacity to the over-all appearance of the pixel. The remaining two voxels should contribute extremely little or nothing (they could contribute something due to the bilinear interpolation process, see Watt (2000, 387) for a detailed explanation, which would go too far at this point) (Watt, 2000, 379 fig. 13.6).

rendered 100% transparent (Watt, 2000, 373). On the other hand we may be interested in a single object (e.g. a stratum) in particular, but also in its relation to the surrounding matter, which in that case should be rendered semi-transparent (Watt, 2000, 375).

For both of these cases it may be preferable to (interpolate and) render a shaded surface—for example, by employing a Phong shader—because it helps in perceiving the shape of the object even though it is unrealistic to light up, e.g. a stratum, from all sides (Watt, 2000, 381 f.).

If all the voxels are to be visible, they all need to be transparent; Watt (2000, 377) calls this the '"semi-transparent gel" option', whereby individual values for transparency and colour can be applied to each set of voxels.

Irrespective of the final voxel colour and opacity values, two different approaches to rendering voxel data exist (Fig. 10.18): ray casting and voxel projection (Watt, 2000, 384), which will be introduced now.

10.3.3.1 Ray casting

From each pixel on the image plane a ray is cast, perpendicular to the plane, into the scene (Fig. 10.19), where it passes through the volume data (Watt, 2000, 385 f.). The voxels which are encountered by the ray have to be noted (Fig. 10.20) including their opacity and colour values (Watt, 2000, 386). If the ray cuts only across a corner of a voxel (resulting in a small distance on the ray's path), the contribution of this voxel to the all-over appearance in colour and opacity of the pixel would be small, in contrast to a voxel which is traversed diagonally by the ray (i.e. a long distance on the ray's path), which would therefore contribute more in colour and opacity to the final appearance of the pixel (Watt, 2000, 386 f.).[67]

In fact the compositing of the pixel's appearance should proceed from the point furthest away from the viewer towards the image plane in order to ensure that voxels with a high opacity, which lie close to the image plane, will occlude those voxels behind them (Watt, 2000, 379 f.).

10.3.3.2 Voxel projection

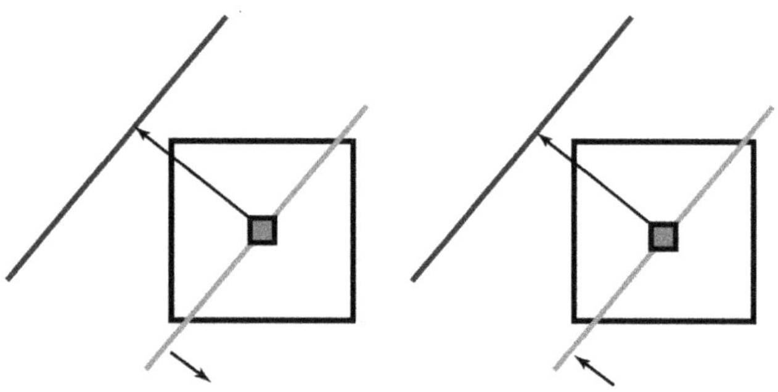

Figure 10.21: Voxel projection: a plane (red) is moved through the data (one voxel is marked in green), either from the front to the back (left) or from the back to the front (right). The results are finally mapped onto the image plane (blue) (after Watt, 2000, 385 fig. 13.12).

In voxel projection a plane (which is parallel to the image plane) is moved through the data (Fig. 10.21) either from the front to the back or from the back to the front. As the plane traverses from the front to the back through the data it accumulates colour *and* opacity values from each voxel it crosses, so that voxels closer to the image plane will occlude voxels further away. If the image plane moves from the back to the front only colour values have to be accumulated.

[67] Fig. 10.20 is far from being an ideal example for illustrating this issue, because the distance the ray traverses through each voxel is the same for all voxels. But assuming that the fifth grey voxel from the left would be minimally crossed by the ray, then its contribution to the final colour and opacity of the pixel on the image plane would be small. The first, second, third and fourth voxel from the left would each contribute more of their own appearance to the appearance of the pixel because of the longer traversal distance.

The result is projected onto the image plane (Watt, 2000, 388). The shape of the footprint which the voxels leave on the image plane, will be identical for all voxels in the data set (Watt, 2000, 385).

10.3.4 Mapping techniques

Texture mapping is used to make objects look more interesting or 'realistic' without adding too much computational cost because textured objects can be rendered with the computationally cheap Phong shader, but manage to avoid its trademark plastic appearance[68] (Watt, 2000, 223 f.).

Texture mapping usually implies mapping a two-dimensional texture onto a three-dimensional surface. The result has to be projected onto the two-dimensional screen space to be visible (Watt, 2000, 226). The simplest method is planar projection mapping (Fig. 10.22, 3–6 and 10) (O'Rourke, 2003, 121 f. cf. Watt, 2000, 228 ff.) which does not always yield the desired results (O'Rourke, 2003, 122).

It is possible to use an intermediate surface (cube, cylinder, sphere) onto which the texture map is projected before being projected onto the object surface, for example, in a 'shrinkwrap' mapping (O'Rourke, 2003, 122 f. Watt, 2000, 230 f.). Results of cubic, cylindrical and spherical projection mapping are illustrated in Fig. 10.22, 2, 7–9, 11.[69] The intermediate surface can be imagined as a surface surrounding the object onto which the texture will be finally mapped (Watt, 2000, 231).

The most frequent application areas for texture mapping are (Watt, 2000, 225):

1. Colour
2. Light mapping
3. Specular colour or environment mapping
4. Normal vector perturbance or bump mapping
5. Displacement mapping
6. Transparency

The different mapping applications and their results will be introduced in short.

[68]The 'plastic look' is, for example, apparent in the simple rendered model of the St. Lawrence's church interior (Ename): http://www.ename974.org/Ndl/pagina/NuhV/990905_1.html, last image on the page.

[69]Note that cylindrical mapping can also yield useful results for a spherical object, depending on the desired outcome: Fig. 10.22, 9 is the result of a shrinkwrap mapping.

10.3.4.1 Colour

Colour mapping is achieved by using a two-dimensional texture map (either a photograph, a procedurally generated map or an image created in a paint program) and combining it with the three-dimensional object in the scene (Watt, 2000, 224). The Phong shader changes the colour of the diffuse component according to the colour values dictated by the texture map when calculating the diffuse reflection coefficients (Watt, 2000, 224). It is also possible to change the specular coefficients, but this change only becomes visible if a highlight falls upon it (Watt, 2000, 225).

Colour mapping has been employed in all the three case studies, for example, photographs were used to texture the stones which were created with PhotoModeler in the Negotiating Avebury project (Gillings, 2000, 61–68), and a combination of photographed and painted textures was used for the walls and floors in the model of the Casa del Centenario (Coralini, 2007, 23). Colour mapping can also be employed to create a non-photorealistic impression as demonstrated by Kieven et al. (2000) (see Fig. 5.6, top left).

If photographs are to be used for texturing it may be necessary to rectify them (Barceló, 2000, 20). A sample image of a wood grain structure used to texture a wooden railing will not require rectification, while photographs of large mosaics which constitute an integral part of the reconstruction should certainly be rectified (cf. Section 6.3.2). It is not always necessary (or feasible) to use one single image for texture mapping. In many cases it is sufficient to use a small image and copy it over and over. This principle, called tiling, was used in the Casa del Centenario project (Coralini, 2007, 24) to create the mosaic floor in the atrium (see Fig. 6.10).

10.3.4.2 Light mapping

Light maps can be used to store pre-calculated lighting of a scene in two-dimensional maps; these can then be mapped onto the objects which thus appear to be illuminated correctly, without recourse to another computationally expensive rendering pass (Watt, 2000, 240).

The pre-calculation should be based on a view-independent global illumination method like radiosity or photon mapping. The amount and colour of

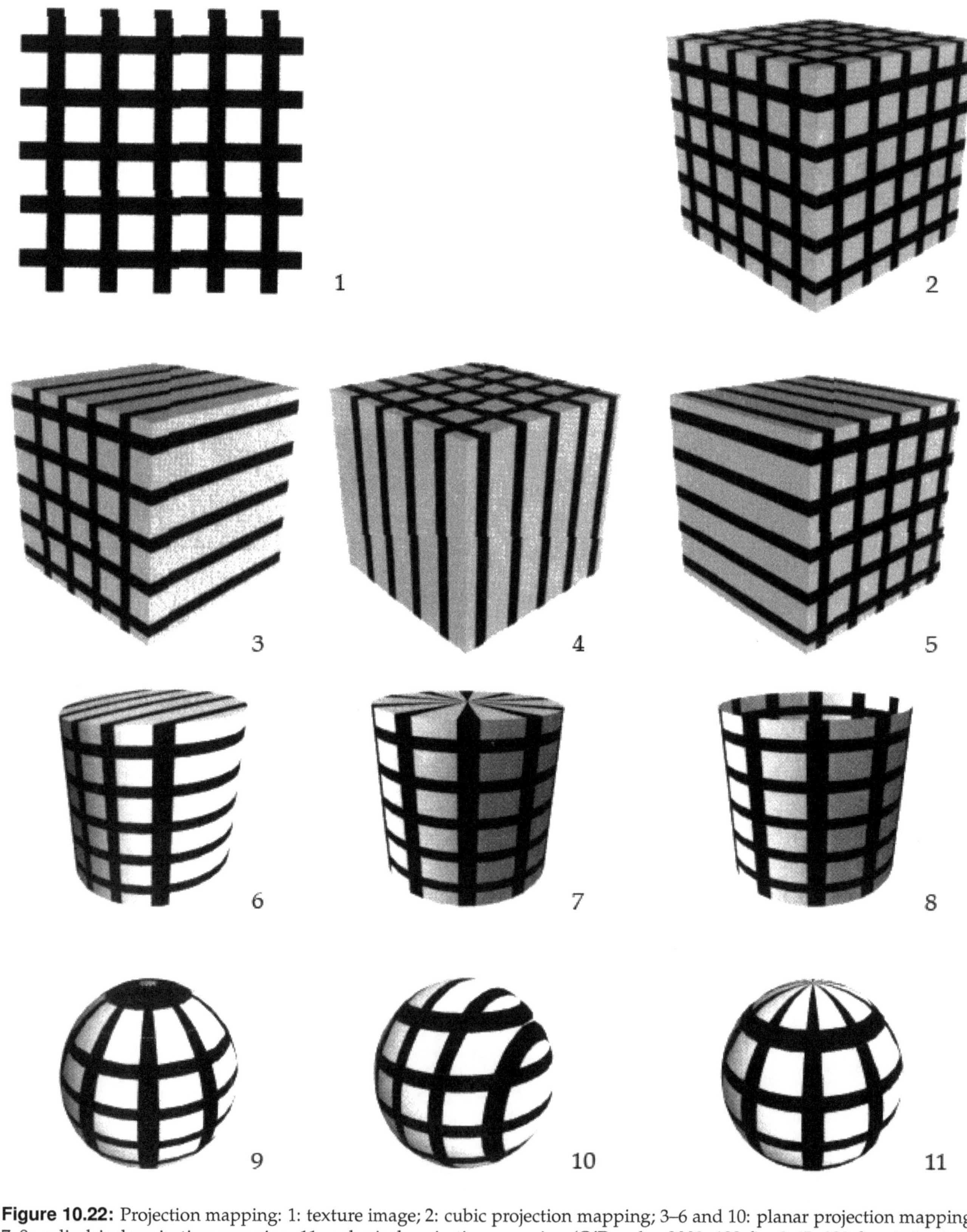

Figure 10.22: Projection mapping: 1: texture image; 2: cubic projection mapping; 3–6 and 10: planar projection mapping; 7–9: cylindrical projection mapping; 11: spherical projection mapping (O'Rourke, 2003, 122 fig. 3-65, 123 figs. 3-66, 3-67, 3-68). (Image copyright Michael O'Rourke, 2003).

Figure 10.23: Shadow map and shadow mapped scene: left: shadow maps were generated for the whole scene. The six images were combined into a shadow depth cube. Right: the shadow maps are applied and the scene is rendered using a Phong shader (which does not calculate shadows). The resultant shadows in the scene seem 'realistic'. The reflections on the teapot were generated through environment mapping (see Section 10.3.4.3) (Watt, 2000, colour plates fig. 18.8).

Figure 10.24: Environment mapped teapot and environment map: left: the environment mapped teapot from the scene in Fig. 10.23, right. Right: the cubic environment map was generated from six images which were rendered from the position of the teapot (Watt, 2000, colour plates figs. 18.8 and 18.9).

the reflected light which falls on the different surfaces in the scene is stored as a set of two-dimensional maps, which also include shadows (Watt, 2000, 240) and caustics if photon mapping is used. Fig. 10.23, left illustrates a shadow map in the form of a shadow depth cube. Fig. 10.23, right depicts the corresponding rendered scene.

The method of creating a two-dimensional light texture from a three-dimensional object in the scene is the reverse of texture mapping (see Section 10.3.4), whereby the desired result is a texture map which spans more than just one polygon (Watt, 2000, 241).

If both a light map and a texture map are to be applied, they can be stored separately or combined into one (Watt, 2000, 240). For the Casa del Centenario model the combination of a texture map and light map has been created by hand, i.e. the shade has been manually painted onto the colour texture maps for the objects (Coralini, 2007, 24).

10.3.4.3 Specular colour or environment mapping

Environment mapping imitates the mirror image of the surroundings on specular objects[70] as they are created by ray tracing. This is managed through pre-calculating a specular reflection map,[71] whereby each specular object needs its own environment map (Watt, 2000, 243).

Just as in light mapping, three-dimensional information has to be mapped onto two dimensions. For this purpose cubic, latitude-longitude and spheric environment mapping can all be employed (Watt, 2000, 243).

To create the environment map, the specular object, which will later receive the map, is removed from the scene and from its former position the incident illumination is calculated for the scene (Watt, 2000, 245). In a cubic mapping, which will suffice as an example here, it means that from a point in space[72] six maps are created, one for each of the four sides and one each for the top and bottom of the cube (Fig. 10.24, right).[73] If an object is situated in one of the corners

Figure 10.25: Procedurally generated bump map applied to a vase. Notice the smooth curve of the vase's outline which is *not* affected by the bump map (after Watt, 2000, colour plates, fig. 8.10 b).

of the cube it will map on three of the cube's map surfaces (Watt, 2000, 245 f.).

The resultant cubic environment map (as mentioned before, spherical environment maps are also possible) is, in a second step, mapped onto the specular object in the scene, which appears to reflect it (Fig. 10.24, left). This reflection can also be combined with, for example, Phong shading, so that only the specular component is influenced by the environment map while the diffuse component is unaffected (Watt, 2000, 249).[74]

Application areas are those scenes in which glass or burnished metal objects are included, and where a walk-through is planned for the presentation.[75]

[70]For instance a chrome sphere will mirror its surroundings.

[71]This method does not work for concave objects (Watt, 2000, 243).

[72]The point is identical with the point where the environment mapped object will later appear. This point is also the centre of the cube (Watt, 2000, 245 f.).

[73]This method can be used to include, i.e. 'matte', a computer generated specular object into a real scene. The six sides of the cube can in this case be created simply by using photographs of the real environment (Watt, 2000, 247).

[74]The effect of this approach can be envisaged by imagining an opaque glass marble which will retain its colour but also reflect the environment.

[75]It has to be remembered that each change of viewing position requires a renewed ray tracing pass, which is computationally very expensive.

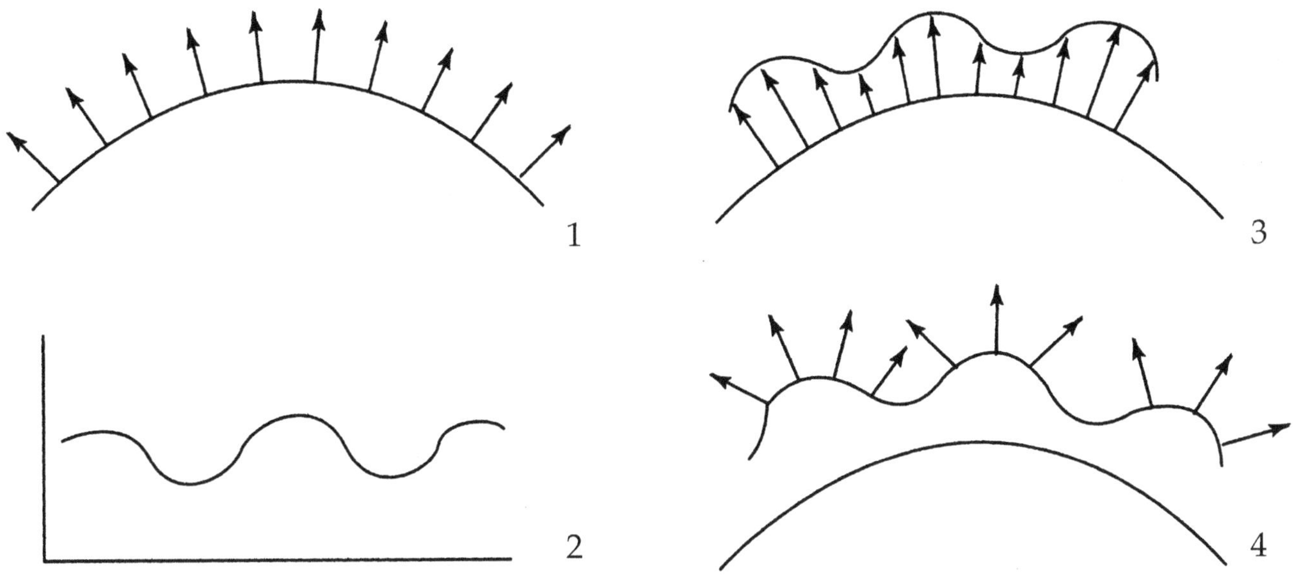

Figure 10.26: Bump mapping: 1: the original surface; 2: the perturbance indicated by a bump map; 3: lengthening or shortening of the surface normals according to the bump map; 4: the 'perturbed' surface with the new surface normals, the original surface still lies underneath (Bump mapping after Blinn (1978, 287 fig. 2, 291 fig. 9) © 1978 Association for Computing Machinery, Inc. Reprinted by permission. Image as published in Watt (2000, 237 fig. 8.11)).

10.3.4.4 Normal vector perturbance or bump mapping

Bump mapping has been developed by James F. Blinn in 1978 (Watt, 2000, 236). The effect achieved is to create the illusion of an uneven surface. The effect is only an illusion because the outline or the cross-section of the object onto which the bump map has been mapped will continue to appear smooth (Fig. 10.25) (Watt, 2000, 236). The bump map affects only the local reflection model and thus creates the impression of roughness, but does not change the geometry of the object (Watt, 2000, 236). This constitutes the main difference from displacement mapping, where the geometry of the object is indeed changed (see Section 10.3.4.5).

Fig. 10.26 illustrates how the process works: image 1 shows the original object surface with its surface normals (Watt, 2000, 236). The two-dimensional bump map (usually a greyscale image whereby white and black constitute the minimal and maximal perturbance respectively) indicates the amount of perturbance. This is depicted in image 2 in form of a wavy line. The normals of the original surface are now lengthened or shortened according to the values indicated by the bump map (image 3) and new normals for the perturbed surface have to be calculated (image 4). The smooth original object surface still lies underneath (Watt, 2000, 237).

Bump mapping was, for example, used in the Casa del Centenario project to create the impression of granularity for the wooden railing in the peristyle (Section 6.3.2).

10.3.4.5 Displacement mapping

Displacement mapping works similarly to bump mapping[76] except that in this case the surface of the object is actually altered by the displacement map, which is not easy to implement (Watt, 2000, 225). As in bump mapping, the starting point is a two-dimensional greyscale texture wherein white or light pixels are interpreted as high displacements, and dark or black pixels as little or no displacement (O'Rourke, 2003, 130). The amount of detail which can be achieved with a displacement map is obviously dependent on the number of the object's vertices because only they can be affected by the displacement (O'Rourke, 2003, 216, 216 fig. 5-24). The fact that the surface has actually been perturbed is visible on the outline of the object which is now not smooth but uneven (O'Rourke, 2003, 130 f.).

[76]This apparently results in similar application areas for these methods.

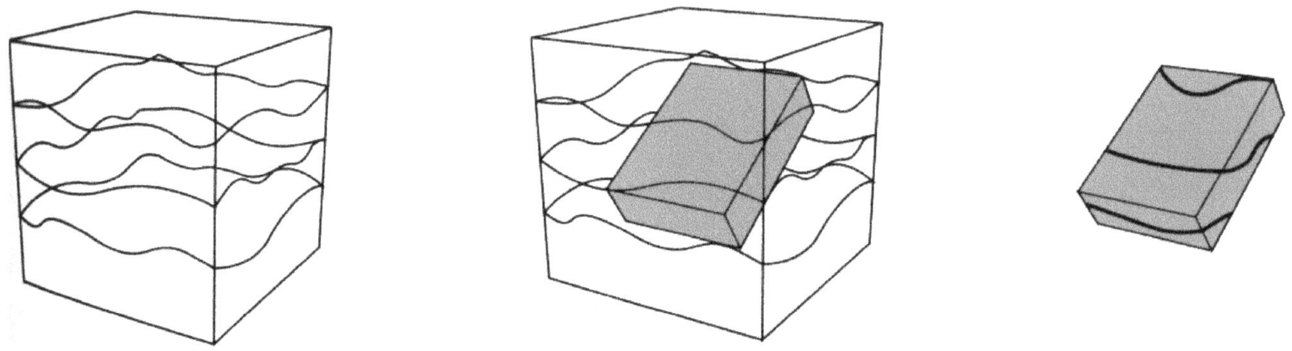

Figure 10.27: Applying a 3D-texture to an object: left: the solid texture; centre: the object 'immersed' in the texture; right: the textured object (O'Rourke, 2003, 135 fig. 3-87). (Image copyright Michael O'Rourke, 2003).

10.3.4.6 Transparency

Transparency mapping is used to differentiate between transparent and opaque areas (for example, an etching on a glass surface creates an opaque area) (Watt, 2000, 225). Once again a greyscale map is used to indicate the degree of transparency (white) or opacity (black) of the surface (O'Rourke, 2003, 128 f.).

10.3.5 Three-dimensional textures

Three-dimensional textures circumvent the mapping problems observed with two-dimensional textures (see Section 10.3.4 and Fig. 10.22). They are especially useful when wood grain (Fig. 10.28) or natural stone textures are to be created. The actual mapping process (Fig. 10.27) can be envisaged as carving the object out of a three-dimensional texture block or as immersing the object into the texture (Watt, 2000, 251; O'Rourke, 2003, 135).

The texture should preferably be procedurally generated, even though there are other ways to create a three-dimensional texture—but they are computationally expensive (Watt, 2000, 251). This prerequisite somewhat limits the number of possible solid textures (Watt, 2000, 251), because they rely on noise, turbulence or fractal mathematics (Watt, 2000, 251–254, O'Rourke, 2003, 138). Three-dimensional textures are not restricted to colour as they can also be used for bump mapping or to control transparency (O'Rourke, 2003, 137).

Archaeological applications lie in the texturing of complex shapes, like statues, with wood grain or marble textures. A three-dimensional texture has been used for the hypothetical thatched roof in the Navan project (Reilly, 1992, 156) (see Section 4.1).

Figure 10.28: Procedurally generated 3D wood grain texture applied to a vase (after Watt, 2000, on CD).

10.3.6 Geometric shadows

Geometric shadows are unlike the shadows which result from global illumination concepts; instead, they are rather comparable to texture maps (Watt, 2000, 263). A shadow map is calculated for each light source in the scene and accessed during rendering (Watt, 2000, 263). Shadows are important for judging the spatial relationship between objects, as well as the position and size of the light source (Watt, 2000, 263). To create shadows we need to compute their shape and their light intensity (they are usually not completely dark due to indirect illumination, so radiosity can be used to calculate this value) (Watt, 2000, 263).

Shadows can be hard-edged (if the light source is a point light) consisting almost only of the umbra. Conversely, large light sources will produce soft shadows with a considerable penumbra area (Watt, 2000, 264).

Hard-edged shadows can be calculated in the following simple ways (Watt, 2000, 265):

By projection: a polygon is projected two-dimensionally onto a surface. The centre of projection is the light source (Watt, 2000, 265 f.).

As hidden area: the shadow is an area which is hidden from the light source, so that hidden surface algorithms can be applied (Watt, 2000, 271 ff.).

For the Casa del Centenario project (see Section 6.3.2) shadows were calculated in this manner and included in the scene in the shape of semi-transparent grey polygons (Coralini, 2007, 24).

To generate soft-edged shadows, shadow volumes can be employed (Watt, 2000, 268). The shadow volume is also a projection of a polygon or object, but this time it has to be imagined as a three-dimensional (infinite) volume in space (Watt, 2000, 268). For this, the silhouette edge of the object has to be found, by separating the polygons which receive light from those which do not (Watt, 2000, 268).

When the shadow volume is generated it is not rendered, but instead used to define whether a point lies in shadow (Watt, 2000, 268). This is accomplished by using a Z-buffer algorithm. From a pixel on the screen a perpendicular ray is sent into the scene and it is noted whether or not it encounters a shadow volume and also if the point which has to be rendered lies within or behind the shadow volume. If the point lies within the volume it is in shadow, if it lies behind the volume it is in the light (Watt, 2000, 269 f.).

10.4 Representation

After the model shape has been created the representation of the model has to be considered.

The decision whether to present the reconstruction as a still image, an animation or an interactive model has influence on how the data should be treated (Section 10.4.1). Further constraints can be added by the way the representation is to be distributed and presented (Section 10.4.2).

Also, general decisions, which have an influence on how the model is perceived, have to be taken. They relate to the coherence of the depicted scene: i.e. which period or even year does the scene represent, which season and what time of day (Hodgson, 2001, 5). These matters can have an influence on people's activities, their appearance as well as the appearance of the flora and fauna (Hodgson, 2001, 11) and not least on the reconstructed building or object itself. Another factor in the creation of a populated scene is whether an event or every-day life is to be depicted (Hodgson, 2001, 5), though it seems more often the case that people and animals (plants being usually only present in the form of grass with the occasional interspersed tree) are not included in the scene, but rather that the reconstructed object or building takes precedence (cf. Krasniewicz, 2000, 164).[77] The creators of the model bring preconceptions with them which influence the appearance of the scene. This issue encompasses the whole bandwidth, ranging from gender roles (Mehling, 2002, 89) to depicting certain periods as tidy whereas others are dominated by filth (Maschek et al., 2010, 450).

General considerations also include whether and how uncertainties are indicated (Pletinckx, 2008, 5, 20 f.); several methods (e.g. transparency, wireframe, augmented reality, colour coding and NPR) have been discussed in Section 5.9.1.

Decisions concerning the final appearance of the model are comprised within the general considerations as well. The rendition can take a very simplistic form: for instance an untextured model which is rendered with a Phong shader (e.g. Daniels, 1997, frontispiece or Maschek et al., 2010, 453, 452 fig. 4, top), but photo-realistic (e.g. Eiteljorg, 2000, fig. 1 or Grellert and Svenshon, 2010, 190 fig. 3) or non-photo-realistic images are possible also (e.g. Isenberg, 1999, 20 fig. 11 or Lehmann and Haarlammert, 2010, 175 fig. 8).

10.4.1 Method of presentation

As mentioned above the finished reconstruction can be presented in several ways, as an *image*, an *anima-*

[77] This is even the case in models which have been created for phenomenological studies (see Chapter 8), where the encounter with other people and the environment should play a predominant role.

tion or an *interactive model*. Assets and drawbacks of various presentation methods have been discussed in Section 5.5.5.

10.4.1.1 Images

An image is the simplest form of output and places the least constraints on the methods used to create it. As always the subject of the image should be chosen carefully and cluttering avoided (Hodgson, 2001, 5).

The rendering methods depend on the model type (surface or solid), but should also be chosen with the objects of the scene in mind: a reliquary with its gold and silver surface, gems and pearls benefits from a ray-tracing approach, a collection of Venetian glass vessels may even be a case for photon mapping to do justice to the caustic effects, while the interior of a cave will probably be best rendered using radiosity.[78] All three approaches are very computationally expensive (though time is no constraint where just a single image is involved) and deliver stunning photo-realistic results. Obviously computationally less expensive shaders (like Gouraud and Phong shading) can be applied just as well as non-photorealistic approaches. For the latter, even 2D-image processing techniques can be employed. The question is which rendering technique will provide the desired results.

The same applies to rendering voxel data: does the primary interest lie in the surfaces or in viewing a volume in relation to other volumes in the dataset? The challenge here might actually be in finding the right angle and position for the clipping planes (see Section 10.3.1) in order to attain the desired visualisation result.

10.4.1.2 Animations

An animation has to provide the viewer with 24 or 25 images per second (Grellert, 2007, 179) and usually includes movement—either the camera moves (as in a walk-through or a fly-through) or parts of the scene move (i.e. animated people, animals or objects) or both. That this is a difficult situation for employing view-dependent and computationally expensive rendering methods, like ray-tracing, is obvious.[79] But even view-independent global illumination concepts (e.g. radiosity) are not ideally suited for all applications. Only one rendering run suffices for simple walk-through or fly-through animations (because of the view-independency), but as soon as parts of the scene are supposed to move (e.g. a door is shut) light and shadow in the scene change, and so a new rendering run becomes necessary for each animated image. This is often not a feasible approach. Instead local illumination concepts (Section 10.3.1.1), which are computationally much cheaper, in combination with several pre-calculated maps (Section 10.3.4), e.g. for light and shadow, could be used.

10.4.1.3 Interactive models

Computational ease is paramount in this application area because the represented images[80] have to be computed in run-time—though there are some exceptions.

The main question is which kind of interactivity is intended: several different approaches will be introduced in the following, progressing from simple to more complex approaches.

Bubble worlds

Bubble worlds provide a view on an object which can be rotated and inspected from all sides or enable views of the environment from a single standpoint.[81] The model is built either from photographs (Krasniewicz, 2000, 164, 168)[82] or computed from 3D-data (Bohuslav et al., 2002, 209).[83] In addition to viewing the object or the scenery it is also possible to include 'hot spots' in the scene,[84] these are links which can be used either to transport the user into the next bubble world or to show additional information about the part of the scene upon which the user has clicked.

[78]For all three approaches see Section 10.3.1.2.

[79]This applies for ray-tracing of CSG and voxel models, too. They should be transformed into more applicable model types, e.g. polygon meshes.

[80]The number of images is equivalent to that for animations, i.e. 24 to 25 images per second.

[81]The bubble world can be envisaged as being similar to a snow globe. The viewer is either able to look at the scene inside the snow globe from all sides or is stuck at the centre of the snow globe (as part of the scene) and can view from this vantage point the scene around him- or herself.

[82]Cummings (2000, 16) uses panoramas created from several photographs. This is approximately the same as a bubble world only with a cylindrical projection instead of a spherical one.

[83]QuickTime VR, which is a bubble world format, was extensively used in the Ename 974 project, see, for example, Callebaut et al. (2000) http://www.ename974.org/Ndl/pagina/NuhV/990905_1.html.

[84]Pletinckx et al. (2003, 227, 228 fig. 3) provide an example of such hotspots in TimeScope 3; clicking on the highlighted areas leads to further information about them.

View-independent and even view-dependent computationally intensive rendering methods are applicable for this kind of model because only a limited number of images are necessary to create the bubble world.[85]

Interactive animated walk-throughs

They give the user the opportunity to explore a model (e.g. a city or a larger building) on predefined routes (Lock, 2003, 153). The user can decide which way he or she wants to go, and movement along the chosen path will be simulated by showing a short pre-calculated animation. At its end the user can choose again in which direction to proceed.

The constraints are the same as for usual animations.

Models with additional information

Letting users access additional information by clicking on model parts is also a kind of interaction.[86] It can be combined with almost all presentation methods, even a 2D-image can provide the necessary backdrop for this kind of interactivity.[87] Only animations pose a problem, because they follow a predetermined course at a given pace and are usually not meant to be interrupted.

There is also the possibility of combining a text-based menu (listing certain periods or monument parts) with images and 3D-data. Selecting an entry from the text menu results in highlighting a certain part in the 3D-model or displaying appropriate 2D-plans (cf. Meyer et al., 2007, 356). The difficulty lies in displaying the selected part in the three-dimensional model in such a way that the user can immediately identify it[88] and comprehend its position in the model as well as its spatial relationship with other parts[89] (Strothotte and Schlechtweg, 2002, 209).

To enable access to additional information, it must be possible to link the model or parts of the presentation to external data[90].

Real-time interactive models

Virtual reality has been defined as being real-time, interactive and immersive (Frischer et al., 2002, 10). The last property depends on the way the model is experienced (see Section 10.4.2). It could be argued that similar constraints exist for augmented reality only that there the real environment is enhanced by virtual additions. Therefore models which are discussed in this section could either fall under the heading of virtual reality, if they are displayed in the appropriate way, or otherwise not (cf. Lock, 2003, 153). Immersive or almost immersive experiences are provided by Head Mounted Displays, CAVEs and Virtual Environment Theatres (see Section 10.4.2).

One of the simplest ways to generate a real-time interactive model is by using VRML or its successor X3D[91].

The constraints placed upon the models are high. The interaction with the model (walking around, using objects) has to be displayed smoothly so that the computer has to calculate each image within 1/25 of a second. Therefore no global illumination concepts can be used; instead, local illumination concepts and texture mapping[92] have to suffice.

10.4.2 Distribution, access and display

To achieve the aims envisaged for the virtual reconstruction, specific presentation media are often required. Several different media have been referred to already in the case studies. They include mobile devices, a virtual theatre[93] and a virtual TV-production set for the Casa del Centenario project (Section 6.3.3), as well as a variety of kiosks in the Ename 974 project (Section 7.3.1). These devices have been specially

[85] Simpler shaders can always be used.

[86] View-dependent pre-selection of additional information by the information system is also possible: the choice of further information can then be made available to the user via a menu, see, for example, the Casa del Centenario project (Section 6.3.3). This way, the generation of interactive images, panoramas, *etc.* is avoided.

[87] Interactive maps were, for example, used in the Ename 974 project (Pletinckx et al., 2004, 235). That bubble worlds can give access to additional information has been mentioned above.

[88] Strothotte and Schlechtweg (2002, 209) recommend highlighting, labelling or enlarging the object in question.

[89] It may be necessary to rotate the model or to cut away other model parts to make the selected object visible. This has to be done in such a manner that the user can comprehend what is happening, e.g. via an animation (Strothotte and Schlechtweg, 2002, 209).

[90] This data can be stored in database, but it is not a prerequisite.

[91] See also Footnote 118 on Page 43.

[92] Whereby the application of multiple textures and 3D-textures is supported in X3D. See http://www.web3d.org/x3d/specifications/ISO-IEC-19775-1.2-X3D-AbstractSpecification/Part01/components/texturing.html#Concepts section 18.2.4 and http://www.web3d.org/x3d/specifications/ISO-IEC-19775-1.2-X3D-AbstractSpecification/Part01/components/texture3D.html#3DTextureconcepts section 33.2.2. (Both accessed: 14th of March 2011.)

[93] This kind of equipment is also called Virtual Environment Theatre (VET) (Lock, 2003, 154); some examples of such theatres are mentioned in Frischer et al. (2002, 9).

built for the purpose of presenting virtual models or for acting as information systems.[94]

Other special equipment for displaying virtual models are CAVEs[95] they consist of a cube-shaped projection room onto whose walls, floor and ceiling the virtual environment is projected (Frischer et al., 2002, 9). CAVEs have been used by Kadobayashi et al. (1999)[96] to visualise a Japanese village, as well as in the ARCHAVE project which is concerned with the great temple of Petra (Vote et al., 2002, 84).

Also, Head Mounted Displays (HMD) (Lock, 2003, 153) have been used to show archaeological models, for example, an early (see Frischer et al., 2002, 11) and a recent[97] reconstruction of Pompeii.

A number of 3D-viewers can be used for this purpose too. For example, the scanned gravestones of the Jewish cemetery in Worms[98] are displayed on a passive stereo board[99] at the IWR in Heidelberg.

Standard PCs are used more often than special equipment for displaying virtual models. This applies to models which are distributed on CD-ROMs[100] or DVDs[101] or are available over the internet[102].

In these cases, the hard- and software of the end-users as well as—with regard to internet applications—the available bandwidth of the internet connection, all place constraints upon the complexity of the scenes and respectively on the size of the models.

10.5 Data preservation

During the process of creating a 3D-visualisation a considerable amount of data is produced and should be archived for later reference. This comprises the data and model basis, the documentation of all the interpretational steps, the geometric model and the final representational output[103], i.e. a rendered image, an animation or an interactive application (Pletinckx, 2008, 5, 23 f.).

All this information has to be stored in a way which guarantees the accessibility and displayability of the data at a later date. This implies a considerable preservational effort:

A running computer system consists of a combination of hardware (the computer together with input and output devices), operating system, software and data (documentation, geometric model and presentation) (Margulies, 2009, 326, 335 f., 368), whereby the data is often stored on an external data storage medium (e.g. CD-ROM, tape or disk) (AHDS, 2002, section 6.2)[104].

Technology preservation, that is keeping and maintaining the original hard- and software, is difficult and rather risky (AHDS, 2002, section 6.2). Broken legacy hardware is difficult, if not impossible, to replace and the replacement may not work with the original software (Margulies, 2009, 338). The same accounts for problems with the original software—in the worst case it becomes impossible to operate the hardware (Margulies, 2009, 339). In both cases the whole system can break down (Margulies, 2009, 338 f.).

Also, the data storage media do deteriorate over time. Magnetic media store data safely up to 5–10 years, optical media up to 30 years (AHDS, 2002, section 6.2). To prevent the eventual loss of the information it is necessary to conduct a data refreshment (AHDS, 2002, section 6.2). *Data refreshment* describes the process of copying the data from one medium to another medium of the same kind (Margulies, 2009, 340, 345). If data is transferred from one medium (e.g. a tape) to another medium (e.g. a CD-ROM) this process is called *data replication*. This necessitates in some cases a *repackaging* of the data to keep it in readable form (Margulies, 2009, 346).[105]

The rapid development of hard- and software creates further problems (Christodoulou et al., 2006, 61). New software versions cannot read data written by

[94]Some of the component parts may be off-the-shelf hardware, especially in the case of the kiosks, but the final result has been tuned to fulfil a specific purpose. (See Section 7.3.1.)

[95]CAVE Automatic Virtual Environment (Lock, 2003, 154).

[96]See also Kadobayashi et al. (2000).

[97]A reconstruction by the ETH Zürich: http://www.g-o.de/wissen-aktuell-8094-2008-04-16.html. (Accessed: 14th of March 2011.)

[98]http://www.iwr.uni-heidelberg.de/groups/ngg/Worms/ (Accessed: 7th of August 2012.).

[99]http://www.iwr.uni-heidelberg.de/groups/ngg/equipment.php?L=0 (Accessed: 7th of August 2012.).

[100]This has been planned for the Castle of Ferrara (Forte and Borra, 2000, 237) and the Ename 974 project (Pletinckx et al., 2000, 48).

[101]The intention to publish the excavation report of Tsoungiza including virtual models on DVD is stated by Sanders (2000, 43).

[102]For instance the Peel Gap model was accessible via the internet (Gillings, 1999, 249).

[103]The latter two may be identical: for instance, a VRML model includes the geometric description and can be interactively explored at the same time.

[104]http://guides.archaeologydataservice.ac.uk/g2gp/Vr_6-2

[105]None of these methods changes the data contents (Margulies, 2009, 345 f.).

older versions. New hardware is incompatible with old data storage media. New software does not run on old hardware and vice versa (Christodoulou et al., 2006, 61; Margulies, 2009, 353, 355). Therefore it may be advisable to establish a two-year maintenance cycle (Pletinckx, 2008, 23).

Two possibilities for assuring that the data can still be accessed in future exist: *data migration* and *emulation*.

10.5.1 Data migration

Data migration is the process of transferring data from one format or structure into another (AHDS, 2002, section 6.2). In particular, binary formats cannot be decoded if the structure of the document is unknown (Margulies, 2009, 355). Therefore proprietary formats should be avoided—they are dependent on a specific software and cannot be easily converted into other formats (Margulies, 2009, 355).[106] Instead international or open standard formats should be preferred (Göldner, 2010, 121; AHDS, 2002, section 6.2). Problems can occur with container formats (e.g. TIFF which can contain data in a variety of formats) or files which include content of different formats (e.g. audio-visual files will contain audio data in one encoding and video content in another) (Margulies, 2009, 356, 358). In some cases the source format and the target format will have different properties so that the data content cannot be displayed in its original mode after data migration (Margulies, 2009, 357).[107] This is particularly problematic for the documentation and the geometric model, less for the presentational output which is more frequently stored in standard formats (Christodoulou et al., 2006, 61). All data migration processes should be documented, so that decisions and changes can be reconstructed (AHDS, 2002, section 6.2).

Data migration is the most suitable method for dealing with a limited number of files (Christodoulou et al., 2006, 62).

10.5.2 Emulation

Emulation mimics the workings of old hardware and software on new hardware and software (Margulies, 2009, 362), so that a data migration process is not required and the data can stay in their original format (Margulies, 2009, 362; Christodoulou et al., 2006, 62). Emulation is more appropriate if a data loss has to be avoided; nevertheless differences in the functionality and display of the data can occur (Margulies, 2009, 362 f.). The creation of an emulation is very costly because new software has to be programmed for the purpose (Margulies, 2009, 367), so that this process is only feasible if a large amount of data profits from the effort (Christodoulou et al., 2006, 62). The amount to which the emulation has to simulate the original system is variable. It can suffice to emulate only the hardware, but there might be cases where the hardware and the original operating system (or even hardware, operating system plus original software) have to be emulated to ensure that the data can be displayed and used appropriately (Margulies, 2009, 368). To guarantee that the emulation does faithfully represent the original system, careful planning and documentation are necessary (AHDS, 2002, section 6.2).

The difficulties which arise from data preservation can be taken as a reminder to think about storage formats at the very beginning of the project so as to ensure sustainability of the created model and the associated data.

[106] Cf. Footnote 147 on page 120, concerning problems with proprietary software at Ename.

[107] In other cases improvements concerning the usability of the data can occur (AHDS, 2002, section 6.2).

Chapter 11

Conclusion

This dissertation approaches applied ethical issues in virtual reconstruction creation from three different perspectives. First, an overview over applied ethical aspects is provided in Chapter 5. Second, Chapters 6, 7 and 8 supply detailed insight into how different projects were approached, whereby the selected case studies were chosen because they represent a wide spectrum of application areas. Finally, the reconstruction process from the sources to the finished model (Chapters 9 and 10) is elaborated, revealing constraints created by the available material as well as the employed methods and methodologies.

There is the danger of overrating or underrating the potential of virtual models. They are neither authentic nor objective (see Section 5.4) and present only a reduced and interpreted reflection of a possible past, but this does not mean that they cannot be fruitfully employed to tackle research questions or be effectively applied in communication tasks. Rather the opposite is the case.[1]

What this thesis reveals is that there is no generally applicable rule for generating an 'ethically correct' model, just as there is no all-purpose model (for structural analysis *and* documentation *and* communication to the general public and specialists, *etc.*).

Instead, several important notions have emerged which are heavily interconnected, i.e. decisions taken at a particular point will have influence on others:

11.1 Definition of a clear goal

Simply 'producing a visually stunning reconstruction model' cannot be seen as a valid goal in an applied ethical context. A goal should include, as a minimum, a clearly defined application area, the target audience and consideration of the steps which are necessary to reach that goal.

That the goals for virtual models can be very different becomes apparent when looking at the three case studies. They range from:

- aiding the restorers in their work and including the reconstruction in an information system intended for both a lay and a professional audience interested in Pompeii (Section 6.2.1), over:
- using the model as a subsidiary part of an information system for the local population, by presenting via many updates the ongoing research and interpretational work at St. Lawrence's church in Ename (Sections 7.3.1 and 7.3.3.1), to:
- letting archaeologists assess different hypotheses on a phenomenological basis for a neolithic monument complex at different stages in time, while at the same time using the model of the extant remains as a documentation of the modern site (Section 8.2.1).

From these examples it becomes apparent that sometimes two or more aims co-exist. In such cases, it is essential that the different goals are compatible.

Useful aims for reconstruction models are in general a) communication and b) research.

Communication can fulfil two aims, either *imparting information* (in which case the recipient is more or less passive) or *provide the basis for a discussion* (where the user has to be active). Imparting information is usually the aim for museum exhibitions, educational software, *etc.*, i.e. in cases where specialists address a lay audience,[2] or a

[1] Model makers should nevertheless be aware of the limitations of the virtual models.

[2] For example, in the Casa del Centenario project and in the Ename 974 project.

specialist audience which is unfamiliar with this particular information. Providing a basis for discussion, on the other hand, is usually aimed at specialists who are acquainted with the site.[3] These aims can also dictate different representation methods. For example, it could be shown that a representation which resembles a quickly drawn sketch made by hand provokes discussion much more than a tidy line drawing.[4] Also, presentation methods (film, still images, interactive model, *etc.*) may depend on the communication goals.

In both cases it can be useful to define the target audience as precisely as possible (or otherwise provide a system which can deal with different target audiences). Important factors can be: age, computer literacy, previous knowledge on the subject, special interests and time constraints.

It is also important to consider the role the reconstruction will play in the communication process. Is it supposed to stand alone, provide access to further information (e.g. in the Casa del Centenario project) or does it play a subsidiary role among other communication modes.[5]

Research can take place either with the model as the only research tool[6] or the reconstruction can be combined with other applications, as, for example, a Geographical Information System[7] or a database.

The type of research has to be considered also: will the model be used for structural analysis or for the visual exploration of a site? These decisions have an effect on the model type required: in the first case a solid model is needed while in the second a surface model suffices (cf. Section 10.2.1). The visual representation can also be influenced by the research goals. While a visual examination may demand a photo-realistic model, a simulation (for example, an analysis of the results an earthquake could have on the excavated ruins of Pompeii) does not need such an approach. In some cases it can be helpful if not only is a visual output produced, but also other senses are addressed by the model[8] (e.g. by audible or haptic output); such endeavours may obviously call for special equipment.

Research models also need to be manipulated in order to visualise different hypotheses, set up new simulation scenarios, *etc.* It is important to consider in this context whether these changes can be effected by the researchers themselves or whether computer experts have to be consulted.

11.2 Careful planning

Careful planning of all the modelling phases[9] is very important in order to reach the envisaged goal. All of the modelling phases, sources, interpretations, decisions and methods (including formats, software and hardware used in the visualisation process) should be documented for future reference. In the beginning, all the available sources concerning the object or site have to be collected and evaluated; during these activities it may become apparent that further information has to be generated.[10] The available sources will make it necessary to involve experts from different fields in their evaluation and interpretation. Further know-how is necessary for generating the computer-visualisation and connecting it with additional applications (GIS, database, *etc.*). This may not only involve the work of computer specialists, but also requires proficiency in presentation issues,[11] particularly (but not exclusively) if the model is not a stand-alone application but is meant to be viewed in combination with other contents.

Concerning the generation and presentation of the model the question of the necessary equipment has to be posed. Presentation methods can range from a simple PC, head mounted displays, to CAVEs, which in turn require more or less sophisticated equipment. Besides models which are designed for being viewed in a special location (e.g. a museum) others are meant to be used at home or in class. Here the way the

[3] Discussing the various hypotheses was certainly one of the aims in the Negotiating Avebury project, but also the models of the Casa del Centenario and St. Lawrence's church in Ename helped in discussions (see Sections 6.4.1 and 7.4.1).

[4] The effect a photo-realistic rendering will induce, can be speculated upon.

[5] For instance in the Ename 974 project, where the main communication modes are frequently texts and images while the reconstructions only provide additional information (see Sections 7.3.3.1, 7.3.4 and 7.3.5).

[6] See, for instance, Section 8.2.2.2: *The West Kennet Avenue at the southern entrance to the henge* in the Negotiating Avebury project.

[7] For example, see Sections 8.2.1.3, 8.2.2.2: *The Beckhampton Avenue*, and 8.3.2.1 in the Negotiating Avebury project.

[8] Multisensory output would be sensible especially for phenomenological approaches, cf. Section 8.2.1.4 in the Negotiating Avebury project.

[9] The necessary phases are discussed in detail in Chapters 9 and 10.

[10] New information can come, for instance, from physicochemical analyses, as in the Casa del Centenario project (Section 6.3.1), or from surveys, as in the Negotiating Avebury project (Section 8.2.2.1).

[11] In the Casa del Centenario project the project members considered involving communication specialists (Section 6.3.3).

model is distributed (e.g. on CD or via the internet) and the associated constraints (download time and potentially limited computing power) have to be considered.

Considerations related to the funding of the project are essential, too. When allotting money to the different project stages, particularly the part the research efforts play should not be underestimated, because they (and not the graphical output or the display equipment) determine the quality of the model. Naturally this work should fall to experts in the respective field. All other approaches will inevitably lead to mediocre and misleading results.

11.3 Collaboration

Another important factor is the quantity and quality of the communication between the different parties within the project.[12] That communication between members of widely different disciplines[13] is (initially) difficult, is undeniable. However, only if the goals of the project can be appropriately communicated and any problems which occur during the interpretation or modelling phase are discussed and resolved jointly, will the final product be satisfying for both sides as well as for the end-users.

It has been shown that it can be beneficial if the reconstruction model is produced while the field work or research is still ongoing because the model can provide stimuli for new enquiries[14] or alter the course of the interpretation[15].

11.4 Making the model comprehensible

Observers are frequently confronted with models which provide only one (often photo-realistic) interpretation of a site or object. They are thus being deceived into believing that everything about it is known. In reality, very often little is known about the subject of the reconstruction. But even if the observer has some doubts about the reliability of the visualisation, the photo-realistic reconstruction usually offers no possibility for him or her to question it.

Showing what lies behind the virtual models enables the observer to judge for him- or herself. There are several methods used to reveal some or all of the information which has gone into the visualisation:

- The marking of uncertain parts in the reconstruction is one of them: in the Casa del Centenario project three levels of certainty have been identified, according to a catalogue concerning the reliability of source types (Section 6.2.1.1). These are presented in the form of the extant remains (certain), together with a reconstruction which shows areas which can reconstructed with confidence in colour, while uncertain parts are depicted in greyscale. This is just one example of the various possibilities of indicating different levels of certainty in reconstructions.[16]
- The provision of alternative reconstructions can also help to indicate uncertainties or to convey diverging interpretations, especially if the latter effect significant changes in the appearance of the object or site. They also have their place in research where they are employed for hypothesis assessment, as, for example, in the Negotiating Avebury project (Section 8.2.1.4).
- A progression of models can show both the field work or research progress and/or the interpretational process. The models on the *Nieuws uit het Verleden* website document the development stages of the reconstruction, based on the latest discoveries during the excavations and building research at St. Lawrence's church in Ename (see Section 7.3.4). This approach enables the user to comprehend on what basis assumptions were initially made and (later) dismissed, as well as the effects they had on the appearance of the model.
- Another way to provide insight into model construction is by presenting the underlying sources. In augmented reality approaches, the extant remains are contrasted with the the reconstruction. This provides at least some insight into the data basis for the model.[17] However,

[12]This obviously applies particularly to groups which include specialists from various fields, as, for example, in the Casa del Centenario or the Ename 974 projects (see Sections 6.3 and 7.3.3.2). In the Negotiating Avebury project only archaeologists with knowledge in computing were involved (see Section 8.2.2).

[13]For instance between archaeologists and computer scientists.

[14]For example, in the Ename 974 project (see Section 7.3.3.3).

[15]For instance an interpretation can become unlikely due to structural constraints (cf. Section 8.1.3 concerning the Sanctuary model by Jennifer Garofalini).

[16]For a more detailed account see Section 5.9.1.

[17]See, for example, the reconstruction of St. Saviour's abbey in the Ename 974 project (Section 7.3.2.1).

often enough additional information besides the extant remains is available and so has been incorporated into the reconstruction—these sources can also be provided for the observer. This has been realised for some of the reconstructed parts in the Casa del Centenario project (Section 6.2.1.1). The best solution would be to display all the available sources for a reconstructed part[18]—the ones supporting the presented interpretation as well as the ones which do not agree with it. On this basis the user of the model can come to his or her own conclusions concerning the reliability of the reconstruction. Arguably the sources alone are not sufficient for comprehending the decisions made by the researchers who created the model. In order to achieve this, more information has to be made available, i.e. (parts of) the project documentation.

- The most profound insight into the generation of the model is provided by the documentation of the project. In most cases the complete documentation will only be interesting to experts in the field, although excerpts from it may provide an interested lay audience with more background information concerning the reconstructed parts (see above).[19] Documenting the interpretation and the modelling process can also help in facilitating later changes to the model and preserving the data for future use because it clearifies the reasons why decisions were made and which parts would be affected if these decisions are revised.

If uncertainties are depicted and the sources or the whole documentation are made available, different questions can be raised by the observer, e.g. 'Does the available material make a reconstruction feasible?' or 'Could the sources be interpreted differently?'. This way the audience members can partake in the interpretation process, instead of being divorced from it.

Incorporating the depiction of uncertainties into the model will require a non-photorealistic approach, because the uncertain parts have to be highlighted in some way. In general, non-photorealistic images have the benefit of informing the observer that he or she is only looking at a model, i.e. an interpretation, and not at a 'photograph taken from a past reality'. Photo-realistic models are usually not well suited to providing this kind of additional information, but they can be the basis for models which disclose their sources (and the underlying interpretations). Nevertheless, further research into the perception of different representation modes is strongly needed.

11.5 Presentation of the model

A number of presentation devices are available and indeed the three case studies here exhibit a variety of presentation equipment.[20] It is very important to keep the initial goal and the target audience in mind when choosing the appropriate presentation equipment for the reconstruction.[21] For example, for the individual exploration of a site like the Casa del Centenario a handheld device is the sensible choice, while an interactive museum display[22] will often take the form of a large (touch-)screen.

Also, funding, upkeep[23] and error-proneness will play a considerable role in the selection of a display medium.

11.6 Sustainability and maintenance

Sustainability and maintenance (see Section 10.5) are often neglected when planning a reconstruction project. Therefore, it is essential to avoid proprietary software and data formats to ensure the future accessibility of the model data. For both communication and research models, it would be desirable if they were kept up to date concerning their informational content, but often not enough funding is allotted to these activities. Additionally, the data should not only be copied to new data carriers from time to time, but should also be migrated into more recent formats in order to preserve the information for the future. Another way is to program an emulation which can

[18] In the Casa del Centenario project only the sources which agree with the reconstruction are presented.

[19] An insight into the value this information holds for the understanding of the decision-making and modelling process, is provided by Coralini (2007, 26–31) (see Section 6.3.2) who describes these processes using several case studies.

[20] See for instance Section 6.2.1 as well as Sections 7.3.2.1, 7.3.2.2, 7.3.3.1 and 7.3.5.

[21] As Owen et al. (2005, 136) remarks: 'It is an easy mistake for those developing technology to assume that the visitor will use the device because it happens to be there. Visitors are trading off their time with using the technology. Therefore the application's benefits and uses must be made immediately apparent to the visitor in order to encourage use.'

[22] Particularly if it has to be suitable for group viewing.

[23] For instance portable devices have to be distributed and recollected. Additionally, their batteries have to be recharged.

be used to mimic legacy hard- and software so that the old data can still be read out, but this is a cost intensive process and only feasible if a considerable amount of data is involved which can otherwise not be preserved.

11.7 Closing remarks

From the above-mentioned it becomes apparent that a reconstruction which heeds applied ethical issues calls for more work, time and funding than the creation of mere 'pretty pictures' such as can often be encountered in museums and on television. Despite being created for just a quick impression of the site or object, the latter are prone to stay in the observer's mind.[24]

In contrast to these, models which were built with regard to applied ethical issues disclose their hypothetical character, and facilitate the accomplishment of research, as well as provide added content which aids understanding and encourages engagement with the visualisation.

Last but not least, by adhering to applied ethical issues, quality assurance for virtual models can be accomplished.

[24]This may be equally true for reconstructions which were built with regard to applied ethical aspects, but it can be hoped that their hypothetical nature is realised and the false impression of a knowable past is avoided.

Bibliography

(AAHRG) Avebury Archaeological & Historical Research Group (Ed.) (2001). *Archaeological Research Agenda for the Avebury World Heritage Site.* Salisbury: Trust for Wessex Archaeology Ltd.

Abernathy, D. and C. Johanson (2005). Evolving Strategies: Projects of the UCLA CVRLab. See Forte (2005), pp. 221–228.

Adamy, R. (1891). *Die Fränkische Thorhalle und Klosterkirche zu Lorsch an der Bergstraße.* Darmstadt: Selbstverlag des Historischen Vereins für das Großherzogthum Hessen. (Facsimile, Verlag Degener, 2000).

Addison, A. (2001). Virtual Heritage - Technology in the Service of Culture. See Arnold et al. (2001), pp. 343–354.

AHDS (2002). Archaeology Data Service / Digital Antiquity Guides to Good Practice. Creating and Using Virtual Reality: A Guide for the Arts and Humanities. http://guides.archaeologydataservice.ac.uk/g2gp/Vr_Toc. (Accessed: 11th of June 2012.).

Ameels, V. (2000). Ondergronds Ename: Handen Wassen en Eten in de Abdij. *Ename 974 12,* 14–15.

Ameels, V. (2003). Ondergronds Ename: Romeinen in Ename. *Ename 974 23,* 12.

Ameels, V. (2005). Werken in Kerken: Muurschilderingen in de Sint-Laurentiuskerk van Ename. *Ename 974 27,* 2.

Ameels, V., J. Bastiaens, D. Callebaut, K. De Groote, N. Lemay, E. Roels, and M. Vandenbruaene (2000). Kroniek van een Restauratie. *Ename 974 14,* 8–9.

Ameels, V., D. Callebaut, A. Ervynck, N. Lemay, and E. Roels (2000). Kroniek van een Restauratie. *Ename 974 13,* 6–7.

Anderson, M. (2008). Putting the "Reality" in Virtual Reality: New Advances through Game Engine Technology. See Posluschny et al. (2008), pp. 1–8, only on CD–ROM.

Andersson, H., B. Scholkmann, and M. S. Kristiansen (2007). Medieval Archaeology at the Outset of the Third Millennium: Research and Teaching. In J. Graham-Campbell and M. Valor (Eds.), *The Archaeology of Medieval Europe: Eighth to Twelfth Centuries AD,* Acta Jutlandica 83,1, Humanities Series 79, Chapter 1, pp. 19–45. Aarhus, Lancaster and Oakville: Aarhus University Press.

Andrén, A. (1998). *Between Artifacts and Texts: Historical Archaeology in Global Perspective.* Contributions to Global Historical Archaeology. New York and London: Plenum Press.

Arnold, C. and J. Huggett (1995). Pre-Norman Rectangular Earthworks in Mid-Wales. *Medieval Archaeology 39,* 171–174.

Arnold, C., J. Huggett, P. Reilly, and C. Springham (1989). Mathrafal: a Case Study in the Application of Computer Graphics. See Rahtz and Richards (1989), pp. 147–155.

Arnold, D. (2002). Economic Reconstructions of Populated Environments: Progress with the Charismatic Project. See Niccolucci (2002b), pp. 203–207.

Arnold, D., A. Chalmers, and D. Fellner (Eds.) (2001). *Proceedings of the 2001 Conference on Virtual Reality, Archaeology, and Cultural Heritage,* New York. ACM Press.

Arnold, D. B. (Ed.) (2003). *VAST 2003: the 4th International Symposium on Virtual Reality, Archaeology, and Intelligent Cultural Heritage, incorporating 1st EUROGRAPHICS Symposia on Graphics and Cultural Heritage,* Aire-la-Ville, Switzerland. Eurographics Association.

Ashbee, P. (2004). Early Ditches: Their Forms and Infills. See Cleal and Pollard (2004), pp. 1–14.

Atkinson, R. J. C. (1967). Silbury Hill. *Antiquity 41,* 259–262.

Atkinson, R. J. C. (1970). Silbury Hill, 1969-70. *Antiquity 44,* 313–314.

Attardi, G., M. Betrò, M. Forte, R. Gori, S. Imboden, and F. Mallegni (2000). 3D Facial Reconstruction and Visualization of Ancient Egyptian Mummies Using Spiral CT Data Soft Tissue Reconstruction and Texture Application. See Barceló et al. (2000), pp. 79–85.

Aubrey, J. (1980). *Monumenta Britannica or A Miscellany of British Antiquities,* compiled mainly between the years 1665 and 1693, eds. Rodney Legg and John Fowles. Sherborne: Dorset Publishing Company.

Avern, G. (2002). Reconstructions of the Excavations of Two Iron Age Chariot Burials from Belgium: Applying Virtual Reality to Old Excavation Data. See Burenhult (2002), pp. 157–162.

Balthazar, H., J.-P. Van Der Meiren, L. Santens, M. De Meulemeester, W. Merchie, L. Van Quickenborne, G. De Boe, and D. Callebaut (1997). Ondergrondst Ename: De Sint-Laurentiuskerk: een Verrassing van Formaat. *Ename 974 5,* 1–2.

Barceló, J. A. (2000). Visualizing What Might Be: An Introduction to Virtual Reality Techniques in Archaeology. See Barceló et al. (2000), pp. 9–35.

Barceló, J. A. (2001). Virtual Reality for Archeological Explanation. Beyond "picturesque" Reconstruction. *Archeologia e Calcolatori 12,* 221–244.

Barceló, J. A., M. Forte, and D. H. Sanders (Eds.) (2000). *Virtual Reality in Archaeology: Computer Applications and Quantitative Methods in Archaeology 1998,* BAR International Series 843, Oxford. BAR Publishing.

Barrett, J. C. (1994). *Fragments from Antiquity: An Archaeology of Social Life in Britain,2900–1200* BC. Social Archaeology. Oxford and Cambridge: Blackwell.

Bastiaens, J. and H. Tency (2001). Floraal Verhaal: Brand in de Graanschuur...eh...Kerk! *Ename 974 15,* 12.

Bateman, J. (2000). Immediate Realities: An Anthropology of Computer Visualisation in Archaeology. *Internet Archaeology 8.*

Batselier, P. (1977). *Monasticon Belge,* Volume 7 Province de Flandre Orientale. Maredsous: Abbaye.

Baudrillard, J. (1983). *Simulations.* Semiotext(e) Foreign Agents. New York: Semiotext(e).

Baumeier, S. (2008). 3D Models in Settlement Research: Potentials and Drawbacks of Digital Models as Scientific Working Tools and for Presentation to Non-Professionals. See Posluschny et al. (2008), pp. 1–6, only on CD–ROM.

Bayliss, A., F. McAvoy, and A. Whittle (2007). The World Recreated: Redating Silbury Hill in its Monumental Landscape. *Antiquity 81,* 26–53.

Beacham, R., F. Niccolucci, H. Denard, S. Hermon, and A. Bentkowska-Kafel (2009). The London Charter: For the Computer-Based Visualisation of Cultural Heritage. http://www.londoncharter.org/. (Accessed: 1st of December 2010.).

Behn, F. (1934a). *Die karolingische Klosterkirche von Lorsch an der Bergstraße. Nach den Ausgrabungen von 1927–1928 und 1932–1933,* Volume Text und Tafeln. Berlin and Leipzig: De Gruyter.

Behn, F. (1934b). *Die karolingische Klosterkirche von Lorsch an der Bergstraße. Nach den Ausgrabungen von 1927–1928 und 1932–1933,* Volume Pläne. Berlin and Leipzig: De Gruyter.

Berings, G. (1994, April). Archivalische Gegevens Betreffende de Bouwgeschiedenis van de Sint-Laurentiuskerk van Ename vóór de Restauratie van 1908–1909 (Vorlopige Tekst). Unpublished.

Bethmann (Ed.), L. C. (1844). Auctarium Affligemense. In *Monumenta Germaniae Historica, Scriptores 6,* pp. 398–405. Hannover: Hahn.

Bethmann (Ed.), L. C. (1846). Gesta Episcoporum Cameracensium. In *Monumenta Germaniae Historica, Scriptores 7,* pp. 393–525. Hannover: Hahn.

Bethmann (Ed.), L. C. (1851). Genealogiae Comitum Flandriae. In *Monumenta Germaniae Historica, Scriptores 9,* pp. 302–336. Hannover: Hahn.

Bewley, R., M. Cole, A. David, R. Featherstone, A. Payne, and F. Small (1996). New Features within the Henge at Avebury, Wiltshire: Aerial and Geophysical Evidence. *Antiquity 70,* 639–646.

Blinn, J. F. (1978, August). Simulations of Wrinkeled Surfaces. *SIGGRAPH Computer Graphics 12*(3), 286–292.

Bobowski, B., K. Walczak, and M. Stawniak (2008). Hybrid 3D Visualisations of Archaeological Sites: Dynamic 3D Visualisations of Harris Matrix Data for Rescue Town Excavations, Gdańsk / Szafarnia Site, Poland. See Posluschny et al. (2008), pp. 1–5, only on CD–ROM.

Bohuslav, P., F. Humer, A. Voigt, and M. Schrenk (2002). Virtual Tour of the Roman City of Carnuntum: An Example for Virtual Reconstruction of Archaeological Finds. See Niccolucci (2002b), pp. 209–212.

Bonazzi, A. (2006). Dati Archeometrici su Malte e Intonaci dell'*Insula* del Centenario a Pompei. See Custodi and Sciortino (2006), pp. 59–78.

Bonde, S., C. Maines, E. Mylonas, and J. Flanders (2009). The Virtual Monastery: Re-Presenting Time, Human Movement, and Uncertainty at Saint-Jean-des-Vignes, Soissons. *Visual Resources 25*(4, special issue), 363–377.

Bonfigli, M. E. and A. Guidazzoli (2000). A WWW Virtual Museum for Improving the Knowledge of the History of a City. See Barceló et al. (2000), pp. 143–147.

Borchert, D. M. (Ed.) (2006). *Encyclopedia of Philosophy* (2nd ed.), Volume Abbagnano - Byzantine philosophy. Detroit, New York, San Francisco, San Diego, New Haven, Waterville, London and Munich: Thomson Gale.

Borra, D. (2000). La Modellazione Virtuale per l'Architettura Antica: Un Metodo verso l'Isomorfismo Percettivo. *Archeologia e Calcolatori 11*, 259–272.

Brandl, U., C. Diessenbacher, and A. Rieche (2002). Colonia Ulpina Triana: Ein multimediales Informationssystem zur Archäologie der römischen Stadt. See Rieche and Schneider (2002), pp. 23–32.

Brehm, T., T. Chafik, B. Oesl, J. Reiche, H. Schäfer, R. Schlief-Ehrismann, A. von Seggern, and H. Weber (1998). *Bilder, die lügen. Begleitbuch zur Ausstellung im Haus der Geschichte der Bundesrepublik Deutschland, Bonn, 27. November 1998 bis 28. Februar 1999*. Bonn: Bouvier Verlag.

Brück, J. (2005). Experiencing the Past? The Development of a Phenomenological Archaeology in British Prehistory. *Archaeological Dialogues 12*(1), 45–72.

Burenhult, G. (Ed.) (2002). *Archaeological Informatics: Pushing the Envelope CAA 2001: Computer Applications and Quantitative Methods in Archaeology*, BAR International Series 1016, Oxford. BAR Publishing.

Burl, A. (1969). Henges: Internal Features and Regional Groups. *The Archaeological Journal 126*, 1–28.

Burl, A. (1979). *Prehistoric Avebury*. New Haven and London: Yale University Press.

Burl, A. (1993). *From Carnac to Callanish: The Prehistoric Stone Rows and Avenues of Britain, Ireland and Brittany*. New Haven and London: Yale University Press.

Burl, A. (2000). *The Stone Circles of Britain, Ireland, and Brittany*. New Haven and London: Yale University Press.

Burridge, J. M., B. M. Collins, B. N. Galton, A. R. Halbert, T. R. Heywood, W. H. Latham, R. W. Phippen, P. Quarendon, P. Reilly, M. W. Ricketts, J. Simmons, S. J. P. Todd, A. G. N. Walter, and J. R. Woodwark (1989). The WINSOM Solid Modeller and Its Application to Data Visualization. *IBM Systems Journal 28*(4), 548–568.

Burton, N. (2000). New Tools at Avebury. *Antiquity 74*, 279–280.

Burton, N., M. Hitchen, and P. Bryan (1999). Virtual Stonehenge: A Fall from Disgrace? See Dingwall et al. (1999), pp. CD.

Callebaut, D. (1985). De portus en abdij van Ename. *Archaeologia Belgica 1*(2), 89–94.

Callebaut, D. (1987). De vroeg-middeleeuwse portus en Benedictijnenabdij van Ename (Stad Oudenaarde). *Archaeologia Belgica 3*, 213–224.

Callebaut, D. (1992). De Sint-Laurentiuskerk van Ename (stad Oudenaarde, prov. Oost-Vlaanderen): een vroeg-11de-eeuws symbol van *stabilitas regni et fidelitas imperatoris*. *Archeologie in Vlaanderen 2*, 435–470.

Callebaut, D. (1996a). Ondergrondst Ename: De Burcht: een Sterk Verhaal. *Ename 974 1*, 4–5.

Callebaut, D. (1996b). Ondergrondst Ename: De Sint-Laurentiuskerk: een Verrassing van Formaat. *Ename 974 2*, 4–5.

Callebaut, D. (1997). Ondergrondst Ename: De Sint-Salvatorskerk Herrijst. *Ename 974 5*, 4–5.

Callebaut, D. (1999). Kijen in de Regio: Nederename: Nostalgie om de Sint-Vedastuskerk. *Ename 974 10*, 14–15.

Callebaut, D. (2002). The Experiences of the Ename 974 Project with New Media: Where and How Do Virtual Reality and Interactivity Fit In? See Niccolucci (2002b), pp. 179–185.

Callebaut, D. and K. De Groote (1998). Ondergrondst Ename: De Handelsnederzetting van Ename: de Grote Uitdaging voor het Onderzoek van de Komende Jaren. *Ename 974 9*, 4–5.

Callebaut, D., K. De Groote, A. Ervynck, and M. J. Y. Van Strydonck (1997/1998). Was het nu '70 of '80? Radiokoolstofdateringen voor het castrum te Ename (Oudenaarde, prov. Oost-Vlaanderen. *Archeologie in Vlaanderen 6*, 231–241.

Callebaut, D., N. Silberman, D. Pletinckx, M.-C. Van der Donckt, and G. Tack (2000, March). Nieuws uit het Verleden: Sint Laurentiuskerk. http://www.ename974.org/Ndl/pagina/NuhV/actueel.html. (Accessed: 14th of January 2008.).

Callebaut, D. and J. Sunderland (1998). Ename: New Technologies Perpetuate the Past. *Museum International 50*(2), 51–54.

Callebaut, D. and G. Tack (1998). Provinciaal Museum t'Ename: Tijdslijn. *Ename 974 8*, 6–7.

Callebaut, D. and P. van der Plaetsen (1992). Castrum, Portus und Abtei von Ename. In H. W. Böhme (Ed.), *Burgen der Salierzeit*, Volume 1, In den nördlichen Landschaften des Reiches of *Römisch-Germanisches Zentralmuseum. Forschungsinstitut für Vor- und Frühgeschichte. Monographien 25*, pp. 291–309. Sigmaringen: Jan Thorbecke Verlag.

Campana, S. and F. Remondino (2008). Fast and Detailed Digital Documentation of Archaeological Excavations and Heritage Artifacts. See Posluschny et al. (2008), pp. 36–42.

Capoferro Cencetti, A. M., A. Coralini, A. Custodi, A. Guidazzoli, N. Santopuoli, S. Santoro, D. Scagliarini Corlàita, L. Seccia, and E. Vecchietti (2001). *L' Alma Mater a Pompei: Le Pitture dell'Insula del Centenario. Mostra Didattica.* Imola: University Press Bologna.

Carver, M. (1986). Contemporary Artefacts Illustrated in Late Saxon Manuscripts. *Archaeologia or Miscellaneous Tracts Relating to Antiquity 108*, 117–145.

Carver, M. (2002). Marriages of True Minds: Archaeology with Texts. In B. Cunliffe, W. Davies, and C. Renfrew (Eds.), *Archaeology: The Widening Debate*, British Academy Centenary Monographs, pp. 465–496. Oxford: Oxford University Press.

Chadwick, A. M. (2004). Geographies of Sentinence: An Introduction to Space, Place and Time. In A. M. Chadwick (Ed.), *Stories from the Landscape: Archaeologies of Inhabitation*, BAR International Series 1238, Oxford, pp. 1–31. BAR Publishing.

Chalmers, A. and S. Stoddart (1996). Photo-Realistic Graphics for Visualizing Archaeological Site Reconstructions. See Higgins et al. (1996), pp. 85–93.

Chalmers, A., S. Stoddart, J. Tidmus, and R. Miles (1995). INSITE: an Interactive Visualisation System for Archaeological Sites. See Huggett and Ryan (1995), pp. 225–228.

Christodoulou, V., A. Chaleplioglou, and I. Papadakis (2006). Developing a Preservation Program for 3D Computer Graphics Cultural Content. See Ioannides et al. (2006a), pp. 59–64.

Cinotti, T. S., G. Raffa, L. Roffia, F. Garzotto, R. Muzii, V. Varlese, M. Malavasi, and S. Galasso (2004). Evaluating Context-Aware Mobile Applications in Museums: Experiences from the MUSE Project. http://www.archimuse.com/mw2004/papers/salmon/salmon.html. (Accessed: 8th of January 2009.).

Cinotti, T. S., L. Roffia, G. Mincolelli, F. Sforza, and M. Malavasi (2007). L'Archeologia Virtuale e la Fruizione sui Siti Archeologici: Il Progetto MUSE. See Coralini and Scagliarini Corlàita (2007), pp. 41–51.

Clark, J. T. and E. M. Hagemeister (Eds.) (2007). *Digital Discovery: Exploring New Frontiers in Human Heritage. CAA 2006. Computer Applications and Quantitative Methods in Archaeology, Proceedings of the 34th Conference, Fargo,United States, April 2006*, Budapest. Archaeolingua.

Cleal, R. and J. Pollard (Eds.) (2004). *Monuments and Material Culture: Papers in Honour of an Avebury Archaeologist: Isobel Smith.* Salisbury: Hobnob Press.

Collins, B., D. Williams, R. Haak, M. Trux, H. Herz, L. Genevriez, P. Nicot, P. Brault, X. Coyere, B. Krause, J. Kluckow, and A. Paffenholz (1995). The Dresden Frauenkiche—Rebuilding the Past. See Wilcock and Lockyear (1995), pp. 19–24.

Coralini, A. (2007). L'Archeologia Attraverso un 3D Virtual Model. See Coralini and Scagliarini Corlàita (2007), pp. 17–39.

Coralini, A. and D. Scagliarini Corlàita (Eds.) (2007). *Ut Natura Ars: Virtual Reality e Archeologia.* Imola: University Press Bologna.

Cornforth, J., C. Davidson, C. J. Dallas, and G. R. Lock (1992). Visualising ancient Greece: computer graphics in the Sacred Way Project. In G. Lock and J. Moffett (Eds.), *Computer Applications and Quantitative Methods in Archaeology 1991*, BAR International Series S577, Oxford, pp. 219–225. BAR Publishing.

Cosmas, J., T. Itegaki, D. Green, E. Grabczewski, F. Weimer, L. van Gool, A. Zalesny, D. Vanrintel, F. Leberl, M. Grabner, K. Schindler, K. Karner, M. Gervautz, S. Hynst, M. Waelkens, M. Pollefeys, R. DeGeest, R. Sablatnig, and M. Kampel (2001). 3D MURALE: A Multimedia System for Archaeology. See Arnold et al. (2001), pp. 297–305.

Craig, E. (Ed.) (1998). *Routledge Encyclopedia of Philosophy*. London: Routledge.

Cripps, P. (2001). Pathways through the Avebury Landscape: A Study of Spatial Relationships Associated with the Beckhampton Avenue, Avebury, Wilts. Master's thesis, University of Southampton.

Cummings, V. (2000). Landscapes in Motion: Interactive Computer Imagery and the Neolithic Landscapes of the Outer Hebrides. In C. Buck, V. Cummings, C. Henley, S. Mills, and S. Trick (Eds.), *U.K. Chapter of the Computer Applications and Quantitative Methods in Archaeology: Proceedings of the Fourth Meeting, Cardiff University, 27 and 28 February 1999*, BAR International Series 844, Oxford, pp. 11–20. BAR Publishing.

Cunnington, M. E. (1912). The Discovery of a Skeleton and "Drinking Cup" at Avebury. *Man 12*, 200–203.

Cunnington, M. E. (1931). The "Sanctuary" on Overton Hill, near Avebury. *The Wiltshire Archaeological and Natural History Magazine 45*, 300–335.

Custodi, A. and L. Sciortino (Eds.) (2006). *Rilievo, Modellazione e Restauro di Murature Antiche: Il Caso dell'Insula del Centenario a Pompei*. Arrone: Edizioni Thyrus.

Custodi, A. and L. Sciortino (2009). Dal Rilievo all'Analisi Strutturale: Consolidamento e Restauro nell'*Insula* del Centenario (IX, 8) a Pompei. In A. Coralini (Ed.), *Vesuviana: Archeologie a Confronto. Atti del Convengno Internazionale (Bologna, 14–16 Gennaio 2008)*, Bologna, pp. 623–635. Ante Quem.

Custodi, A., L. Sciortino, and G. Castellazzi (2006). Sito Web e Banca Dati Relativi all'Insula del Centenario a Pompei. See Custodi and Sciortino (2006), pp. 189–208.

Custodi, A., L. Sciortino, G. Castellazzi, L. Govoni, and M. Rivola (2006). Rilievo, Modellazione e Analisi Strutturale della Murature dell'Insula del Centenario a Pompei. See Custodi and Sciortino (2006), pp. 37–57.

D'Andrea, A. and F. Niccolucci (2001). L'informatica dell'archeologo: alcune istruzioni per l'uso. *Archeologia e Calcolatori 12*, 199–220.

Daniels, R. (1997). The Need for the Solid Modelling of Structure in the Archaeology of Buildings. *Internet Archaeology 2*.

Davies, S. R. (2009). Digital Avebury: New 'Avenues' of Research. *Internet Archaeology 27*.

DCMI (2010). The Dublin Core Metadata Initiative: Making It Easier to Find Information. http://dublincore.org/. (Accessed: 5th of May 2011.).

De Groote, K. (2000). Pottenkijker: Goud- en Bronsbewerking in de Middeleeuwse Handelsnederzetting van Ename. *Ename 974 12*, 6.

De Groote, K. (2001). Pottenkijker: Van een Leien Dakje? *Ename 974 17*, 6.

Decaudin, P. (1996, June). Cartoon Looking Rendering of 3D Scenes. Research Report 2919, INRIA.

Delooze, K. and J. Wood (1991). Furness Abbey Survey Project—The Application of Computer Graphics and Data Visualisation to Reconstruction Modelling of an Historic Monument. See Lockyear and Rahtz (1991), pp. 141–148.

Devereux, P. (1991). Three-dimensional Aspects of Apparent Relationships between Selected Natural and Artificial Features within the Topography of the Aveburx Complex. *Antiquity 65*, 894–898.

Devliegher, L. (1969). Enkele Nota's over de Sint-Laurentiuskerk te Ename. *Bulletin de la Commission Royale des Monuments et des Sites 18*(1), 83–90.

Di Filippo, M., M. Di Nezza, and B. Toro (2007). La Prospezione Microgravimetrica nell'Insula del Centenario. See Santoro (2007), pp. 33–49.

Dikaiakou, M., A. Efthymiou, and Y. Chrysanthou (2003). Modelling the Walled City of Nicosia. See Arnold (2003), pp. 61–69.

Dingwall, L., S. Exon, V. Gaffney, S. Lafin, and M. Van Leusen (Eds.) (1999). *Archaeology in the Age of the Internet: Proceedings of the 25th Anniversary Conference of CAA. Computer Applications and Quantitative Methods in Archaeology*, BAR International Series 750, Oxford. BAR Publishing.

Ditmar-Trauth, G. (2006). *Alltag und Sachkultur des Mittelalters in Bildquellen von 800 bis zum Anfang des 14. Jh.* Münster: self-published.

Dobbins, J. J. and P. W. Foss (Eds.) (2007). *The World of Pompeii*. London and New York: Routledge.

Doerr, M. and A. Sarris (Eds.) (2003). *CAA 2002: The Digital Heritage of Archaeology: Computer Applications and Quantitative Methods in Archaeology*. Archive of Monuments and Publications Hellenic Ministry of Culture.

Dovey, K. (1985). The Quest for Authenticity and the Replication of Environmental Meaning. In D. Seamon and R. Mugerauer (Eds.), *Dwelling, Place and Environment: Towards a Phenomenology of Person and World*, pp. 33–49. Dordrecht, Boston and Lancaster: Martinus Nijhoff Publishers.

Earl, G. (1999). Visualising Danebury: Modelled Approaches to Spatial Theory. See Dingwall et al. (1999), pp. 259–263.

Earl, G. (2005). Video Killed Engaging VR? Computer Visualizations on the TV Screen. See Smiles and Moser (2005), Chapter 11, pp. 204–221.

Earl, G. and D. Wheatley (2002). Virtual Reconstruction and the Interpretative Process: A Case-study from Avebury. In D. Wheatley, G. Earl, and S. Poppy (Eds.), *Contemporary Themes in Archaeological Computing*, pp. 5–15. Oxford: Oxbow Books.

Edwards, B. (2003). Changing Avebury. *3rd Stone Magazine 47*, 44–58.

Eiteljorg, H. (2000). The Compelling Computer Image: A Double-edged Sword. *Internet Archaeology 8*.

Ervynck, A. and W. V. Neer (1992). De Voedselvoorziening in de Sint-Salvatorsabdij te Ename (stad Oudenaarde, prov. Oost-Vlaanderen) I. Beenderen onder een Keukenvloer (1450-1550 A.D.). *Archeologie in Vlaanderen 2*, 419–434.

Evans, J. G. (1972). *Land Snails in Archaeology: With Special Reference to the British Isles*. Studies in archaeological science. London and New York: Seminar Press.

Fehring, G. P. (2000). *Die Archäologie des Mittelalters: Eine Einführung* (3rd ed.). Darmstadt: Wissenschaftliche Buchgesellschaft.

Felgenhauer-Schmiedt, S. (1993). *Die Sachkultur des Mittelalters im Lichte der archäologischen Funde*. Europäische Hochschulschriften Reihe 38 Archäologie 42. Frankfurt am Main, Berlin, Bern, New York, Paris and Wien: Peter Lang.

Ferschin, P., I. Kulitz, A. Jonas, and D. Raue (2008). Spatial and Temporal Visualization in Archaeology: Examples from the Excavation on Elephantine, Egypt. See Posluschny et al. (2008), pp. 1–7, only on CD–ROM.

Flack, P. A., J. Willmott, S. P. Browne, D. B. Arnold, and A. M. Day (2001). Scene Assembly for Large Scale Urban Reconstructions. See Arnold et al. (2001), pp. 227–234.

Florenzano, M., J.-Y. Blaise, and P. Drap (1999). Close Range Photogrammetry and Architectural Models. See Dingwall et al. (1999), pp. only on CD–ROM.

Foley, J. D., A. van Dam, S. K. Feiner, and J. F. Hughes (1996). *Computer Graphics: Principles and Practice* (2nd ed.). The Systems Programming Series. Reading, Menlo Park, New York, Don Mills, Wokingham, Amsterdam, Bonn, Sydney, Singapore, Tokyo, Madrid, San Juan, Milan, Paris: Addison-Wesley.

Forster, C. (2005). Wormser Formengut in der Lorscher Bauskulptur des 12. Jahrhunderts. *Zeitschrift für Archäologie des Mittelalters 33*, 215–222.

Forte, M. (2000). About *Virtual Archaeology*: Disorders, Cognitive Interactions and Virtuality. See Barceló et al. (2000), pp. 247–259.

Forte, M. (Ed.) (2005). *The Reconstruction of Archaeological Landscapes through Digital Technologies: Proceedings of the 2nd Italy-United States Workshop, Rome, Italy, November 3–5, 2003, Berkeley, USA, May 2005*, BAR International Series 1379, Oxford. BAR Publishing.

Forte, M. (2008). Virtual Archaeology: Communication in 3D and Ecological Thinking. See Frischer and Dakouri-Hild (2008), pp. 21–35, 135–138.

Forte, M. and D. Borra (2000). The Estense Castle of Ferrara (Italy): Multimedia Project and Virtual Reconstruction. See Barceló et al. (2000), pp. 237–244.

Forte, M., S. Kay, C. Perlingieri, and R. Perlingieri (2003). Remote Sensing Technologies and Virtual Reconstruction of Archaeological Landscapes: New Developments of the Aksum Project. See Doerr and Sarris (2003), pp. 49–55.

Foucault, M. (2009). *Archaeology of Knowledge* (1st publication (1972) reprint ed.). Routledge classics. London and New York: Routledge.

Franz, H. (1940). Die Marken Valenciennes, Eename und Antwerpen im Rahmen der kaiserlichen Grenzsicherungspolitik an der Schelde im 10.–11. Jahrhundert. *Rheinische Vierteljahrsblätter 10*, 229–276.

Freudenberg, B., M. Masuch, N. Röber, and T. Strothotte (2001). The Computer-Visualistik-Raum: Veritable and Inexpensive Presentation of a Virtual Reconstruction. See Arnold et al. (2001), pp. 97–102.

Frischer, B. and A. Dakouri-Hild (Eds.) (2008). *Beyond Illustration: 2D and 3D Digital Technologies as Tools for Discovery in Archaeology*, BAR International Series 1805, Oxford. BAR Publishing.

Frischer, B. and D. Favro (2005). Digital Roman Forum. http://dlib.etc.ucla.edu/projects/Forum/. (Accessed: 3rd of January 2009.).

Frischer, B., D. Favro, P. Liverani, S. De Blaauw, and D. Abernathy (2000). Virtual Reality and Ancient Rome: The UCLA Cultural VR Lab's Santa Maria Maggiore Project. See Barceló et al. (2000), pp. 155–162.

Frischer, B., F. Niccolucci, N. S. Ryan, and J. A. Barceló (2002). From CVR to CVRO: The Past, Present, and Future of Cultural Virtual Reality. See Niccolucci (2002b), pp. 7–18.

Fritz, F., A. Susperregui, and M. T. Linaza (2005). Enhancing Cultural Tourism Experiences with Augmented Reality Technologies. See Mudge et al. (2005).

Fronza, V., A. Nardini, and M. Valenti (2003). An Integrated Information System for Archaeological Data Management: Latest Developments. See Doerr and Sarris (2003), pp. 147–153.

Gaskell, I. (1991). History of Images. In P. Burke (Ed.), *New Perspectives on Historical Writing*, Chapter 8, pp. 168–192. Cambridge: Polity Press.

Gibson, A. (2004). Round in Circles: Timber Circles, Henges and Stone Circles: Some Possible Relationships and Transformations. See Cleal and Pollard (2004), pp. 70–82.

Gillings, M. (1999). Engaging Place: a Framework for the Integration and Realisation of Virtual-Reality Approaches in Archaeology. See Dingwall et al. (1999), pp. 247–254.

Gillings, M. (2000). Plans, Elevations and Virtual Worlds: The Development of Techniques for Routine Construction of Hyperreal Simulations. See Barceló et al. (2000), pp. 59–69.

Gillings, M. (2002). Virtual Archaeologies and the Hyper-Real: Or, What Does It Mean to Describe Something As *Virtually*-Real? In D. Unwin and P. Fisher (Eds.), *Virtual Reality in Geography*, pp. 17–34. London and New York: Taylor & Francis.

Gillings, M. (2005). The Real, the Virtually Real, and the Hyperreal: The Role of VR in Archaeology. See Smiles and Moser (2005), Chapter 12, pp. 223–239.

Gillings, M. and G. Goodrick (1996). Sensous and Reflexive GIS: Exploring Visualisation and VRML. *Internet Archaeology 1*.

Gillings, M., R. Peterson, and J. Pollard (2004). The Destruction of the Avebury Monuments. See Cleal and Pollard (2004), pp. 139–163.

Gillings, M. and J. Pollard (1999). Non-Portable Stone Artefacts and Contexts of Meaning: The Tale of Grey Wether (www.museums.ncl.ac.uk/Avebury/stone4.htm). *World Archaeology 31*(2), 179–193.

Gillings, M. and J. Pollard (2004). *Avebury*. Duckworth Archaeological Histories. London: Duckworth.

Gillings, M., J. Pollard, and D. Wheatley (2000). The Beckhampton Avenue and a 'New' Neolithic Enclosure near Avebury: An Interim Report on the 1999 Excavations. *Wiltshire Archaeological & Natural History Magazine 93*, 1–8.

Gillings, M., J. Pollard, and D. Wheatley (2002). Excavations at the Beckhampton Enclosure, Avenue and Cove, Avebury: An Interim Report on the 2000 Season. *Wiltshire Archaeological & Natural History Magazine 95*, 249–258.

Gillings, M., J. Pollard, D. Wheatley, and R. Peterson (2008). *Landscape of the Megaliths: Excavation and Fieldwork on the Avebury Monuments, 1997–2003*. Oxford: Oxbow books.

Goetz, H.-W. (2006). *Proseminar Geschichte: Mittelalter* (3rd ed.). UTB 1719 Geschichte. Stuttgart: Ulmer.

Göldner, R. (2010). Preserving Digital Data Step by Step: Small Solutions for Small Institutions. In K. Fischer Ausserer (Ed.), *Proceedings of the 14th International Congress "Cultural Heritage and New Technologies" Vienna, 2009*, pp. 117–125.

Gooch, A., B. Gooch, P. Shirley, and E. Cohen (1998). A Non-photorealistic Lighting Model for Automatic Technical Illustration. In *Proceedings of the*

25th *Annual Conference on Computer Graphics and Interactive Techniques*, SIGGRAPH '98, New York, pp. 447–452. ACM.

Goodrick, G. and G. Earl (2004). A Manufactured Past: Virtual Reality in Archaeology. *Internet Archaeology 15*.

Goodrick, G. and M. Gillings (2000). Constructs, Simulations and Hyperreal Worlds: The Role of Virtual Reality (VR) in Archaeological Research. In G. Lock and K. Brown (Eds.), *On the Theory and Practice of Archaeological Computing*, pp. 41–58. Oxford: Oxford University Committee for Archaeology.

Goodrick, G. and J. Harding (2000). Virtual Reality at the Neolithic Monument Complex of Thornborough, North Yorkshire. See Barceló et al. (2000), pp. 115–119.

Gorissen, P. (1952). *Sigeberti Gemblacensis Chronographiae Auctarium Affligemense.* Verhandelingen van de Koninklijke Vlaamse Academie voor Wetenschappen, Letteren en Schone Kunsten van België, 15. Bruxelles: Paleis der Academiën.

Gray, H. S. G. (1935). The Avebury Excavations, 1908–1922. *Archaeologia or Miscellaneous Tracts Relating to Antiquity 84*, 99–162.

Grellert, M. (2007). *Immaterielle Zeugnisse. Synagogen in Deutschland. Potentiale digitaler Technologien für das Erinnern zerstörter Architektur*. Kultur- und Museumsmanagement. Bielefeld: Transcript.

Grellert, M. and H. Svenshon (2010). Rekonstruktion ohne Befund? *Mitteilungen der Deutschen Gesellschaft für Archäologie des Mittelalters und der Neuzeit 22*, 189–198.

Griffiths, N., A. Jenner, and C. Wilson (1990). *Drawing Archaeological Finds: A Handbook.* Number 13 in Occasional Paper of the Institute of Archaeology, University College London. London: Archetype Publications.

Guidazzoli, A. (2007). L'Esperienza del CINECA nel Campo della Virtual Archaeology. See Coralini and Scagliarini Corlàita (2007), pp. 81–89.

Gysseling, M. and A. C. F. Koch (Eds.) (1950). *Diplomata Belgica ante Annum Millesimum Centesimum Scripta*, Volume Teksten of *Bouwstoffen en Studien voor de Geschiedenis en de Lexicografie van het Nederlands, 1*. Brussel: Belgisch Inter-Universitair Centrum voor Neerlandistiek.

Haack, J. (2002). Interaktivität als Kennzeichen von Multimedia und Hypermedia. See Issing and Klimsa (2002), pp. 127–136.

Hävernick, W. (1973–1975). Münzen als Grabbeigaben 750–1815 (Auswertung des Fundkatalogs der Numismatischen Kommission der Länder in der Bundesrepublik Deutschland). *Hamburger Beiträge zur Numismatik 27–29*, 27–51.

Hermon, S. (2008). Reasoning in 3D: A Critical Appraisal of the Role of 3D Modelling and Virtual Reconstruction in Archaeology. See Frischer and Dakouri-Hild (2008), pp. 36–45.

Hermon, S. and P. Fabian (2002). Virtual Reconstruction of Archaeological Sites Some Archaeological Scientific Considerations: Avdat Roman Military Camp as a Case-study. See Niccolucci (2002b), pp. 103–108.

Hermon, S. and F. Niccolucci (2004). A Fuzzy Logic Approach to Reliability in Archaeological Virtual Reconstruction. http://public-repository.epoch-net.org/articles/caa2004-fuzzy.pdf. (Unpublished CAA 2004 Proceedings. Accessed: 27th of December 2010.).

Hermon, S. and J. Nikodem (2008). 3D Modelling as a Scientific Research Tool in Archaeology. See Posluschny et al. (2008), pp. 1–6, only on CD–ROM.

Herrmann, V. (2010). Archäologische RuhrZeiten: Chancen und Möglichkeiten virtueller Rekonstruktion in GoogleEarth. *Mitteilungen der Deutschen Gesellschaft für Archäologie des Mittelalters und der Neuzeit 22*, 199–210.

Higgins, T., P. Main, and J. Lang (Eds.) (1996). *Imaging the Past: Electronic Imaging and Computer Graphics in Museums and Archaeology.* British Museum Occasional Papers Number 114. London: British Museum Press.

Hodder, I. (1999). *The Archaeological Process: An Introduction.* Oxford and Malden: Blackwell.

Hodder, I. (2000). Symbolism, Meaning and Context. See Thomas (2000a), pp. 86–96.

Hodder, I. and S. Hutson (2003). *Reading the Past: Current Approaches to Interpretation in Archaeology* (3rd ed.). Cambridge, New York, Port Melbourne, Madrid and Cape Town: Cambridge University Press.

Hodgson, J. (2001). *Archaeological Reconstruction: Illustrating the Past.* IFA Paper 5. Reading: Institute of Field Archaeologists.

Hoffsummer, P. (1992). Typologie et Dendrochronologie des Toitures de l'Eglise Saint-Laurent á Ename (Oudenaarde, prov. de Flandre Orientale). *Archeologie in Vlaanderen 2*, 471–480.

Holder-Egger (Ed.), O. (1888). Ex Vita S. Macharii Priore. In *Monumenta Germaniae Historica, Scriptores 15,2*, pp. 614–616. Hannover: Hahn.

Holst, M. K. (2001). Formalizing Fact and Fiction in Four Dimensions: A Relational Description of Temporal Structures in Settlements. See Stančič and Veljanovski (2001), pp. 159–163.

Huggett, J. and C. Guo-Yuan (2000). 3D Interpretative Modelling of Archaeological Sites: A Computer Reconstruction of a Medieval Timber and Earthwork Castle. *Internet Archaeology 8*.

Huggett, J. and N. Ryan (Eds.) (1995). *Computer Applications and Quantitative Methods in Archaeology 1994*, BAR International Series 600, Oxford. BAR Publishing.

Ioannides, M. J., D. Arnold, F. Niccolucci, and K. Mania (Eds.) (2006a). *The e-volution of Information Communication Technology in Cultural Heritage. Where Hi-Tech Touches the Past: Risks and Challenges for the 21st Century. Project papers from the joint event CIPA / VAST / EG / EuroMed 2006, 30. October — 4. November 2006 Nicosia, Cyprus*, Volume 1, Budapest. Epoch: Archaeolingua.

Ioannides, M. J., D. Arnold, F. Niccolucci, and K. Mania (Eds.) (2006b). *The e-volution of Information Communication Technology in Cultural Heritage. Where Hi-Tech Touches the Past: Risks and Challenges for the 21st Century. Project papers from the joint event CIPA / VAST / EG / EuroMed 2006, 30. October — 4. November 2006 Nicosia, Cyprus*, Volume 2, Budapest. Epoch: Archaeolingua.

Isenberg, T. (1999). Visualisierung von Modellierungsentscheidungen und Unsicherheiten in virtuellen Rekonstruktionen. Technical report, Department of Simulation and Graphics, Otto-von-Guericke-Universität Magdeburg.

Issing, L. J. and P. Klimsa (Eds.) (2002). *Information und Lernen mit Multimedia und Internet - Lehrbuch für Studium und Praxis* (3rd ed.). Weinheim: Beltz.

Jablonka, P. (2004). Reconstructing Sites and Archives: Information and Presentation Systems at Troy. See Magistrat der Stadt Wien, Referat Kulturelles Erbe, Stadtarchäologie Wien (2004), pp. on CD, 281–285 in book. (Formatting and content of the articles on CD and in the book are different).

Jablonka, P., S. Kirchner, and J. Serangeli (2003). Troy VR: A Virtual Reality Model of Troy and the Troad. See Doerr and Sarris (2003), pp. 13–18.

James, S. (1997). Drawing Inferences: Visual Reconstructions in Theory and Practice. See Molyneaux (1997), pp. 22–48.

Jaritz, G. (1996). 'Et est ymago ficta non veritas'. Sachkultur und Bilder des späten Mittelalters. In G. Jaritz (Ed.), *Pictura quasi fictura: die Rolle des Bildes in der Erforschung von Alltag und Sachkultur des Mittelalters und der frühen Neuzeit. Internationales Round-Table-Gespräch, Krems an der Donau, 3. Oktober 1994*, Number 1 in Forschungen des Instituts für Realienkunde des Mittelalters und der Frühen Neuzeit 1, Wien, pp. 9–13. Österreichische Akademie der Wissenschaften Wien/Philosophisch-Historische Klasse: Verlag der Österreichischen Akademie der Wissenschaften.

Jarosz, W., H. W. Jensen, and C. Donner (2008). Advanced Global Illumination Using Photon Mapping. In *ACM SIGGRAPH 2008 classes*, SIGGRAPH '08, New York, pp. 2:1–2:112. ACM.

Jobst, M. (2004). Presenting Archaeological Information with the Help of Multimedia Cartography. See Magistrat der Stadt Wien, Referat Kulturelles Erbe, Stadtarchäologie Wien (2004), pp. on CD, 187–191 in book. (Formatting and content of the articles on CD and in the book are different).

Johnsen, H. and B. Olsen (2000). Hermeneutics and Archaeology: On the Philosophy of Contextual Archaeology. See Thomas (2000a), pp. 97–117.

Johnson, D. (2008). Architectural Drafting Standards in Archaeological Computer Modeling: Reconstructions from Drawings and Surveys of the Metropolitan Museum of Art Egyptian Expedition. See Posluschny et al. (2008), pp. 1–5, only on CD–ROM.

Jope, E. M. (1999). The Saxon and Medieval Pottery from Alexander Keiller's Excavations at Avebury. *The Wiltshire Archaeological and Natural History Magazine 92*, 60–91.

Jud, P. and G. Kaenel (Eds.) (2002). *Lebensbilder – Scènes de vie: Actes du colloque de Zoug (13-14 mars 2001)*, Documents du GPS, 2, Zug. Kantonales Museum für Urgeschichte Zug.

Jungblut, D., S. Karl, H. Mara, S. Krömker, and G. Wittum (2012). Automated GPU-based Surface Morphology Reconstruction of Volume Data for Archaeology. In *Proceedings of SCCH09, Scientific Computing in the Cultural Heritage*, Heidelberg. Springer.

Kadobayashi, R., E. Neeter, K. Mase, and R. Nakatsu (1999). VisTA: An Interactive Visualization Tool for Archaeological Data. See Dingwall et al. (1999), pp. CD.

Kadobayashi, R., K. Nishimoto, and K. Mase (2000). Immersive Walk-throught Experience of Japanese Ancient Villages with the VisTA-walk System. See Barceló et al. (2000), pp. 135–142.

Kanter, J. (2000). Realism vs. Reality: Creating Virtual Reconstructions of Prehistoric Architecture. See Barceló et al. (2000), pp. 47–52.

Keen, D., S. Rahtz, and P. Ucko (1992). Visualization and Analysis of the Antiquarian Record in Archaeology. *Bulletin of the John Rylands University Library of Manchester 74*(3), 121–137.

Keiller, A. (1939). Avebury: Summary of Excavations, 1937–1938. *Antiquity 13*, 223–233.

Keiller, A. and S. Piggott (1936). The Recent Excavations at Avebury. *Antiquity 10*, 417–427.

Kemp, D. (1995). Personal Computer-Based Three-Dimensional Reconstruction Modelling of Standing Buildings. See Wilcock and Lockyear (1995), pp. 249–254.

Kensek, K. M. (2007). A Survey of Methods for Showing Missing Data, Multiple Alternatives, and Uncertainty in Reconstructions. *CSA Newsletter 19*(3).

Kieven, E., H. Schlimme, and G. Eger (2000). Visualisation of Filippo Juvarra's Palace Project and Garden Plan (1705): Case Study in the Project: 3D-Bridge - Transferring Cultural Heritage with New Technology. http://wissensgeschichte.biblhertz.it/3d-bridge-html/index3D.html. (Accessed: 21st of November 2011.).

Kirchner, S. and P. Jablonka (2001). Virtual Archaeology - VR Based Knowledge Management and Marketing in Archaeology First Results – Nexts Steps. See Arnold et al. (2001), pp. 235–239.

Klein, U., G. Isenberg, and M. Untermann (1999). Das Tübinger Kolloquium der Arbeitsgemeinschaft im November 1995: Archäologie des Mittelalters und der Neuzeit in Mitteleuropa 'Vorträge und Berichte'. Arbeitsgruppe I: Der Befund und seine Deutungsprobleme: Quellenanalyse und Quellenkritik. *Mitteilungen der Arbeitsgemeinschaft für Archäologie des Mittelalters und der Neuzeit 10*, 18–34.

Knoll, W., P. Braumann, and I. Frels (2004). Enter the Past – the E-Way into the Four Dimensions of Cultural Heritage: CAA 2003: Computer Applications and Quantitative Methods in Archaeology. See Magistrat der Stadt Wien, Referat Kulturelles Erbe, Stadtarchäologie Wien (2004), pp. on CD, 558–561 in book. (Formatting and content of the articles on CD and in the book are different).

Koob, M. (2000). Die dreidimensionale Computer-Rekonstruktion des Burcharddoms und der Johanneskirche zu Worms. In G. Bönnen and I. Spille (Eds.), *Bischof Burchard 1000–1025: Tausend Jahre Romanik in Worms*, pp. 71–81. Worms: Verlag Stadtarchiv Worms.

Krasniewicz, L. (2000). Immersive Imaging Technologies for Archaeological Research. See Barceló et al. (2000), pp. 163–169.

Kuratorium Weltkulturdenkmal Kloster Lorsch e.V. (Ed.) (2007). *Das verlorene Kloster: Studien zur Archäologie und Baugeschichte*. Lorscher Studien 1. Insingen: Degener & Co.

Lambers, K. and F. Remondino (2008). Optical 3D Measurement Techniques in Archaeology: Recent Developments and Applications. See Posluschny et al. (2008), pp. 27–35.

Lancaster, L. (2005). Virtual Reality within the Humanities. See Forte (2005), pp. 1–8.

Lehmann, T. and U. Haarlammert (2010). Die Entwicklung der digitalen Rekonstruktion des spätantiken Pilgerheiligtums in Cimitile/Nola. *Mitteilungen der Deutschen Gesellschaft für Archäologie des Mittelalters und der Neuzeit 22*, 167–176.

Levy, R. and P. Dawson (2008). Structural Analysis: A Tool for Testing 3D Computer Reconstructions of Thule Whalebone Houses. See Posluschny et al. (2008), pp. 134–139.

Li, B., S. Tao, R. W. Dawson, J. Cao, and K. Lam (2002). A GIS Based Road Traffic Noise Prediction Model. *Applied Acoustics 63*(6), 679–691.

Lieberwirth, U. (2008a). 3D GIS Voxel-Based Model Building in Archaeology. See Posluschny et al. (2008), pp. 1–8, only on CD–ROM.

Lieberwirth, U. (2008b). Voxel-Based 3D GIS: Modelling and Analysis of Archaeological Stratigraphy. See Frischer and Dakouri-Hild (2008), pp. 85–94, 157–158.

Lock, G. R. (2003). *Using Computers in Archaeology: Towards Virtual Pasts*. London and New York: Routledge.

Lockyear, K. and S. Rahtz (Eds.) (1991). *Computer Applications and Quantitative Methods in Archaeology 1990*, BAR International Series 565, Oxford. BAR Publishing.

Lucet, G. L. (1997). The Virtual Reconstruction of the Mesoamerican Site of Cacaxtla: A Methodological Point of View. *Archeologia e Calcolatori 8*, 129–138.

Lucet, G. L. (2000). Visualisation of Virtual Environments of Ancient Architecture: The Problem of Illumination. See Barceló et al. (2000), pp. 87–95.

Ludwig, T. (2004). Ein authentisches Bild des karolingischen Klosters Lorsch? Beiträge zu einer Theorie der Rekonstruktion. In I. Ericsson and M. Sanke (Eds.), *Aktuelle Forschungen zum ehemaligen Reichs- und Königskloster Lorsch*, Bamberger Beiträge zur Archäologie des Mittelalters und der Neuzeit 1, pp. 17–34. Darmstadt: Hessische Historische Kommission.

Lukesh, S. S. (1996). Expanding the Archaeologist's Toolkit: Scientific Visualisation of Archaeological Data. See Higgins et al. (1996), pp. 245–257.

Luyten, K., H. Van Loon, D. Teunkens, K. Gabriëls, K. Coninx, and E. Manshoven (2006). ARCHIE: Disclosing a Museum by a Socially-aware Mobile Guide. See Ioannides et al. (2006a), pp. 221–226.

Maas, P. (1963). *Textual Criticism* (Reprint ed.). Oxford: University Press.

Magistrat der Stadt Wien, Referat Kulturelles Erbe, Stadtarchäologie Wien (Ed.) (2004). *Enter the Past – the E-Way into the Four Dimensions of Cultural Heritage: CAA 2003: Computer Applications and Quantitative Methods in Archaeology*, BAR International Se-ries 1227. BAR Publishing. (Formatting and content of the articles on CD and in the book are different).

Mainka-Mehling, A. (2008). *LebensBilder: Zur Darstellung des ur- und frühgeschichtlichen Menschen in der Archäologie*, Volume 1 and 2 of *Frühgeschichtliche Studien 1*. Remshalden: Greiner.

Mallory, J. P. and C. J. Lynn (2002). Recent Excavations and Speculations on the Navan Complex. *Antiquity 76*, 532–541.

Margulies, S. B. (2009). *Digitale Daten als Quelle der Geschichtswissenschaft. Eine Einführung*. Number 2 in Kölner Beiträge zu einer geisteswissenschaftlichen Fachinformatik 2. Hamburg: Verlag Dr. Kovač.

Martens, B. and H. Peter (2004). Virtual Reconstruction of Viennese Synagogues: Sustainable 3D Models. See Magistrat der Stadt Wien, Referat Kulturelles Erbe, Stadtarchäologie Wien (2004), pp. on CD, 204–207 in book. (Formatting and content of the articles on CD and in the book are different).

Martens, F., J. Legrand, P. Legrand, L. Loots, and M. Waelkens (2000). Computer-Aided Design and Archaeology at Sagalassos: Methodology and Possibilities of CAD Reconstructions of Archaeological Sites. See Barceló et al. (2000), pp. 205–212.

Martinez, P. (2001). Digital Realities and Archaeology: A Difficult Relationship or a Fruitful Marriage? See Arnold et al. (2001), pp. 9–15.

Martins, M. and P. Bernandes (2000). A Multi-Disciplinary Approach for Research and Presentation of Bracara Augusta's Archaeological Heritage . *Archeologia e Calcolatori 11*, 347–357.

Maschek, D., M. Schneyder, and M. Tschannerl (2010). Virtual 3D Reconstructions: Benefit or Danger for Modern Archaeology? In K. Fischer Ausserer (Ed.), *Proceedings of the 14th International Congress "Cultural Heritage and New Technologies" Vienna, 2009*, pp. 447–460.

Masuda, T., Y. Yamada, N. Kuchitsu, and K. Ikeuchi (2004). Sunlight Illumination Simulation for Archaeological Investigation: Case Study of the Fugoppe Cave. http://www.cvl.iis.u-tokyo.ac.jp/papers/all/687.pdf. (Accessed: 3rd of March 2011.).

Mau, A. (1881). Scavi di Pompei. Isola ad Oriente di IX 5 e IX 6. *Bullettino dell'Instituto di Corrispondenza Archeologica*, 113–128, 169–175, 221–238.

Mau, A. (1882a). *Geschichte der decorativen Wandmalerei in Pompeji*, Volume Text and Tafeln. Berlin: Reimer.

Mau, A. (1882b). Scavi di Pompei. Isola ad Oriente di IX 5 e IX 6. *Bullettino dell'Instituto di Corrispondenza Archeologica*, 23–32, 47–53, 87–91, 104–116.

Mayr-Harting, H. (1991). *Ottonian Book Illumination*, Volume Themes. London: Harvey Miller.

McGregor, C. C. (2000). A Virtual Reconstruction of the Inner Enclosure of the Temple of Tutu, Dakhleh Oasis, Egypt. *Archaeological Computing Newsletter 56*, 7–16.

Mehling, A. (2002). Zur Struktur und Rolle von Lebensbildern in der Archäologie. See Jud and Kaenel (2002), pp. 87–94.

Messika, N. (2002). Reviving Armageddon (Tel Megiddo). See Niccolucci (2002b), pp. 147–150.

Meyer, E., P. Grussenmeyer, J.-P. Perrin, A. Durand, and P. Drap (2006). Intra-site Level Cultural Heritage Documentation: Combination of Survey, Modeling and Imagery Data in a Web Information System. See Ioannides et al. (2006b), pp. 129–134.

Meyer, E., P. Grussenmeyer, J.-P. Perrin, A. Durand, and P. Drap (2007). Integration of Heterogeneous Cultural Heritage Data in a Web-based Information System: A Case Study from Vianden Castle, Luxembourg. See Clark and Hagemeister (2007), pp. 352–363.

Milis, L. and D. Callebaut (1990). Ename: burcht en "pre-stedelijke" nederzetting. In *La genèse et les premiers siècles des villes médiévales dans les Pays-Bas méridionaux: un problème archéologique et historique: 14e colloque international, Spa, 6 – 8 septembre 1988: actes. Ontstaan en vroegste geschiedenis van de middeleeuwse steden in de Zuidelijke Nederlanden*, Collection Histoire 83, Bruxelles, pp. 459–497. Crèdit Communal - Gemeentekrediet.

Miller, P. and J. Richards (1995). The Good, the Bad and the Downright Misleading: Archaeological Adoption of Computer Visualization. See Huggett and Ryan (1995), pp. 19–22.

Mlekuz, D. (2004). Listening to Landscapes: Modelling Past Soundscapes in GIS. *Internet Archaeology 16*.

Mols, S. T. A. M. (1999). *Wooden Furniture in Herculaneum. Form, Technique and Function*. Amsterdam: Gieben.

Mols, S. T. A. M. (2002). Identification of the Woods Used in the Furniture at Herculaneum. In W. M. F. Jashemski (Ed.), *The natural history of Pompeii*, pp. 225–234. Cambridge: Cambridge University Press.

Molyneaux, B. L. (1992). From Virtuality to Actuality: the Archaeological Site Simulation Environment. See Reilly and Rahtz (1992), Chapter 19, pp. 312–322.

Molyneaux, B. L. (Ed.) (1997). *The Cultural Life of Images*. Theoretical Archaeology Group 7. London and New York: Routledge.

Monod, E. and H. K. Klein (2005). From E-Heritage to Interpretive Archaeology Systems (IAS): A Research Framework for Evaluating Cultural Heritage Communication in the Digital Age. In *Proceedings of the 13th European Conference on Information Systems, Information Systems in a Rapidly Changing Economy, ECIS 2005, Regensburg, Germany, May 26-28, 2005*.

Moreland, J. (2001). *Archaeology and Text*. Duckworth debates in archaeology. London: Duckworth.

Moser, S. (1998, November). Seeing the Past in Standard Images. *British Archaeology 39*, 10–11.

Mudge, M., N. Ryan, and R. Scopigno (Eds.) (2005). *VAST 2005: 6th International Symposium on Virtual Reality, Archaeology and Cultural Heritage. 3rd Eurographics Workshop on Graphics and Cultural Heritage. November 8–11 2005, ISTI-CNR Pisa, Italy*, Eurographics symposium proceedings, Aire-la-Ville. Eurographics Association.

Murphy, J. T. (2007). The Hohokam Water Management Simulation: A Collaborative Model for Exploring Alternative Pasts. See Clark and Hagemeister (2007), pp. 18–21.

Nerdinger, W. (Ed.) (2010). *Geschichte der Rekonstruktion, Konstruktion der Geschichte. Publikation zur Ausstellung des Architekturmuseums der TU München in der Pinakothek der Moderne, 22. Juli bis 31. Oktober 2010*. München, Berlin, London, New York: Prestel.

Newell, A. and H. A. Simon (1972). *Human Problem Solving*. Englewood Cliffs: Prentice-Hall.

Niccolucci, F. (2002a). Virtual Archaeology: An Introduction. See Niccolucci (2002b), pp. 3–6.

Niccolucci, F. (Ed.) (2002b). *Virtual Archaeology: Proceedings of the VAST Euroconference, Arezzo 24-25 November 2000*, BAR International Series 1075, Oxford. BAR Publishing.

Niccolucci, F. and F. Cantone (2003). Legend and Virtual Reconstruction: Porsenna's Mausoleum in X3D. See Doerr and Sarris (2003), pp. 57–62.

Nigro, J. D., W. F. Limp, K. K. Kvamme, D. J. de Ruyter, and L. R. Berger (2002). The Creation and Potential Application of a 3-Dimensional GIS for the Early Hominin Site of Swartkrans, South Africa. See Burenhult (2002), pp. 113–124.

Nonn, U. (1983). *Pagus und Comitatus in Niederlothringen. Untersuchung zur politischen Raumgliederung in frühen Mittelalter*. Bonner Historische Forschungen, 49. Bonn: Ludwig Röhrscheid Verlag.

OED (2010). Oxford English Dictionary Online. http://dictionary.oed.com/. (Accessed: 4th of September 2010.).

O'Flaherty, B. (1988). The Southampton-York Archaeological Simulation System. In S. Rahtz (Ed.), *Computer and Quantitative Methods in Archaeology 1988*, Volume 2 of *BAR International Series 446 (ii)*, Oxford, pp. 491–497. BAR.

O'Rourke, M. (2003). *Principles of Three-Dimensional Computer Animation: Modeling, Rendering, and Animating with 3D Computer Graphics* (3rd ed.). New York and London: W. W. Norton & Company.

Ottaway, B. S., L. Sawyer, and A. Miller (1986). Archaeology Gets Graphic. *Nature 323*, 651–652.

Owen, R., D. Buhalis, and D. Pletinckx (2005). Visitors' Evaluations of ICTs Used in Cultural Heritage. See Mudge et al. (2005), pp. 129–136.

Ozawa, K. (1993). Reconstruction of Japanese Ancient Tombs. In J. Andresen, T. Madsen, and I. Scollar (Eds.), *Computing the Past: Computer Applications and Quantitative Methods in Archaeology*, pp. 415–423. Aarhus University Press.

Ozawa, K. (2006). Computer-Assisted Estimation of the Original Shape of a Japanese Ancient Tomb Mound Based on Its Present Contour Map. See Ioannides et al. (2006b), pp. 143–146.

Parker Pearson, M. and Ramilisonina (1998). Stonehenge for the Ancestors: The Stones Pass on the Message. *Antiquity 72*, 308–326.

Pásztor, E., Ákos Juhász, M. Dombi, and C. Roslund (2000). Computer Simulation of Stonehenge. See Barceló et al. (2000), pp. 111–113.

Pavelka, K., J. Svatušková, J. Preisler, V. Němcová, M. Nečesánek, J. Pazdera, and R. Balík (2006). Complex Documentation of Charles Bridge in Prague by Using Laser Scanning, Photogrammetry and GIS Technology. See Ioannides et al. (2006b), pp. 163–168.

Peterson, P., F. D. Fracchia, and B. Hayden (1995). A Virtual Computer Imaging Technique for Archaeological Research. *Society for American Archaeology Bulletin 13*(4), 30–33.

Pietroni, E. (2005). 3D Data Aquisition and 3D Modelling Applied to Cultural Heritage: From Laser Scanner to Virtual Reality Applications. See Forte (2005), pp. 241–249.

Piggott, S. (1935). Stukeley, Avebury and the Druids. *Antiquity 9*, 22–32.

Piggott, S. (1940). Timber Circles: A Re-examination. *The Archaeological Journal 96*(2), 193–222.

Piggott, S. (1964). Excavations at Avebury, 1960. *The Wiltshire Archaeological and Natural History Magazine 59*, 28–29.

Piro, S. (2007). Il Indagini Georadar nella Domus del Centenario. See Santoro (2007), pp. 51–62.

Pitts, M. (2001a). Excavating the Sanctuary: New Investigations on Overton Hill, Avebury. *The Wiltshire Archaeological and Natural History Magazine 94*, 1–23.

Pitts, M. (2001b). *Hengeworld* (2 ed.). Arrow.

Pitts, M. and A. Whittle (1992). The Development and Date of Avebury. *Proceedings of the Prehistoric Society 58*, 203–212.

Platz, T. (2005). Neue Forschungen zum Kirchenrest in Lorsch. *Zeitschrift für Archäologie des Mittelalters 33*, 207–214.

Platz, T. (2007a). Bemerkungen zu den Grabungen Friedrich Behns in Lorsch. See Kuratorium Weltkulturdenkmal Kloster Lorsch e.V. (2007), pp. 11–13.

Platz, T. (2007b). Neue Forschungen zum Kirchenrest in Lorsch. See Kuratorium Weltkulturdenkmal Kloster Lorsch e.V. (2007), pp. 38–39.

Pletinckx, D. (2004). EPOCH: Een Tweede Europees Project in Ename. *Ename 974 24*, 5.

Pletinckx, D. (2008). Interpretation Management: How to Make Sustainable Visualisations of the Past. Stockholm.

Pletinckx, D., D. Callebaut, A. E. Killebrew, and N. A. Silberman (2000, April-June). Virtual-Reality Heritage Presentation at Ename. *Multimedia, IEEE 7*(2), 45–48.

Pletinckx, D., L. D. Jaegher, T. Helsen, I. Langen, N. Silberman, M.-C. V. der Donckt, and J. Stobbe (2004). Telling the Local Story: An Interactive Cultural Presentation System for Community and Regional Settings. In *Proceedings of the VAST2004 Conference, Brussels: Conscience-auditorium /Oudenaarde: Ename Center*, pp. 233–240. unpublished.

Pletinckx, D. and E. Servaes (2001). De Ename 974-Website. *Ename 974 17*, 3–4.

Pletinckx, D., N. Silberman, and D. Callebaut (2001). Presenting a Monument in Restoration: The Saint Laurentius Church in Ename and Its Role in the Francia Media Heritage Initiative. See Arnold et al. (2001), pp. 197–204.

Pletinckx, D., N. Silberman, and D. Callebaut (2003). Heritage Presentation through Interactive Storytelling: A New Multimedia Database Approach. *The Journal of Visualization and Computer Animation 14*(4), 225–231.

Pohl, W. (2005). Von Nutzen und Methodik des Edierens. In B. Merta, A. Sommerlechner, and H. Weigl (Eds.), *Vom Nutzen des Edierens. Akten des Internationalen Kongresses zum 150-jährigen Bestehen des Instituts für Österreichische Geschichtsforschung, Wien, 3.–5. Juni 2004*, Mitteilungen des Instituts für Österreichische Geschichtsforschung, Supplement 47, Wien and München, pp. 349–354. R. Oldenbourg Verlag.

Pollard, J. (1992). The Sanctuary, Overton Hill, Wiltshire: A Re-examination. *Proceedings of the Prehistoric Society 58*, 213–226.

Pollard, J. and R. Cleal (2004). Dating Avebury. See Cleal and Pollard (2004), pp. 120–129.

Pollard, J. and M. Gillings (1998). Romancing the Stones: Towards a Virtual and Elemental Avebury. *Archaeological Dialogues 5*(2), 143–164.

Pollard, J. and A. Reynolds (2002). *Avebury: the Biography of a Landscape*. Stroud and Charleston: Tempus.

Pollefeys, M., R. Koch, M. Vergauwen, and L. V. Gool (2000). Automated Reconstruction of 3D Scenes from Sequences of Images. *ISPRS Journal of Photogrammetry and Remote Sensing 55*(4), 251–267.

Pollefeys, M., M. Proesmans, R. Koch, M. Vergauwen, and L. van Gool (2000). Aquisition of Detailed Models for Virtual Reality. See Barceló et al. (2000), pp. 71–77.

Pope, J. and A. Chalmers (2000). Pre-rendering Acoustics and Illumination for Archeological Reconstructions. See Barceló et al. (2000), pp. 105–110.

Posluschny, A., K. Lambers, and I. Herzog (Eds.) (2008). *Layers of Perception: Proceedings of the 35th International Conference on Computer Applications and Quantitative Methods in Archaeology (CAA) Berlin, Germany, April 2–6, 2007*, Kolloquien zur Vor- und Frühgeschichte, 10, Bonn. Habelt.

Pringle, M. and M. Moulding (1997). Applications for Virtual Reality, and Associated Information Technology, in the Illustration of Archaeological Material. *Graphic Archaeology 4*, 22–34.

Ragia, L. and E. Leopold (2003). Matching Ancient Texts with Geographical Data. See Doerr and Sarris (2003), pp. 103–107.

Rahtz, S. and J. Richards (Eds.) (1989). *Computer Applications and Quantitative Methods in Archaeology 1989*, BAR S548, Oxford. British Archaeological Reports.

Reilly, P. (1988). *Computer Analysis of an Archaeological Landscape – Medieval Land Divisions in the Isle of Man*. BAR British Series 190. Oxford: BAR.

Reilly, P. (1989). Data Visualization in Archaeology. *IBM Systems Journal 28*(4), 569–579.

Reilly, P. (1991). Towards a Virtual Archaeology. See Lockyear and Rahtz (1991), pp. 133–139.

Reilly, P. (1992). Three-dimensional Modelling and Primary Archaeological Data. See Reilly and Rahtz (1992), Chapter 12, pp. 147–173.

Reilly, P. (1996). Access to Insights: Stimulating Archaeological Visualisation in the 1990s. In I. Gedai (Ed.), *The Future of our Past '93–'95*, Budapest, pp. 38–51. Hungarian National Museum.

Reilly, P. and S. Rahtz (Eds.) (1992). *Archaeology and the Information Age: A Global Perspective*. One World Archaeology. London and New York: Routledge.

Reilly, P. and S. Shennan (1989). Applying Solid Modelling and Animated Three-Dimensional Graphics. See Rahtz and Richards (1989), pp. 157–165.

Rieche, A. and B. Schneider (Eds.) (2002). *Archäologie virtuell: Projekte, Entwicklungen,Tendenzen seit 1995. Beiträge zum Colloquium in Köln, 5.–6. Juni 2000, während der Archäologischen Landesausstellung 2000 „Fundort Nordrhein-Westfalen - Millionen Jahre Geschichte" im Römisch-Germanischen Museum der Stadt Köln*, Schriften zur Bodendenkmalpflege in Nordrhein-Westfalen 6, Bonn. Dr. Rudolf Halbelt GmbH.

Riedel, A. and T. Bauer (2008). Pretty and Useful? Three-dimensional Computer Models as a Working Tool for Documentation and Investigation in Building Archaeology. See Posluschny et al. (2008), pp. 1–8, only on CD–ROM.

Röber, R. (2010). Historische Ansichten als Quelle zur Rekonstruktion von Baubefunden—Hilfsmittel oder Illusion? Das Beispiel Kloster Petershausen in Konstanz. *Mitteilungen der Deutschen Gesellschaft für Archäologie des Mittelalters und der Neuzeit 22*, 101–110.

Roberts, J. C. and N. Ryan (1997). Alternative Archaeological Representations within Virtual Worlds. http://www.cs.kent.ac.uk/people/staff/nsr/arch/vrsig97/vrsig.html. (Accessed: 12th of January 2012.).

Rosenberg, R., J. Betz, and C. Klein (2008). Augensprünge. *Bildwelten des Wissens 6*(1), 127–129.

Roussou, M. (2002). Virtual Heritage: From the Research Lab to the Broad Public. See Niccolucci (2002b), pp. 93–100.

Roussou, M. and G. Drettakis (2003). Photorealism and Non-Photorealism in Virtual Heritage Representation. See Arnold (2003), pp. 51–60.

Ryan, N. (1996). Computer Based Visualisation of the Past: Technical 'Realism' and Historical Credibility. See Higgins et al. (1996), pp. 95–108.

Ryan, N. (2001). Documenting and Validating Virtual Archaeology. *Archeologia e Calcolatori 12*, 245–273.

Saito, T. and T. Takahashi (1990). Comprehensible Rendering of 3-D Shapes. In *Proceedings of the 17th Annual Conference on Computer Graphics and Interactive Techniques*, SIGGRAPH '90, New York, pp. 197–206. ACM.

Sanders, D. H. (1999). Virtual Worlds for Archaeological Research and Education. See Dingwall et al. (1999), pp. CD.

Sanders, D. H. (2000). Archaeological Applications Using Virtual Reality: Case Studies and Caveats. See Barceló et al. (2000), pp. 37–46.

Sanders, D. H. (2007). Why Do Virtual Heritage? See Clark and Hagemeister (2007), pp. 563–572.

Santoro, S. (2006). Archeologia, archeometria e ingegneria nello studio dell'Insula del Centenario. See Custodi and Sciortino (2006), pp. 23–35.

Santoro, S. (Ed.) (2007). *Indagini Diagnostiche Geofisiche e Analisi Archeometriche.* Pompei. Insula del Centenario (IX, 8) 1. Bologna: Ante Quem.

Santoro, S., D. S. Corlàita, B. Sassi, C. Ciarlantini, E. Vitali, G. Bigliardi, G. Guiducci, L. Cattani, L. M. Saracino, M. L. Sciarratta, and S. Morsiani (2005). Progetto Insula del Centenario (IX, 8). Saggi di Scavo 1999-2004. *Rivista di Studi Pompeiani 16*, 221–256.

Scagliarini, D. (2006). Il progetto Insula del Centenario a Pompei. See Custodi and Sciortino (2006), pp. 7–21.

Scagliarini, D., A. Coralini, E. Vecchietti, T. S. Cinotti, L. Roffia, S. Galasso, M. Malavasi, M. Pigozzi, E. Romagnoli, and F. Sforza (2001). Exciting Understanding in Pompeii through On-site Parallel Interaction with Dual Time Virtual Models. See Arnold et al. (2001), pp. 83–90, 364.

Scagliarini Corlàita, D., A. Coralini, A. Guidazzoli, T. S. Cinotti, G. Raffa, L. Roffia, C. Taboni, M. Malavasi, F. Sforza, and E. Vecchietti (2003). Archeologia virtuale e supporti informatici nella ricostruzione di una *domus* di Pompei. *Archeologia e Calcolatori 14*, 237–274.

Scaife, M. and Y. Rogers (1996). External Cognition: How Do Graphical Representations Work? *International Journal of Human–Computer Studies 45*, 185–213.

Schedl, B. (2003). ...von Burgen, Kapellen und Klosterkirchen ...: Virtuelle Mediaevistik als Wissensvermittlung. In W. Börner and L. Dollhofer (Eds.), *Archäologie und Computer 2002*, Workshop Archäologie und Computer 2002. Forschungsgesellschaft Wiener Stadtarchäologie 7, Wien, pp. 1–11. Phoibos Verlag.

Schnotz, W. (2002). Wissenserwerb mit Texten, Bildern und Diagrammen. See Issing and Klimsa (2002), pp. 65–81.

Scholkmann, B. (2003). Die Tyrannei der Schriftquellen? Überlegungen zum Verhältnis materieller und schriftlicher Überlieferung in der Mittelalterarchäologie. In M. Heinz and M. Eggert (Eds.), *Zwischen Erklären und Verstehen? Beiträge zu den erkenntnistheoretischen Grundlagen archäologischer Interpretation. Beiträge einer Tagung in Freiburg 1998*, Tübinger archäologische Taschenbücher 2, pp. 239–257. Münster: Waxmann.

Shanks, M. (1997). Photography and Archaeology. See Molyneaux (1997), Chapter 5, pp. 73–107.

Shanks, M. and I. Hodder (1995). Processual, Postprocessual and Interpretive Archaeologies. In I. Hodder, M. Shanks, A. Alexandri, V. Buchli, J. Carman, J. Last, and G. Lucas (Eds.), *Interpreting Archaeology: Finding Meaning in the Past*, Chapter 1, pp. 3–29. London and New York: Routledge.

Shennan, S. (1997). *Quantifying Archaeology* (2nd ed.). Edinburgh: Edinburgh University Press.

Sifniotis, M., K. Mania, P. L. Watten, and M. White (2006). Presenting Uncertainty in Archaeological Reconstructions Using Possibility Theory and Information Visualisation Schemes. See Ioannides et al. (2006b), pp. 198–202.

Sims, D. (1997, January-February). Archaeological Models: Pretty Pictures or Research Tools? *IEEE Computer Graphics & Applications*, 13–15.

Smiles, S. and S. Moser (Eds.) (2005). *Envisioning the Past: Archaeology and the Image*. New Interventions in Art History 5. Malden and Oxford and Victoria: Blackwell.

Smith, I. (1965). *Windmill Hill and Avebury: Excavations by Alexander Keiller 1925–1939*. Oxford: Clarendon Press.

Solmi, S. and E. Vecchietti (2001-2002). "Dalla Fotogrammetria Digitale alla Restituzione Virtuale: Due Esempi di Applicazione sul Campo". *Ocnus 9-10*, 185–204.

Spence, R. (2001). *Information Visualization*. ACM Press books. Harlow and Munich: Addison-Wesley.

Stančič, Z. and T. Veljanovski (Eds.) (2001). *Computing Archaeology for Understanding the Past CAA 2000: Computer Applications and Quantitative Methods in Archaeology*, BAR International Series 931, Oxford. BAR Publishing.

Stork, D. (2009). Computer Graphics for the Analysis of Realist Master Art: Current Methods and Future Challenges. In R. Sablatnig, M. Kampel, and M. Lettner (Eds.), *VSMM 2009: Proceedings of the 15th International Conference on Virtual Systems and Multimedia*, Los Alamitos, Washington, Tokyo, pp. 3–5. IEEE Computer Society Conference Pubishing Services.

Strothotte, T., M. Masuch, and T. Isenberg (1999). Visualizing Knowledge about Virtual Reconstructions of Ancient Architecture. In *Proceedings Computer Graphics International 1999 (CGI 99, June 7–11, 1999, Canmore, Alberta, Canada)*, Los Alamitos, pp. 36–43. The Computer Graphics Society: IEEE Computer Society.

Strothotte, T. and S. Schlechtweg (2002). *Non-Photorealistic Computer Graphics: Modeling, Rendering, and Animation*. Morgan Kaufmann Series in Computer Graphics and Geometric Modeling. Amsterdam, Boston, London, New York, Oxford, Paris, San Diego, San Francisco, Singapore, Sydney and Tokyo: Morgan Kaufman Publishers.

Stukeley, W. (1743). *Abury, a Temple of the British Druids, with Some Others, Described*, Volume 2. London: self-published.

Tergan, S.-O. (2002). Information und Lernen mit Multimedia und Internet - Lehrbuch für Studium und Praxis. See Issing and Klimsa (2002), pp. 99–112.

Terras, M. M. (1999). A Virtual Tomb for Kelvingrove: Virtual Reality, Archaeology and Education. *Internet Archaeology 7*.

Theuerkauf, G. (1991). *Einführung in die Interpretation historischer Quellen. Schwerpunkt: Mittelalter*. Uni-Taschenbücher 1554. Paderborn, München, Wien and Zürich: Ferdinand Schöningh.

Thomas, J. (1993). The Politics of Vision and the Archaeologies of Landscape. In B. Bender (Ed.), *Landscape: Politics and Perspectives*, Explorations in Anthropology, pp. 19–48. Providence and Oxford: Berg.

Thomas, J. (Ed.) (2000a). *Interpretative Archaeology: A Reader*. London and New York: Leicester University Press.

Thomas, J. (2000b). Introduction: The Polarities of Post-Processual Archaeology. See Thomas (2000a), pp. 1–18.

Thomas, J. (2004). *Archaeology and Modernity*. London and New York: Routledge.

Tilley, C. Y. (1994). *A Phenomenology of Landscape: Places, Paths and Monuments*. Explorations in Anthropology. Oxford and Providence: Berg.

Tobler, R. F., S. Maierhofer, K. Karner, and M. Sormann (2006). Creative Histories: The Josefsplatz Experience. See Ioannides et al. (2006a), pp. 318–324.

Tost, L. P. (2008). Does Virtual Archaeology Exist? See Posluschny et al. (2008), pp. 101–107.

Traser, F. (2007). Resemblance of the Once Existing. See Clark and Hagemeister (2007), pp. 579–592.

Ucko, P., M. Hunter, A. Clark, and A. David (1991). *Avebury Reconsidered: From the 1660s to the 1990s*. London: Unwin Hyman.

Van Acker, K. G. (1987). Ida van Ename. *Het Land van Aalst 39*(6), 281–291.

Van de Walle, A. L. J. (1952). De St Vedastuskerk te Nederename, een X[e] eeuwse praestedelijke moederkerk. *Handelingen van de Geschied- en Oudheidkundige Kring van Oudenaarde, van zijn Kastelnij en van den Lande tusschen Maerck en Ronne 10*, 19–32.

Van der Donckt, M.-C., D. Pletinckx, and D. Callebaut (2002). Projekt Ename 974: Neue Technologien in der Präsentation von Archäologie und Stadtgeschichte. See Rieche and Schneider (2002), pp. 50–56.

Van Der Meiren, J.-P. and D. Callebaut (1999). Nieuw Tijdsvenster Maakt Restauratie Toegankelijk voor Iedereen. *Ename 974 11*, 2.

Van Der Meiren, K. and W. Derde (2006). Unieke Kart van Ename uit 1596. *Ename 974 30*, 4–5.

Van Der Meiren (Ed.), J.-P. (1999). De Sint-Laurentiuskerk: De Geschiedenis van een Restauratie. *Ename 974 11*, 10–11.

Van Dijck, L. (1995, June). Ename, Sint-Laurentiuskerk: Restauratie van muurschilderingen in het boogveld boven een dichtgemetste arcade op de oostelijke muur van het middenschip. Verslag.

Van Strydonck, M. J. Y. (1997). Radiocarbon Dating and Medieval Europe. In G. De Boe and F. Verhaeghe (Eds.), *Method and Theory in Historical Archaeology*, Papers of the 'Medieval Europe Brugge 1997' Conference, 10, Zellik.

Vatanen, I. (2003). Deconstructing the (Re)Constructed: Issues in Conceptualising the Annotation of Archaeological Virtual Realities. See Doerr and Sarris (2003), pp. 69–74.

Vatanen, I. (2004). Argumentation Paths in Information Infrastructure of the Archeological Virtual Realities. See Magistrat der Stadt Wien, Referat Kulturelles Erbe, Stadtarchäologie Wien (2004), pp. on CD, 494–499 in book. (Formatting and content of the articles on CD and in the book are different).

Veryovka, O. (1999). *Texture Control in Digital Halftoning*. Alberta: University of Alberta.

Viti, S. (2004). Between Reconstruction and Reproduction: The Role of Virtual Models in Archaeological Research. See Magistrat der Stadt Wien, Referat Kulturelles Erbe, Stadtarchäologie Wien (2004), pp. on CD, 525–528 in book. (Formatting and content of the articles on CD and in the book are different).

Vlahakis, V., J. Karigiannis, M. Tsotros, M. Gounaris, L. Almeida, D. Stricker, T. Gleue, I. T. Christou, R. Carlucci, and N. Ioannidis (2001). ARCHEOGUIDE: First results of an Augmented Reality, Mobile Computing System in Cultural Heritage Sites. See Arnold et al. (2001), pp. 131–140.

Vote, E., D. Acevedo, D. Laidlaw, and M. S. Joukowsky (2001). ARCHAVE: A Virtual Environment for Archaeological Research. See Stančič and Veljanovski (2001), pp. 313–316.

Vote, E. L., D. Acevedo, D. H. Laidlaw, and M. S. Joukowsky (2002). What's Virtual Reality Good for? The ARCHAVE System – Problems and Possibilities. See Niccolucci (2002b), pp. 83–86.

Walter-Karydi, E. (1998). Sachliche Bilddokumente und Phantasiekompositionen: Frühe Bilder des Naxier Apollon. *Antike Welt 29*(4), 339–344.

Ware, C. (2004). *Information Visualization: Perception for Design*. Morgan Kaufmann Series in Interactive Technologies. San Francisco: Morgan Kaufman Publishers.

Watson, A. (2001). Composing Avebury. *World Archaeology 33*(2), 296–314.

Watson, A. (2004). Monuments That Made the World: Performing the Henge. See Cleal and Pollard (2004), pp. 83–97.

Watt, A. H. (2000). *3D Computer Graphics* (3rd ed.). Harlow, London, New York, Reading, San Francisco, Toronto, Don Mills, Sydney, Tokyo, Singapore, Hong Kong, Seoul, Taipei, Cape Town, Madrid, Mexico City, Amsterdam, Munich, Paris and Milan: Addison-Wesley.

Weidenmann, B. (2002a). Abbilder in Multimediaanwendungen. See Issing and Klimsa (2002), pp. 83–96.

Weidenmann, B. (2002b). Multicodierung und Multimodalität im Lernprozess. See Issing and Klimsa (2002), pp. 45–62.

West, M. L. (1973). *Textual Criticism and Editorial Technique: Applicable to Greek and Latin Texts*. Stuttgart: B. G. Teubner.

Wheatley, D. (1996). The Use of GIS to Understand Regional Variation in Neolithic Wessex. In D. G. Maschner (Ed.), *New Methods, Old Problems: Geographic Information Systems in Modern Archaeological Research*, pp. 75–103. Southern Illinois University Press.

Wheatley, D. (1999). So Wm Stukeley Esq Was Right after All. *British Archaeology 48*, 18.

Wheatley, D. and M. Gillings (2002). *Spatial Technology and Archaeology: The Archaeological Applications of GIS*. London and New York: Taylor & Francis.

Whittle, A. (1991). A Late Neolithic Complex at West Kennet, Wiltshire, England. *Antiquity 65*, 256–262.

Whittle, A. (1993). The Neolithic of the Avebury Area: Sequence, Environment, Settlement and Monuments. *Oxford Journal of Archaeology 12*, 29–53.

Whittle, A. W. R. (1997). *Sacred Mound, Holy Rings: Silbury Hill and the West Kennet Palisade Enclosures: a Later Neolithic Complex in North Wiltshire.* Oxbow Monographs 74. Oxford: Oxbow Books.

Wilcock, J. and K. Lockyear (Eds.) (1995). *Computer Applications and Quantitative Methods in Archaeology 1993*, BAR International Series 598, Oxford. BAR Publishing.

Wittur, J. and S. Krömker (2009). Lorsch Abbey and the London Charter: An Information System for a World Heritage Site. In R. Sablatnig, M. Kampel, and M. Lettner (Eds.), *VSMM 2009: Proceedings of the 15th International Conference on Virtual Systems and Multimedia*, Los Alamitos, Washington, Tokyo, pp. 85–90. IEEE Computer Society Conference Publishing Services.

Wood, J. and G. Chapman (1992). Three-dimensional Computer Visualization of Historic Buildings — with Particular Reference to Reconstruction Modelling. See Reilly and Rahtz (1992), Chapter 11, pp. 123–146.

Woodwark, J. (1991, January). Reconstructing History with Computer Graphics. *IEEE Computer Graphics & Applications*, 18–20.

Zanfini, M. and E. Vecchietti (2004). Fotogrammetria e *Virtual Reality*: La Casa del Centenario (IX, 8, 3.6.a) a Pompei. In C. Agnelli (Ed.), *Atti del IX Colloquio dell'Associazione Italiana per lo Studio e la Conservazione del Mosaico*, Ravenna, pp. 849–856. Edizioni del Girasole.

Zhukovsky, M. (2001). Virtual 3D Reconstruction of the Kiafar Site, North Caukasus, Russia. See Stančič and Veljanovski (2001), pp. 297–301.

Zuk, T., S. Carpendale, and W. D. Glanzman (2005). Visualizing Temporal Uncertainty in 3D Virtual Reconstructions. See Mudge et al. (2005), pp. 99–106.

www.ingramcontent.com/pod-product-compliance
Lightning Source LLC
LaVergne TN
LVHW072250300926
849160LV00018B/1629
* 9 7 8 1 4 0 7 3 1 0 7 1 8 *